HUMAN FACTORS METHODS

Human Factors Methods

A Practical Guide for Engineering and Design

2nd Edition

NEVILLE A. STANTON

PAUL M. SALMON

LAURA A. RAFFERTY

GUY H. WALKER

CHRIS BABER

DANIEL P. JENKINS

CRC Press
Taylor & Francis Group
Boca Raton London New York

CRC Press is an imprint of the
Taylor & Francis Group, an **informa** business

CRC Press
Taylor & Francis Group
6000 Broken Sound Parkway NW, Suite 300
Boca Raton, FL 33487-2742

© 2013 by Neville A. Stanton, Paul M. Salmon, Laura A. Rafferty, Guy H. Walker, Chris Baber and Daniel P. Jenkins
CRC Press is an imprint of Taylor & Francis Group, an Informa business

No claim to original U.S. Government works

International Standard Book Number-13: 978-1-4094-5753-4 (Hardback) 978-1-4094-5754-1 (Paperback)

Visit the Taylor & Francis Web site at
http://www.taylorandfrancis.com

and the CRC Press Web site at
http://www.crcpress.com

Contents

List of Figures

List of Tables

About the Authors

Neville A. Stanton holds a Chair in Human Factor Engineering at the University of Southampton. He has published over 160 peer-reviewed journal papers and 20 books on Human Factors and Ergonomics. In 1998, he was awarded the Institution of Electrical Engineers Divisional Premium Award for a co-authored paper on engineering psychology and system safety. The Institute of Ergonomics and Human Factors awarded him the Otto Edholm medal in 2001, the President's Medal in 2008 and the Sir Frederic Bartlett Medal in 2012 for his substantial and original contribution to basic and applied ergonomics research. In 2007, the Royal Aeronautical Society awarded him and his colleagues the Hodgson Medal and Bronze Award for their work on flight deck safety. He is an editor of the *Ergonomics* journal and on is the editorial boards of the *Theoretical Issues in Ergonomics Science* and the *Human Factors and Ergonomics in Manufacturing and Service Industries* journals. He consults for a wide variety of organisations on topics such as Human Factors, safety cases, safety culture, risk assessment, human error, product design, warning design, system design and operation. He has also acted as an expert witness in accident cases. He is a Fellow and Chartered Occupational Psychologist registered with the British Psychological Society and a Fellow of the Ergonomics Society. He has a BSc (Hons) in Occupational Psychology from the University of Hull, a well as an MPhil in Applied Psychology and a PhD in Human Factors from Aston University in Birmingham.

Paul M. Salmon is an Associate Professor in Human Factors and leader of the USCAR (University of the Sunshine Coast Accident Research) team at the University of the Sunshine Coast. Paul holds an Australian National Health and Medical Research Council (NHMRC) post-doctoral training fellowship in the area of Public Health and has over 12 years experience in applied Human Factors research in a number of domains, including the military, aviation, and road and rail transport. Paul has co-authored 10 books, over 70 peer reviewed journal articles, and numerous conference articles and book chapters. He has received various accolades for his research to date, including the 2007 Royal Aeronautical Society Hodgson Prize for best research and best paper, and the 2008 Ergonomics Society's President's Medal. Paul was also recently named as one of three finalists in the 2011 Scopus Young Australian Researcher of the Year Award.

Paul Salmon's contribution to this book was funded through the Australian National Health and Medical Research Council post doctoral training fellowship scheme.

Laura A. Rafferty completed her undergraduate studies in 2007, graduating with a BSc in Psychology (Hons) from Brunel University. In the course of this degree, she completed two industrial placements, the second of which was working as a research assistant in the Ergonomics Research Group. During this seven-month period, she helped to design, run and analyse a number of empirical studies being run for the Human Factors Integration Defence Technology Centre (HFI DTC) at Brunel. Within this time, she also completed her dissertation, which explored the qualitative and quantitative differences between novices and experts within military command and control. She is currently in her third year of part-time PhD studies, focusing on team work and decision-making associated with fratricide. Since April 2009 she has been employed in the Transportation Research Group at the University of Southampton as a project assistant for the HFI DTC working on projects including Naturalistic Decision-Making in Teams and compiling a Human Factors Methods Database.

Guy H. Walker is a Lecturer at the School of the Built Environment at Heriot-Watt University, Edinburgh and his research focuses on human factors issues in infrastructure and transport. He is a recipient, with his colleagues, of the Ergonomics Society's President's Medal for original research. He is also author/ co-author of nine books on diverse topics in human factors, including a major text on human factors methods, and is author/co-author of over 50 international peer-reviewed journals.

Chris Baber graduated with a BA (Hons) in Psychology/English at Keele University before joining Aston University's Applied Psychology Unit in 1987. He was awarded a PhD in Human Factors of Speech Technology and joined Birmingham University in 1990. He taught on the MSc Work Design and Ergonomics course for 12 years before joining the School of Electronic, Electrical and Computer Engineering (where he was appointed Head of School in 2010). His research concerns human interaction with novel technology and ranges from theories of distributed cognition to social network analysis to the design and evaluation of wearable computers.

Daniel P. Jenkins leads the Human factors and usability team at DCA Design. DCA is one of Europe's leading product design and development consultancies, working across the Medical & Scientific, Transport, Commercial and Industrial and Consumer sectors. DCA offers an integrated approach to product development with services covering: applied product research and planning; design strategy; industrial design; interaction design; mechanical engineering; electronic hardware and software engineering; usability and ergonomics; prototyping and production support.

Dan started his career as an automotive engineer, graduating in 2004, with an M.Eng (Hons) in Mechanical Engineering and Design, receiving the 'University Prize' for the highest academic achievement in the school. During his time in the car industry, Dan developed a great interest in ergonomics and human factors. In 2005, Dan returned to Brunel University taking up the full-time role of Research Fellow in the Ergonomics Research Group. Dan studied part-time for his PhD in Human Factors and interaction design - graduating in 2008, and receiving the 'Hamilton Prize' for the Best Viva in the School of Engineering and Design. In 2009, Dan started his own consultancy (Sociotechnic Ltd) with the aim of developing industrial experience across a wide range of domains. In July 2012, Dan joined the team at DCA, seizing the opportunity to work as part of an integrated product development team. Dan has developed experience of applied research in domains including medical, defence, nuclear facilities, automotive, submarines, maritime, aviation, policing, and control room design. Dan has co-authored nine books and over 45 peer-reviewed journal papers, alongside numerous conference articles and book chapters. Dan and his colleagues were awarded the Ergonomics Society's President's Medal in 2008 for contribution to basic and applied ergonomics research.

Preface to the Second Edition

This is the revised and updated version of our original book entitled *Human Factors Methods: A Practical Guide for Engineering and Design*. We have been delighted by the praise heaped on the first edition by researchers and practitioners alike. Not all reviews have been glowing however, so we have sought to improve the weaknesses and have revisited all of the methods. The result is that we have removed a few and added many. We have also updated references to methods and had plenty of opportunities to use the methods ourselves over the past seven years.

What people have liked is the 'cookbook' approach to helping people get to grips with the methods. Surprisingly, perhaps, quite a lot of methods books don't actually tell you how to use the methods. We have sought to cover over 100 methods, which meant that we need to give a brief overview of the approach with the advantages and disadvantages, the procedure, and worked example and references to more information. This worked very well last time.

Despite this being a multi-authored book, we have tried to convey the information in a consistent manner, as this is meant to be a book that people will pick up and flick through to find the methods they want. We would certainly encourage readers to try new methods and to use the methods in combination. Often new insights can be gained by looking at the information in a different way or from multiple perspectives.

We would suggest those new to Human Factors methods start by examining a smaller, more self-contained problem first before moving up to the larger, unconstrained, more ambiguous problems. We have found that when using methods to examine problems, they can help reframe the problem in new ways – to show that the original problem was really just a symptom and that the real underlying problems have more fundamental causes which need to be tackled. Solving problems can be immensely satisfying. In this work we have analysed many systems in military and civilian domains. We have also developed design specifications for new systems.

Human Factors methods provide a useful structure when investigating problems in order to help the analyst understand the domain and the issue relatively quickly. When working with subject-matter experts, these methods can also help reflect the usefulness of the discipline of Human Factors and gain acceptance by the domain expert of the approach. Successful design projects often require working with domain experts, and Human Factors methods can provide the conduit for useful discussion and collaboration.

Those expert in Human Factors methods would probably only need this book as an aide-memoire, to help guide them through familiar methods, but even they are unlikely to know all of these methods thoroughly. We would also point this group to the final chapter covering the EAST method, which integrates several well-known methods into a teamwork analysis approach.

To those developing new methods, or adapting existing ones, we recommend the structure used in this book as a useful way of conveying the method to the newcomer. To those training people in Human Factors methods, we have found that the book supports workshops and lectures. To those analysing and designing systems, this book should help you to select and use the approaches of choice. In short, we hope that everyone will find this book useful and will continue to explore the use of Human Factors methods to solve practical problems.

Neville A. Stanton
University of Southampton

Acknowledgments

This work from the Human Factors Integration Defence Technology Centre was partially funded by the Human Sciences Domain of the UK Ministry of Defence Scientific Research Programme.

For more information about the Human Factors Integration Defence Technology Centre, please see: www.hfidtc.com

Chapter 1

Introduction to Human Factors Methods

What is a Human Factors Method?

Most readers will be able to identify an example of a Human Factors (HF) problem from their own experience of work or study. An HF problem will more than likely possess some, or all, of the following attributes. It will be a problem which impacts negatively on overall system performance. It will involve humans in systems not behaving as they were expected to. It will be a problem which existing methods of design, evaluation and procurement have somehow not captured, despite indepth analysis. Above all, it will usually be frustratingly resistant to a whole range of purely technical interventions. HF problems therefore have the very real capability to reduce overall system performance to a level that is substantially less effective than was originally intended.

This creates an interesting situation, again one that many readers may find resonates with their own experience. Faced with a 'Moore's Law' of accelerating technological progress, some authors have noted how much more difficult it is becoming to compete on functionality, reliability or manufacturing costs (Green and Jordan, 1999). What they mean is that technology is not the only precursor to the success of a system and that functionality, reliability and manufacturing costs relate more to what a system 'is' (in terms of its technical/engineering content) rather than to what a system 'does' (in terms of harnessing that technology to enable people to accomplish meaningful real-world tasks). HF problems are troublesome because they do not reside exclusively within the purview of engineering; nor are they the exclusive domain of human scientists. HF problems reside at the interface of both, and HF methods are the means by which they can be tackled.

HF and Design

Much has been made about the timeliness of HF input into projects and it is true to say that considerable time, effort and expense can often be saved by early intervention rather than being faced with a completed system which requires considerable re-design. Unfortunately, this is a common problem. That being said, the appropriateness of a particular HF analysis will of course depend on a number of factors, including which stage of design the project is at, how much time and resources are available, the skills of the analyst, access to the end-user population and what kind of data are required (Stanton and Young, 1999). Fortunately, many HF methods are flexible with regard to the design stage they could be applied to, even if the system itself is no longer as flexible in terms of subsequent changes. Figure 1.1 provides an illustration of this, showing that, in terms of overall effort, HF input can add most value early on in the design process when it is still relatively inexpensive to carry out modifications to mock-ups and prototypes.

There are many methods explained in this book which lend themselves well to being applied at the very early stages of design. In addition to this, there are many methods explained which may be used in a predictive as well as an evaluative manner. Called 'analytical prototyping', this is the process of applying HF insights to systems which do not yet exist in physical form. Figure 1.2 presents a generic design process in which different HF methods become applicable and useful at different stages. At the start, we begin with methods that are suited to 'analytical prototyping' and to modelling the constraints of a particular problem domain to reveal opportunities for unexpected behaviours. The analysis would

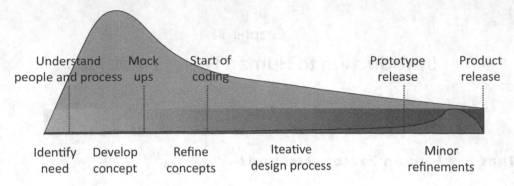

Understand people and process · Mock ups · Start of coding · Prototype release · Product release

Identify need · Develop concept · Refine concepts · Iteative design process · Minor refinements

Figure 1.1   Illustration to show that HF effort is better placed in the early stages of the design process

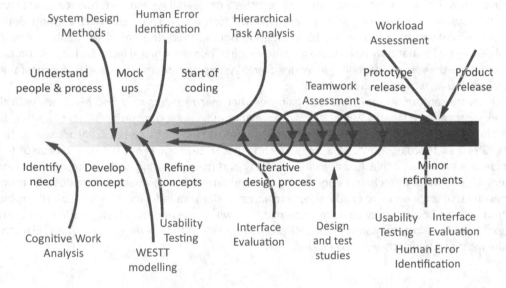

System Design Methods · Human Error Identification · Hierarchical Task Analysis · Workload Assessment

Understand people & process · Mock ups · Start of coding · Teamwork Assessment · Prototype release · Product release

Identify need · Develop concept · Refine concepts · Iterative design process · Minor refinements

Cognitive Work Analysis · Usability Testing · WESTT modelling · Interface Evaluation · Design and test studies · Usability Testing · Interface Evaluation · Human Error Identification

Figure 1.2   Application of HF methods by phase of the design process

then proceed forward with analyses of human error, usability and interface evaluation, amongst others. Each method would be chosen to suit the particular stage of the design life-cycle. For example, in the early stages, methods would be chosen to enable designers and engineers to diagnose important HF dimensions of their proposed systems. In later stages, methods would be chosen which reflect the fact that a physical manifestation of the system now exists and that users themselves can start interacting with it.

Given that most HF problems emerge from unexpected interactions at the boundary between people and systems, the need to engage in an evolutionary, iterative, 'design-test-design' process emerges as a consistent theme in projects within which the authors have worked. This book is not about the design process per se, but suffice to say that HF problems can be avoided with a systems approach to which HF methods lend themselves very well.

Human Factors Integration

From Figures 1.1 and 1.2, it is clear that the flexibility of application to the various design stages bodes well for HF methods. Something else that bodes well is ISO-13407, 'Human-Centered Design of Systems'. This international standard codifies the requirement to ensure adequate focus on the potential users of systems at all stages in the design and development process. In support of this, a practice called Human Factors Integration (HFI) represents the 'business process' which seeks to meet this aim. HFI is about:

> providing a balanced development of both the technical and human aspects of equipment procurement. It provides a process that ensures the application of scientific knowledge about human characteristics through the specification, design and evaluation of systems. (Ministry of Defence, 2000: 6)

HFI exists in various forms within certain sectors of industry. In the UK Ministry of Defence, for example, the HFI process covers six domains: Manpower, Personnel, Training, HF Engineering, System Safety and Health Hazards. The HFI process is intended to be seen as an activity which supports attention to all six domains during the entire system design life-cycle, from design to manufacture, to use and to maintenance and disposal. This book contains methods which can be used to support all these domains. We cover methods that are essential to System Safety and to Manpower, and that can support Training and Personnel. Issues relating to Health Hazards relate in turn to risk analysis, which requires additional knowledge and techniques beyond the scope of this book, yet the methods presented interface well with this partnering domain.

Scientist or Practitioner?

HF methods are built on a foundation of robust human science, with actionable methods forming a major part of the HF discipline. The *International Encyclopaedia of Human Factors and Ergonomics* (Karwowski, 2001) has an entire section devoted to methods and techniques. Many of the other sections of the *Encyclopaedia* also make reference to, if not provide actual examples of, HF methods. HF consultants provide services to industry in which HF methods are routinely deployed. In short, the importance of HF methods cannot be overstated. These methods offer the engineer, the designer and the specialised HF practitioner a structured approach to the analysis and evaluation of practical problems. The overall approach can be described using the scientist-practitioner model (Stanton, 2005). As a scientist, the process of applying these methods is as follows:

- extending the work of others;
- testing theories of human-system performance;
- developing hypotheses;
- questioning everything;
- using rigorous data collection and analysis techniques;
- ensuring the repeatability of results;
- disseminating the findings of studies.

As a practitioner, the application of these methods is as follows:

- addressing real-world problems;
- seeking the best compromise under difficult circumstances;
- looking to offer the most cost-effective solution;
- developing demonstrators and prototype solutions;
- analysing and evaluating the effects of change;
- developing benchmarks for best practice;
- communicating findings to interested parties.

In applying the methods contained in this book, you will work somewhere between the poles of scientist and practitioner, varying the emphasis of your approach depending upon the problems that you face.

HF methods are useful in the scientist-practitioner model because of the structure and potential for repeatability that they offer. There is an implicit guarantee in the use of methods that, provided they are used properly, they will produce certain types of useful products. It has been suggested that HF methods are a route to making the discipline more accessible to all (Diaper, 1989; Wilson, 1995; Stanton and Young, 2003). This is entirely appropriate given the multi-disciplinary nature of the problems in which HF professionals and engineers/designers will encounter each other. However, despite the rigour offered by methods, there is still plenty of scope for the role of experience. The most frequently asked questions raised by users of ergonomics methods are the following:

- How deep should the analysis be?
- Which methods of data collection should be used?
- How should the analysis be presented?
- Where is the use of the method appropriate?
- How much time/effort does each method require?
- How much, and what type, of expertise is needed to use the method(s)?
- What tools are there to support the use of the method(s)?
- How reliable and valid is/are the method(s) (Stanton and Annett, 2000)?

This book will help to answer some of these questions.

Reliability and Validity

To the engineer or designer, the human sciences in general (and possibly HF in particular) may fall victim to the popular, albeit wholly inaccurate, perception of being a rather 'woolly' field. This is not so. HF methods certainly deal with problems which, to engineers or designers, may seem alarmingly loose in definition and which do not conform to any readily identifiable chain of cause and effect. However, in facing such problems, HF methods will provide a welcome source of structure and rigour.

Researchers have identified a dichotomy of Ergonomics methods: analytical methods and evaluative methods (Annett, 2002). They argue that analytical methods (i.e. those methods that help the analyst gain an understanding of the mechanisms underlying the interaction between human and machines) require construct validity, whereas evaluative methods (i.e. those methods that estimate parameters of selected interactions between human and machines) require predictive validity. Construct and criterion-referenced validity play a role in the development of HF theory itself. There is a difference between construct validity (how acceptable the underlying theory is), predictive validity (the usefulness and efficiency of the approach in predicting the behaviour of an existing or future system) and reliability (the repeatability of the results). This distinction is made in Table 1.1.

This presents an interesting question for HF. Are the methods really mutually exclusive? Some methods appear to have dual roles (i.e., both analytical and evaluative, such as Task Analysis For Error

Table 1.1   Annett's dichotomy of ergonomics methods (adapted from Annett, 2002)

	Analytic	Evaluative
Primary purpose	Understand a system.	Measure a parameter.
Examples	Task analysis, training needs analysis, etc.	Measures of workload, usability, comfort, fatigue, etc.
Construct validity	Based on an acceptable model of the system and how it performs.	Is consistent with theory and other measures of parameters.
Predictive validity	Provides answers to questions, e.g. structure of tasks.	Predicts performance.
Reliability	Data collection conforms to an underlying model.	Results from independent samples agree.

Identification), which implies that they must satisfy both criteria. It is plausible, however, as Baber (2005a) argues in terms of evaluation, that the approach taken will influence which of the purposes one might wish to emphasise. The implication is that the way in which one approaches a problem – or, in other words, where on the scientist-practitioner continuum one places oneself – could well have a bearing on how a method is employed. At first glance (particularly from a 'scientist' perspective), such a 'pragmatic' approach appears highly dubious: if we are selecting methods piecemeal in order to satisfy contextual requirements, how can be certain that we are producing useful, valid, reliable output? While it may be possible for a method to satisfy three types of validity: construct (i.e. theoretical validity), content (i.e. face validity) and predictive (i.e. criterion-referenced empirical validity), it is not always clear whether this arises from the method itself or from the manner in which it is applied. The solution, simply stated, is that care needs to be taken before embarking on any application of methods to make sure that one is attempting to use the method in the spirit in which it was originally designed.

Which Method?

How do you decide which of the 107 methods contained in this book to apply to a particular problem? Many HF problems require only a basic level of HF insight and a correspondingly basic methodological intervention. Many other problems require greater levels of sophistication, and determining an appropriate set of methods (because individual methods are rarely used alone in such cases) requires some planning and preparation. Stanton and Young (1999a) have devised a process model to guide the selection of methods, as shown in Figure 1.3 below. This is a valuable first step, but increasingly complex systems require you to have a flexible strategy, so pilot studies are often helpful in scoping out the problem before a detailed study is undertaken. This may mean that there can be several iterations through the criterion development and methods selection process. From a practitioner perspective, the time taken to carry out pilot studies might simply be unavailable. However, we would argue that there is no harm in running through one's selection of methods as a form of 'thought-experiment' in order to ascertain what type of output each method is likely to produce, and deciding whether or not to include a method in the battery of methods that will be applied. While it is important not to rely too heavily on a single approach, there is no guarantee that simply throwing a lot of methods at a problem will guarantee useful results. An informed approach is needed.

As shown in Figure 1.3, the initial method selection can be viewed as a closed-loop process with three feedback loops. The first feedback loop validates the selection of the methods against the selection criteria. The second feedback loop validates the methods against the adequacy of the ergonomic intervention. The third feedback loop validates the initial criteria against the adequacy of the intervention. There could be errors in the development of the initial criteria, the selection of the methods and the appropriateness of the intervention. Each should be checked. The main stages in the process are identified as: determine criteria (where the criteria for assessment are identified); compare methods against criteria (where the pool of methods are compared for their suitability); application of methods

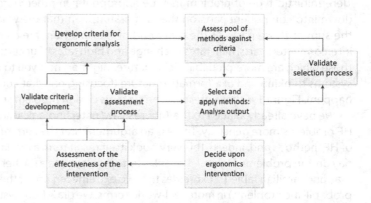

Figure 1.3   **Validating the selection of the methods and the ergonomics intervention process (adapted from Stanton and Young, 1999a)**

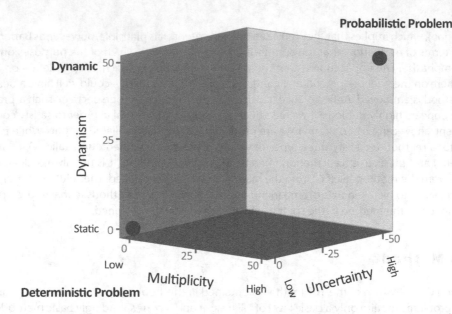

Figure 1.4 The systems design problem space is a device that can be used to shape thinking as to what method might be appropriate to what problem

(where the methods are applied); implementation of ergonomics intervention (where an ergonomics programme is chosen and applied); and evaluation of the effectiveness of the intervention (where the assessment of change brought about by the intervention is assessed). This process represents a highly useful and pragmatic first step, helping you to generate a defensible set of appropriate methods. As noted above, however, HF problems can be alarmingly unpredictable and the systems within which they arise can be similarly complex. Added to this, a degree of flexibility in method selection still exists, so we proceed down into the next level of sophistication in order to continue probing the question 'which method?'.

Faced with particularly complex and seemingly intractable HF issues, a fundamental question to ask is 'what is the nature of the HF problem that my selection of HF methods is aiming to resolve?'. The notion of a 'HF problem space' could serve as a useful device in shaping your thinking. If your particular problem can be defined as having low levels of change over time, a small number of interconnected parts, and the principles of the system's operation are well understood, then methods suited to this more 'deterministic' type of problem may be appropriate; in other words, methods which break a problem down into component parts on the tacit assumption that the whole can be no more, or no less, than the sum of those parts. At the other end of the spectrum are complex problems, those with multiple interconnected parts, high rates of change and high rates of uncertainty. Faced with problems like these, those which are more 'probabilistic' in nature might prompt you to use methods more closely aligned to systems thinking and of a 'formative' nature (focusing on what 'could' happen rather than what 'should' happen). Figure 1.4 above presents the systems design problem space.

We have already alluded to the 'Moore's Law' of technological advance and to the troubling nature of HF problems more generally. There is an argument that the current drive towards HFI, the proactive use of HF methods and, indeed, this very book itself result from a fundamental shift in the nature of present-day design problems. Figure 1.1 enables us to direct you to a further trend which, as before, many readers may find familiar. Table 1.2 expresses the two extreme regions of the problem space (deterministic versus probabilistic problems) in more real-world terms. We draw from systems engineering to identify facets of systems that HF methods are increasingly required to operate within.

At first glance, it seems rather improbable to suggest that the domains of health care, air traffic control, railway operations, nuclear power and many more besides have become significantly more

Table 1.2   **Wider trends associated with opposite corners of the ergonomics problem space (Boehm, 2006)**

Deterministic problems	Probabilistic problems
A focus on specialisation.	An increasing integration of disciplines, specialisms and expertise.
A focus on what a system 'is' (i.e. requirements and functionality).	An increased emphasis on what a system 'does' (i.e. end value, effects and capabilities).
An increasing level of criticality and dependability required of less complex systems.	An increasing level of criticality and dependability required of more complex systems.
A focus on constraining dynamism and imposing stable behaviour.	A focus on responding and adapting to dynamism and change.
A focus on stand-alone systems.	An increased emphasis on interoperability.
An emphasis on controlling complexity.	An increasing emphasis on ever more complex systems and systems of systems.
A focus on end products often based on new technology to replace obsolete equipment.	An increasing trend towards through-life capability, integration of legacy systems and reuse.
An increase in computational power and the ability of entities and artifacts to exhibit complex but predictable behaviour.	An increase in computational power and the ability of entities and artifacts to exhibit complex emergent behaviour.

or less complex than they were in the 1950s (O'Brian and O'Hare, 2007). The development of research tools which allow aspects of complex problems to be confronted more directly (such as computer simulations), a shift in the frame of reference within which ergonomics problems are viewed and a wider transition from 'industrial age' to 'information age' problems mean that knowledge of the region that a particular HF problem occupies becomes important. So too does the type of method most appropriate to that region. What this means, simply stated, is that one methodological size does not fit all and that the question of 'which method?' is a non-trivial one. Whilst it is not possible to provide rules which apply to all situations, what the discussion so far enables us to do is to offer the following advice.

Think about the design life-cycle and the potential role of HFI and HF methods throughout it. What role could HF methods play early on in shaping and guiding the developing programme of work? What impact would this have on the design process itself?

Think about the stage of the design life-cycle that you are currently operating within. Does the system exist in physical form and, if so, can it be easily changed? Who will be participating in the analysis? End users? Analysts? Both?

Think about the nature of the problem and/or the system you are working with. Is it comprised of multiple components? Is its behaviour well understood? Is it stable?

Use this thought process to then develop criteria for the analysis and proceed with method selection using the flowchart shown in Figure 1.3 as a guide. Iterate this method with the aim of identifying specifically what parameters you wish to analyse and which methods would provide best access to them.

If time and resources permit, a pilot study with your selected methods can be extremely useful in identifying issues before conducting a larger analysis. In addition, for particularly complex problems, it will provide you with insight into the predictive efficiency of your selected methods.

Of course, you may not merely be a user of HF methods; you may also be a consumer. This guidance can in turn help you to become a more informed customer, able to better judge 'why' a particular method or methods have been chosen, and how adequate they may really be in a given set of circumstances.

Using the Book

This book includes the phrase 'a practical guide' in its title for a reason. It does not dwell extensively on theory; rather, it focuses on the more pragmatic question of what HF methods are, their advantages and disadvantages, and step-by-step guidance on how to carry them out yourself. In putting the book together, we reviewed a very large collection of contemporary HF methods over three stages. First, an initial review of existing HF methods and techniques was conducted. Second, a screening process was employed in order to remove any duplicated methods or any methods which require more than paper and pencil to conduct. The reason for this latter criterion was not to disparage any of the computer-based tools on the market, but to focus on those techniques that the practitioner could use without recourse to specialised equipment and/or providers. Third, the methods selected for review were then analysed using a set of pre-determined criteria. To give you confidence in the breadth and depth of the methods selected for inclusion herein, each stage of the review is briefly described below.

Stage 1: Initial Literature Review of the Existing HF Methods

The initial literature review was based upon a survey of standard ergonomics textbooks, relevant scientific journals and existing HF method reviews. At this stage, none of the HF methods were subjected to any further analysis and were simply recorded by name, author(s) or source(s), and class of method (e.g. mental workload assessment, Human Error Identification, Data Collection and Task Analysis). In order to make the list as comprehensive as possible, any method discovered in the literature was recorded and added to a database. The result of this initial literature review was a database of over 300 HF methods and techniques grouped into 11 categories, as shown in Table 1.3:

Table 1.3   HF method categories

Method category	Description
Data collection techniques.	Data collection techniques are used to collect specific data regarding a system or scenario. According to Stanton (2003), the starting point for designing future systems is a description of a current or analogous system.
Task analysis techniques.	Task analysis techniques are used to represent human performance in a particular task or scenario under analysis. Task analysis techniques break down tasks or scenarios into the required individual task steps, in terms of the required human–machine and human–human interactions.
Cognitive task analysis techniques.	Cognitive task analysis (CTA) techniques are used to describe and represent the unobservable cognitive aspects of task performance. CTA is used to describe the mental processes used by system operators in completing a task or set of tasks.
Charting techniques.	Charting techniques are used to depict graphically a task or process using standardised symbols. The output of charting techniques can be used to understand the different task steps involved in a particular scenario and also to highlight when each task step should occur and which technological aspect of the system interface is required.
HEI/HRA techniques.	Human error identification (HEI) techniques are used to predict any potential human/operator error that may occur during a man–machine interaction. Human reliability analysis (HRA) techniques are used to quantify the probability of error occurrence.
Situation awareness assessment techniques.	Situation awareness (SA) refers to an operator's knowledge and understanding of the situation that he or she is placed in. According to Endsley (1995a), SA involves a perception of appropriate goals, comprehending their meaning in relation to the task and projecting their future status. SA assessment techniques are used to determine a measure of operator SA in complex, dynamic systems.
Mental workload assessment techniques.	Mental workload (MWL) represents the proportion of operator resources demanded by a task or set of tasks. A number of MWL assessment techniques exist, which allow the HF practitioner to evaluate the MWL associated with a task or set of tasks.
Team performance analysis techniques.	Team performance analysis techniques are used to describe, analyse and represent team performance in a particular task or scenario. Various facets of team performance can be evaluated, including communication, decision-making, awareness, workload and coordination.
Interface analysis techniques.	Interface analysis techniques are used to assess the interface of a product or systems in terms of usability, error, user satisfaction and layout.

Stage 2: Initial Methods Screening

Before the HF techniques were subjected to further analysis, a screening process was employed to remove any techniques that were not suitable for review with respect to their use in the design and evaluation of systems. Techniques were deemed unsuitable for review if they fell into the following categories:

- *Unavailable* – to be included, the method should be freely available in the public domain.
- *Inapplicable* – those methods deemed unsuitable for use in the design of systems were rejected. In addition, anthropometric, physiological and biomechanical techniques were not reviewed, and the reader is referred to Stanton et al. (2005) for a comprehensive pre-existing account.
- *Duplication* – HF methods are often reiterated and presented in a new format. Any methods that were very similar to other methods already chosen for review were rejected (the version in most common usage being selected instead).
- *Limited use* – the field of HF is not short of methods and quite often a method is developed and not used by anyone other than the developer. Any methods that had not been applied in an analysis of some sort were rejected.

As a result of the method screening procedure, a list of 107 HF methods suitable for use in the system design and evaluation process was created. This list was then circulated within the professional HF community to ensure the suitability of the methods chosen for review and also to check the comprehensiveness of the list.

Stage 3: Methods Review

The 107 HF design and evaluation methods were then analysed using the set of pre-determined criteria outlined in Table 1.4. The criteria were designed not only to establish which of the techniques was the most suitable for use in the design and evaluation of systems, but to provide a standardised, simplified way of communicating 'how to' perform the method in practice. Therefore, the output is designed to act as a manual, with the headings in Table 1.4 serving as the structure for each of the methods contained in this book.

Table 1.4   Descriptions of method review criteria

Criteria	Description of criteria
Name and acronym.	The name of the technique or method and its associated acronym.
Author(s), affiliations(s) and address(es).	The names, affiliations and addresses of the authors are provided to assist with citation and requesting any further help in using the technique.
Background and applications.	This section introduces the method, its origins and development, the domain of application of the method and also application areas in which it has been used.
Domain of application.	This describes the domain that the technique was originally developed for and applied in.
Procedure and advice.	This section describes the procedure for applying the method as well as general points of expert advice.
Flowchart.	A flowchart is provided, depicting the method's procedure.
Advantages.	Lists the advantages associated with using the method in the design of systems.
Disadvantages.	Lists the disadvantages associated with using the method in the design of systems.
Example.	An example (or examples) of the application of the method is provided to show the method's output.
Related methods.	Any closely related methods are listed, including contributory and similar methods.
Approximate training and application times.	Estimates of the training and application times are provided to give the reader an idea of the commitment required when using the technique.
Reliability and validity.	Any evidence on the reliability or validity of the method is cited.
Tools needed.	This describes any additional tools required when using the method.
Bibliography.	A bibliography lists recommended further reading on the method and the surrounding topic area.

Six years have now passed since the first edition of this book and the continual growth of the HF domain brings with it an increasingly diverse range of methodologies. Within this second edition, an additional 107 methods are included, spanning 12 categories. In addition to the new methods, the book also contains updated examples and references for the pre-existing methods, outlining recent applications and adaptations to the techniques. The propagation of technology is reflected in this second edition, Chapter 2 on data collection has been revised to illustrate the increasing level of reliance upon technology with 'technological advances' sections for each method, and applicable software tools have been reviewed and are discussed in the appropriate methods sections.

In summary, this book has been designed for you to consult for advice and guidance on which methods have potential application to your problem, how to actually use any given method in practice and what questions you might wish to ask as a consumer of HF methods. The book is also designed to help you understand the interdependencies between methods and which method outputs are required to act as inputs for other methods. An HF methods matrix has been constructed with this aim in mind, highlighting that the integration of HF techniques is often the only way to ensure that all relevant issues within a multi-faceted problem are captured. As it is, HF methods represent a surprisingly underused resource for engineers, designers, procurers and testers. As a cursory glance at the methods contained in this book will show, most if not all are fully compatible with systems engineering and are highly relevant to quality assurance processes as well as the aims and aspirations of intelligent, more informed procurement. HF methods enable engineers, designers and specialist HF practitioners to add considerable value to current and future systems. This book presents an actionable set of methods which can be put to immediate use in achieving this aim.

Chapter 2
Data Collection Methods

The starting point of any Human Factors (HF) analysis will be the scoping and definition of expected outcomes; for example, this might mean defining hypotheses or determining which questions the analysis is intended to answer. Following this stage, effort normally involves collecting specific data regarding the system, activity and personnel that the analysis effort is focused upon. In the design of novel systems, information regarding activity in similar, existing systems is required. This allows the design team to evaluate existing or similar systems in order to determine existing design flaws and problems, and also to highlight efficient aspects that may be carried forward into the new design. The question of what constitutes a 'similar' system is worth considering at this juncture. If we concentrate solely on the current generation of systems (with a view to planning the next generation), then it is likely that any design proposals would simply be modifications to current technology or practice. While this might be appropriate in many instances, it does not easily support original design (which might require a break with current systems). An alternative approach is to find systems that reflect some core aspect of current work and then attempt to analyse the activity within these systems. Thus, in designing novel technology to support newspaper editing, production and layout planning, Bødker (1988) focused on manual versions of the activities rather than on the contemporary word processing or desktop publishing systems. An obvious reason for doing this is that the technology (particularly at the time of her study) would heavily constrain the activity that people could perform, and these constraints might be appropriate for the limitations of the technology but not supportive of the goals and activities of the people working within the system. In a similar manner, Stanton and Baber (2002), in a study re-designing a medical imaging system, decided to focus their analysis on cytogeneticists using conventional microscopes rather than analysts using the sophisticated imaging equipment. Thus, it can be highly beneficial to look at activity away from the technology for several reasons:

1. avoiding the problems of technology constraining possible activity;
2. allowing appreciation of the fundamental issues relating to the goals of people working with the system (as opposed to understanding the manner in which particular technology needs to be used); and
3. allowing (often) rapid appreciation of basic needs without the need to fully understand complex technology.

The evaluation of existing, operational systems (e.g. usability, error analysis and task analysis) also requires that specific data regarding task performance in the system under analysis is collected, represented and analysed accordingly. Data collection methods therefore represent the cornerstone of any HF analysis effort. Such methods are used by the HF practitioner to collect specific information regarding the system, activity or artefact under analysis, including the nature of the activity conducted within the system, the individuals performing the activity, the component task steps and their sequence, the technological artefacts used by the system and its personnel in performing the tasks (controls, displays, communication technology, etc.), the system environment and also the organisational environment. In terms of HFI, therefore, the methods can readily contribute to our understanding of personnel, training, HF engineering and system safety.

The importance of an accurate representation of the system or activity under analysis cannot be underestimated and is a necessary prerequisite for any further analysis efforts. As we noted above, the starting point for designing future systems is a description of the current or analogous system, and any inaccuracies within the description could potentially hinder the design effort. Data collection methods are used to collect the relevant information that is used to provide this description of the system or activity under analysis. There are a number of different data collection methods available to the HF practitioner, including observation, interviews, questionnaires, analysis of artefacts, usability metrics and the analysis of performance. Often, data collected through the use of these methods can be utilised as the starting point or input for another HF method, such as human error identification (HEI), task analysis and charting techniques.

The main advantage associated with the application of data collection methods is the high volume and utility of the data that is collected. The analyst(s) using the methods also have a high degree of control over the data collection process and are able to direct the data collection procedure as they see fit. Despite the usefulness of data collection methods, there are a number of potential problems associated with their use. For example, one problem associated with the utilisation of data collection methods such as interviews, observational study and questionnaires is the high level of resource usage incurred, particularly during the design of data collection procedures. The design of interviews and questionnaires is a lengthy process, involving numerous pilot runs and reiterations. In addition to this, large amounts of data are typically collected, and lengthy data analysis procedures are common. For example, analysing the data obtained during observational study efforts is particularly laborious and time-consuming, even with the provision of supporting computer software such as Observer™, and can last weeks rather than hours or days. In addition to the high resource usage incurred, data collection techniques also require access to the system and personnel under analysis, which is often very difficult and time-consuming to obtain. If the data needs to be collected during operational scenarios, getting the required personnel to take part in interviews is also difficult, and questionnaires often have very low return rates i.e. typically 10 per cent for a postal questionnaire. Similarly, institutions do not readily agree to personnel being observed whilst at work, and often access is rejected on this basis. A brief description of each of the data collection methods is given below. A summary of the data collection techniques review is presented in Table 2.1 on the following page.

Interviews

Interviews offer a flexible approach to data collection and have consequently been applied for a plethora of different purposes. They can be used to collect a wide variety of data, ranging from user perceptions and reactions to usability and error-related data. There are three types of interview available to the HF practitioner: structured, semi-structured and unstructured or open interviews. Typically, participants are interviewed on a one-to-one basis and the interviewer uses pre-determined probe questions to elicit the required information. A number of interview-based techniques have been developed, including the critical decision method (CDM: Klein and Armstrong, 2004) and the applied cognitive task analysis technique (ACTA: Militello and Hutton, 2000). Both are semi-structured interviews based on cognitive task analysis approaches that are used to elicit information regarding operator decision-making in complex, dynamic environments.

Questionnaires

Questionnaires offer a very flexible means of quickly collecting large amounts of data from large participant populations. They have been used in many forms to collect data regarding numerous issues within HF design and evaluation, and can be used to collect information regarding almost anything at all, including usability, user satisfaction, opinions and attitudes. More specifically, they can be employed throughout the design process to evaluate design concepts and prototypes, to probe user perceptions and reactions, and to evaluate existing systems. Established questionnaires such as the system usability scale (SUS), the questionnaire for user interface satisfaction (QUIS) and the software usability

Table 2.1   Summary of the data collection technique

Method	Type of Method	Domain	Training Time	App Time	Related Methods	Tools Needed	Validation Studies	Advantages	Disadvantages
Interviews	Data collection.	Generic.	Med–high.	High.	Critical Decision Method CASI.	Pen and paper. Audio recording equipment.	Yes.	1) Flexible technique that can be used to assess anything from usability to error. 2) Interviewer can direct the analysis. 3) Can be used to elicit data regarding cognitive components of a task.	1) Data analysis is time consuming and laborious. 2) Reliability is difficult to assess. 3) Subject to various source of bias.
Method	**Type of Method**	**Domain**	**Training Time**	**App Time**	**Related Methods**	**Tools Needed**	**Validation Studies**	**Advantages**	**Disadvantages**
Questionnaires.	Data collection.	Generic.	Low.	High.	SUMI. QUIS. SUS. CSAQ.	Pen and paper. Video and audio recording equipment.	Yes.	1) Flexible technique that can be used to assess anything from usability to error. 2) A number of established HF questionnaire techniques already exist, such as SUMI and SUS. 3) Easy to use, requiring minimal training.	1) Data analysis is time consuming and laborious. 2) Subject to various source of bias. 3) Questionnaire development is time consuming and requires a large amount of effort on behalf of the analyst(s).
Method	**Type of Method**	**Domain**	**Training Time**	**App Time**	**Related Methods**	**Tools Needed**	**Validation Studies**	**Advantages**	**Disadvantages**
Observation.	Data collection.	Generic.	Low.	High.	Acts as an input to various HF methods e.g. HTA.	Pen and paper. Video and audio recording equipment.	Yes.	1) Can be used to elicit specific information regarding decision-making in complex environments. 2) Acts as the input to numerous HF techniques such as HTA. 3) Suited to the analysis of C4i activity.	1) Data analysis procedure is very time consuming. 2) Coding data is also laborious. 3) Subject to bias.
Mouse tracking.	Data collection.	Generic.	Low.	Low.	VPA. Eye tracking.	Computer and software.	Yes.	1) The method is unobtrusive, simple with minimal costs. 3) A link between cognition and physical movement has been established.	1) Only provides information on physical movements not cognition.

measurement inventory (SUMI) are available for practitioners to apply to designs and existing systems. Alternatively, specific questionnaires can be designed and administered during the design process.

Observation

Observation (and observational studies) is used to gather data regarding activity conducted in complex, dynamic systems. In its simplest form, it involves observing an individual or group of individuals performing work-related activity. A number of different types of observational study exist, such as direct observation, covert observation and participant observation. Observation is attractive due to the volume and utility of the data collected, and also the fact that the data is collected in an operational context. Although, at first glance, simply observing an operator at work seems to be a very simple technique to employ, it is evident that this is not the case, and that careful planning and execution are required (Stanton 2003). Observational techniques also require the provision of technology, such as video and audio recording equipment. The output from an observational analysis is used as the primary input for most HF techniques, such as task analysis, error analysis and charting techniques.

Mouse Tracking

Mouse tracking is used to collect data regarding a user's interaction with a computer interface. A piece of software automatically records all mouse movement and mouse clicks during task performance, providing both a high-level summary and detailed descriptions of action. It is deemed to be a cheap, convenient alternative to eye tracking (Atterer, Wnuk and Schmidt, 2006; Chen, Anderson and Sohn, 2001; Rodden and Fu, 2007) and although the method only collects information regarding users' physical interaction with the interface, a link has been established between movement and cognition (Arroyo, Selker and Wei, 2006; Chen, Anderson and Sohn, 2001; Rodden and Fu, 2007). The technique is argued to be a valuable resource in the collection of data regarding interface design as well as for the development of user profiles (Atterer, Wnuk and Schmidt, 2006).

Interviews

Background and Applications

Interviews provide the HF practitioner with a flexible means of gathering large amounts of specific information regarding a particular subject. Due to the flexible nature of interviews, they have been used extensively to gather information on a plethora of topics, including system usability, user perceptions, reactions and attitudes, job analysis, cognitive task analysis, error and many more. As well as designing their own interviews, HF practitioners also have a number of specifically designed interview techniques at their disposal. For example, the CDM (Klein and Armstrong, 2004) is a cognitive task analysis technique that provides the practitioner with a set of cognitive probes designed to elicit information regarding decision-making during a particular scenario (see the relevant section for a full CDM description. The three generic interview 'types' typically employed by the HF practitioner are outlined below:

- *Structured*: in a structured interview, the interviewer probes the participant using a set of pre-defined questions designed to elicit specific information regarding the subject under analysis. The content of the interview (questions and their order) is pre-determined and no scope for further discussion is permitted. Due to their rigid nature, structured interviews are the least popular type of interview. They are only used when the type of data required is rigidly defined, and no additional data is required.

- *Semi-structured*: when using a semi-structured interview, some of the questions and their order are pre-determined. However, semi-structured interviews are flexible in that the interviewer can direct the focus of the interview and also use further questions that were not originally part of the planned interview structure. As a result, information surrounding new or unexpected issues is often uncovered during semi-structured interviews. Due to this flexibility, the semi-structured interview is the most commonly applied type of interview.
- *Unstructured*: when using an unstructured interview, there is no pre-defined structure or questions and the interviewer goes into the interview 'blind' so to speak. This allows the interviewer to explore, on an ad hoc basis, different aspects of the subject under analysis. Whilst their flexibility is attractive, unstructured interviews are infrequently used, as their lack of structure may result in crucial information being neglected or ignored.

Focus Group

While many interviews concentrate on the one-to-one elicitation of information, group discussions can provide an efficient means of canvassing consensus opinion from several people. Ideally, the focus group would contain around five people with similar backgrounds and the discussion would be managed at a fairly high-level, i.e. rather than asking specific questions, the analyst would introduce topics and would facilitate their discussion. A useful text for exploring focus groups is Langford and McDonagh (2002).

Question Types

An interview involves the use of questions or probes designed to elicit information regarding the subject under analysis. An interviewer typically employs three different types of question during the interview process. These are closed questions, open-ended questions and probing questions. A brief description of each interview question type is presented below:

- *Closed*: closed questions are used to gather specific information and typically permit yes or no answers. An example of a closed question would be: 'Do you think that system X is usable?' The question is designed to gather a yes or no response and the interviewee does not elaborate on their chosen answer.
- *Open-ended*: an open-ended question is used to elicit more than the simple yes/no information that a closed question gathers. It allows the interviewees to answer in whatever way they wish and also to elaborate on their answer. For example, an open-ended question approach to the topic of system X's usability would be something like: 'What do you think about the usability of system X?' By allowing the interviewee to elaborate upon answers given, open-ended questions typically gather more pertinent data than closed questions. However, open-ended question data requires more time to analyse than closed question data, and so closed questions are more commonly used.
- *Probing*: a probing question is normally used after an open-ended or closed question to gather more specific data regarding the interviewee's previous answer. Typical examples of a probing question would be 'Why did you think that system X was not usable?' or 'How did it make you feel when you made that error with the system?'.

Stanton and Young (1999a) recommend that interviewers should begin with a specific topic and probe it further until the topic is exhausted; then move on to a new topic. They advocate that the interviewer should begin by focusing on a particular topic with an open-ended question, and then, once the interviewee has answered, use a probing question to gather further information. A closed question should then be used to gather specific information regarding the topic. This cycle of open-ended, probe and closed questions should be maintained throughout the interview. An excellent general text on interview design is Oppenheim (2000).

Domain of Application

Generic.

Procedure and Advice (Semi-structured Interview)

There are no set rules to adhere to during the construction and conduction of an interview. The following procedure is intended to act as a set of flexible guidelines for the HF practitioner.

Step 1: Define the Interview Objective
First, before any interview design takes place, the analyst should clearly define the objective of the interview. Without a clearly defined objective, the focus of the interview is unclear and the data gathered during the interview may lack specific content. For example, when interviewing a civil airline pilot for a study into design-induced human error on the flight deck, the objective of the interview would be to discover which errors the pilot had made or had seen being made in the past, in which part of the interface, and during which task. A clear definition of the interview objectives ensures that the interview questions used are wholly relevant and that the data gathered is of optimum use.

Step 2: Question Development
Once the objective of the interview is clear, the development of the questions to be used during the interview can begin. The questions should be developed based upon the overall objective of the interview. In the design-induced pilot error case, examples of pertinent questions would be: 'What sort of design-induced errors have you made in the past on the flight deck?' This would then be followed by a probing question such as 'Why do you think you make this error?' or 'What task were you performing when you made this error?'. Once all of the relevant questions are developed, they should be put into some sort of coherent order or sequence. The wording of each question should be very clear and concise, and the use of acronyms or confusing terms should be avoided. An interview transcript or data collection sheet should then be created, containing the interview questions and spaces for demographic information (name, age, sex, occupation, etc.) and interviewee responses.

Step 3: Piloting the Interview
Once the questions have been developed and ordered, the analyst should then perform a pilot or trial run of the interview procedure. This allows any potential problems or discrepancies to be highlighted. Typical pilot interview studies involve submitting the interview to colleagues or even performing a trial interview with real participants. This process is very useful in shaping the interview into its most efficient form and allows any potential problems in the data collection procedure to be highlighted and eradicated. The analyst is also given an indication of the type of data that the interview may gather and can change the interview content if appropriate.

Step 4: Re-design Interview Based upon Pilot Run
Once the pilot run of the interview is complete, any changes highlighted should be made. This might include the removal of redundant questions, the rewording of existing questions or the addition of new questions.

Step 5: Select Appropriate Participants
Once the interview has been thoroughly tested and is ready for use, the appropriate participants should be selected. Normally, a representative sample from the population of interest is used. For example, in an analysis of design-induced human error on the flight deck, the participant sample would comprise airline pilots with varying levels of experience.

Step 6: Conduct and Record the Interview

According to Stanton and Young (1999a), the interviewee should use a cycle of open-ended, probe and closed questions. He or she should persist with one particular topic until it is exhausted and should then move on to a new topic. General guidelines for conducting an interview prescribe that the interviewer is confident and familiar with the topic in question, communicates clearly and establishes a good rapport with the interviewee. The interviewer should avoid being over-bearing and should not mislead, belittle, embarrass or insult the interviewee. The use of technical jargon or acronyms should also be avoided. It is recommended that the interview be recorded using either audio or visual recording equipment.

Step 7: Transcribe the Data

Once the interview is completed, the analyst should proceed to transcribe the data. This involves replaying the initial recording of the interview and transcribing fully everything that is said during the interview, both by the interviewer and the interviewee. This is typically a lengthy and laborious process and requires much patience on behalf of the analyst involved. It might be worth considering paying someone to produce a word-processed transcription, e.g. by recruiting someone from a Temp Agency for a week or two, or employing a dedicated transcription company. Specialist software (for example, NCH© Software's Express Scribe version 5.0.1) and transcription foot pedals (for example, VEC's Infinity In-USB-2 transcription pedals) can also be utilised to speed up the transcription process.

Step 8: Data Gathering

Once the transcript of the interview is complete, the analyst should examine the interview transcript, looking for the specific data that was required by the objective of the interview. This is known as the 'expected' data. Once all of the 'expected data' is gathered, the analyst should re-examine the interview in order to gather any 'unexpected data', that is, any extra data (not initially outlined in the objectives) that is unearthed.

Step 9: Data Analysis

Finally, the analysts should then examine the data using appropriate statistical tests, graphs, etc. The form of analysis used is dependent upon the aims of the analysis, but typically involves converting the words collected during the interview into numerical form in readiness for statistical analysis. A good interview will always involve planning, so that the data is collected with a clear understanding of how subsequent analysis will be performed. In other words, it is not sufficient to have piles of handwritten notes following many hours of interviewing and then have no idea what to do with them. A good starting point is to take the transcribed information and then perform some 'content analysis', i.e. divide the transcription into specific concepts. Then it can be determined whether the data collected from the interviews can be reduced to some numerical form, e.g. counting the frequency with which certain concepts are mentioned by different individuals or the frequency with which concepts occur together.

Alternatively, the content of the interview material might not be amenable to reduction to numerical form, and so it is not possible or sensible to consider statistical analysis. In this case, it is common practice to work through the interview material and look for common themes and issues. These can be separated out and (if possible) presented back to the interviewees, using their own words. This can provide quite a powerful means of presenting opinion or understanding. If the interview has been video-taped, then it can be useful to edit the video down in a similar manner, i.e. to select specific themes and use the video of the interviewees to present and support these themes.

Advantages

- Interviews can be used to gather data regarding a wide range of subjects.
- Interviews offer a very flexible way of gathering large amounts of data.
- The data gathered is potentially very powerful.
- The interviewer has full control over the interview and can direct the interview in any way.
- Response data can be treated statistically.
- A structured interview offers consistency and thoroughness (Stanton and Young, 1999a).
- Interviews have been used extensively in the past for a number of different types of analysis.
- Specific, structured HF interview techniques already exist, such as the CDM (Klein and Armstrong, 2004).

Disadvantages

- The construction and data analysis process ensure that the interview technique is a time-consuming one.
- The reliability and validity of the technique is difficult to address.
- Interviews are susceptible to both interviewer and interviewee bias.
- Transcribing the data is a laborious, time-consuming process.
- Conducting an interview correctly is quite difficult and requires great skill on behalf of the interviewer.
- The quality of the data gathered is based entirely upon the skill of the interviewer and the quality of the interviewee.

Approximate Training and Application Times

In a study comparing 12 HF techniques, Stanton and Young (1999a) reported that interviews took the longest to train of all the methods, due to the fact that the technique is a refined process requiring a clear understanding on the analyst's behalf. In terms of application times, a normal interview could last anything between 10 and 60 minutes. Kirwan and Ainsworth (1992) recommend that an interview should last a minimum of 20 minutes and a maximum of 40 minutes. Whilst this represents a low application time, the data analysis part of the interview technique can be an extremely lengthy one (e.g. data transcription, data gathering and data analysis). Transcribing the data is a particularly time-consuming process. For this reason, the application time for interviews is estimated as very high.

Reliability and Validity

Although the reliability and validity of interview techniques is difficult to address, Stanton and Young (1999a) report that in a study comparing 12 HF techniques, a structured interview technique scored poorly in terms of reliability and validity.

Tools Needed

An interview requires a pen and paper and an audio recording device, such as a digital voice recorder. A PC with a word processing package such as Microsoft Word™ is also required in order to transcribe the data and statistical analysis packages such as SPSS™ may be required for data analysis procedures.

Flowchart

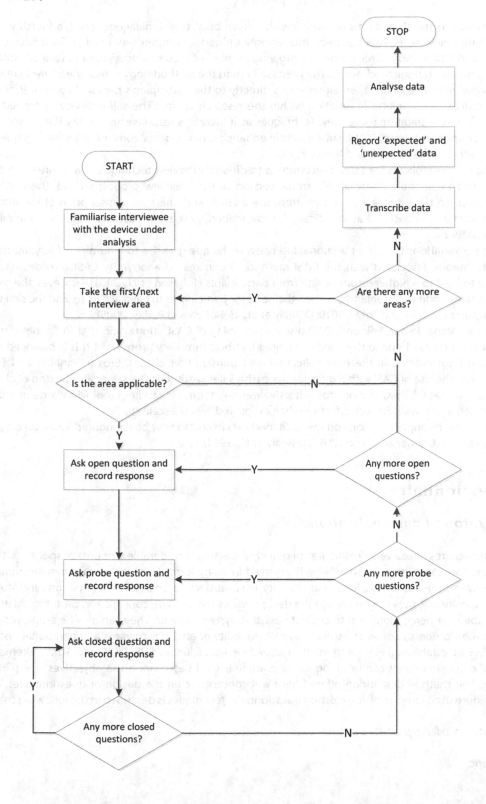

Interviews in the Twenty-First Century

In addition to the three forms of interview that have been traditionally employed, a fourth variation of the interview technique has been increasingly utilised – Computer-Assisted Self-Interviews (CASI: Evans and Miller, 1969). CASI is one of a growing number of methods developed to take advantage of emergent technological advances. The premise behind the methodology is that the analyst creates an interview which he or she then either sends directly to the participant's personal computer or which the participant completes in isolation within the research setting. The self-interview is argued to be preferable over traditional interview techniques as it increases response anonymity (Evans and Miller, 1969; Couper and Rowe, 2010), subsequently enhancing participants' comfort in answering questions on sensitive topics (Couper and Rowe, 2010).

The method follows the same procedure as traditional interview techniques, with three exceptions: first, the analyst must create an electronic version of the interview probes; second, the participant will complete the interview separately from the analyst; and, third, no transcription of the interview is required as the data will automatically be transcribed by the computer used (Brown, Vanable and Eriksen, 2008).

These variations from the traditional interview technique give rise to a number of advantages and disadvantages associated with the CASI method. The increased anonymity CASI provides has been agued to lead to a higher response rate from participants (Johnson et al., 2001). However, the method also creates additional demands, such as the need for participants to have access to, and the skill to use, a computer (Couper and Rowe, 2010; O'Reilly et al., 1994; Rea and Parker, 2005).

Brown, Vanable and Eriksen (2008) discuss the utility of CASI, suggesting that initial development times are increased due to the need to create electronic interview probes, yet this is balanced by the increased automation in the data collection and transcription phases. Brown, Vanable and Eriksen compared the use of CASI with a questionnaire (the Safety Attitudes Questionnaire: Sexton et al., 2006) and found that CASI was a more cost-effective method for use in studies employing large numbers of participants, but was less cost-effective when employed in small-scale studies.

There are multiple variations on the CASI method which can now be configured into a text, audio or video format (Couper and Rowe, 2010; O'Reilly et al., 1994).

Questionnaires

Background and Applications

Questionnaires offer a very flexible way of quickly collecting considerable amounts of specific data from a large population sample. They have been used in many forms to collect data regarding numerous issues within HF and design, including usability, user satisfaction, error, and user opinions and attitudes. More specifically, they can be used in the design process to evaluate concept and prototypical designs, to probe user perceptions and to evaluate existing system designs. They can also be employed in the evaluation process in order to evaluate system usability or attitudes towards an operational system. A number of established HF questionnaires already exist, including SUS, QUIS and SUMI. Alternatively, specific questionnaires can be designed and administered based upon the objectives of a particular study. The method description offered here will concentrate on the design of questionnaires, as the procedure used when applying existing questionnaire techniques is described in the following chapters.

Domain of Application

Generic.

Procedure and Advice

There are no set rules for the design and administration of questionnaires. The following procedure is intended to act as a set of guidelines to consider when constructing a questionnaire.

Step 1: Define the Study Objectives

The first step involves clearly defining the objectives of the study, i.e. what information is wanted from the questionnaire data that is gathered. Before any effort is put into the design of the questions, the objectives of the questionnaire must be clearly defined. It is recommended that the analyst should go further than merely describing the goal of the research. For example, when designing a questionnaire in order to gather information on the usability of a system or product, the objectives should contain precise descriptions of different usability problems already encountered and descriptions of the usability problems that are expected. In addition, the different tasks involved in the use of the system in question should be defined and the different personnel should be categorised. What the results are supposed to show and what they could show should also be specified, as well as the types of questions (closed, multiple-choice, open-ended, rating, ranking, etc.) to be used. This stage of questionnaire construction is often neglected, and consequently the data obtained normally reflects this (Wilson and Corlett, 1995).

Step 2: Define the Population

Once the objectives of the study are clearly defined, the analyst should define the sample population, i.e. the participants whom the questionnaire will be administered to. Again, the definition of the participant population should go beyond simply describing an area of personnel, such as 'control room operators', and should be as exhaustive as possible, including defining age groups, different job categories (control room supervisors, operators, management, etc.) and different organisations. The sample size should also be determined at this stage. Sample size is dependent upon the scope of the study and also the amount of time and resources available for data analysis.

Step 3: Construct the Questionnaire

A questionnaire is typically comprised of four parts: an introduction, a participant information section, an information section and an epilogue. The introduction should contain information that lets the participant know who you are, what the purpose of the questionnaire is and what the results are going to be used for. One must be careful to avoid putting information in the introduction that may bias the participant in any way. For example, describing the purpose of the questionnaire as 'determining usability problems with existing C4i interfaces' may lead the participant before the questionnaire has begun. The classification part of the questionnaire normally contains multiple-choice questions requesting information about the participant, such as age, sex, occupation and experience. The information part of the questionnaire is the most crucial part, as it contains the questions designed to gather the required information related to the initial objectives. There are numerous categories of questions that can be used in this part of the questionnaire. Which type of question to be used is dependent upon the analysis and the type of data required. Where possible, the type of question used in the information section of the questionnaire should be consistent, i.e. if the first few questions are multiple choice, then all of the questions should be kept as multiple choice. The different types of questions available are displayed in Table 2.2. Each question used in the questionnaire should be short in length and worded clearly and concisely, using relevant language. Data analysis should be considered when constructing the questionnaire. For instance, if there is little time available for the data analysis process, then the use of open-ended questions should be avoided, as they are time-consuming to collate and analyse. If time is limited, then closed questions should be used, as they offer specific data that is quick to collate and analyse. The size of the questionnaire is also important. If it is too large, the participants will not complete the questionnaire, yet a very small questionnaire may seem worthless and could suffer the same fate.

The optimum questionnaire length is dependent upon the participant population, but it is generally recommended that questionnaires should be no longer than two pages (Wilson and Corlett, 1995).

Step 4: Piloting the Questionnaire
Wilson and Corlett (1995) recommend that once the questionnaire construction stage is complete, a pilot run of the questionnaire is required. This is a crucial part of the questionnaire design process, yet it is often neglected by HF practitioners due to various factors, such as time and financial constraints. During this step, the questionnaire is evaluated by its potential user population, domain experts and other HF practitioners. This allows any problems with the questionnaire to be removed before the critical administration phase. Typically, numerous problems are encountered during the pilot stage, such as errors within the questionnaire, redundant questions and questions that the participants simply do not understand or find confusing. Wilson and Corlett (1995) recommend that the pilot stage should comprise the following three stages:

- *Individual criticism*: the questionnaire should be administered to several colleagues who are experienced in questionnaire construction, administration and analysis. These colleagues should be encouraged to offer criticisms of the questionnaire.
- *Depth interviewing*: once the individual criticisms have been attended to and any changes have been made, the questionnaire should be administered to a small sample of the intended population. Once they have completed the questionnaire, the participants should be subjected to an interview regarding the answers that they provided. This allows the analyst to ensure that the questions were fully understood and that the correct (required) data is obtained.
- *Large sample administration*: the re-designed questionnaire should then be administered to a large sample of the intended population. This allows the analyst to ensure that the correct data is being collected and also that sufficient time is available to analyse the data. Redundant questions can also be highlighted during this stage. The likely response rate can also be predicted based upon the returned questionnaires in this stage.

Table 2.2   Types of questions used in questionnaire design

Type of question	Example question	When to use
Multiple choice.	On approximately how many occasions have you witnessed an error being committed with this system? (0–5, 6–10, 11–15, 16–20, more than 20).	When the participant is required to choose a specific response.
Rating scales.	I found the system unnecessarily complex (Strongly Agree (5), Agree (4), Not sure (3), Disagree (2), Strongly Disagree (1)).	When subjective data regarding participant opinions is required.
Paired associates (bipolar alternatives).	Which of the two tasks A + B subjected you to the most mental workload? (A or B).	When two alternatives are available to choose from.
Ranking.	Rank, on a scale of 1 (Very Poor Usability) to 10 (Excellent Usability), the usability of the device.	When a numerical rating is required.
Open-ended questions.	What did you think of the system's usability?	When data regarding participants' own opinions about a certain subject is required, i.e. subjects compose their own answers.
Closed questions.	Which of the following errors have you committed or witnessed whilst using the existing system (action omitted, action on wrong interface element, action mistimed, action repeated, action too little, action too much).	When the participant is required to choose a specific response.
Filter questions.	Have you ever committed an error whilst using the current system interface? (Yes or No; if Yes, go to question 10, if No, go to question 15).	To determine whether participant has specific knowledge or experience; to guide participant past redundant questions.

Step 5: Questionnaire Administration
Once the questionnaire has been successfully piloted, it is ready to be administered. Exactly how the questionnaire is administered is dependent upon the aims and objectives of the analysis, and also the target population. For example, if the target population can be gathered together at a certain time and place, then the questionnaire could be administered at this time, with the analyst(s) present. This ensures that the questionnaires are completed. However, gathering the target population in one place at the same time can be problematic and so questionnaires are often administered by post. Although this is quick and cheap, requiring little input from the analyst(s), the response rate is very low, typically 10 per cent. Procedures to address poor responses rates are available, such as offering payment on completion, the use of encouraging letters, offering a donation to charity upon return, contacting non-respondents by telephone and sending shortened versions of the initial questionnaire to non-respondents. All these methods have been shown in the past to improve response rates, but almost all involve substantial extra costs.

Step 6: Data Analysis
Once all (or a sufficient amount) of the questionnaires have been returned or collected, the data analysis process should begin. This is a lengthy process, the exact time required being dependent upon a number of factors (e.g. number of question items, sample size, required statistical techniques and data reduction). Questionnaire data is normally computerised and analysed statistically.

Step 7: Follow-up Phase
Once the data is analysed sufficiently and conclusions are drawn, the participants who completed the questionnaire should be informed regarding the outcome of the study. This might include a thank-you letter and an associated information pack containing a summary of the research findings.

Advantages

- Questionnaires offer a very flexible way of collecting large volumes of data from large participant samples.
- When the questionnaire is properly designed, the data analysis phase should be quick and very straightforward.
- Very few resources are required once the questionnaire has been designed.
- A number of HF questionnaires already exist (QUIS, SUMI, SUS, etc.), allowing the analyst to choose the most appropriate for the purposes of the study. This also removes the time associated with the design of the questionnaire. The use of pre-existing questionnaires also enables a comparison with past results obtained using the same questionnaire.
- Questionnaires are very easy to administer to large numbers of participants.
- Skilled questionnaire designers can use the questions to direct the data collection.

Disadvantages

- Designing, piloting, administering and analysing a questionnaire is time-consuming.
- Reliability and validity of questionnaires is questionable.
- The questionnaire design process is taxing, requiring great skill on the analyst's part.
- Typically, response rates are low (around 10 per cent for postal questionnaires).
- The answers provided in questionnaires are often rushed and non-committal.
- Questionnaires are prone to a number of different biases, such as prestige bias.
- Questionnaires can offer a limited output.

Flowchart

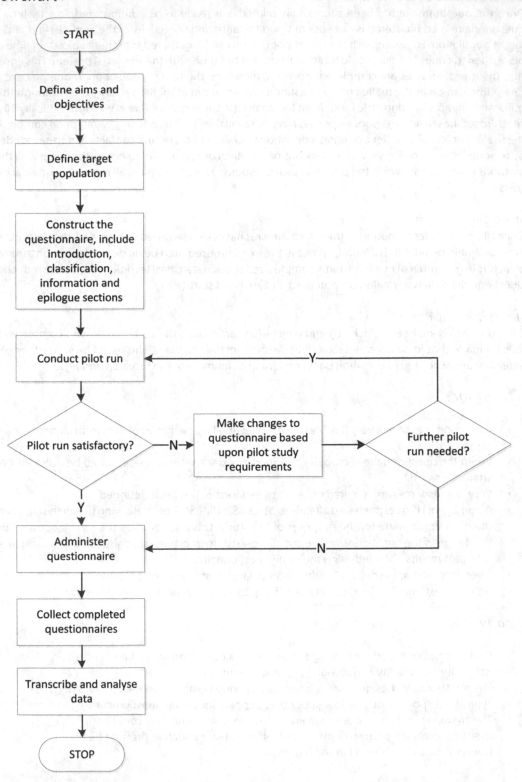

Example

Marshall et al. (2003) conducted a study designed to investigate the prediction of design-induced error on civil flight decks. The human error template (HET) technique was developed and used to predict potential design-induced errors on the flight deck of aircraft X during the flight task 'Land aircraft X at New Orleans Airport using the autoland system'. In order to validate the error predictions made, a database of error occurrence for the flight task under analysis was required. A questionnaire was developed based upon the results of an initial study using the Systematic Human Error Reduction and Prediction Approach (SHERPA) (Embrey, 1986) technique to predict design-induced error during the flight task under analysis. The questionnaire was based upon the errors identified using the SHERPA technique and included a question for each error identified. Each question was worded to ask respondents whether they had ever made the error in question or whether they knew of anyone else who had made the error. The questionnaire contained 73 questions in total. A total of 500 questionnaires were sent out to civil airline pilots and 46 (9.2 per cent) were completed and returned (Marshall et al., 2003). An extract of the questionnaire is presented below.

Aircraft Pilot Error Questionnaire Extract

The questionnaire aims to establish mistakes or errors that you have made or that you know have been made when completing approach and landing. For the most part, it is assumed that the task is carried out using the Flight Control Unit for most of the task. We are hoping to identify the errors that are made as a result of the design of the flight deck, what are termed 'Design Induced Errors'.

Position: _____

Total Flying Hours: _____

Hours on Aircraft Type: _____

This questionnaire has been divided broadly into sections based upon the action being completed. In order to be able to obtain the results that we need, the questionnaire may appear overly simplistic or repetitive but this is necessary for us to break down the possible problems into very small steps that correspond to the specific pieces of equipment or automation modes being used.

Some of the questions may seem to be highly unlikely events that have not been done as far as you are aware but please read and bypass these as you need to.

Next to each statement, there are two boxes labelled 'Me' and 'Other'. If it is something that you have done personally then please tick 'Me'. If you know of colleagues who have made the same error, then please tick 'Other'. If applicable, please tick both boxes.

Q	Error	Me	Other
	Failed to check the speed brake setting at any time.	☐	☐
	Intended to check the speed brake setting and checked something else by mistake.	☐	☐
	Checked the speed brake position and misread it.	☐	☐
	Assumed that the lever was in the correct position and later found that it was in the wrong position.	☐	☐
	Set the speed brake at the wrong time (early or late).	☐	☐
	Failed to set the speed brake (at all) when required.	☐	☐
	Moved the flap lever instead of the speed brake lever when intended to apply the speed brake.	☐	☐

Q	Error	Me	Other
	Started entering an indicated air speed on the Flight Control Unit and found that it was in MACH mode or vice versa.	☐	☐
	Misread the speed on the Primary Flight Display.	☐	☐
	Failed to check airspeed when required to.	☐	☐
	Initially, dialled in an incorrect airspeed on the Flight Control Unit by turning the knob in the wrong direction.	☐	☐
	Found it hard to locate the speed change knob on the Flight Control Unit.	☐	☐
	Having entered the desired airspeed, pushed or pulled the switch in the opposite way to the one that you wanted.	☐	☐
	Adjusted the heading knob instead of the speed knob.	☐	☐
	Found the Flight Control Unit too poorly lit at night to be able to complete actions easily.	☐	☐
	Found that the speed selector knob is easily turned too little or too much, i.e. speed is set to fast/slow.	☐	☐
	Turned any other knob when intending to change speed.	☐	☐
	Entered an airspeed value and accepted it but it was different to the desired value.	☐	☐

Q	Error	Me	Other
	Failed to check that the aircraft had established itself on the localiser when it should have been checked.	☐	☐
	Misread the localiser on the ILS.	☐	☐
	If not on localiser, started to turn in wrong direction to re-establish localiser.	☐	☐
	Incorrectly adjusted heading knob to regain localiser and activated the change.	☐	☐
	Adjusted the speed knob by mistake when intending to change heading.	☐	☐
	Turned heading knob in the wrong direction but realised before activating it.	☐	☐
	Pulled the knob when you meant to push it and vice versa.	☐	☐

Q	Error	Me	Other
	Misread the glideslope on the ILS.	☐	☐
	Failed to monitor the glideslope and found that the aircraft had not intercepted it.	☐	☐

Q	Error	Me	Other
	Adjusted the speed knob by mistake when intending to change heading.	☐	☐
	Turned heading knob in the wrong direction but realised before activating it.	☐	☐
	Turned the knob too little or too much.	☐	☐
	Entered a heading on the Flight Control Unit and failed to activate it at the inappropriate time.	☐	☐

Q	Error	Me	Other
	Misread the altitude on the Primary Flight Display.	☐	☐
	Maintained the wrong altitude.	☐	☐
	Entered the wrong altitude on the Flight Control Unit but realised before activating it.	☐	☐
	Entered the wrong altitude on the Flight Control Unit and activated it.	☐	☐
	Not monitored the altitude at the necessary time.	☐	☐
	Entered an incorrect altitude because the 100/1,000 feet knob wasn't clicked over.	☐	☐
	Believed that you were descending in FPA and found that you were in fact in V/S mode or vice versa.	☐	☐
	Having entered the desired altitude, pushed or pulled the switch in the opposite way to the one that you wanted.	☐	☐

If you would like to tell us anything about the questionnaire or you feel that we have missed out some essential design induced errors, please feel free to add them below and continue on another sheet if necessary.

Please continue on another sheet if necessary.

If you would be interested in the results of this questionnaire then please put the address or email address below that you would like the Executive Summary sent to.

_____

_____

I would be interested in taking part on the expert panel of aircraft X pilots. ☐
Thank you very much for taking the time to complete this questionnaire.

Related Methods

There are numerous questionnaire techniques available to the HF practitioner. Different types of questionnaires include rating scale questionnaires, paired comparison questionnaires and ranking questionnaires. A number of established questionnaire techniques exist, such as the cockpit management attitude questionnaire (CMAQ), SUMI, QUIS and SUS.

Approximate Training and Application Times

Wilson and Corlett (1995) suggest that questionnaire design is more of an art than a science. Practice makes perfect, and practitioners normally need to make numerous attempts at questionnaire design before becoming proficient at the process (see Oppenheim, 2000). Similarly, although the application time associated with questionnaires is at first glance minimal (i.e. the completion phase), when one considers the time expended in the construction and data analysis phases, it is apparent that the total application time is high.

Reliability and Validity

The reliability and validity of questionnaire techniques is questionable. Questionnaire techniques are prone to a number of biases and often suffer from 'social desirability', whereby the participants are merely 'giving the analyst(s) what they want'. Questionnaire answers are also often rushed and non-committal. In a study comparing 12 HF techniques, Stanton and Young (1999) report that questionnaires demonstrated an acceptable level of inter-rater reliability, but unacceptable levels of intra-rater reliability and validity.

Tools Needed

Questionnaires are normally paper-based and completed using pen and paper. Questionnaire design normally requires a computer, along with a word processing package such as Microsoft Word. In the analysis of questionnaire data, a spreadsheet package such as Microsoft Excel™ is required, and a statistical software package such as SPSS™ is also required to treat the data statistically.

Questionnaires in the Twenty-First Century

In the section on interviews above, the CASI method was discussed as a technique which utilises technological advances to improve data collection. A variation on the CASI method is the Computerised Self-Administered Questionnaire (CSAQ: Ramos, Sedivi and Sweet, 1998), a questionnaire that requests participants to enter information electronically using their computer.

The utility attached to employing technology to questionnaire data collection is a common feature in experimental studies today. A recent study by Stanton, Rafferty and Forster (2012) utilised a bespoke online questionnaire application (hosted by the University of Southampton) in order to gain responses from military personnel on a series of questionnaires. The online host enabled data on demographic information, mental workload, usability and situation awareness to be automatically collected, transcribed, entered into a Microsoft Excel template and assigned a unique, anonymous identification number. The utilisation of the online host enabled data to be collected from 186 participants in an entirely automated, error-free and anonymous manner.

Online questionnaires have been used to explore a host of topics in a variety of domains, including mountain sport accidents (Chamarro and Fernandez-Castro, 2009), driving (Tay, 2009), skydiving (Westman et al., 2010) and the safety climate within grocery stores (Kath, Magley and Marmet, 2010).

In addition to bespoke applications, a number of commercial online questionnaire hosts are available, such as Survey Monkey and Zoomerang.

Observation

Background and Applications

Observational techniques are used to gather data regarding the physical and verbal aspects of a task or scenario. These include tasks catered for by the system, the individuals performing the tasks, the tasks themselves (task steps and sequence), errors made, communications between individuals, the technology used by the system in conducting the tasks (controls, displays, communication technology, etc.), the system environment and the organisational environment. Observation has been extensively used and typically forms the starting point of an analysis effort. The most obvious and widely used form of observational technique is direct observation, whereby an analyst visually records a particular task or scenario. However, a number of different forms of observation exist, including direct observation as well as participant observation and remote observation. Drury (1990) suggests that there are five different types of information that can be elicited from observational techniques: the sequence of activities; the duration of activities; the frequency of activities; the amount of time spent in states; and spatial movement. As well as physical (or visually recorded) data, verbal data is also recorded, in particular verbal interactions between the agents involved in the scenario under analysis. Observational techniques can be used at any stage of the design process in order to gather information regarding existing or proposed designs.

Domain of Application

Generic.

Procedure and Advice

There is no set procedure for carrying out an observational analysis. The procedure would normally be determined by the nature and scope of analysis required. A typical observational analysis procedure can be split into the following three phases: the observation design stage, the observation application stage and the data analysis stage. The following procedure provides the analyst with a general set of guidelines for conducting a 'direct'-type observation.

Step 1: Define the Objective of the Analysis
The first step in observational analysis involves clearly defining the aims and objectives of the observation. This should include determining which product or system is under analysis, in which environment the observation will take place, which user groups will be observed, what type of scenarios will be observed and what data is required. Each point should be clearly defined and stated before the process continues.

Step 2: Define the Scenario(s)
Once the aims and objectives of the analysis are clearly defined, the scenario(s) to be observed should be defined and described further. For example, when conducting an observational analysis of control room operation, the type of scenario required should be clearly defined. Normally, the analyst(s) has a particular type of scenario in mind. For example, operator interaction and performance under emergency situations may be the focus of the analysis. The exact nature of the required scenario(s) should be clearly defined by the observation team. It is recommended that a Hierarchical Task Analysis (HTA) is then conducted for the task or scenario under analysis.

Step 3: Observation Plan
Once the aim of the analysis is defined and the type of scenario to be observed is determined, the analysis team should proceed to plan the observation. The team should consider what they are hoping to observe, what they are observing and how they are going to observe it. Depending upon the nature of the observation, access to the system in question should be gained first. This may involve holding meetings with the organisation or establishment in question, and is typically a lengthy process. Any recording tools should be defined and the length of observations should also be determined. In addition, placement of video and audio recording equipment should be considered. To make things easier, a walkthrough of the system/environment/scenario under analysis is recommended. This allows the analyst(s) to become familiar with the task in terms of the activity conducted, the time taken, location and also the system under analysis.

Step 4: Pilot Observation
In any observational study, a pilot or practice observation is crucial. This allows the analysis team to assess any problems with the data collection, such as noise interference or problems with the recording equipment. The quality of data collected can also be tested, as well as any effects upon task performance that may result from the presence of observers. If major problems are encountered, the observation may have to be re-designed. Steps 1–4 should be repeated until the analysis team are happy that the quality of the data collected will be sufficient for their study requirements.

Step 5: Conduct Observation
Once the observation has been designed, the team should proceed with the observation(s). Typically, data is recorded using video and audio recording equipment. An observation transcript is also created during the observation. An example of an observation transcript is presented in Table 2.3. Observation length and timing is dependent upon the scope and requirements of the analysis and also the scenario(s) under analysis. The observation should end only when the required data is collected.

Step 6: Data Analysis
Once the observation is complete, the data analysis procedure begins. Typically, the starting point of the analysis phase involves typing up the observation notes or transcript made during the observation. This is a very time-consuming process, but is crucial to the analysis. Depending upon the analysis requirements, the team should then proceed to analyse the data in the format that is required, such as frequency of tasks,

verbal interactions and sequence of tasks. When analysing visual data, typically user behaviours are coded into specific groups. The software package Observer is frequently used to aid the analyst in this process.

Step 7: Further Analysis

Once the initial process of transcribing and coding the observational data is complete, further analysis of the data begins. Depending upon the nature of the analysis, observation data is used to inform a number of different HF analyses, such as task analysis, error analysis and communications analysis. Typically, observational data is used to develop a task analysis (e.g. HTA) of the task or scenario under investigation.

Step 8: Participant Feedback

Once the data has been analysed and conclusions have been drawn, the participants involved should be provided with feedback of some sort. This could be in the form of a feedback session or a letter to each participant. The type of feedback used is determined by the analysis team.

Example

Stanton et al. (2010a) conducted an observation of five helicopter crews as part of an analysis of distributed decision-making within the military. A team of six analysts observed the performance of five helicopter crews within a series of connected, high-fidelity simulators. The helicopter crews were tasked with planning and executing the evacuation of a number of civilians from a country engulfed by a volatile political situation. During the planning phase of the mission, the helicopter crews were mainly collocated and were observed by the analytic team in a large planning room. Each flight crew was assigned one observer from the research team, who carried a voice recorder and clipboard with an observational transcript – as shown below in Table 2.3. The 'Tag name' refers to the decision-making style label the analyst assigned to each decision.

During the execution phase of the mission, analysts were located inside the helicopter simulators, in a rear compartment to avoid interference, or from an exercise control room where voice communication and video data from the simulators was available in addition to a 'god's eye' view of events. The data for the analysis was used to compare information dissemination, team processes and decision styles utilised within the collocated and distributed environments.

Table 2.3   Extract from the observation transcript of the military decision-making scenario

Time	Actor	Event/conversation	Decision	Context	Tag name
8:56	S	Just use 2. Can we get 4 pallets in each?	Using 2 tanks to take the pallets.	Planning how to take load.	LH
	L	Yes.			
	S	We should be alright then.			
9:01	S	Can you check HTA. What is the northern route on HTA? We look to get clearance – failing that, just go around it.	The route to the HTA.	Planning routes.	CHL
Chinook 2 approaches Chinook 1					
9:15	J	Is that alright?	Refuelling location and agreeing to refuel.	Planning refuelling.	
	P	We'll fuel up there.			
	J	So we can do the whole thing without refuel?			
	P	I would fuel up there.			
Chinook 1 approaches Chinook 2					
9:41	J	We haven't been told that – I made an assumption.	Refuelling time.	Planning refuelling.	HX (assumptive)
9:44	S	Best get fuel there and straight up to there You might want to do that first and then come back and fuel.			

Flowchart

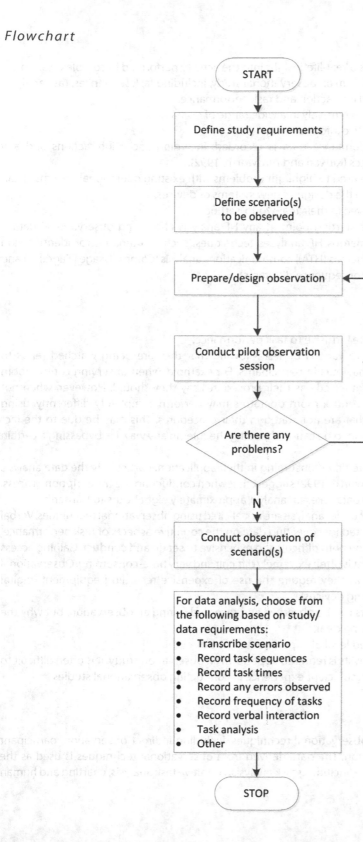

Advantages

- Observational data provides a 'real-life' insight into the activity performed in complex systems.
- Various data can be elicited from an observational study, including task sequences, task analysis, error data, task times, verbal interaction and task performance.
- Observation has been used extensively in a wide range of domains.
- Observation provides objective information.
- Detailed physical task performance data is recorded, including social interactions and any environmental task influences (Kirwan and Ainsworth, 1992).
- Observation analysis can be used to highlight problems with existing operational systems. It can be used in this way to inform the design of new systems or devices.
- Specific scenarios are observed in their 'real-world' setting.
- Observation is typically the starting point in any HF analysis effort, and observational data is used as the input into numerous HF analyses techniques, such as human error identification techniques (SHERPA), task analysis (HTA), communications analysis (Comms Usage Diagrams) and charting techniques (operator sequence diagrams).

Disadvantages

- Observational techniques are intrusive to task performance.
- Observation data is prone to various biases. Knowing that they are being watched tends to elicit new and different behaviours in participants. For example, when observing control room operators, they may perform exactly as their procedures say they should. However, when not being observed, the same control room operators may perform completely differently, using shortcuts and behaviours that are not stated in their procedures. This may be due to the fact that the operators do not wish to be caught bending the rules in any way, i.e. bypassing a certain procedure.
- Observational techniques are time-consuming in their application, particularly the data analysis procedure. Kirwan and Ainsworth (1992) suggest that when conducting the transcription process, one hour of recorded audio data takes an analyst approximately eight hours to transcribe.
- Cognitive aspects of the task under analysis are not elicited using observational techniques. Verbal protocol analysis is more suited for collecting data on the cognitive aspects of task performance.
- An observational study can be both difficult and expensive to set up and conduct. Gaining access to the required establishment is often extremely difficult and very time-consuming. Observational techniques are also costly, as they require the use of expensive recording equipment (digital video camera, audio recording devices).
- Causality is a problem. Errors can be observed and recorded during an observation, but why the errors occur may not always be clear.
- The analyst has only a limited level of experimental control.
- In most cases, a team of analysts is required to perform an observation study. It is often difficult to acquire a suitable team with sufficient experience in conducting observational studies.

Related Methods

There are a number of different observational techniques, including indirect observation, participant observation and remote observation. The data derived from observational techniques is used as the input to a plethora of HF techniques, including task analysis, cognitive task analysis, charting and human error identification techniques.

Approximate Training and Application Times

Whilst the training time for an observational analysis is low (Stanton and Young, 1999a), the application time is typically high. The data analysis phase in particular is extremely time-consuming. The data analysis phase in particular is extremely time-consuming and so too is the transcription phase, as highlighted above by Kirwan and Ainsworth (1992).

Reliability and Validity

Observational analysis is beset by a number of problems that can potentially affect the reliability and validity of the technique. According to Baber and Stanton (1996), problems with causality, bias (in a number of forms), construct validity, external validity and internal validity can all arise unless the correct precautions are taken. Whilst observational techniques possess a high level of face validity (Drury, 1990) and ecological validity (Baber and Stanton, 1996), analyst or participant bias can adversely affect their reliability and validity.

Tools Needed

For a thorough observational analysis, the appropriate visual and audio recording equipment is necessary. Simplistic observational studies can be conducted using pen and paper only; however, for observations in complex, dynamic systems, more sophisticated equipment is required, such as video and audio recording equipment. For the purposes of data analysis, a computer with the Observer software is required.

Observation in the Twenty-First Century

In the discussion on observation above, the utilisation of recording equipment was touched upon. Within this section, technological advances that have impacted the method of observation are outlined. Erlandsson and Jansson (2007) suggest that the use of video recording equipment is becoming an increasingly frequent occurrence in data collection. Additional equipment advances include the use of head-mounted cameras, enabling a first-person account of the scenario under analysis to be recorded. Revell, Stanton and Bessell (2010) employed head-mounted cameras to provide insights into military planning scenarios. Participants wore these cameras during a three-day Battle Group planning mission in which tripod-mounted video cameras were also employed to gain an overall appreciation of the scenario and capture data from video screens and whiteboards. Post-mission, the participants were each given their first-person recordings, along with the tripod-mounted video footage, to provide a high level of detail on which to base their development of an HTA on their role in the mission. The video footage was also used by the analytic team to create sequences of action and metaphorical views of task performance, and to populate the abstraction hierarchy phase of cognitive work analysis. This data fed into the development of a demonstrator to illustrate technological improvements to the mission planning system employed in the task.

Although the use of recording equipment in observations does provide the ability to accurately record a high degree of task performance, it is possible that problems may arise through planning oversights, resulting in some aspect of the scenario not being captured due to problems with camera angles or audio recording ranges. In addition to this, the wealth of data collected can be time-consuming to collate and analyse, and deciphering relevant data may be difficult.

In the three previous sections, the benefits attached to utilising technological developments in data collection have been outlined. Within the following section, a new form of data collection, rather than a variation on already-existing data collection techniques, is discussed.

Mouse Tracking

Background and Applications

Mouse tracking involves the utilisation of software to capture data relating to a user's interaction with an interface, specifically based upon mouse movement and mouse clicks (Arroyo, Selker and Wei, 2006; Atterer, Wnuk and Schmidt, 2006). A user's interaction with the interface is clearly presented through the use of a heat map to illustrate areas of high- and low-frequency mouse travel. The software also provides a summary table representing all mouse movements and clicks that occurred during the task scenario.

The method is based on the hypothesis that mouse travel is correlated with mental processing in that a user will move the mouse to areas of consideration (Chen, Anderson and Sohn, 2001; Rodden and Fu, 2007). Research by Arroyo, Selker and Wei (2006) argues that mouse tracking represents an indication of visual attention. In this way, it can map the different pathways considered by a user when moving through a system interface. Rodden and Fu (2007) and Arroyo, Selker and Wei (2006) posit that the technique can elicit data regarding the appeal of different interface elements, providing valuable information for interface design. A study by Atterer, Wnuk and Schmidt (2006) hypothesised that mouse-tracking data could be employed in both the evaluation of user interfaces and in the development of user profiles.

Atterer, Wnuk and Schmidt (2006), Chen, Anderson and Sohn (2001) and Rodden and Fu (2007) suggest that mouse tracking is a promising alternative to eye tracking, offering similar data with lower financial costs, the ability to collect data remotely and the removal of set up costs (financial and time).

Domain of Application

According to Atterer, Wnuk and Schmidt (2006), the main area of application for mouse tracking has previously been in the evaluation of web-based applications. The method is generic and could be utilised in the evaluation of any system interface; for example, Stanton and McIlroy (2012) applied the method in the comparison of military mission planning interfaces.

Procedure and Advice

Step 1: Define Task
The first stage of the method involves clearly defining the task under analysis. The analyst must be aware of exactly what tasks need to be observed and what information is required. Without a clear definition of the task under analysis, the information gathered may be inappropriate.

Step 2: Design Task
Once the analyst has a clear understanding of the task under analysis, an experimental task needs to be designed in order to elicit the relevant information from the participants.

Step 3: Recruit Participants
At this stage, the analyst must recruit appropriate participants for the analysis. The choice of participants will be dependent upon the focus of analysis; if the analysis is examining expert users, then experts should be recruited, while if the analysis is exploring the way in which novices interact with a system, then novices should be recruited.

Step 4: Brief Participants
Once recruited, the participants should be given a brief on the task they are being asked to undertake, the objectives of the study and exactly what the mouse-tracking software is capturing.

Step 5: Conduct a Pilot
In order to ensure that the correct data is collected, a pilot study should be run, and any amendments required based upon this should be made before the full-scale study. Rodden and Fu (2007) also suggest that providing participants with an example and practice run before the main trial may be advantageous.

Step 6: Participants Complete Task
The next stage of the analysis involves the participants undertaking the task whilst the mouse-tracking software records their movements.

Step 7: Analyse Data
Once the task is complete, the analyst should examine the heat map and summary data in order to deduce the user's cognition whilst interacting with the system interface.

Advantages

- A simple and easily comprehensible data summary (Arroyo, Selker and Wei, 2006).
- No interference with task performance (Arroyo, Selker and Wei, 2006; Atterer, Wnuk and Schmidt, 2006).
- Requires no training or expertise (Rodden and Fu, 2007).
- Removes the need for participants to travel to research trial as trials can be conducted remotely (Atterer, Wnuk and Schmidt, 2006; Chen, Anderson and Sohn, 2001) and in naturalistic settings (Arroyo, Selker and Wei, 2006).
- Lower costs than eye tracking (Atterer, Wnuk and Schmidt, 2006; Chen, Anderson and Sohn, 2001; Rodden and Fu, 2007).
- The data derived can be utilised to develop user profiles (Atterer, Wnuk and Schmidt, 2006), categorise user behaviour (Arroyo, Selker and Wei, 2006) and identify potential usability problems (Arroyo, Selker and Wei, 2006).

Disadvantages

- The method provides information on physical movements alone, and the link between physical movements and cognition is only supported by a relatively low number of studies (Chen, Anderson and Sohn, 2001).

Reliability and Validity

The reliability of the method is supported by studies such as that by Rodden and Fu (2007), which explores the correlation between mouse movement and eye movement when navigating through an online search engine results page. They concluded that mouse movement provides a reliable measure of the interface options considered by a user when making a mouse click. However, the reliability and validity of this method is not certain, as highlighted in research by Chen, Anderson and Sohn (2001), which discusses the limited level of empirical data supporting the link between mouse movement and option consideration.

Tools Needed

In order to conduct this method, the interface under analysis is required in addition to mouse-tracking software. Bespoke software can be developed or free software can be utilised, such as Fridgesoft Odo Plus (see www.fridgesoft.de/odoplus.php) and MouseTracker (see http://mousetracker.jbfreeman.net/index.htm).

Application and Training

The method is associated with low training and application times.

Related Methods

Mouse tracking is similar to other data collection procedures such as observation and eye tracking. The method has also been employed in combination with verbal protocol analysis. In a study exploring users' interaction with websites, Arroyo, Selker and Wei (2006) played back the mouse movement data to participants and asked them to provide verbalisations of their thought processes.

Example

In order to illustrate the outputs of the mouse-tracking method, Odo Plus was applied to the creation of a Microsoft Word document using web resources over a time period of two-and-a-half hours. Figure 2.1 illustrates the mouse movements of the user mapped onto a full-size screen.

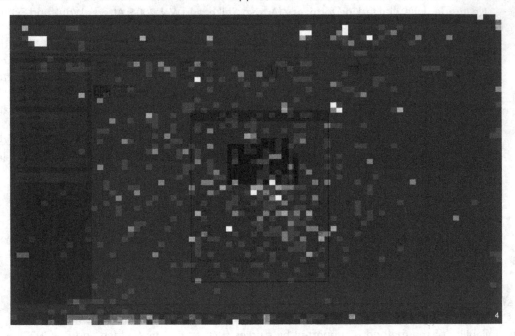

Figure 2.1   Example output of Odo Plus

From this figure, the exact movement of the mouse can be seen; for example, it is clear that the user was familiar with the task bar along the bottom of the screen and frequently used this to navigate between tasks. In addition, there appears to be a high level of activity in the top-right-hand corner where the minimise and close tabs are positioned in the window. The method also provides a summary box, as shown in Figure 2.2.

This box presents a smaller version of the screenshot of mouse movements and also contains information regarding the metres the mouse has travelled, the number of mouse clicks (divided into left, centre and right mouse clicks), the number of wheel ticks and the time the software has been running for.

Clearly, such software has great benefits for data collection and interface evaluation, as well as providing a wealth of information for interface design.

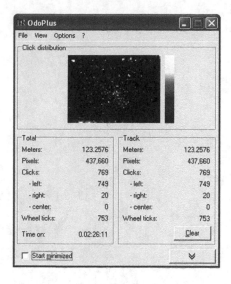

Figure 2.2   Example summary output

Flowchart

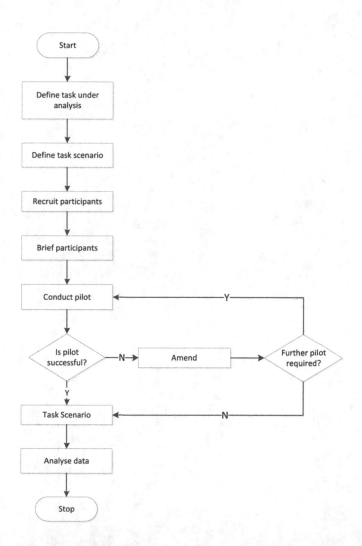

Chapter 3
Task Analysis Methods

Whilst data collection techniques are used to collect specific data regarding the activity performed in complex systems, task analysis techniques describe and represent the activity performed. Another emphatically used group of HF techniques, task analysis techniques are used to understand and represent human and system performance in a particular task or scenario under analysis. Task analysis involves identifying tasks, collecting task data, analysing the data so that tasks are understood and then producing a documented representation of the analysed tasks (Annett et al., 1971). According to Diaper and Stanton (2004), there are, or at least have been, over 100 task analysis techniques described in the literature. Typical task analysis techniques are used, in terms of the required human–machine and human–human interactions, to break down tasks or scenarios into component task steps or physical operations. According to Kirwan and Ainsworth (1992), task analysis can be defined as the study of what an operator (or team of operators) is required to do (their actions and cognitive processes) in order to achieve system goals.

The use of task analysis techniques is widespread, with applications in a wide range of domains, including military operations, aviation (Marshall et al., 2003), air traffic control, driving (Walker, Stanton and Young, 2001a), public technology (Stanton and Stevenage, 1998), product design and nuclear petrochemical domains to name a few. According to Annett (2004), a survey of defence task analysis studies demonstrated its use in system procurement, manpower analysis, interface design, operability assessment and training specification. Diaper (2004) suggests that task analysis is potentially the most powerful technique available to human–computer interaction (HCI) practitioners, and it has potential application at each stage in the system design and development process. Stanton (2004) also suggests that task analysis is the central method for the design and analysis of system performance, involved in everything from design concept to system development and operation. Stanton also highlights the role of task analysis in task allocation, procedure design, training design and interface design.

A task analysis of the task(s) and system under analysis is the next logical step after the data collection process. Specific data is used to conduct a task analysis, allowing the task to be described in terms of the individual task steps required, the technology used in completing the task (controls, displays, etc.) and the sequence of the task steps involved. The task description offered by task analysis techniques is then typically used as the input to further analysis techniques, such as human error identification (HEI) techniques and process charting techniques. For example, SHERPA (Embrey, 1986) and the human error template (HET: Marshall et al., 2003) are both HEI techniques that are applied to the bottom-level task steps identified in a hierarchical task analysis (HTA). In doing so, the task under analysis can be scrutinised to identify potential errors that might occur during the performance of that task. Similarly, an operations sequence diagram (OSD) is another example of a method that is based upon an initial task analysis of the task or process in question.

The popularity of task analysis techniques is a direct function of their usefulness and flexibility. Typically, a task analysis of some sort is required in any HF analysis effort, be it usability evaluation, error identification or performance evaluation. Task analysis outputs are particularly useful, providing a step-by-step description of the activity under analysis. Also, analysts using task analysis approaches often develop a (required) deep understanding of the activity under analysis.

However, task analysis techniques are not without their flaws. The resource usage incurred when using such approaches is often considerable. The data collection phase is time-consuming and often requires the provision of video and audio recording equipment. Such techniques are also typically time-consuming in their application, and many reiterations are needed before an accurate representation of the activity under analysis is produced. Task analysis techniques also suffer from reliability problems, and different analysts may produce entirely different representations of the same activity. Similarly, analysts may produce different representations of the same activity on different occasions.

There are a number of different approaches to task analysis available to the HF practitioner, including HTA, tabular task analysis (TTA), verbal protocol analysis (VPA), goals, operators, methods and selection rules (GOMS) and the Sub-Goal Template (SGT) method. A brief summary description of the task analysis techniques reviewed is given below.

The most commonly used and well-known task analysis technique is HTA (Annett, 2004; Stanton, 2006). HTA involves breaking down the task under analysis into a nested hierarchy of goals, operations and plans. GOMS (Card, Moran and Newell, 1983) attempts to define the user's goals, break down these goals into sub-goals and demonstrate how the goals are achieved through user interaction. Verbal protocol analysis (VPA) is used to derive the processes (cognitive and physical) that an individual uses to perform a task. VPA involves creating a written transcript of operator behaviour as he or she performs the task under analysis. Task decomposition (Kirwan and Ainsworth, 1992) can be used to create a detailed task description using specific categories to exhaustively describe actions, goals, controls, error potential and time constraints. The SGT method is a development of HTA that is used to specify information requirements to system designers. The output of the SGT method provides a re-description of HTA for the task(s) under analysis in terms of information handling operations (IHOs), SGT task elements and the associated information requirements.

Task analysis techniques have evolved in response to increased levels of complexity and the increased use of teams within work settings. A wide variety of task analysis procedures now exist, including techniques designed to consider the cognitive aspects of decision-making and activity in complex systems (cognitive task analysis) and also collaborative or team-based activity (team task analysis). Cognitive task analysis techniques, such as the CDM (Klein and Armstrong, 2004), and applied cognitive task analysis (ACTA; Militello and Hutton, 2000) use probe interview techniques in order to analyse, understand and represent the unobservable cognitive processes associated with tasks or work. TTA techniques attempt to describe the process of work across teams or distributed systems. A summary of the task analysis techniques reviewed is presented in Table 3.1.

Hierarchical Task Analysis (HTA)

Background and Applications

HTA (Annett, 2004; Stanton, 2006) is the most popular task analysis technique and has become perhaps the most widely used of all available HF techniques. Originally developed in response to the need for greater understanding of cognitive tasks (Annett, 2004), HTA involves describing the activity under analysis in terms of a hierarchy of goals, sub-goals, operations and plans. The end result is an exhaustive description of task activity. One of the main reasons for the enduring popularity of the technique is its flexibility and the scope for further analysis that it offers to the HF practitioner.

The majority of HF analysis methods either require an initial HTA of the task under analysis as their input or at least are made significantly easier through the provision of an HTA. HTA acts as an input into numerous HF analyses techniques, such as HEI, allocation of function, workload assessment, interface design and evaluation and many more. In a review of ergonomics texts, Stanton (2004) highlights at least 12 additional applications to which HTA has been put, including interface design and evaluation, training, allocation of functions, job description, work organisation, manual design, job aid design, error prediction and analysis, team task analysis, workload assessment and procedure design. Consequently, HTA has been applied across a wide spectrum of domains, including the process control and power generation industries (Annett, 2004), emergency services, military applications (Ainsworth and Marshall, 1998; Kirwan and Ainsworth, 1992), civil aviation (Marshall et al., 2003), driving (Walker, Stanton and

Table 3.1 Summary of task analysis techniques

Method	Type of method	Domain	Training time	App time	Related methods	Tools needed	Validation studies	Tools needed	Disadvantages
Hierarchical Task Analysis (HTA).	Task analysis.	Generic.	Med.	Med.	HEI task analysis	Pen and paper.	Yes.	1) HTA output feeds into numerous HF techniques. 2) Has been used extensively in a variety of domains. 3) Provides an accurate description of task activity.	1) Provides mainly descriptive information. 2) Cannot cater for the cognitive components of task performance. 3) Can be time-consuming to conduct for large, complex tasks.
Goals, Operators, Methods (GOMS) and Selection Rules.	Task analysis.	HCI.	Med–High.	Med–High.	NGOMSL. CMN-GOMS. KLM. CPM-GOMS.	Pen and paper.	Yes (but not outside of HCI).	1) Provides a hierarchical description of task activity.	1) May be difficult to learn and apply for non-HCI practitioners. 2) Time-consuming in its application. 3) Remains to be validated outside of HCO domain.
Verbal Protocol Analysis (VPA).	Task analysis.	Generic.	Low.	High.	Walkthrough analysis.	Audio recording equipment, observer software, PC.	Yes.	1) Rich data source. 2) Verbalisations can give a genuine insight into cognitive processes. 3) Easy to conduct, providing the correct equipment is used.	1) The data analysis process is very time-consuming and laborious. 2) It is often difficult to verbalise cognitive behaviour. 3) Verbalisations intrude upon primary task performance.
Task decomposition.	Task analysis.	Generic.	High.	High.	HTA. Observation, interviews questionnaire, walkthrough.	Pen and paper, video recording equipment.	No.	1) A very flexible technique, allowing the analyst to direct the analysis as he or she wishes. 2) Potentially very exhaustive. 3) Can cater for numerous aspects of the interface under analysis, including error, usability, interaction time, etc.	1) Very time-consuming and laborious to conduct properly.
Sub-goal template method.	Task analysis.	Generic.	Med.	High.	HTA.	Pen and paper.	No.	1) The output is very useful. 2) Information requirements for the task under analysis are specified.	1) Techniques require further testing regarding reliability and validity. 2) Can be time-consuming in its application.
Tabular Task Analysis.	Task analysis.	Generic.	Low.	High.	HTA, interface surveys, task decomposition.	Pen and paper.	No.	1) A very flexible technique, allowing the analyst to direct the analysis as he or she wishes. 2) Can cater for numerous aspects of the interface under analysis. Potentially very exhaustive.	1) Time-consuming to conduct properly. 2) Used infrequently.

Young, 2001a), public technology (Stanton and Stevenage, 1998) and retail (Shepherd, 2002) to name but a few.

The use of HTA is still increasing and modern applications include the analysis of helicopter mission planning systems (Salmon et al., 2010), friendly fire (Rafferty, Stanton and Walker, 2012); drug administration (Lane, Stanton and Harrison, 2006), software design (Mills, 2007) and product sales figure predictions (Asimakopoulos, Dix and Fildes, 2011). Asimakopoulos, Dix and Fildes (2011) employed HTA to develop a formal model of prescriptive behaviour that should occur in the prediction of product sales figures. They compare this formal HTA to observations of actual processes and process data derived from interviews with subject-matter experts. The resulting data is used to inform the development of a piece of forecasting software. They use the method both to analyse tasks and to elicit information to refine task analyses.

Domain of Application

HTA was originally developed for the chemical processing and power generation industries (Annett, 2004). However, the technique is generic and can be applied in any domain.

Procedure and Advice

Step 1: Define Task(s) under Analysis
The first step in conducting an HTA is to clearly define the task (or tasks) under analysis. As well as identifying the task under analysis, the purpose of the task analysis effort should also be defined. For example, Marshall et al. (2003) conducted an HTA of a civil aircraft landing task in order to predict design-induced error for the flight task in question.

Step 2: Data Collection Process
Once the task (or tasks) under analysis is clearly defined, specific data regarding the task should be collected. The data collected during this process is used to inform the development of the HTA. Data regarding the task steps involved, the technology used, interaction between man and machine and team members, decision-making and task constraints should be collected. There are a number of ways to collect this data, including observations, interviews with subject matter experts (SMEs), questionnaires and walkthroughs. The techniques used are dependent upon the analysis effort and the various constraints imposed, such as time and access constraints. Once sufficient data regarding the task under analysis is collected, the development of the HTA should begin.

Step 3: Determine the Overall Goal of the Task
The overall goal of the task under analysis should first be specified at the top of the hierarchy, i.e. 'Land aircraft X at New Orleans Airport using the autoland system' (Marshall et al., 2003), 'Boil kettle' or 'Listen to in-car entertainment' (Stanton and Young, 1999a).

Step 4: Determine Task Sub-goals
Once the overall task goal has been specified, the next step is to break this overall goal down into meaningful sub-goals (usually four or five, but this is not rigid), which together form the tasks required to achieve the overall goal. In the task 'Land aircraft X at New Orleans Airport using the autoland system' (Marshall et al., 2003), the overall goal of landing the aircraft was broken down into the following sub-goals: 'Set up for approach', 'Line up aircraft for runway' and 'Prepare aircraft for landing'. In an HTA of a Ford in-car radio (Stanton and Young, 1999a) the overall task goal, 'Listen to in-car entertainment', was broken down into the following sub-goals: 'Check unit status', 'Press on/off button', 'Listen to the radio', 'Listen to cassette' and 'Adjust audio preferences'.

Step 5: Sub-goal Decomposition

Next, the analyst should break down the sub-goals identified during step four into further sub-goals and operations, according to the task step in question. This process should go on until an appropriate operation is reached. The bottom level of any branch in an HTA should always be an operation. Whilst everything above an operation specifies goals, operations actually say what needs to be done. Therefore, operations are actions to be made by an agent in order to achieve the associated goal. For example, in the HTA of the flight task 'Land aircraft X at New Orleans Airport using the autoland system' (Marshall et al., 2003), the sub-goal 'Reduce airspeed to 210 Knots' is broken down into the following operations: 'Check current airspeed' and 'Dial the Speed/Mach selector knob to enter 210 on the IAS/Mach display'.

Step 6: Analysis of Plans

Once all of the sub-goals and operations have been fully described, the plans need to be added. Plans dictate how the goals are achieved. A simple plan would say do 1, then 2, and then 3. Once the plan is completed, the agent returns to the super-ordinate level. Plans do not have to be linear and exist in many forms, such as do 1, or 2 and 3. The

Table 3.2   Example HTA plans

Plan	Example
Linear.	Do 1, then 2, then 3.
Non-linear.	Do 1, 2 and 3 in any order.
Simultaneous.	Do 1, then 2 and 3 at the same time.
Branching.	Do 1; if X present, then do 2 then 3, but if X is not present, then EXIT.
Cyclical.	Do 1, then 2, then 3 and repeat until X.
Selection.	Do 1, then 2 or 3.

different types of plans used are presented in Table 3.2. The output of an HTA can either be a tree diagram (see Figure 3.1) or a tabular diagram (see Table 3.3).

Advantages

- HTA requires minimal training and is easy to implement.
- The output of an HTA is extremely useful and forms the input for numerous HF analyses, such as error analysis, interface design, and evaluation and allocation of function analysis.
- It is an extremely flexible technique that can be applied in any domain for a variety of purposes.
- It is quick to use in most instances.
- The output provides a comprehensive description of the task under analysis.
- It has been used extensively in a wide range of contexts.
- Conducting an HTA gives the user considerable insight into the task under analysis. Salmon et al. (2010) propose that the process of creating an HTA enables key insights to be gained in addition to the results of the analysis.
- It is an excellent technique to use when requiring a task description for further analysis. If performed correctly, the HTA should depict everything that needs to be done in order to complete the task in question.
- The technique is generic and can be applied to any task in any domain.
- Tasks can be analysed to any required level of detail, depending on the purpose.

Disadvantages

- HTA provides mainly descriptive information rather than analytical information.
- It contains little that can be used directly to provide design solutions.
- It does not cater for the cognitive components of the task under analysis.
- The technique may become laborious and time-consuming to conduct for large, complex tasks.

- The initial data collection phase is time-consuming and requires the analyst to be competent in a variety of HF techniques, such as interviews, observations and questionnaires.
- The reliability of the technique may be questionable in some instances. For example, for the same task, different analysts may produce very different task descriptions.
- Conducting an HTA is more of an art than a science, and much practice is required before an analyst becomes proficient in the application of the technique.
- An adequate software version of the technique has yet to emerge.
- There are few prescriptive guidelines on how to apply it (Stanton, 2006).

Related Methods

HTA is widely used in HF and often forms the first step in a number of analyses, such as HEI, HRA and mental workload assessment. Stanton (2006) conducted a comprehensive review outlining a variety of applications of HTA, including interface design, error prediction, workload analysis, team performance assessment and training requirement identification.

Mills (2007) argues that HTA is at its best when used alongside other methods; for example, Mills (2007) used HTA alongside usability context analysis and Salmon et al. (2010) have developed a piece of software which integrates HTA with additional HF methods, including methods providing insights into workload (the NASA Task Load Index (NASA-TLX)) error (SHERPA) and design (key stroke level model, HCI analysis).

Approximate Training and Application Times

According to Annett (2004), a study by Patrick, Gregov and Halliday (2000) gave students a few hours training with not entirely satisfactory results on the analysis of a very simple task, although performance improved with further training. A survey by Ainsworth and Marshall (1998) found that the more experienced practitioners produced more complete and acceptable analyses.

Stanton and Young (1999a) report that the training and application time for HTA is substantial. The application time associated with HTA is dependent upon the size and complexity of the task under analysis. For large, complex tasks, the application time for HTA would be high. Salmon et al. (2010) also suggest that HTA application times are high, stating that the high fidelity of information captured in an HTA can increase application times to almost double those of other methods such as cognitive work analysis. However, they also argue that increased application times are correlated with a greater granularity of detail, suggesting that if cognitive work analysis was conducted to the same level of granularity, its application time would be far higher than HTA.

Harvey and Stanton (2012) applied HTA to the analysis of multi-modal in-vehicle car interfaces along with SHERPA, critical path analysis (CPA), heuristic analysis and layout analysis. They concluded that HTA required a significant level of time, suggesting that two to four hours were needed to collect and six to eight hours to analyse the data.

Reliability and Validity

According to Annett (2004), the reliability and validity of HTA is not easily assessed. From a comparison of 12 HF techniques, Stanton and Young (1999a) reported that the technique achieved an acceptable level of validity but a poor level of reliability. The reliability of the technique is certainly questionable. It seems that different analysts with different levels of experience may produce entirely different analyses for the same task (intra-analyst reliability). Similarly, the same analyst may produce different analyses on different occasions for the same task (inter-analyst reliability).

Tools Needed

HTA can be carried out using pencil and paper only. The HTA output can be developed and presented in a number of software applications, such as Microsoft Visio, Microsoft Word and Microsoft Excel. A number of HTA software tools also exist, such as the C@STTA HTA tool.

Example

An example HTA for the 'Boil kettle' task is presented in Figure 3.1. The same HTA is presented in tabular format in Table 3.1. This is typically the starting point in the training process of the method and is presented in order to depict a simplistic example of the method's output. An extract of the HTA for the flight task 'Land aircraft X at New Orleans Airport using the autoland system' is presented in Figure 3.2.

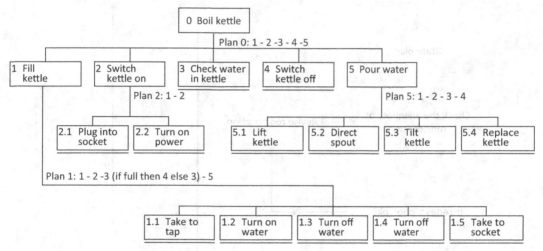

Figure 3.1   HTA of the task 'Boil kettle'

Table 3.3   Tabular HTA for the 'Boil kettle' task

0. Boil kettle Plan 0: Do 1, then 2, then 3, then 4, then 5.
1. Fill kettle Plan 1: Do 1, then 2, then 3 (if full, then 4, otherwise 3), then 5.

Take to tap.
    Turn on water.
    Check level.
    Turn off water.
    Take to socket.

2. Switch kettle on. Plan 2: Do 1, then 2.
2.1 Plug into socket.   2.2 Turn on power.
3. Check water in kettle.
4. Switch kettle off.
5. Pour water. Plan 5: Do 1, then 2, then 3.

5.1 Lift kettle.
    5.2 Direct spout.
    5.3 Tilt kettle.
    5.4 Replace kettle.

Flowchart

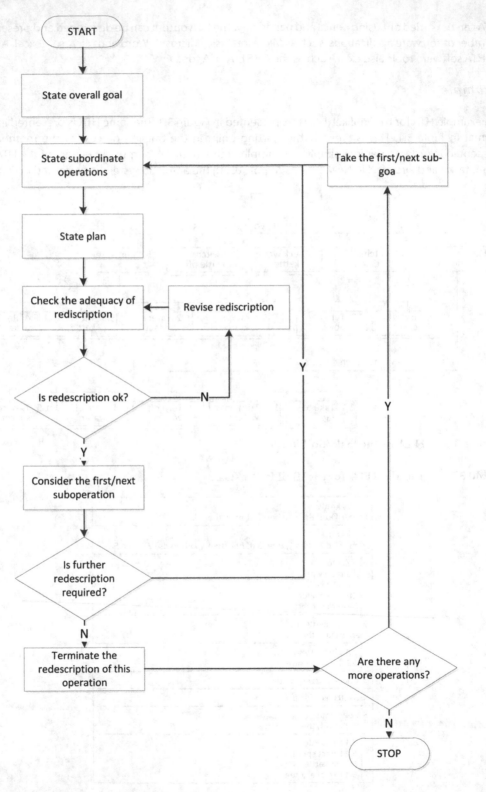

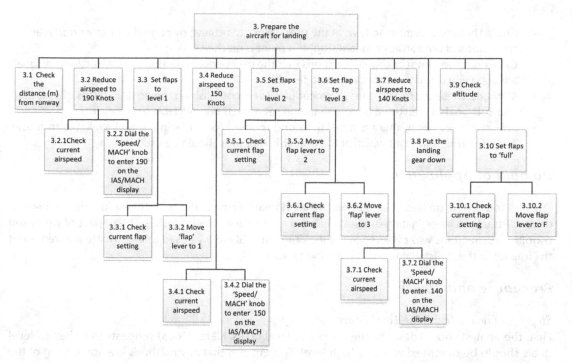

Figure 3.2   **HTA extract for the landing task 'Land aircraft X at New Orleans Airport using the autoland system**

Source: Marshall et al., 2003.

Goals, Operators, Methods and Selection Rules (GOMS)

Background and Applications

The GOMS (Card, Moran and Newell, 1983) technique is part of a family of HCI-based techniques that is used to provide a description of human performance in terms of user goals, operators, methods and selection rules. GOMS attempts to define the user's goals, break these goals down into sub-goals and demonstrate how the goals are achieved through user interaction. GOMS can be used to provide a description of how a user performs a task, to predict performance times and to predict human learning. Whilst the GOMS techniques are most commonly used for the evaluation of existing designs or systems, it is also feasible that they could be used to inform the design process, particularly to determine the impact of a design concept on the user. Within the GOMS family, there are four techniques, which are described below:

1.   NGOMSL (Natural GOMS Language).
2.   KLM (Keystroke Level Model).
3.   CMN-GOMS (Card, Moran and Newell – GOMS).
4.   CPM-GOMS (Cognitive Perceptual Model – GOMS).

The GOMS techniques are based upon the assumption that the user's interaction with a computer is similar to solving problems. Problems are broken down into sub-problems, which are then broken down further and so on. The GOMS technique focuses upon four basic components of human interaction: goals, operators, methods and selection rules. These components are described below.

- *Goals*: these represent exactly what the user wishes to achieve through the interaction. Goals are decomposed until an appropriate stopping point is reached.
- *Operators*: the motor or cognitive actions that the user performs during the interaction. The goals are achieved through performing the operators.
- *Methods*: these describe the user's procedures for accomplishing the goals in terms of operators and sub-goals. Often there is more than one set of methods available to the user.
- *Selection rules*: when there is more than one method for achieving a goal available to a user, selection rules highlight which of the available methods should be used.

Domain of Application

The method can be utilised to explore HCI in any domain. Liang and Lee (2010) explore the problem of driver distraction, investigating its causal influence on motor-vehicle crashes. The impact of visual and cognitive distractions was explored in combination, and GOMS was used to investigate and represent the impact of these distractions on the driving task.

Procedure and Advice

Step 1: Define the User's Top-Level Goals
First, the analyst should describe the user's top-level goals. Kieras (2003) suggests that the top-level goals should be described at a very high level. This ensures that any methods are not left out of the analysis.

Step 2: Goal Decomposition
Once the top-level goal or set of goals has been specified, the next step is to break down the top-level goal into a set of sub-goals.

Step 3: Determine and Describe Operators
Operators are actions executed by the user to achieve a goal or sub-goal. The next phase of a GOMS analysis involves describing the operators required for the achievement of the sub-goals specified during step 2. Each high-level operator should be replaced with another goal/method set until the analysis is broken down to the level desired by the analyst (Kieras, 2003).

Step 4: Determine and Describe Methods
Methods describe the procedures or set of procedures used to achieve the goal (Kirwan and Ainsworth, 1992). In the next phase of the GOMS analysis, the analyst should describe each set of methods that the user could use to achieve the task. Often there are a number of different methods available to the user, and the analyst is encouraged to include all possible methods.

Step 5: Describe Selection Rules
If there is more than one method for achieving a goal, the analyst should determine selection rules for the goal. Selection rules predict which of the available methods will be used by the user to achieve the goal.

Advantages

- GOMS can be used to provide a hierarchical description of task activity.
- The methods part of a GOMS analysis allows the analyst to describe a number of different potential task routes.
- GOMS analysis can aid designers in choosing between systems, as performance and learning times can be specified.
- It has been applied extensively in the past and has a wealth of associated validation evidence.

Disadvantages

- GOMS is a difficult technique to apply. Far simpler task analysis techniques are available.
- It can be time-consuming to apply.
- The GOMS technique appears to be restricted to HCI. As it was developed specifically for use in HCI, most of the language is HCI-orientated. Reported use of GOMS outside of the HCI domain is limited.
- A high level of training and practice would be required.
- GOMS analysis is limited as it only models error-free, expert performance.
- Context is not taken into consideration.
- The GOMS methods remain largely unvalidated outside of HCI.

Related Methods

There are four main techniques within the GOMS family. These are NGOMSL, KLM, CMN-GOMS and CPM-GOMS.

Approximate Training and Application Times

For non-HCI experienced practitioners, it is expected that the training time would be medium to high. The application time associated with the GOMS technique is dependent upon the size and complexity of the task under analysis. For large, complex tasks involving many operators and methods, the application time for GOMS would be very high. However, for small, simplistic tasks, the application time would be minimal.

Reliability and Validity

Within the HCI domain, the GOMS technique has been validated extensively. According to Salvendy (1997), Card, Moran and Newell (1983) reported that for a text-editing task, the GOMS technique predicted the user's methods 80–90 per cent of the time and also the user's operators 80–90 per cent of the time. However, evidence of the validation of the GOMS technique in applications outside of the HCI domain is limited.

Tools Needed

GOMS can be conducted using pen and paper. Access to the system, programme or device under analysis is also required.

Flowchart

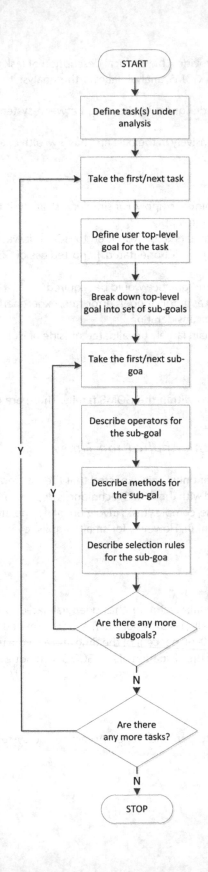

Verbal Protocol Analysis (VPA)

Background and Applications

As mentioned earlier, VPA is used to derive descriptions of the processes (cognitive and physical) that an individual uses to perform a task. It involves creating a written transcript of operator behaviour as he or she performs the task or scenario under analysis. The transcript is based upon the operator 'thinking aloud' as he or she conducts the task under analysis. VPA has been used extensively as a means of gaining an insight into the cognitive aspects of complex behaviours. It has been used in many domains, ranging from investigating expertise in nursing (Hoffman, Aitken and Duffield, 2009), to exploring the reliability of personality questionnaires (Robie, Brown and Beaty, 2007), to the examination of the compatibility of motorcyclists' and car drivers' mental representations of the road (Walker, Stanton and Salmon, 2011) and usability evaluations (Wu et al., 2008).

Domain of Application

Generic.

Procedure and Advice

The following procedure is adapted from Walker (2004).

Step 1: Define Scenario under Analysis
First, the scenario under analysis should be clearly defined. It is recommended that an HTA is used to describe the task under analysis.

Step 2: Instruct/Train the Participant
Once the scenario is clearly defined, the participant should be briefed regarding what is required of him or her during the analysis. What he or she should report verbally is clarified here. According to Walker (2004), it is particularly important that the participant is informed that he or she should continue talking even when what he or she is saying does not appear to make much sense. A small demonstration should also be given to the participant at this stage. A practice run may also be undertaken, although this is not always necessary.

Step 3: Begin Scenario and Record Data
The participant should begin to perform the scenario under analysis. The whole scenario should be audio recorded (at least) by the analyst. It is also recommended that a video recording be made.

Step 4: Verbalisation of Transcript
Once collected, the data should be transcribed into a written form. An Excel spreadsheet is normally used. This aspect of VPA is particularly time-consuming and laborious.

Step 5: Encode Verbalisations
The verbal transcript (written form) should then be categorised or coded. Depending upon the requirements of the analysis, the data is coded into one of the following five categories; words, word senses, phrases, sentences or themes. The encoding scheme chosen should then be encoded according to a rationale determined by the aims of the analysis. Walker (2004) suggests that this involves attempting to ground the encoding scheme according to some established theory or approach, such as mental workload or SA. The analyst should also develop a set of written instructions for the encoding scheme. These instructions should be strictly adhered to and constantly referred to during the encoding process (Walker, 2004). Once the encoding type, framework and instructions are completed, the analyst should

proceed to encode the data. Various computer software packages are available to aid the analyst with this process, such as General Enquirer and Nvivo.

Step 6: Devise Other Data Columns
Once the encoding is complete, the analyst should devise any 'other' data columns. This allows the analyst to note any mitigating circumstances that may have affected the verbal transcript.

Step 7: Establish Inter- and Intra-rater Reliability
Reliability of the encoding scheme then has to be established (Walker, 2004). In VPA, reliability is established through reproducibility, i.e. independent raters need to encode previous analyses.

Step 8: Perform Pilot Study
The protocol analysis procedure should now be tested within the context of a small pilot study. This will demonstrate whether the verbal data collected is useful, whether the encoding system works, and whether inter- and intra-rater reliability are satisfactory. Any problems highlighted through the pilot study should be refined before the analyst conducts the VPA for real.

Step 9: Analyse Structure of Encoding
Finally, the analyst can study the results from the VPA. During any VPA analysis, the responses given in each encoding category require summing, and this is achieved simply by adding up the frequency of occurrence noted in each category. Walker (2004) suggests that for a more fine-grained analysis, the structure of encodings can be analysed contingent upon events that have been noted in the 'other data' column(s) of the worksheet or in light of other data that has been collected simultaneously.

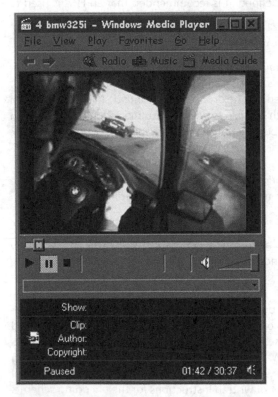

Figure 3.3   Digital audio/video recording of protocol – analysis scenario

Example

The following example is a VPA taken from Walker (2004). This digital video image (Figure 3.3) is taken from the study reported by Walker, Stanton and Young (2001b) and shows how the Protocol Analysis was performed with normal drivers. The driver in Figure 3.3 is providing a concurrent verbal protocol whilst being simultaneously videoed. The driver's verbalisations and other data gained from the visual scene are transcribed into the data sheet in Figure 3.4, which illustrates the two-second incremental time index, the actual verbalisations provided by the driver's verbal commentary, the encoding categories, the events column and the protocol structure. In this study, three encoding groups were defined: behaviour, cognitive processes and feedback. The behaviour group defined the verbalisations as referring to the driver's own behaviour (OB), behaviour of the vehicle (BC), behaviour of the road environment (RE) and behaviour of other traffic (OT). The cognitive processes group was sub-divided into perception (PC), comprehension (CM), projection (PR) and action execution (AC). The feedback category offered an opportunity for vehicle feedback to be further categorised according to whether it

TIME mm:ss	VERBALISATIONS	ENCODING BEHAV.				COG.				F/B			EVENTS
		OB	BC	RE	OT	PC	CM	PR	AC	SD	CD	IN	
01:34	70mph, 5th gear		1			1					1	1	Glances at gear lever
01:36	2800 rpm		1			1						1	
01:38	that is quite smooth		1				1						
01:40	he's slowing down				1	1							Other car crossing from lane 3 over
01:42	don't know what is wrong with him	1											to hard shoulder in front of driver
01:44													
01:46													
01:48													
01:50													
01:52													
01:54													
01:56													
01:58													
02:00													
02:02													
02:04	It's all clear ahead				1	1							
02:06													
02:08	chap behind has eased off a bit luckly				1	1	1						
02:10													
02:12	make my intention clear that I'm going right	1							1				Indicating right
02:14	so I'll stick to the right side of this lip lane	1		1		1							
02:16													
02:18													
02:20	bit worried about overtaking him	1			1		1						passing other vehicle
02:22													
02:24													
02:26													
Section Frequency Counts		4	3	1	5	5	5	0	1	0	1	2	
02:28													

Figure 3.4   Transcription and encoding sheet

referred to system or control dynamics (SD or CD), or vehicle instruments (IN). The cognitive processes and feedback encoding categories were couched in relevant theories in order to establish a conceptual framework. The events column was for noting road events from the simultaneous video log, and the protocol structure was colour-coded according to the road type being travelled upon. In this case, the shade corresponds to a motorway and would permit further analysis of the structure of encoding contingent upon road type. The section frequency counts simply sum the frequency of encoding for each category for that particular road section.

Advantages

- VPA provides a rich data source.
- It is particularly effective when used to analyse sequences of activities.
- Verbalisations can provide a genuine insight into cognitive processes.
- Domain experts can provide excellent verbal data.
- It has been used extensively in a wide variety of domains.
- It is simple to conduct with the right equipment.

Disadvantages

- Data analysis (encoding) can become extremely laborious and time-consuming.
- VPA is a very time-consuming method to apply (data collection and data analysis).

- It is difficult to verbalise cognitive behaviour. Researchers have been cautioned in the past for relying on verbal protocol data (Militello and Hutton, 2000).
- Verbal commentary can sometimes serve to change the nature of the task.
- Complex tasks involving high demand can often lead to a reduced quantity of verbalisations (Walker, 2004).
- Strict procedure is often not adhered to fully.
- VPA is prone to bias on the participant's behalf.

Related Methods

VPA is related to observational techniques such as walkthroughs and direct observation. Task analysis techniques such as HTA are often used in constructing the scenario under analysis. VPA is frequently used alongside other HF methods; for example, Hoffman, Aitken and Duffield (2009) employed both VPA and retrospective interviewing in their exploration of the differences in decision-making between experts and novices within the medical nursing domain. The method can also act as an input for numerous other HF methods; for example, Walker et al. (2012) used verbal protocols as an input to an automatic concept map tool to develop illustrations of car drivers' and motorcyclists' mental representations of driving. VPA is also used for various purposes, including SA measurement, mental workload assessment and task analysis.

Approximate Training and Application Times

Although the technique is very easy to train, the VPA procedure is time-consuming to implement. According to Walker (2004), if transcribed and encoded by hand, 20 minutes of verbal transcript data at around 130 words per minute can take between six and eight hours to transcribe and encode.

Reliability and Validity

Walker (2004) reports that the reliability of the technique is reassuringly good. For example, Walker, Stanton and Young (2001b) used two independent raters and established inter-rater reliability at Rho=0.9 for rater 1 and Rho=0.7 for rater 2. Intra-rater reliability during the same study was also high, being in the region of Rho=0.95.

Hoffman, Aitken and Duffield (2009) argue that the use of both concurrent and retrospective protocols increases the reliability and validity of the method as one measures long-term memory and the other short-term memory. They propose that using researchers with domain experience and maintaining consistency of researchers throughout analysis also increase levels of reliability and consistency.

Tools Needed

A VPA can be conducted using pen and paper, a digital audio recording device and a video recorder if required. The device or system under analysis is also required. For the data analysis part of VPA, Excel is normally required, although this can be done using pen and paper. A number of software packages can also be used by the analyst, including Observer, General Enquirer, TextQuest, Nvivo and Wordstation.

Task Decomposition

Background and Applications

Kirwan and Ainsworth (1992) describe the task decomposition methodology that can be used to gather detailed information regarding a particular task or scenario. Task decomposition involves describing the task or activity under analysis and then using specific task-related information to decompose the task in terms of specific statements regarding the task. The task can be broken down to describe a variety of task-related features, including the devices and interface components used, the time taken, errors made, feedback and decisions required. The categories used to decompose the task steps should be chosen by the analyst based on the requirements of the analysis. There are numerous decomposition categories that can be used and new categories can be developed if required by the analysis. According to Kirwan and Ainsworth (1992), Miller (1953) was the first practitioner to use the task decomposition technique. Miller recommended that each task step should be decomposed around the following categories:

- Description.
- Subtask.
- Cues initiating action.
- Controls used.
- Decisions.
- Typical errors.
- Response.
- Criterion of acceptable performance.
- Feedback.

However, further decomposition categories have since been defined (e.g. Kirwan and Ainsworth, 1992). It is recommended that the analyst develops a set of decomposition categories based upon the analysis requirements.

Domain of Application

Generic.

Procedure and Advice

Step 1: HTA
The first step in a task decomposition analysis involves creating an initial description of the task or scenario under analysis. It is recommended that an HTA is conducted for this purpose, as a goal-driven, step-by-step description of the task is particularly useful when conducting a task decomposition analysis.

Step 2: Create Task Descriptions
Once an initial HTA for the task under analysis has been conducted, the analyst should create a set of clear task descriptions for each of the different task steps. These descriptions can be derived from the HTA developed during step 1. The task description should give the analyst enough information to determine exactly what has to be done to complete each task element. The detail of the task descriptions should be determined by the requirements of the analysis.

Step 3: Choose Decomposition Categories
Once a sufficient description of each task step is created, the analyst should choose the appropriate decomposition categories. Kirwan and Ainsworth (1992) suggest that there are three types of decomposition categories: descriptive, organisation-specific and modelling. Table 3.4 presents a taxonomy of descriptive decomposition categories that have been used in various studies.

Step 4: Information Collection
Once the decomposition categories have been chosen, the analyst should create a data collection pro-forma for each decomposition category. The analyst should then work through each decomposition

Table 3.4   Task decomposition categories

Description of task	Task difficulty
Description	Task criticality
Type of activity/behaviour	Amount of attention required
Task/action verb	Performance on the task
Function/purpose	Performance
Sequence of activity	Time taken
Requirements for undertaking task	Required speed
Initiating cue/event	Required accuracy
Information	Criterion of response adequacy
Skills/training required	Other activities
Personnel requirements/manning	Subtasks
Hardware features	Communications
Location	Coordination requirements
Controls used	Concurrent tasks
Displays used	Outputs from the task
Critical values	Output
Job aids required	Feedback
Nature of the task	Consequences/problems
Actions required	Likely/typical errors
Decisions required	Errors made/problems
Responses required	Error consequences
Complexity/task complexity	Adverse conditions/hazards

Source: Kirwan and Ainsworth, 1992.

category, recording task descriptions and gathering the additional information required for each of the decomposition headings. To gather this information, Kirwan and Ainsworth (1992) suggest that there are many possible methods to use, including observation, system documentation, procedures, training manuals and discussions with system personnel and designers. Interviews, questionnaires, VPA and walkthrough analysis can also be used.

Step 5: Construct Task Decomposition
The analyst should then put the data collected into a task decomposition output table. The table should comprise all of the decomposition categories chosen for the analysis. The amount of detail included in the table is also determined by the scope of the analysis.

Advantages

• Task decomposition is a very flexible approach. By selecting which decomposition categories to use, the analyst can determine the direction and focus of the analysis.
• A task decomposition analysis has the potential to provide a very comprehensive analysis of a particular task.
• Task decomposition techniques are easy to learn and use.
• The method is generic and can be used in any domain.
• It provides a much more detailed description of tasks than traditional task analysis techniques do.
• As the analyst has control over the decomposition categories used, potentially any aspect of a task can be evaluated. In particular, the technique could be adapted to assess the cognitive components associated with tasks (goals, decisions, SA).

Disadvantages

• As the task technique is potentially so exhaustive, it is a very time-consuming technique to apply and analyse. The HTA only serves to add to the lengthy application time. Furthermore, obtaining information about the tasks (observation, interview, etc.) creates even more work for the analyst.
• Task decomposition can be laborious to perform, involving observations, interviews, etc.

Example

A task decomposition analysis was performed on the landing task 'Land Aircraft X at New Orleans Airport using the autoland system' (Marshall et al., 2003). The purpose of the analysis was to ascertain

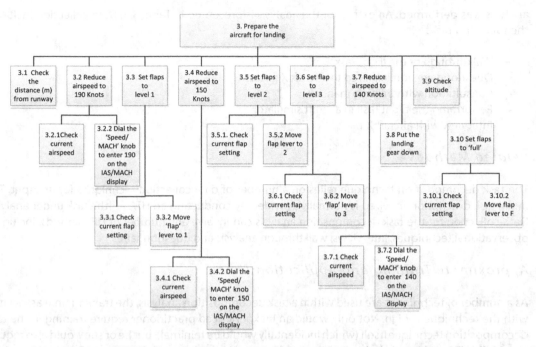

Figure 3.5   Extract of HTA 'Land aircraft X at New Orleans Airport using the autoland system'
Source: Marshall et al., 2003.

Table 3.5   Extract of task decomposition analysis for flight task 'Land aircraft X at New Orleans Airport using the autoland system'

Task step description: 3.2.2 Dial the speed/MACH knob to enter 190 knots on the IAS/MACH display	Complexity: Medium. The task involves a number of checks in quick succession and also the use of the Speed/MACH knob, which is very similar to the HDG/Track knob
Initiating cue/event: Check that the distance from the runway is 15 miles	Difficulty: Low
Displays used: Captain's Primary Flight display IAS/MACH window (Flight Control Unit) Captain's navigation display	Criticality: High. The task is performed in order to reduce the aircraft's speed so that the descent and approach can begin
Controls used: IAS/MACH knob	Feedback provided: Speed/MACH window displays current airspeed value. CPFD displays airspeed
Actions required: Check distance from runway on CPFD Dial in 190 using the IAS/MACH display Check IAS/MACH window for speed value	Probable errors: a) Using the wrong knob, i.e. the HDG/Track knob b) Failing to check the distance from runway c) Failing to check current airspeed d) Dialling in the wrong speed value e) Fail to enter new airspeed
Decisions required: Is distance from runway 15 miles or under? Is airspeed over/under 190 knots? Have you dialled in the correct airspeed (190 knots)? Has the aircraft slowed down to 190 knots?	Error consequences: a) Aircraft will change heading to 190 b) Aircraft may be too close or too far way from the runway c) Aircraft travelling at the wrong airspeed d) Aircraft may be travelling too fast for the approach

how suitable the task decomposition technique was for the prediction of design-induced error on civil flight decks. An HTA of the flight task was constructed (Figure 3.5) and a task decomposition analysis was performed. An extract of the analysis is presented in Table 3.5. Data collection included the following tasks:

- Walkthrough of the flight task.
- Questionnaire administered to aircraft X pilots.
- Consultation with training manuals.
- Performing the flight task in aircraft simulator
- Interview with aircraft X pilot.

Related Methods

The task decomposition technique relies on a number of data collection techniques for its input. The initial task description required is normally provided by conducting an HTA for the task under analysis. Data collection for the task decomposition analysis can involve any number of HF methods, including observational techniques, interviews, walkthrough analysis and questionnaires.

Approximate Training and Application Times

As a number of techniques are used within a task decomposition analysis, the training time associated with the technique is high. Not only would an inexperienced practitioner require training in the task decomposition technique itself (which incidentally would be minimal), but he or she would also require training in HTA and any techniques that would be used in the data collection part of the analysis. Also, due to the exhaustive nature of a task decomposition analysis, the associated application time is also very high. Kirwan and Ainsworth (1992) suggest that task decomposition can be a lengthy process and that its main disadvantage is the huge amount of time associated with collecting the required information.

Reliability and Validity

At present, no data regarding the reliability and validity of the technique is offered in the literature. It is apparent that such a technique may suffer from reliability problems, as a large proportion of the analysis is based upon the analyst's subjective judgment.

Tools Needed

The tools needed for a task decomposition analysis are determined by the scope of the analysis and the techniques used for the data collection process. Task decomposition can be conducted using just pen and paper. However, it is recommended that for the data collection process, visual and audio recording equipment should be employed. The system under analysis is also required in some form – either in mock-up, prototype or operational form.

Flowchart

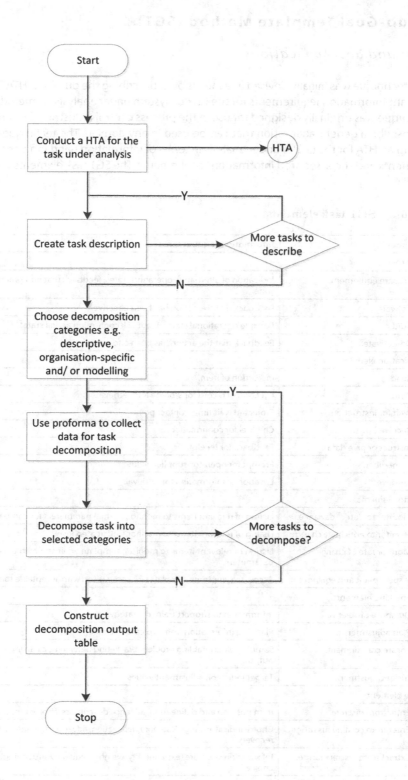

The Sub-Goal Template Method (SGT)

Background and Applications

The SGT technique was initially devised as a means of re-describing the output of HTA in order to specify the relevant information requirements for the task or system under analysis (Ormerod, 2000). Although the technique was originally designed for use in the process control industries, Ormerod and Shepherd (2003) describe a generic adaptation that can be used in any domain. The technique itself involves re-describing an HTA for the task(s) under analysis in terms of information handling operations (IHOs), SGT task elements and the associated information requirements. The SGT task elements used are presented in Table 3.6.

Table 3.6   SGT task elements

Code	Label	Information requirements
Action elements		
A1	Prepare equipment.	Indication of alternative operating states, feedback that equipment is set to required state.
A2	Activate.	Feedback that the action has been effective.
A3	Adjust.	Possible operational states, feedback confirming actual state.
A4	De-activate.	Feedback that the action has been effective.
Communication elements		
C1	Read.	Indication of item.
C2	Write.	Location of record for storage and retrieval.
C3	Wait for instruction.	Projected wait time, contact point.
C4	Receive instruction.	Channel for confirmation.
C5	Instruct or give data.	Feedback for receipt.
C6	Remember.	Prompt for operator-supplied value.
C7	Retrieve.	Location of information for retrieval.
Monitoring elements		
M1	Monitor to detect deviance.	Listing of relevant items to monitor, normal parameters for comparison.
M2	Monitor to anticipate change.	Listing of relevant items to monitor, anticipated level.
M3	Monitor rate of change.	Listing of relevant items to monitor, template against which to compare observed parameters.
M4	Inspect plant and equipment.	Access to symptoms, templates for comparison with acceptable tolerances if necessary.
Decision-making elements		
D1	Diagnose problems.	Information to support trained strategy.
D2	Plan adjustments.	Planning information from typical scenarios.
D3	Locate containment.	Sample points enabling problem bracketing between a clean input and a contaminated output.
D4	Judge adjustment.	Target indicator, adjustment values.
Exchange elements		
E1	Enter from discrete	Item position and delineation, advance descriptors, choice recovery.
E2	Enter from continuous range.	Choice indicator, range/category delineation, advance descriptors, end of range, range recovery.
E3	Extract from discrete range.	Information structure (e.g. criticality, weight, frequency structuring), feedback on current choice.
E4	Extract from continuous range.	Available range, information structure (e.g. criticality, weight, frequency structuring), feedback on current choices.

Table 3.6 Continued

Code	Label	Information requirements
Navigation elements		
N1	Locate a given information set.	Organisation structure cues (e.g. screen set/menu hierarchy, catalogue), choice descriptor conventions, current location, location relative to start, selection indicator.
N2	Move to a given location.	Layout structure cues (e.g. screen position, menu selection, icon), current position, position relative to information coordinates, movement indicator.
N3	Browse an information set.	Information (e.g. screen/menu hierarchy, catalogue), organisation cues, information scope, choice points, current location, location relative to start, selection indicator.

Source: Ormerod, 2000.

Ormerod and Shepherd (2003) describe a modified set of task elements, presented in Table 3.7.

Table 3.7 Modified SGT task elements

SGT	Task elements	Context for assigning SGT and task element	Information requirements
Act		Perform as part of a procedure or subsequent to a decision made about changing the system.	Action points and order, current, alternative, and target states, preconditions, outcomes, dependencies, halting, recovery indicators.
	A1 Activate.	Make subunit operational: switch from off to on.	Temporal/stage progression, outcome activation level.
	A2 Adjust.	Regulate the rate of operation of a unit maintaining 'on' state.	Rate of state of change.
	A3 Deactivate.	Make subunit non-operational: switch from on to off.	Cessation descriptor.
Exchange		To fulfil a recording requirement To obtain or deliver operating value.	Indication of item to be exchanged, channel for confirmation.
	E1 Enter.	Record a value in a specified location.	Information range (continuous, discrete).
	E2 Extract.	Obtain a value of a specified parameter.	Location of record for storage and retrieval; prompt for operator.
Navigate		To move an informational state for exchange, action or monitoring.	System/state structure, current relative location.
	N1 Locate.	Find the location of a target value or control.	Target information, end location relative to start.
	N2 Move.	Go to a given location and search it.	Target location, directional descriptor.
	N3 Explore.	Browse through a set of locations and values.	Current/next/previous item categories.
Monitor		To be aware of system states that determine need for navigation, exchange and action.	Relevant items to monitor; record of when actions were taken; elapsed time from action to the present.
	M1 Monitor to detect deviance.	Routinely compare system state against target state to determine need for action.	Normal parameters for comparison.
	M2 Monitor to anticipate cue.	Compare system state against target state to determine readiness for known action.	Anticipated level.
	Monitor transition.	Routinely compare state of change during state transition.	Template against which to compare observed parameters.

Source: Ormerod and Shepherd, 2003.

Domain of Application

The SGT technique was originally developed for use in the process control industries.

Procedure and Advice

Step 1: Define the Task(s) under Analysis
The first step in a SGT analysis involves defining the task (or tasks) or scenario under analysis. The analyst should specify the task that is to be subjected to the SGT analysis. A task or scenario list should be created, including the task, system, environment and personnel involved.

Step 2: Collect Specific Data Regarding the Task under Analysis
Once the task under analysis is defined, the data that will inform the development of the HTA should be collected. Specific data regarding the task should be collected, including the task steps involved, the task sequence, the technology used, the personnel involved and the communications made. There are a number of ways to collect this data, including observations, interviews and questionnaires. It is recommended that a combination of observation of the task under analysis and interviews with the personnel involved should be used when conducting a task analysis.

Step 3: Conduct an HTA for the Task under Analysis
Once sufficient information regarding the task under analysis is collected, an HTA for the task under analysis should be conducted.

Step 4: Assign SGT to HTA Sub-goals
Each bottom-level task from the HTA should then be assigned a SGT. SGT sequencing elements are presented as an example in Table 3.8.

Table 3.8   SGT sequencing elements

Code	Label	Syntax
S1	Fixed.	S1, then X.
S2	Choice/contingent.	S2 if Z, then X; if not Z, then Y.
S3	Parallel.	S3, then do X and Y together.
S4	Free.	S4 then do X and Y in any order.

Source: Ormerod, 2000.

Step 5: Specify Sequence
The order in which the tasks should be carried out is specified next using the SGT sequencing elements presented in Table 3.8.

Step 6: Specify Information Requirements
Once an SGT has been assigned to each bottom level operation in the HTA and the appropriate sequence of the operations has been derived, the information requirements should be derived. Each SGT has its own associated information requirements, and so this involves merely looking up the relevant SGTs and extracting the appropriate information requirements.

Advantages

- The SGT technique can be used to provide a full information requirements specification to system designers.
- It is based upon the widely used HTA technique.
- Once the initial concepts are grasped, it is easy to apply.

Disadvantages

- There is no data available relating to the reliability and validity of the technique.
- The initial requirement of an HTA for the task/system under analysis creates further work for the analyst(s).
- Further categories of SGT may require development, depending upon the system under analysis.
- One might argue that the output of an HTA would suffice.

Flowchart

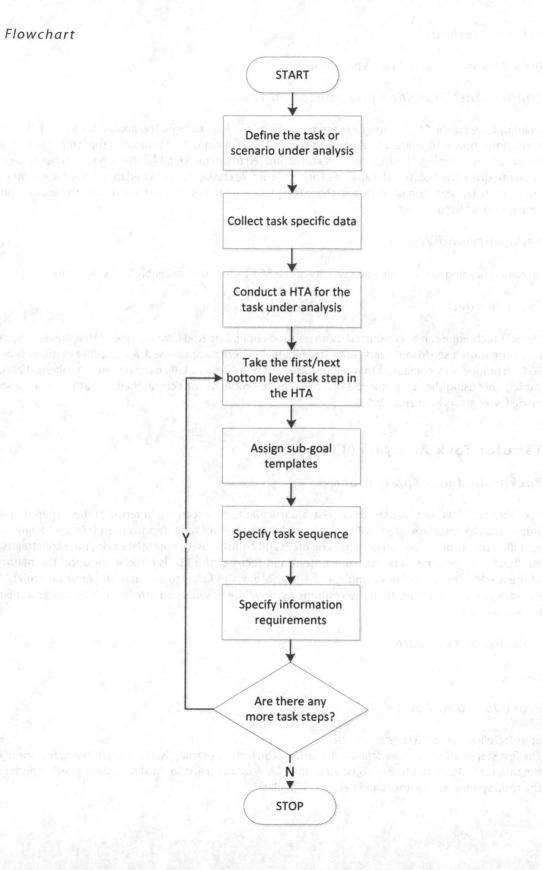

Related Methods

The SGT technique uses HTA as its primary input.

Approximate Training and Application Times

Training time for the SGT technique is estimated to be medium to high. The analyst is required to fully understand how HTA works and then to grasp the SGT technique. It is estimated that this may take a couple of days' training. The application is also estimated to be considerable, although this is dependent upon the size of the task(s) under analysis. For large, complex tasks, it is estimated that the SGT application time is high. For small, simple tasks and those tasks where an HTA is already constructed, the application time is estimated to be low.

Reliability and Validity

No data regarding the reliability and validity of the SGT technique is available in the literature.

Tools Needed

The SGT technique can be conducted using pen and paper. Ormerod (2000) suggests that the technique would be more user-friendly and easier to execute if it were computerised. A computer version of the SGT technique was compared to a paper-based version (Ormerod, Richardson and Shepherd, 1998). Participants using the computer version solved more problems correctly at their first attempt and also made fewer errors (Ormerod, 2000).

Tabular Task Analysis (TTA)

Background and Applications

TTA (Kirwan, 1994) can be used to analyse a particular task or scenario in terms of the required task steps and the interface used. A TTA takes each bottom-level task step from an HTA and analyses specific aspects of the task step, such as displays and controls used, potential errors, time constraints, feedback, triggering events, etc. The content and focus of the TTA is dependent upon the nature of the analysis required. For example, if the purpose of the TTA is to evaluate the error potential of the task(s) under analysis, then the columns used will be based upon errors, their causes and their consequences.

Domain of Application

Generic.

Procedure and Advice

Step 1: Define the Task(s) under Analysis
The first step in a TTA involves defining the task or scenario under analysis. The analyst should first specify the task (or tasks) that is to be subjected to the TTA. A task or scenario list should be created, including the task, system, environment and personnel involved.

Step 2: Collect Specific Data Regarding the Task under Analysis
Once the task under analysis is defined, the data that will inform the development of the TTA should be collected. Specific data regarding the task should be collected, including the task steps involved, the task sequence, the technology used, the personnel involved and the communications made. There are a number of ways to collect this data, including observations, interviews and questionnaires. It is recommended that a combination of observation of the task under analysis and interviews with the personnel involved should be employed when conducting a TTA.

Step 3: Conduct an HTA for the Task under Analysis
Once sufficient data regarding the task under analysis is collected, an initial task description should be created. For this purpose, it is recommended that an HTA is used. The data collected during step 2 should be used as the primary input to the HTA.

Step 4: Convert the HTA into Tabular Format
Once an initial HTA for the task under analysis has been conducted, the analyst should put the HTA into a tabular format. Each bottom-level task step should be placed in a column running down the left-hand side of the table. An example of an initial TTA is presented in Table 3.9.

Table 3.9   Extract of initial TTA

Task no.	Task description	Controls and displays used	Required action	Feedback	Possible errors	Error consequences	Error remedies
3.2.1	Check current airspeed.						
3.2.2	Dial in 190 knots using the speed/MACH selector knob.						
3.3.1	Check current flap setting.						
3.3.2	Set the flap lever to level '3'.						

Step 5: Choose Task Analysis Categories
Next the analyst should select the appropriate categories and enter them into the TTA. The selection of categories is dependent upon the nature of the analysis. The example in Table 3.9 was used to investigate the potential for design-induced error on the flight deck, and so the categories used are based upon error identification and analysis.

Step 6: Complete TTA Table
Once the categories are chosen, the analyst should complete the columns in the TTA for each task. How this is achieved is not a strictly defined process. A number of techniques can be used, such as walkthrough analysis, heuristic evaluation, observations or interviews with SMEs. Typically, the TTA is based upon the analyst's subjective judgment.

Advantages

- TTA is flexible technique, allowing any factors associated with the task to be assessed.
- A TTA analysis has the potential to provide a very comprehensive analysis of a particular task or scenario.

- It is easy to learn and use.
- It is generic and can be used in any domain.
- It provides a much more detailed description of tasks than traditional task analysis techniques do.
- As the analyst has control over the TTA categories used, potentially any aspect of a task can be evaluated.
- It is potentially exhaustive if the correct categories are used.

Disadvantages

- As the TTA is potentially so exhaustive, it is a very time-consuming technique to apply. The initial data collection phase and the development of an HTA for the task under analysis also add considerably to the overall application time.
- Data regarding the reliability and validity of the technique is not available in the literature. It is logical to assume that the technique may suffer from problems surrounding the reliability of the data produced.
- An HTA for the task under analysis may suffice in most cases.

Example

A TTA was performed on the landing task 'Land aircraft X at New Orleans Airport using the autoland system' (Marshall et al., 2003). The purpose of the analysis was to ascertain how suitable the TTA technique was for the prediction of design-induced error on civil flight decks. An HTA of the flight task was constructed (Figure 3.6) and a TTA analysis was performed (Table 3.10). Data collection included the following:

- Walkthrough of the flight task.
- Questionnaire administered to aircraft X pilots.
- Consultation with training manuals.
- Performing the flight task in aircraft simulator
- Interview with aircraft X pilot.

Related Methods

TTA is one among many task analysis techniques. It relies on a number of data collection techniques for its input. The initial task description required is normally provided by conducting an HTA for the task under analysis. Data collection for the TTA can involve any number of HF methods, including observational techniques, interviews, walkthrough analysis and questionnaires. The TTA technique is very similar to the task decomposition technique (Kirwan and Ainsworth, 1992).

Approximate Training and Application Times

The training time for the TTA technique is minimal, provided the analyst in question is competent in the use of HTA. The application time is considerably longer. It is estimated that each task step in an HTA requires up to 10 minutes for further analysis. Thus, for large, complex tasks, the TTA application time is estimated to be high. A TTA for the flight task 'Land aircraft X at New Orleans Airport using the autoland system', which consisted of 32 bottom-level task steps, took around four hours to complete.

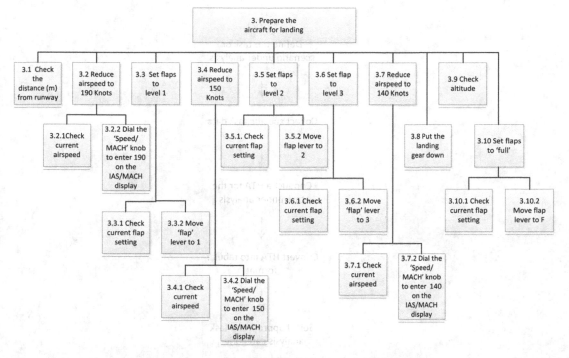

Figure 3.6   Extract of HTA for the task 'Land aircraft X at New Orleans Airport using the autoland system'

Source: Marshall et al., 2003.

Table 3.10   Extract of TTA analysis for the task 'Land aircraft X at New Orleans Airport using the autoland system'

Task no.	Task description	Controls/displays used	Required action	Feedback	Possible errors
3.2.1	Check current airspeed	Captains primary flight display Speed/MACH window	Visual check		Misread, check wrong display, fail to check
3.2.2	Dial in 190 Knots using the speed/MACH selector knob	Speed/MACH selector knob Speed/MACH window Captains primary flight display	Rotate Speed/MACH knob to enter 190 Visual check of speed/MACH window	Speed change in speed/MACH window and on CPFD, aircraft changes speed	Dial in wrong speed, use the wrong knob (e.g. heading knob)
3.3.1	Check current flap setting	Flap lever Flap display	Visual check		Misread, check wrong display, fail to check
3.3.2	Set the flap lever to level '3'	Flap lever Flap display	Move flap lever to '3' setting	Flaps change, Aircraft lifts and slows	Set flaps to wrong setting

Flowchart

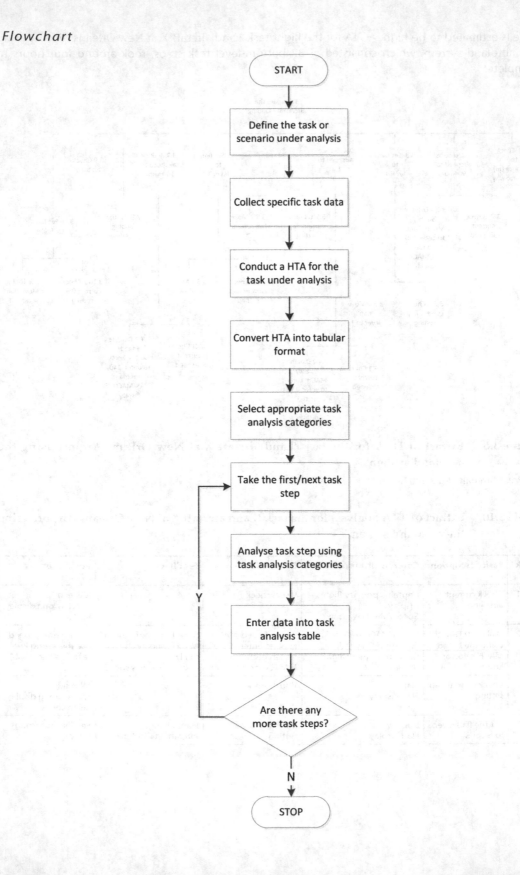

Chapter 4
Cognitive Task Analysis Methods

In contrast to traditional task analysis techniques, which provide a physical description of the activity performed within complex systems, cognitive task analysis (CTA) techniques are used to determine and describe the cognitive processes used by agents. Agents undertaking activity in today's complex systems face increasing demands upon their cognitive skills and resources. As system complexity increases, agents require training in specific cognitive skills and processes in order to keep up. System designers require an analysis of the cognitive skills and demands associated with the operation of these systems in order to propose design concepts, allocate tasks, develop training procedures and work processes, and to evaluate performance. The design, development and evaluation of the artefacts, procedures and training employed within such systems require the analysis of the cognitive skills and demands imposed on operators. Traditional task analysis techniques outputs can be used to develop physical, step-by-step descriptions of agent activity during task performance. Whilst this is useful, it does not explicitly consider the cognitive processes associated with the activity. For some analysts, the detail provided by traditional task analysis can be used as the basis for consideration of more 'cognitive' aspects – for example, the 'plans' in HTA could be taken to reflect the manner in which information is used to guide activity. However, it can be argued that assuming an equivalence between mental processes and the information needed to guide physical tasks can often lead to misunderstanding cognition (or at least requires a view of 'cognition' which is so restricted as to be at odds with what the term usually means).

The past three decades have seen the emergence of CTA, and a number of techniques now exist that can be used to determine, describe and analyse the cognitive processes employed during task performance. According to Schraagen, Chipman and Shalin (2000), CTA represents an extension of traditional task analysis techniques used to describe the knowledge, thought processes and goal structures underlying observable task performance. Militello and Hutton (2000) describe CTA techniques as those that focus upon describing and representing the cognitive elements that underlie goal generation, decision-making and judgments. CTA outputs are used, amongst other things, for interface design and evaluation, the design of procedures and processes, the allocation of functions, the design and evaluation of training procedures and interventions, and the evaluation of individual and team performance within complex systems.

Flanagan (1954) first probed the decisions and actions made by pilots in near-accidents using the critical incident technique (CIT). However, the term 'cognitive task analysis' did not appear until the early 1980s, when it began to be used in research texts. According to Hollnagel (2003), the term was first used in 1981 to describe approaches to the understanding of the cognitive activities required in man-machine systems. Since then, the focus on the cognitive processes employed by system operators has increased, and CTA applications are now on the increase, particularly in complex, dynamic environments such as those seen in the nuclear power, defence and emergency services domains. Various CTA techniques have been in widespread use over the past two decades, with applications in a number of domains, such as fire-fighting (Militello and Hutton, 2000), aviation (O'Hare et al., 2000), emergency services (O'Hare et al., 2000), command and control (Salmon et al., 2004), military operations (Klein, 2000), naval maintenance (Schaafstal and Schraagen, 2000) and even white-water rafting (O'Hare et al., 2000). Consequently, there are a large number of CTA approaches available. In 2006, Crandall, Klein and Hoffman published *Working Minds: A Practitioner's Guide to Cognitive Task Analysis*, a text dedicated to elaborating the 'how to' of CTA techniques. In this work, the authors describe over 1,000 applications of CTA to over 100 domains, including the military, the fire service and mathematics, and provide indepth examinations of a smaller number of techniques. A broader review of CTA and associated techniques is highlighted by

Crandall et al. (2006) and was conducted by the MITRE Corporation, details of which can be found online at the following address: www.mentalmodels.mitre.org. Crandall et al. present CTA as a methodology to 'understand how cognition makes it possible for humans to get things done and then turning that understanding into aids-low or high tech-for helping people get things done better' (2006: 2).

According to Roth, Patterson and Mumaw (2002), there are three different approaches to CTA. The first approach involves analysing the domain in question in terms of goals and functions in order to determine the cognitive demands imposed by the tasks performed. The second approach involves the use of empirical techniques, such as observation and interview techniques, in order to determine how the users perform the task(s) under analysis, allowing a specification of the knowledge requirements and the strategies involved. The third and more recent approach involves developing computer models that can be used to simulate the cognitive activities required during the task under analysis. It is beyond the scope of this book to review all of the CTA techniques available to the HF practitioner. Instead, a review of selected approaches based upon popularity and previous applications is presented. A brief description of the CTA approaches reviewed is presented below.

The cognitive work analysis (CWA) framework (Vicente, 1999) is currently receiving the most attention from the HF community. The CWA approach was originally developed at the Risø National Laboratory in Denmark (Rasmussen, Pejtersen and Goodstein, 1994) and offers a comprehensive framework for the design, evaluation and analysis of complex socio-technical systems. Rather than offer a description of the activity performed within a particular system, the CWA framework provides methods that can be used to develop an indepth analysis of the constraints that shape agent activity within the system. The CDM (Klein and Armstrong, 2004) is a semi-structured interview approach that uses pre-defined probes to elicit information regarding expert decision-making during complex activity. The CDM procedure is perhaps the most commonly used CTA technique and has been employed in a wide variety of domains. Applied cognitive task analysis (ACTA) (Militello and Hutton, 2000) offers a toolkit of semi-structured interview techniques that can be used to analyse the cognitive demands associated with a particular task or scenario. The cognitive walkthrough technique is used to evaluate interface usability. Based upon traditional design walkthrough techniques and a theory of exploratory learning (Polson et al., 1992), the technique focuses upon the usability, particularly from an ease of learning perspective. The CIT (Flanagan, 1954) is a semi-structured interview approach that uses a series of probes designed to elicit information regarding pilot decision-making during non-routine tasks. Object-oriented cognitive task analysis and design (OOCTAD: Wei and Salvendy, 2006) is a CTA methodology based upon Wickens' (1987b) human information processing theory. The method categorises cognition into a hierarchy of standard modules, packages and classes, with links to represent the relationships between them. The Concurrent Observer Narrative Technique (CONT; McIlroy and Stanton, 2011) is an elicitation technique that utilises the principles of both observation and VPA to provide insights into expert knowledge, reasoning strategies and cues that guide behaviour. Finally, collegial verbalisation (Erlandsson and Jansson, 2007; Jansson, Olsson and Erlandsson, 2006) is a knowledge elicitation technique in which two experts are utilised to provide insights into task performance. Expert one conducts a task whilst being video recorded and expert two watches the video, providing verbalisations about the cognitive processes occurring.

CTA techniques are useful in evaluating individual and team performance, in that they offer an analysis of cognitive processes surrounding decisions made and choices taken. This allows the HF practitioner to develop guidelines for effective performance and decision-making in complex environments. The main problem associated with the use of CTA techniques is the considerable amount of resources required. CTA methods are commonly based upon interview and observational data, and therefore require considerable time and effort to conduct. Access to SMEs is also required, as is great skill on the analyst's behalf. CTA techniques are also criticised for their reliance upon the recall of events or incidents from the past. Klein and Armstrong (2004) suggest that methods which analyse retrospective incidents are associated with concerns of data reliability due to memory degradation. A summary of the CTA techniques reviewed is presented in Table 4.1.

Table 4.1 Summary of CTA techniques

Method	Type of method	Domain	Training time	App time	Related methods	Tools needed	Validation studies	Advantages	Disadvantages
ACTA.	Cog task analysis.	Generic.	Med–high.	High.	Interviews, CDM.	Pen and paper, audio recording equipment.	Yes.	1) Requires fewer resources than traditional CTA techniques. 2) Provides the analyst with a set of probes.	1) Great skill is required on behalf of the analyst for the technique to achieve its full potential. 2) Consistency/reliability of the technique is questionable. 3) Time-consuming in its application.
Cognitive Walkthrough.	Cog task analysis.	Generic.	High.	High.	HTA.	Pen and paper, video and audio recording equipment.	Yes.	1) Has a sound theoretical underpinning (Norman's Action Execution model). 2) Offers a very useful output.	1) Requires further validity and reliability testing. 2) Time-consuming in its application. 3) Great skill is required on behalf of the analyst for the technique to achieve its full potential.
Cognitive Work Analysis.	Cog task analysis.	Generic.	High.	High.	Abstraction hierarchy, decision ladder, information flow maps, SRK framework, interviews, observation.	Pen and paper, video and audio recording equipment.	Yes.	1) Extremely flexible approach that can be used for a number of different purposes. 2) Has been used extensively in a number of different domains for the design, development, representation and evaluation of systems and technologies. 3) Based on sound underpinning theory.	1) CWA analyses are typically resource-intensive. 2) Only limited guidance is given to analysts and the techniques within the framework may be difficult to grasp for novice analysts. 3) The latter phases of the framework have previously received only limited attention.
Critical Decision Method.	Cog task analysis.	Generic.	Med–High.	High.	Critical Incident Technique.	Pen and paper, audio recording equipment.	Yes.	1) Can be used to elicit specific information regarding decision-making in complex environments. 2) Seems suited to C4i analysis. 3) Various cognitive probes are provided.	1) Reliability is questionable. 2) There are numerous problems associated with recalling past events, such as memory degradation. 3) Great skill is required on behalf of the analyst for the technique to achieve its full potential.

Table 4.1   Continued

Method	Type of method	Domain	Training time	App time	Related methods	Tools needed	Validation studies	Advantages	Disadvantages
Critical Incident Technique.	Cog task analysis.	Generic.	Med–High.	High.	CDM.	Pen and paper, audio recording equipment.	Yes.	1) Can be used to elicit specific information regarding decision-making in complex environments. 2) Seems suited to C4i analysis.	1) Reliability is questionable. 2) There are numerous problems associated with recalling past events, such as memory degradation. 3) Great skill is required on behalf of the analyst for the technique to achieve its full potential.
Collegial Verbalisation.	Cog task analysis.	Generic.	Low.	Med–High.	VPA, elicitation by critiquing, CONT, video cued recall.	Video and audio recording equipment.	No.	1) Does not interfere with task performance. 2) Avoids rationalisation of performance. 3) Exploration of mental models.	1) Video recording and knowledge of evaluation may affect performance. 2) Assumes a link between colleagues' cognitive processes.
Object-Oriented Cognitive Task Analysis and Design.	Cog task analysis.	Generic.	Med–High.	Med–High.		Pen and paper.	No.	1) Generic categories that can be reused lower costs and provide a common language. 2) Based on sound HF theory.	1) Complex. 2) Generic nature may mean that specific details are lost.
Concurrent Observer Narrative Technique.	Cog task analysis.	Generic.	Low.	Med–High.	VPA, collegial verbalisation, elicitation by critiquing.	Video and audio recording equipment.	No.	1) Does not interfere with task performance. 2) No domain expertise required; minimal resources required.	1) Requires access to multiple experts. 2) Assumes a link between colleagues' cognitive processes.

Cognitive Work Analysis (CWA)

Background and Applications

CWA (Jenkins et al., 2009a; Rasmussen, Pejtersen and Goodstein, 1994; Vicente, 1999) is a framework that was developed to model complex socio-technical work systems. The framework models different types of constraints, building a model of how work could proceed within a given work system. The focus on constraints separates the technique from other approaches to analysis that aim to describe how work is actually conducted, or prescribe how it should be conducted.

CWA was originally developed at the Risø National Laboratory in Denmark (Rasmussen, Pejtersen and Goodstein, 1994). In the years that have followed, attempts have been made to add additional detail and clarification to the framework proposed by Rasmussen et al.; however, the underlying framework remains largely unchanged.

The CWA approach can be used to describe the constraints imposed by purpose of a system, its functional properties, the nature of the activities that are conducted, the roles of the different actors, and their cognitive skills and strategies. Rather than offer a prescribed methodology, the CWA framework instead acts as a set of tools that can be used either individually or in combination with one another, depending upon the analysis needs. These tools are divided between phases. The exact names and scopes of these phases differ slightly depending on the scope of the analysis; however, the overall scope remains largely the same. As defined by Vicente (1999), the CWA framework comprises five different phases: work domain analysis, control task (or activity) analysis, strategies analysis, social organisation and cooperation analysis, and worker competencies analysis.

The different tools within the CWA framework have been used for a plethora of different purposes, including system modelling (Chin, Sanderson and Watson, 1999), system design (Bisantz et al., 2003; Rasmussen, Pejtersen and Goodstein, 1994), process design (Olsson and Lee, 1994), training needs analysis (Naikar and Sanderson, 1999), training design and evaluation, interface design and evaluation (Dinadis and Vicente, 1999; Salmon et al., 2004), information requirements specification (Stoner, Wiese and Lee, 2003), tender evaluation (Naikar and Sanderson, 2001), team design (Naikar et al., 2003) and error management training design (Naikar and Saunders, 2003). Despite the CWA's origins within the nuclear power domain, the CWA applications referred to above have taken place in a wide range of different domains, including naval, military, aviation, driving, and health-care domains.

Domain of Application

The CWA framework was originally developed for the nuclear power domain; however, the generic nature of the methods within the framework allows it to be applied in a wide range of domains. The framework is best suited to domains that can be described as complex (i.e. exuding some or all of the following qualities: high risk, dynamic, uncertain, with interconnected parts).

Procedure and Advice

It is especially difficult to prescribe a strict procedure for the CWA framework. In its true form, the framework is used to provide a description of the constraints within a domain. This description can then be used to address specific research and design aims. For example, work domain analysis (WDA) is commonly used to support interface design and evaluation purposes, but it can also be used to inform training design and evaluation. It would also be beyond the scope of this review to describe the procedure fully. The following procedure is intended to act as a broad set of guidelines for each of the phases defined by the CWA framework. A more complete description can be found in Jenkins et al. (2009a) and Jenkins (2012).

Step 1: Define the Nature of Analysis
The first step in a CWA is to clearly define the purpose of the analysis. The exact aims of the analysis should be clearly specified to allow for an appropriate description of the system.

Step 2: Select Appropriate CWA Phases and Methods
Once the nature and desired outputs of the analysis are clearly defined, the analysis team carefully select the most appropriate CWA phases and methods to be employed during the analysis. For example, when using the framework for the design of a novel interface, it may be that only the WDA component is required.
    Conduct steps 3–8 as appropriate.

Step 3: Work Domain Analysis
The initial phase within the CWA framework, WDA, provides a description of the constraints that govern the purpose and the function of the systems under analysis. The Abstraction Hierarchy (Rasmussen, 1985; Vicente, 1999; see also Figure 4.2) is used to provide a context-independent description of the domain. The analysis, and the resultant set of diagrams, is not specific to any particular technology; rather, they represent the entire domain. The top three levels of the diagrams consider the overall objectives of the domain, and what it can achieve, whereas the bottom two levels concentrate on the physical components and their affordances. Through a series of 'means-ends' links, it is possible to model how individual components can have an impact on the overall domain purpose. The abstraction hierarchy (shown in Figure 4.2) is constructed by considering the work system's objectives (top-down) and the work system's capabilities (bottom-up). The diagram is constructed based upon a range of data collection opportunities. The exact data collection procedure is dependent on the domain in question and the availability of data. In most cases, the procedure commences with some form of document analysis. Document analysis allows the analyst to gain a basic domain understanding, forming the basis for semi-structured interviews with domain experts. Wherever possible, observation of the work in context is highly recommended.
    The abstraction hierarchy consists of five levels of abstraction, ranging from the most abstract level of purposes to the most concrete level of form (Vicente, 1999). The labels used for each of the levels of the hierarchy tend to differ, depending on the aims of the analysis. In this case, the labels used by Xiao et al. (2008) are adopted. It is felt that the use of the word 'domain' in the top three levels and the use of the word 'physical' in the bottom two levels draws a fitting distinction:

- Domain purpose: the domain purpose, displayed at the very top of the diagram, represents the reason why the work system exists. This purpose is independent of any specific situation and is also independent of time – the system purpose exists as long as the system does.
- Domain values: the domain values level of the hierarchy is used to capture the key values that can be used to assess how well the work system is performing its domain purpose(s). These values are likely to be conflicting.
- Domain functions: the middle layer of hierarchy lists the functions that can be performed by the combined work system. These functions are expressed in terms of the domain in question.
- Physical functions: the physical functions that the objects can perform are listed. These are listed generically and are independent of the domain purpose.
- Physical objects: the key physical objects within the work system are listed at the base of the hierarchy. These objects represent the sum of the relevant objects from all of the component technologies. This level of the diagram is independent of purpose; however, analyst judgment is required to limit the object list to a manageable size.

The structure of the abstraction hierarchy framework acts as a guide to acquiring the knowledge necessary to understand the domain. The framework helps to direct the search for deep knowledge about the work domain, providing structure to the document analysis process, particularly for the domain novice. While the output may initially appear overbearing, its value to the analysis cannot be overstated. The abstraction hierarchy defines the systemic constraints at the highest level.

The WDA can also be described using an abstraction decomposition space (ADS). The decomposition hierarchy (the top row in the abstraction-decomposition space) typically comprises five levels of resolution, ranging from the coarsest level of total system to the finest level of component (Vicente, 1999). According to Vicente (1999), each of the five levels represents a different level of granularity with respect to the system in question and moving from left to right across the decomposition hierarchy is the equivalent of zooming into the system, as each level provides a more detailed representation of the system in question. The ADS also employs structural means-ends relationships in order to link the different representations of the system within the ADS. This means that every node in the ADS should be the end that is achieved by all of the nodes below it, and also the means that can be used to achieve all of the nodes above it.

Table 4.2   Abstraction decomposition space template

Decomposition / Abstraction	Total system	Sub-system	Function unit	Sub-assembly	Component
Domain purpose.	Purpose of the entire system.				
Domain values.					
Domain functions.					
Physical functions.					
Physical objects.					Material form of individual components.

Step 4: Conduct Activity Analysis in Work Domain Terms

Up until this point, the analysis has deliberately not considered the constraints that are imposed by specific situations. One tool for considering such constraints is the 'Contextual Activity Template' (Naikar, Moylan and Pearce, 2006). This tool plots the functions identified in the abstraction hierarchy against a number of specific 'situations'. At this stage, the analysis remains independent of the actor. The first stage of the process is to define the situations. Situations can be characterised by either time or location (or a combination of the two). In many cases, it is appropriate to explore more than one set of situations using multiple contextual activity template representations to meet a range of analytic goals.

The contextual activity templates take an output from the abstraction hierarchy and add information on additional constraints. Thus, the products provide a more context-specific description of the domain. The contextual activity templates can be used to inform information exchange requirements by indicating situations where information may be required. Likewise, by adding in additional constraints, the list of possible information requirements is reduced in certain situations.

Step 5: Conduct Activity Analysis in Decision-Making Terms

Continuing with the theme of describing additional constraint, key function-situation cells within the contextual activity template can be explored in terms of decision-making. The decision ladder (Rasmussen, 1974; see Figure 4.1) is the tool most commonly used within CWA to describe decision-making activity. Its focus is on the entire decision-making activity rather than the moment of selection between options. It is not specific to any single actor; instead, it represents the decision-making process of the combined work system. In many cases, the decision-making process may be collaborative, distributed between a range of human and technical decision-makers.

The ladder contains two different types of node: the rectangular boxes represent data-processing activities and the circles represent resultant states of knowledge. Novice users (to the situation) are

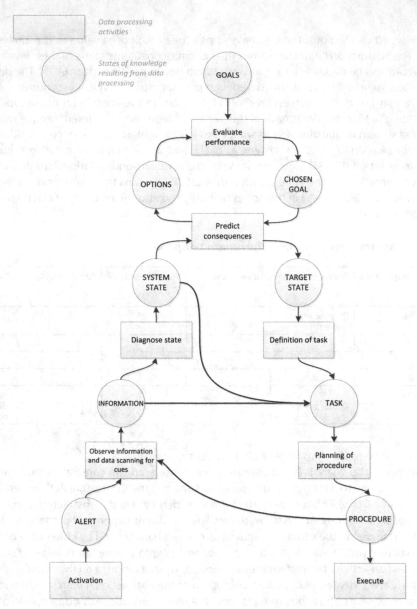

Figure 4.1   Decision ladder

Source: Vicente, 1999.

expected to follow the decision ladder in a linear fashion, whereas expert users are expected to link the two halves by shortcuts. The left side of the decision ladder represents the observation of the current system state, whereas the right side represents the planning and execution of tasks and procedures to achieve a target system state. Decision ladders can be populated based on semi-structured interviews with subject-matter experts (see Jenkins et al., 2010b). Ostensibly, in its raw form, the decision ladder models provide a list of the information requirements for making a decision triggered by a number of presupposed events. At this stage of the analysis, the relative importance of these information elements is not considered.

Step 6: Conduct Strategies Analysis

Activity analysis in terms of mental strategies, or simply 'strategies analysis', addresses the constraints influencing the way in which activity can be conducted. In keeping with the remainder of the framework, it introduces additional detail to the analyses in the previous phases. Ostensibly, the aim of a strategies analysis is to describe the constraints that dictate how a work system can be (rather than how it should be or currently is) moved from one state to another. As Jenkins et al. (2009b) point out, this phase of the analysis is particularly useful for exploring the flexibility within a system. Information flow maps are typically used for the strategies analysis component of CWA.

Step 7: Conduct Social Organisation and Cooperation Analysis

The social organisation and cooperation analysis phase of a CWA involves identifying how the control tasks can be distributed between agents and artefacts within the system. The social organisation and cooperation analysis component of CWA reuses the abstraction decomposition space, decision ladders and information flow maps developed during the preceding phases for this purpose.

Step 8: Conduct Worker Competencies Analysis

The final stage of a CWA involves identifying the cognitive skills required for control task performance in the system under analysis. Worker competencies analysis uses Rasmussen's Skills, Rules and Knowledge (SRK) framework in order to classify the cognitive activities employed by agents during control task performance.

Example

Jenkins (2012) used CWA to describe the role of unmanned air vehicles (UAVs) in military operations. The analysis used the abstraction hierarchy (see Figure 4.2) to describe the purpose of the wider air operations system and the functions that it could perform. The Contextual Activity Template was then used to explore the limitations imposed by certain situations. Building on the idea of successive layers of constraints, the decision ladder (see Figure 4.3) was used to describe the information requirements that support the decision-making processes involved in operating in the military air operations domain. Additional detail was then added by describing the different strategies that could be used to achieve set tasks within this domain. Finally, these products were reproduced and coded to indicate the constraints that influenced whether autonomous UAVs could perform the task or whether human intervention was required.

This analysis of the work system emphasised the inherent complexity within the military air operations domain. From the models produced, it was apparent that the work system components were capable of performing a wide range of functions in an array of situations. These components are required to adjust frequently to balance conflicting domain values. The military air operations work system is expected to balance the completion of mission objectives with minimising casualties and collateral damage. Furthermore, decision-makers are expected to consider the full gamut of imposed constraints (such as rules of engagement and legal), as well as the impact of their actions on media perception, and the hearts and minds of the wider population.

The study of decision-making revealed that decision rights could be allocated across the work system. A range of information elements were identified as having the potential to inform the complex decisions required. While mandated procedures are commonplace within military work systems, the strategies analysis section revealed the variability in the way operators move between even closely related system states. Exploration of the social and organisational aspects of the work system revealed considerable overlap between the functionality of the autonomous UAV and the human. At a superficial level, the autonomous UAV can perform nearly all the functions that a manned aircraft can. However, the key difference, highlighted by the study of the decision ladder, is that the automated UAV lacks the ability to make complicated value judgments in ambiguous situations. In situations where activity can be broken down and controlled by rule-based reasoning, the automated UAV performs well. However, autonomous UAVs lack the ability of the human to truly consider the impact of their actions on the wide range of domain values identified in the abstraction hierarchy.

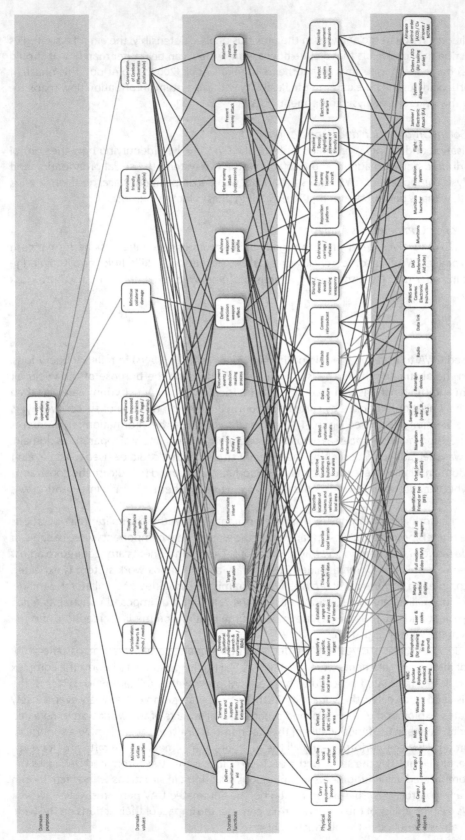

Figure 4.2   Abstraction hierarchy for the air operations domain

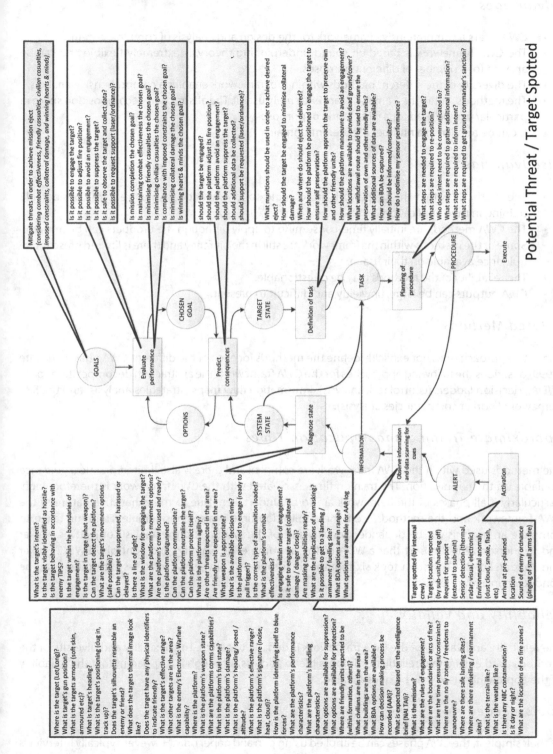

Figure 4.3   Decision ladder showing the process of mitigating potential threats

Advantages

- CWA offers a comprehensive framework for the design and analysis of complex systems.
- The CWA framework is based on sound underpinning theory, is extremely flexible and can be applied for a number of different purposes.
- The diversity of the different methods within the framework enables a comprehensive analysis.
- The methods within the framework are extremely useful. The abstraction decomposition space in particular can be used for a wide range of purposes.
- It can be applied in a number of different domains.

Disadvantages

- The methods within the framework are complex and practitioners may require considerable training in their application.
- The CWA methods are initially time-consuming to apply (although the products can be reused).
- Some of the methods within the framework are still in their infancy and there is limited prescriptive guidance available on their usage.
- The reliability of the methods may be questionable.
- CWA outputs can be large, unwieldy and difficult to present.

Related Methods

The CWA approach does not explicitly define the methods for each of the different CWA phases. Vicente (1999) describes the following approaches for the CWA framework: the abstraction decomposition space (WDA), decision ladders (control task analysis), information flow maps (strategies analysis) and the SRK framework (worker competencies analysis).

Approximate Training and Application Times

The methods used within the CWA framework are complex and there is also limited practical guidance available on their application. The training time associated with the CWA framework is therefore high, particularly if all phases of the framework are to be undertaken. Due to the exhaustive nature of the CWA framework and the methods used, the application time is also considerable. The time taken to complete an analysis will be dependent on the size of the system and the analysis objectives. Naikar and Sanderson (2001) report that a WDA of the airborne early warning and control (AEW&C: Naikar and Sanderson, 2001) system took six months to complete. Conversely, smaller-scale studies can be conducted in a few weeks.

Reliability and Validity

The reliability and validity of the CWA framework is difficult to assess. The flexibility and diversity of the methods used ensure that reliability is impossible to address, although it is apparent that the reliability of the approaches used may be questionable.

Tools Needed

At their simplest, the CWA phases can be applied using pen and paper only. However, typically interviews and observations are required, and so audio and video recording equipment may be required. CWA outputs are also typically large and require software support in their construction. For example, Microsoft Visio is particularly useful in the construction of the abstraction hierarchy. A dedicated software tool for the approach has been produced the Human Factors Integration Defence Technology Centre (HFI DTC) and can be requested via its website (see www.hfidtc.com).

Flowchart

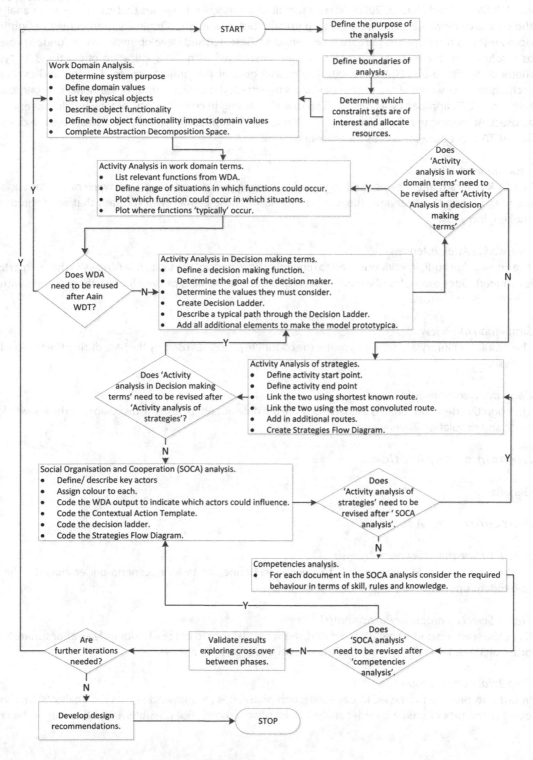

Applied Cognitive Task Analysis (ACTA)

Background and Applications

ACTA (Militello and Hutton, 2000) offers a toolkit of interview techniques that can be used to analyse the cognitive demands associated with a particular task or scenario. Originally used in the fire-fighting domain, it was developed as part of a Navy Personnel Research and Development Centre-funded project as a solution to the inaccessibility and difficulty associated with the application of existing CTA-type methods (Militello and Hutton, 2000). The overall goal of the project was to develop and evaluate techniques that would allow system designers to extract the critical cognitive elements of a particular task. The ACTA approach was designed so that no training in cognitive psychology would be required to use it. According to Militello and Hutton (2000), ACTA outputs are typically used to aid system design. The ACTA procedure comprises the following.

Task Diagram Interview
The task diagram interview is used to provide the analyst with an indepth overview of the task under analysis. During the interview, the analyst highlights those elements of the task that are cognitively challenging.

Knowledge Audit Interview
The knowledge audit interview is used to highlight those parts of the task under analysis where expertise is required. Once examples of expertise are highlighted, the SME is probed for specific examples within the context of the task.

Simulation Interview
The simulation interview is used to probe the cognitive processes used by the SME during the task under analysis.

Cognitive Demands Table
The cognitive demands table is used to integrate the data obtained from the task diagram, the knowledge audit and simulation interviews.

Domain of Application

Generic.

Procedure and Advice

Step 1: Define the Task under Analysis
The first part of an ACTA analysis is to select and define the task or scenario under analysis. This is dependent upon the nature and focus of the analysis.

Step 2: Select Appropriate Participant(s)
Once the scenario under analysis is defined, the analyst(s) should proceed to identify an appropriate SME or set of SMEs. Typically, operators of the system under analysis are used.

Step 3: Task Observation
In order to prepare for the ACTA data collection phase, it is recommended that the analyst(s) involved observe the task or scenario under analysis. If an observation is not possible, a walkthrough of the task

may suffice. This allows the analyst to fully understand the task and the participant's role during task performance.

Step 4: Task Diagram Interview
The purpose of the task diagram interview is to elicit a broad overview of the task under analysis in order to focus the knowledge audit and simulation interview parts of the analysis. Once the task diagram interview is complete, the analyst should have created a diagram representing the component task steps involved and those task steps that require the most cognitive skill. According to Militello and Hutton (2000), the SME should first be asked to break down the task into the relevant task steps. The analyst should use questions like 'Think about what you do when you perform the task [under analysis]. Can you break this task down into less than six, but more than three steps?'. Once the task is broken down into a number of separate task steps, the SME should then be asked to identify which of the task steps require cognitive skills. Militello and Hutton (2000) define cognitive skills as judgments, assessments, problem solving and thinking skills.

Step 5: Knowledge Audit
Next, the analyst should proceed with the knowledge audit interview. This allows the analyst to identify instances during the task under analysis where expertise is used and also what sort of expertise is used. The knowledge audit interview is based upon the following knowledge categories that characterise expertise (Militello and Hutton, 2000):

- diagnosing and predicting;
- situation awareness;
- perceptual skills;
- developing and knowing when to apply tricks of the trade;
- improvising;
- meta-cognition;
- recognising anomalies;
- compensating for equipment limitations.

Once a probe has been administered, the analyst should then query the SME for specific examples of critical cues and decision-making strategies. Potential errors should then be discussed. The list of knowledge audit probes is presented below (Militello and Hutton, 2000).

Basic Probes

- Past and future: is there a time when you walked into the middle of a situation and knew exactly how things got there and where they were headed?
- Big picture: can you give me an example of what is important about the big picture for this task? What are the major elements you have to know and keep track of?
- Noticing: have you had experiences where part of a situation just 'popped' out at you, where you noticed things going on that others didn't catch? What is an example?
- Job smarts: when you do this task, are there ways of working smart or accomplishing more with less that you have found especially useful?
- Opportunities/improvising: can you think of an example when you have improvised in this task or noticed an opportunity to do something better?
- Self-monitoring: can you think of a time when you realised that you would need to change the way you were performing in order to get the job done?

Optional Probes

- Anomalies: can you describe an instance when you spotted a deviation from the norm or knew that something was amiss?
- Equipment difficulties: have there been times when the equipment pointed in one direction but your own judgment told you to do something else? Or when you had to rely on experience to avoid being led astray by the equipment?

Step 6: Simulation Interview

The simulation interview allows the analyst to determine the cognitive processes involved during the task under analysis. The SME is presented with a typical scenario. Once the scenario is completed, the analyst should prompt the SME to recall any major events, including decisions and judgments, that occurred during the scenario. Each event or task step in the scenario should be probed for situation awareness, actions, critical cues, potential errors and surrounding events. Militello and Hutton (2000: 96) present the following set of simulation interview probes: For each major event, elicit the following information:

- As the job you are investigating in this scenario, what actions, if any, would you take at this point in time?
- What do you think is going on here? What is your assessment of the situation at this point in time?
- What pieces of information led you to this situation assessment and these actions?
- What errors would an inexperienced person be likely to make in this situation?

Any information elicited here should be recorded in a simulation interview table. An example simulation interview table is shown in Table 4.3.

Table 4.3   Example simulation interview table

Events	Actions	Assessment	Critical cues	Potential errors
On-scene arrival.	Account for people (names). Ask neighbours. Must knock on or knock down to make sure people aren't there.	Its a cold night, need to find place for people who have been evacuated.	Night time. Cold > 15°. Dead space. Add on floor. Poor materials, metal girders. Common attic in whole building.	Not keeping track of people (could be looking for people who are not there).
Initial attack.	Watch for signs of building collapse. If signs of building collapse, evacuate and throw water on it from outside.	Faulty construction, building may collapse.	Signs of building collapse include: what walls are doing: cracking; what floors are doing: groaning; what metal girders are doing; clicking, popping. Cable in old buildings holds walls together.	Ventilating the attic, which draws the fire up and spreads it through the pipes and the electrical system.

Source: Militello and Hutton, 2000.

Step 7: Construct Cognitive Demands Table

Once the knowledge audit and simulation interview are completed, it is recommended that a cognitive demands table is used to integrate the data collected (Militello and Hutton, 2000). This table is used to help the analyst focus on the most important aspects of the data obtained. The analyst should prepare the cognitive demands table based upon the goals of the particular project involved. An example of a cognitive demands table is shown in Table 4.4.

Table 4.4   Example cognitive demands table

Difficult cognitive element	Why difficult?	Common errors	Cues and strategies used
Knowing where to search after an explosion.	Novices may not be trained in dealing with explosions. Other training suggests you should start at the source and work outwards.	Novice would be likely to start at the source of the explosion. Starting at the source is a rule of thumb for most other kinds of incidents.	Start where you are most likely to find victims, keeping in mind safety considerations. Refer to material data sheets to determine where dangerous chemicals are likely to be. Consider the type of structure and where victims are likely to be. Consider the likelihood of further explosions. Keep in mind the safety of your crew.
Finding victims in a burning building.	There are lots of distracting noises. If you are nervous or tired, your own breathing makes it hard to hear anything else.	Novices sometimes don't recognise their own breathing sounds; they mistakenly think they hear a victim breathing.	Both you and your partner stop, hold your breath and listen. Listen for crying, victims talking to themselves, victims knocking things over, etc.

Source: Militello and Hutton, 2000.

Advantages

- The technique offers a structured approach to CTA.
- The use of three different interview approaches ensures the comprehensiveness of the technique.
- Analysts using the technique do not require training in cognitive psychology.
- Militello and Hutton (2000) reported that in a usability questionnaire focusing on the use of the ACTA techniques, ratings were very positive. The data indicated that participants found the ACTA techniques easy to use and flexible, and that the output of the interviews was clear and the knowledge representations were useful.
- Probes and questions are provided for the analyst, facilitating relevant data extraction.
- Prasanna, Yang and King (2009) argue that there is a high level of guidance regarding the incorporated methods, making it easier to apply than other CTA methods.

Disadvantages

- The quality of data obtained is very much dependent upon the skill of the analyst involved and also the quality of the SMEs used.
- The reliability of such a technique is questionable.
- The technique appears to be time-consuming in its application. In a validation study (Militello and Hutton, 2000), participants using the ACTA techniques were given three hours to perform the interviews and four hours to analyse the data.
- The training time for the ACTA techniques is also considerable. Militello and Hutton (2000) gave participants an initial two-hour workshop introducing CTA and then a six-hour workshop on the ACTA techniques.
- The analysis of the data appears to be a laborious process.
- As with most CTA techniques, ACTA requires further validation. At the moment there is little evidence of validation studies associated with the ACTA techniques.
- It is often difficult to gain sufficient access to appropriate SMEs for the task under analysis.
- Parasanna, Yang and King (2009) argue that the method is more complicated and subjective than other CTA methods.
- There is no guidance on extending ACTA for information requirements analysis, although the authors of the method suggest that it is possible, and little guidance on the integration of the component methods (Prasanna, Yang and King, 2009).

Related Methods

The ACTA technique is an interview-based CTA technique. There are other interview-based CTA techniques, such as the CDM (Klein and Armstrong, 2004). The ACTA technique also employs various data collection techniques, such as walkthrough and observation.

Approximate Training and Application Times

In a validation study (Militello and Hutton, 2000), participants were given eight hours of training, consisting of a two-hour introduction to CTA and a six-hour workshop on the ACTA techniques. This represents a medium training time for the ACTA techniques. In the same study, the total application time for each participant was seven hours, consisting of three hours conducting the interviews and four hours analysing the data. This represents a medium application time for the ACTA techniques.

Reliability and Validity

Militello and Hutton (2000) suggest that there are no well-established metrics that exist in order to establish the reliability and validity of CTA techniques. However, Militello and Hutton (2000) made a number of attempts to establish the reliability and validity of the ACTA techniques. In terms of validity, three questions were addressed:

- does the information gathered address cognitive issues?;
- does the information gathered deal with experience-based knowledge as opposed to classroom-based knowledge?; and
- do the instructional materials generated contain accurate information that is important for novices to learn?

The study by Militello and Hutton (2000) examined each item in the cognitive demands table for its cognitive content. They found that 93 per cent of the items were related to cognitive issues.

To establish the level of experience-based knowledge elicited, participants were asked to subjectively rate the proportion of information that only highly experienced SMEs would know. The study was divided into two parts: the fire-fighting study in which experienced fire commanders were interviewed; and the electronic warfare (EW) study in which experienced EW technicians were interviewed. In the fire-fighting study, the average was 95 per cent, and in the EW study, the average was 90 per cent. The importance of the instructional materials generated was validated via domain experts rating the importance and accuracy of the data elicited. The findings indicated that the instructional materials generated in the study contained important information for novices (70 per cent fire-fighting, 95 per cent EW). The reliability of the ACTA techniques was assessed by determining whether the participants using the techniques generated similar information. It was established that participants using the ACTA techniques were able to consistently elicit relevant cognitive information.

Tools Needed

ACTA can be applied using pen and paper only, providing the analyst has access to the ACTA probes required during the knowledge audit and simulation interviews. An audio recording device may also be useful to aid the recording and analysis of the data.

Flowchart

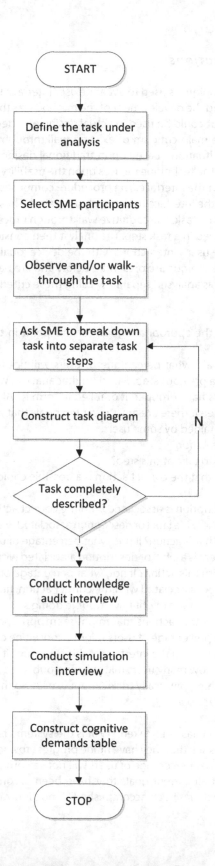

START

Define the task under analysis

Select SME participants

Observe and/or walk-through the task

Ask SME to break down task into separate task steps

Construct task diagram

Task completely described?

N

Y

Conduct knowledge audit interview

Conduct simulation interview

Construct cognitive demands table

STOP

Cognitive Walkthrough

Background and Applications

The cognitive walkthrough technique is used to evaluate user interface usability. According to Polson et al. (1992), the main driver behind the development of the method was the goal to provide a theoretically based design methodology that could be used in actual design and development situations. Polson et al. go on to comment that the main criticism of existing walkthrough techniques suggests that they are unusable in actual design situations. Based upon traditional design walkthrough techniques and a theory of exploratory learning, the technique focuses upon the usability of an interface, in particular the ease of learning associated with the interface. The procedure comprises a set of criteria that the analyst uses to evaluate each task and the interface under analysis against. These criteria focus on the cognitive processes required to perform the task. The cognitive walkthrough process involves the analyst 'walking' through each user action involved in a task step. The analyst then considers each criteria and the effect that the interface has upon the user's interactions with the device (goals and actions). The criteria used in the cognitive walkthrough technique are presented below and are based on the work of Polson et al. (1992). Each task step or action is analysed separately using these criteria.

Goal Structure for a Step
- Correct goals: what are the appropriate goals for this point in the interaction? Describe as for initial goals.
- Mismatch with likely goals: what percentage of users will not have these goals, based on the analysis at the end of the previous step. Based on that analysis, will all users have the goal at this point, or may some users have dropped it or failed to form it? Also check the analysis at the end of the previous step to see if there are any unwanted goals that are not appropriate for this step that will be formed or retained by some users.

Choosing and Executing the Action
- Is this the correct action to take at this step?
- Availability: is it obvious that the correct action is a possible choice here? If not, what percentage of users might miss it?
- Label: what label or description is associated with the correct action?
- Link of label to action: if there is a label or description associated with the correct action, is it obvious and is it clearly linked with this action? If not, what percentage of users might have trouble?
- Link of label to goal: if there is a label or description associated with the correct action, is it obvious and is it clearly linked with this action? If not, what percentage of users might have trouble?
- No label: if there is no label associated with the correct action, how will users relate this action to a current goal? What percentage might have trouble doing so?
- Wrong choices: are there other actions that might seem appropriate to some current goal? If so, what are they, and what percentage of users might choose one of these?
- Time out: if there is a time out in the interface at this step, does it allow time for the user to select the appropriate action? How many users might have trouble?
- Hard to do: is there anything physically tricky about executing the action? If so, what percentage of users will have trouble?

Modification of Goal Structure
- Assume the correct action has been taken. What is the system's response?
- Quit or backup: will users see that they have made progress towards some current goal? What will indicate this to them? What percentage of users will not see progress and try to quit or backup?
- Accomplished goals: list all current goals that have been accomplished. Is it obvious from the system response that each has been accomplished? If not, indicate for each how many users will not realise it is complete.

- Incomplete goals that look accomplished: are there any current goals that have not been accomplished, but might appear to have based upon the system response? What might indicate this? List any such goals and the percentage of users who will think that they have actually been accomplished.
- 'And-then' structures: is there an 'and-then' structure and does one of its sub-goals appear to be complete? If the sub-goal is similar to the super-goal, estimate how many users may prematurely terminate the 'and-then' structure.
- New goals in response to prompts: does the system response contain a prompt or cue that suggests any new goal or goals? If so, describe the goals. If the prompt is unclear, indicate the percentage of users that will not form these goals.
- Other new goals: are there any other new goals that users will form given their current goals, the state of the interface and their background knowledge? Why? If so, describe the goals and indicate how many users will form them. Note that these goals may or may not be appropriate, so forming them may be bad or good.

Domain of Application

Generic. Although originally developed for use in the software engineering domain, it is apparent that the technique could be used to evaluate an interface in any domain.

Procedure and Advice (Adapted from Polson et al., 1992)

The cognitive walkthrough procedure comprises two phases: the preparation phase and the evaluation phase. The preparation phase involves selecting the set of tasks to analyse and determining the task sequence. The evaluation phase involves the analysis of the interaction between the user and the interface, using the criteria outlined above.

Step 1: Select Tasks to be Analysed
First, the analyst should select the set of tasks that are to be the focus of the analysis. In order to ensure that the user interface in question is subjected to a thorough examination, an exhaustive set of tasks should be used. However, if time is limited, the analyst should try to select set of tasks that are as representative of the tasks that can be performed with the interface under analysis as possible.

Step 2: Create Task Descriptions
Each task selected by the analyst must be described fully. Although there are a number of ways of doing this, it is recommended that an HTA describing the general operation of the user interface under analysis is used. An exhaustive HTA should provide a description of each task identified during step 1.

Step 3: Determine the Correct Sequence of Actions
For each of the selected tasks, the appropriate sequence of actions required to complete the task must be specified. Again, it is recommended that the analyst uses the HTA for this purpose.

Step 4: Identify User Population
Next, the analyst should determine the potential users of the interface under analysis. A list of user groups should be created.

Step 5: Describe the User's Initial Goals
The final part of the cognitive walkthrough analysis preparation phase involves identifying and recording the user's initial goals. The analyst should record what goals the user has at the start of the task. This is based upon the analyst's subjective judgment. Again, it is recommended that the HTA output is used to generate the goals required for this step of the analysis.

Step 6: Analyse the Interaction between User and Interface
The second and final phase of the cognitive walkthrough procedure, the evaluation phase, involves analysing the interaction between the user and the interface under analysis. To do this, the analyst should 'walk' through each task, applying the criteria outlined above as he or she goes along. The cognitive walkthrough evaluation concentrates on three key aspects of the user interface interaction (Polson et al., 1992):

1.   The relationship between the required goals and the goals that the user actually has.
2.   The problems in selecting and executing an action.
3.   Changing goals due to action execution and system response.

The analyst should record the results for each task step. This can be done via video, audio or pen and paper techniques.

Advantages

- The cognitive walkthrough technique presents a structured approach to user interface analysis.
- It is used early in the design lifecycle of an interface. This allows any design flaws highlighted in the analysis to be eradicated.
- It is designed to be used by non-cognitive psychology professionals.
- It is based upon sound underpinning theory, including Norman's model of action execution.
- It is easy to learn and apply.
- The output from a cognitive walkthrough analysis appears to be very useful.

Disadvantages

- The cognitive walkthrough technique is limited to cater only for ease of learning of an interface.
- It requires validation.
- It may be time-consuming for more complex tasks.
- A large part of the analysis is based upon the analyst's subjective judgment. For example, the percentage estimates used with the walkthrough criteria require a 'best guess'. As a result, the reliability of the technique may be questionable.
- It requires access to the personnel involved in the task(s) under analysis.

Related Methods

The cognitive walkthrough technique is a development of traditional design walkthrough methods (Polson et al., 1992). HTA or tabular task analysis could also be used when applying the cognitive walkthrough technique in order to provide a description of the task under analysis.

Approximate Training and Application Times

No data regarding the training and application time for the technique is offered by the authors. It is estimated that the training time for the technique would be quite high. It is also estimated that the application time for the technique would be high, particularly for large, complex tasks.

Reliability and Validity

Lewis et al. (1990) reported that in a cognitive walkthrough analysis of four answering machine interfaces, about half of the actual observed errors were identified. More critically, the false alarm rate (errors predicted in the cognitive walkthrough analysis but not observed) was extremely high, at almost 75 per cent. In a study on voicemail directory, Polson et al. (1992) reported that half of all observed errors were picked up in the cognitive walkthrough analysis. It is apparent that the cognitive walkthrough technique requires further testing in terms of the reliability and validity of the technique.

Tools Needed

The cognitive walkthrough technique can be applied using pen and paper only. For larger analyses, the analyst may wish to record the process using video or audio recording equipment. The device or interface under analysis is also required.

Flowchart

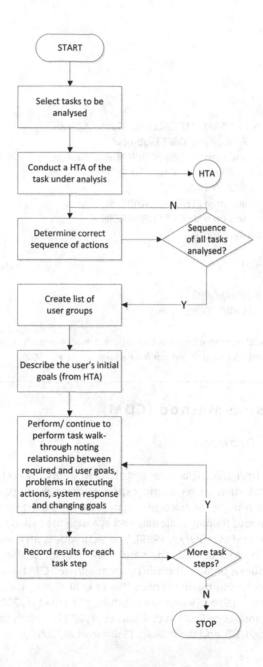

Example

The following example is an extract of a cognitive walkthrough analysis of a phone system task presented in Polson et al. (1992).

Task – Forward all my calls to 492-1234
Task list

1.   Pick up the handset

2.   Press ##7

3.   Hang up the handset

4.   Pick up the handset

5.   Press **7

6.   Press 1234

7.   Hang up the handset

Goals:
75 per cent of users will have FORWARD ALL CALLS TO 492-1234 (Goal)
                                       PICK UP HANDSET (Sub-goal)
                                       and then SPECIFY FORWARDING (Sub-goal)
25 per cent of users will have FORWARD ALL CALLS TO 492-1234
                                       PICK UP HANDSET
                                       and then CLEAR FORWARDING
                                       and then SPECIFY FORWARDING
Analysis of ACTION 1: pick up the handset
Correct goals
FORWARD ALL CALLS TO 492-1234
         PICK UP HANDSET
         and then CLEAR FORWARDING
         and then SPECIFY FORWARDING

75 per cent of the users would therefore be expected to have a goal mismatch at this step, due to the required clear forwarding sub-goal that is required but not formed (Polson et al., 1992).

The Critical Decision Method (CDM)

Background and Applications

The CDM (Klein and Armstrong, 2004) is a semi-structured interview technique that uses cognitive probes in order to elicit information regarding expert decision-making. According to the authors, the technique can serve to provide knowledge engineering for expert system development, identify training requirements, generate training materials and evaluate the task performance impact of expert systems (Klein, Calderwood and MacGregor, 1989). The technique is an extension of the CIT (Flanagan, 1954) and was developed in order to study the naturalistic decision-making strategies of experienced personnel. The CDM procedure is perhaps the most commonly used CTA technique and has been applied in a number of domains, including the fire service (Baber et al., 2004), military and paramedics (Klein, Calderwood and MacGregor, 1989) and white-water rafting (O'Hare et al., 2000), with recent applications including emergency response coordination (Salmon et al., 2011), health care (Fackler et al., 2009), air traffic control (Walker et al., 2010), and the military (Salmon et al., 2009b).

Domain of Application

Generic.

Procedure and Advice

Step 1: Define the Task or Scenario under Analysis
The first part of a CDM analysis is to define the incident that is to be analysed. CDM normally focuses on non-routine incidents, such as emergency incidents, or highly challenging incidents. If the scenario under analysis is not already specified, the analyst(s) may identify an appropriate incident via interview with an appropriate SME by asking them to describe a recent highly challenging (i.e. high workload) or non-routine incident in which they were involved. The interviewee involved in the CDM analysis should be the primary decision-maker in the chosen incident.

Step 2: Select CDM Probes
The CDM technique works by probing SMEs using specific probes designed to elicit pertinent information regarding the decision-making process during key points in the incident under analysis. In order to ensure that the output is compliant with the original aims of the analysis, an appropriate set of CDM probes should be defined prior to the analysis. The probes used are dependent upon the aims of the analysis and the domain in which the incident is embedded. Alternatively, if there are no adequate probes available, the analyst(s) can develop novel probes based upon the analysis needs. A set of CDM probes defined by O'Hare et al. (2000) is presented in Table 4.5.

Step 3: Select Appropriate Participant
Once the scenario under analysis and the probes to be used are defined, an appropriate participant or set of participants should be identified. The SMEs used are typically the primary decision-maker in the task or scenario under analysis.

Step 4: Gather and Record Account of the Incident
The CDM procedure can be applied to an incident observed by the analyst or to a retrospective incident described by the participant. If the CDM analysis is based upon an observed incident, then this step involves first observing the incident and then recording an account of the incident. Alternatively, the incident can be described retrospectively from memory by the participant. The analyst should ask the SME for a description of the incident in question, from its starting point to its end point.

Step 5: Construct Incident Timeline
The next step in the CDM analysis is to construct a timeline of the incident described in step 4. The aim of this is to give the analyst(s) a clear picture of the incident and its associated events, including when each event occurred and what the duration of each event was. According to Klein, Calderwood and MacGregor (1989), the events included in the timeline should encompass any physical events, such as alarms sounding, and also 'mental' events, such as the thoughts and perceptions of the interviewee during the incident.

Step 6: Define Scenario Phases
Once the analyst has a clear understanding of the incident under analysis, the incident should be divided into key phases or decision points. It is recommended that this is done in conjunction with the SME. Normally, the incident is divided into four or five key phases.

Step 7: Use CDM Probes to Query Participant Decision-Making
For each incident phase, the analyst should probe the SME using the CDM probes selected during step 2 of the procedure. The probes are used in an unstructured interview format in order to gather pertinent

information regarding the SME's decision-making during each incident phase. The interview should be recorded using an audio recording device.

Step 8: Transcribe Interview Data
Once the interview is complete, the data should be transcribed accordingly.

Step 9: Construct CDM Tables
Finally, a CDM output table for each scenario phase should be constructed. This involves simply presenting the CDM probes and the associated SME answers in an output table. The CDM output tables for an energy distribution scenario are presented in Tables 4.6–4.9.

Advantages

- The CDM analysis procedure can be used to elicit specific information regarding the decision-making strategies used by agents in complex, dynamic systems.
- The technique is normally quick to apply.
- Once familiar with the technique, CDM is relatively easy to apply.
- It is a popular procedure and has been applied in a number of domains.
- Its output can be used to construct propositional networks which describe the knowledge or SA objects required during the scenario under analysis.

Disadvantages

- The reliability of such a technique is questionable. Klein and Armstrong (2004) suggest that methods that analyse retrospective incidents are associated with concerns of data reliability, due to evidence of memory degradation.
- The data is obtained is highly dependent upon the skill of the analyst conducting the CDM interview and also the quality of the participants used.
- A high level of expertise and training is required in order to use the CDM to its maximum effect (Klein and Armstrong, 2004).
- It relies upon interviewee verbal reports in order to reconstruct incidents. How far a verbal report accurately represents the cognitive processes of the decision-maker is questionable. Facts could easily be misrepresented by the participants involved.
- It is often difficult to gain sufficient access to appropriate SMEs in order to conduct a CDM analysis.

Example

The following example is taken from a CDM analysis that was conducted in order to analyse C4i activity in the civil energy distribution domain (Salmon et al., 2005). The scenario under analysis involved the switching out of three circuits at three substations. Circuit SGT5 was being switched out for the installation of a new transformer for the nearby channel tunnel rail link and SGT1A and 1B were being switched out for substation maintenance. For the CDM analysis, the control room operator coordinating the activity and the senior authorised person (SAP) at the substation who conducted the activity were interviewed. The set of CDM probes used are presented in Table 4.5. The scenario was divided into four key phases:

1. First issue of instructions.
2. Deal with switching requests.
3. Perform isolation.
4. Report back to network operations centre.

Flowchart

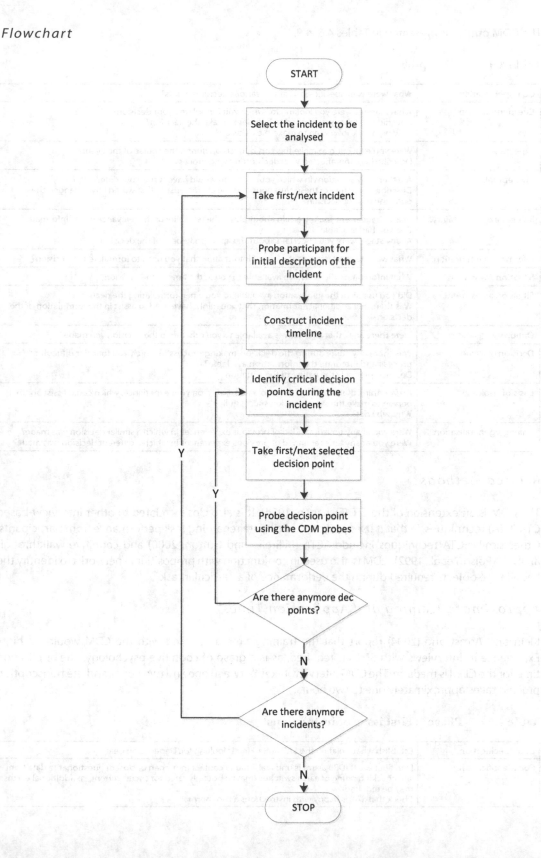

The CDM output is presented in Tables 4.6–4.9.

Table 4.5   CDM probes

Goal specification.	What were your specific goals at the various decision points?
Cue identification.	What features were you looking for when you formulated your decision? How did you what you needed to in order to make the decision? How did you know when to make the decision?
Expectancy.	Were you expecting to make this sort of decision during the course of the event? Describe how this affected your decision-making process
Conceptual.	Are there any situations in which your decision would have turned out differently? Describe the nature of these situations and the characteristics that would have changed the outcome of your decision
Influence of uncertainty.	At any stage, were you uncertain about either the reliability or the relevance of the information that you had available? At any stage, were you uncertain about the appropriateness of the decision?
Information integration.	What was the most important piece of information that you used to formulate the decision?
Situation Awareness.	What information did you have available to you at the time of the decision?
Situation assessment.	Did you use all of the information available to you when formulating the decision? Was there any additional information that you might have used to assist in the formulation of the decision?
Options.	Were there any other alternatives available to you other than the decision you made?
Decision blocking – stress.	Was there any stage during the decision-making process in which you found it difficult to process and integrate the information available? Describe precisely the nature of the situation
Basis of choice.	Do you think that you could develop a rule, based on your experience, which could assist another person to make the same decision successfully? Why/why not?
Analogy/generalisation.	Were you at any time reminded of previous experiences in which a similar decision was made? Were you at any time reminded of previous experiences in which a different decision was made?

Related Methods

The CDM is an extension of the CIT (Flanagan, 1954). It is also closely related to other interview-based CTA (CTA) techniques, in that it uses probes to elicit data regarding task performance from participants. Other similar CTA techniques include ACTA (Militello and Hutton, 2000) and cognitive walkthrough analysis (Polson et al., 1992). CDM is also used in conjunction with propositional networks to identify the knowledge objects required during the performance of a particular task.

Approximate Training and Application Times

Klein and Armstrong (2004) report that the training time associated with the CDM would be high. Experience in interviews with SMEs is required, as is a grasp of cognitive psychology. The application time for the CDM is medium. The CDM interview takes between one and two hours, and the transcription process takes approximately one to two hours.

Table 4.6   Phase 1: First issue of instructions

Goal specification.	Establish what isolation the SAP at Barking is looking for. Depends on gear?
Cue identification.	Don't Believe It (DBI) alarm is unusual – faulty contact (not open or closed) questionable data from site checking rating of earth switches (maybe not fully rated for circuit current, so additional earths may be required). Check that SAP is happy with instructions as not normal.

Table 4.6   Continued

Expectancy.	Decision expected by DBI is not common.
Conceptual model.	Recognised instruction but not stated in WE1000 – as there are not too many front and rear shutters metal clad switch gear.
Uncertainty.	Confirm from field about planned instruction – make sure that SAP is happy with the instruction.
Information.	Reference to front and rear busbars.
Situation Awareness.	WE1000 procedure. Metal clad switchgear. Barking SGT1A/1B substation screen. SAP at Barking.
Situation assessment.	Ask colleagues if needed to.
Options.	No alternatives.
Stress.	N/A
Choice.	WE1000 – need to remove what does not apply. Could add front and rear busbar procedures.
Analogy.	Best practice guide for metal clad EMS switching.

Table 4.7   Phase 2: Deal with switching requests

Goal specification.	Obtain confirmation from NOC that planned isolation is still required.
Cue identification.	Approaching time for planned isolation. Switching phone rings throughout building. Airblast circuit breakers (accompanied by sirens) can be heard to operate remotely (more so in Barking 275 than Barking C 132).
Expectancy.	Yes – routine planned work according to fixed procedures.
Conceptual model.	Wokingham have performed remote isolations already. Circuit configured ready for local isolation.
Uncertainty.	Physical verification of apparatus always required (DBI – don't believe it).
Information.	Proceduralised information from NOC – circuit, location, time, actions required, etc. Switching log.
Situation Awareness.	Switching log. Physical status of apparatus. Planning documentation. Visual or verbal information from substation personnel.
Situation assessment.	Planning documentation used only occasionally.
Options.	Refusal of switching request. Additional conditions to switching request.
Stress.	Some time pressure.
Choice.	Yes – highly proceduralised anyway.
Analogy.	Yes – routine activity.

Reliability and Validity

Both the intra- and inter-analyst reliability of the CDM approach are questionable. It is apparent that such an approach may elicit different data from similar incidents when applied by different analysts on separate participants. Klein and Armstrong (2004) suggest that there are also concerns associated with the reliability of the CDM due to evidence of memory degradation.

Table 4.8   Phase 3: Perform isolation

Goal specification.	Ensure it is safe to perform local isolation. Confirm circuits/equipment to be operated.
Cue identification.	Telecontrol displays/circuit loadings. Equipment labels. Equipment displays. Other temporary notices.
Expectancy.	Equipment configured according to planned circuit switching. Equipment will function correctly.
Conceptual model.	Layout/type/characteristics of circuit. Circuit loadings/balance. Function of equipment.
Uncertainty.	Will equipment physically work as expected (will something jam, etc.)? Other work being carried out by other parties (e.g. EDF).
Information.	Switching log. Visual and verbal information from those undertaking the work.
Situation Awareness.	Physical information from apparatus and telecontrol displays.
Situation assessment.	All information used.
Options.	Inform NOC that isolation cannot be performed/other aspects of switching instructions cannot be carried out.
Stress.	Some time pressure. Possibly some difficulties in operating or physically handling the equipment.
Choice.	Yes – proceduralised within equipment types. Occasional non-routine activities required to cope with unusual/unfamiliar equipment or equipment not owned by NGT.
Analogy.	Yes – often. Except in cases with unfamiliar equipment.

Table 4.9   Phase 4: Report back to Network Operations Centre (NOC)

Goal specification.	Inform NOC of isolation status.
Cue identification.	Switching telephone. NOC operator answers.
Expectancy.	NOC accepts.
Conceptual model.	Manner in which circuit is now isolated. Form of procedures.
Uncertainty.	No – possibly further instructions, possibly mismatches local situation and remote displays in NOC.
Information.	Switching log.
Situation Awareness.	Verbal information from NOC. Switching log.
Situation assessment.	Yes – all information used.
Options.	No (raise or add on further requests, etc. to the same call?).
Stress.	No.
Choice.	Yes – highly proceduralised.
Analogy.	Yes – frequently performed activity.

Tools Needed

When conducting a CDM analysis, pen and paper could be sufficient. However, to ensure that data collection is comprehensive, it is recommended that video or audio recording equipment is used. A set of relevant CDM probes, such as those presented in Table 4.5, is also required. The type of probes used is dependent upon the focus of the analysis.

Critical Incident Technique (CIT)

Background and Applications

The CIT (Flanagan, 1954) is an interview technique that is used to retrospectively analyse operator decision-making. The technique was first used to analyse aircraft incidents that 'almost' led to accidents and has since been used extensively and redeveloped in the form of CDM (Klein and Armstrong, 2004). It involves the use of semi-structured interviews to facilitate operator recall of critical events or incidents, including the actions and decisions made by themselves and colleagues, and the reasons behind these actions and decisions. The analyst uses a set of probes designed to elicit pertinent information surrounding the participant's decision-making during the scenario under analysis. A set of probes used by Flanagan (1954) is presented below:

- Describe what led up to the situation.
- Exactly what did the person do or not do that was especially effective or ineffective?
- What was the outcome or result of this action?
- Why was this action effective or what more effective action might have been expected?

The CIT is a generic method that can be employed in a variety of domains. Recent applications of the method include insights into the limitations associated with weather surveillance radars (Newman, LaDue and Heinselman, 2009); teamwork issues in healthcare (Kvarnstrom, 2008); incivility from university students during lectures (Goodyear, Reynolds and Both Gragg, 2010); and dental students' perception of learning (Zakariasen Victoroff and Hogan, 2005).

Domain of Application

Generic. Although the technique was originally developed for use in analysing pilot decision-making in non-routine (e.g. near-miss) incidents, the technique can be applied in any domain.

Procedure and Advice

Step 1: Select the Incident to be Analysed
The first part of a CIT analysis is to select the incident (or group of incidents) that is to be analysed. Depending upon the purpose of the analysis, the type of incident may already be selected. CIT normally focuses on non-routine incidents, such as emergency scenarios, or highly challenging incidents. If the type of incident is not already known, CIT analysts may select the incident via interview with system personnel, probing the interviewee for recent high-risk, highly challenging emergency situations. The interviewee involved in the CIT analysis should be the primary decision-maker in the chosen incident. CIT can also be conducted on groups of operators.

Step 2: Gather and Record an Account of the Incident
Next, the interviewee(s) should be asked to provide a description of the incident in question, from its starting point (i.e. alarm sounding) to its end point (i.e. when the incident was classed as 'under control').

Step 3: Construct Incident Timeline
The next step in the CIT analysis is to construct an accurate timeline of the incident under analysis. The aim of this is to give the analysts a clear picture of the incident and its associated events, including when each event occurred and what the duration of each event was. According to Klein, Calderwood and MacGregor (1989), the events included in the timeline should encompass any physical events, such

as alarms sounding, and also 'mental' events, such as the thoughts and perceptions of the interviewee during the incident.

Step 4: Select Required Incident Aspects
Once the analyst has an accurate description of the incident, the next step is to select specific incident points that are to be analysed further. The points selected are dependent upon the nature and focus of the analysis. For example, if the analysis is focusing upon team communication, then aspects of the incident involving team communication should be selected.

Step 5: Probe Selected Incident Points
Each incident aspect selected in step 4 should be analysed further using a set of specific probes. The probes used are dependent upon the aims of the analysis and the domain in which the incident is embedded. The analyst should develop specific probes before the analysis begins. In an analysis of team communication, the analyst would use probes such as 'Why did you communicate with team member B at this point?', 'How did you communicate with team member B?' and 'Was there any miscommunication at this point?'.

Advantages

- The CIT can be used to elicit specific information regarding decision-making in complex systems.
- Once learned, the technique requires relatively little effort to apply.
- The incidents which the technique concentrates on have already occurred, removing the need for time-consuming incident observations.
- It has been used extensively in a number of domains and has the potential to be used anywhere.
- It is a very flexible technique – this flexibility is said to enable innovative use (Butterfield et al., 2005).
- It has high face validity (Kirwan and Ainsworth, 1992).
- Kvarnstrom (2008) argues that it provides more detailed information than other comparable techniques.
- It can be used to study both behavioural acts and cognitive states and constructs (Butterfield et al., 2005).

Disadvantages

- The reliability of such a technique is questionable. Methods that analyse retrospective incidents are associated with concerns of data reliability, due to memory degradation.
- A high level of expertise in interview techniques is required.
- The method relies upon the accurate recall of events.
- Operators may not wish to recall events or incidents in which there performance is under scrutiny.
- The data obtained is dependent upon the skill of the analyst and also the quality of the SMEs used.
- The original CIT probes are dated and the technique has effectively been replaced by the CDM.
- There is a lack of formal guidelines on how to establish validity (Butterfield et al., 2005).
- There is a lack of formal structure, which can cause confusion (Butterfield et al., 2005).
- There are no reliability evaluations on the use of CIT to explore cognition (Butterfield et al., 2005).

Related Methods

CIT was the first interview-based technique designed to focus upon past events or incidents. A number of techniques have since been developed as a result of the CIT, such as the critical decision method (Klein, Calderwood and McGregor, 1989).

Flowchart

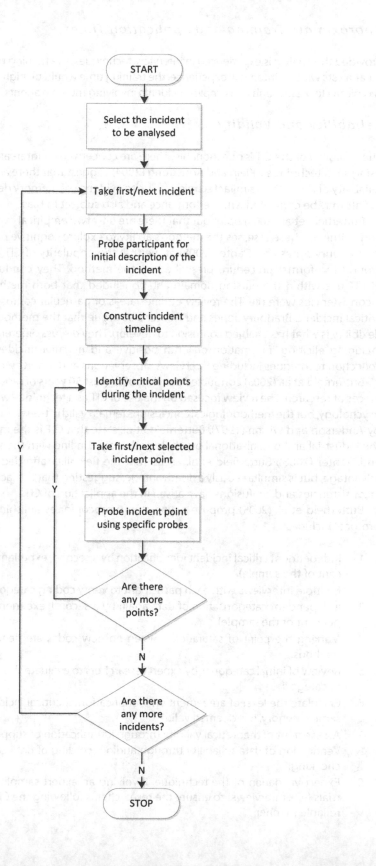

Approximate Training and Application Times

Provided the analyst is experienced in interview techniques, the training time for CIT is minimal. However, for analysts with no interview experience, the training time would be high. The application time for the CIT is typically low, although for complex incidents involving multiple agents, it could increase considerably.

Reliability and Validity

The reliability of the CIT is questionable. There are concerns over inter- and intra-analyst reliability when using such techniques. Klein and Armstrong (2004) suggest that there are concerns associated with the reliability of the CDM (a similar technique) due to evidence of memory degradation. In addition, recalled events may be correlated with performance and also subject to bias.

Butterfield et al. (2005) point out that there are only two empirical evaluations of the reliability of CIT and neither of these assesses the method's ability to explore cognitive processes.

Bradbury-Jones and Tranter (2008) argue that the popularity of CIT has led to a level of ambiguity around the formal procedure prescribed by the method. They conducted a comprehensive review of CIT use within the nursing domain and concluded that both methodological and terminological inconsistencies were rife. The review outlines areas of particular confusion, such as what constitutes a critical incident. Bradbury-Jones and Tranter also argue that the method is highly flexible, but that this flexibility is what has enabled confusion to develop. They discuss different adaptations of the technique, including eliciting information on both positive and negative incidents, and utilising various data collection techniques including interviews, observations and reports written by participants.

Butterfield et al. (2005) conducted a review of the last 50 years of research with CIT studying over 125 pieces of research. The review focused on the use of CIT as a technique within the domain of counselling psychology, but the methodological conclusions remain valid in the wider literature. Referencing research by Anderson and Wilson (1997), Butterfield et al. state that CIT is the most frequently cited method in the industrial and organisational psychology domains. In line with the suppositions of Bradbury-Jones and Tranter (2008), Butterfield et al. argue that the flexibility afforded by the method is not only an advantage, but is simultaneously a disadvantage, suggesting that the lack of a rigid structure means that inconsistencies and confusions have arisen in the application of CIT.

Butterfield et al. (2005) propose a nine-step protocol for establishing reliability in CIT studies. This protocol includes:

1. Independent critical incident identification by a second, experienced researcher (normally 25 per cent of the sample).
2. Multiple interviews with each participant to verify coding categories.
3. Independent categorisation of data again by a second, experienced researcher (again using 25 per cent of the sample).
4. Marking the point of saturation – when no new codes are generated from CIT data collection methods.
5. Review of initial categories by expert researchers to evaluate the usefulness and appropriateness of categories.
6. Calculate the level of agreement on identification of critical incidents (if 25 per cent identify the same category, it is deemed valid).
7. Assessment of theoretical validity through identification of supporting literature.
8. Verification of data reliability through audio recording of data collection and participant cross-checking.
9. Expert validation of the technique, involving an expert sampling a number of data collection trials (e.g. interviews) to ensure the researcher is following the CIT procedure in an accurate and reliable manner.

Tools Needed

CIT can be conducted using pen and paper. However, it is recommended that the analysis is recorded using video and audio recording equipment.

The Concurrent Observer Narrative Technique (CONT)

Background and Applications

CONT (McIlroy and Stanton, 2011) is similar to VPA (Ericsson & Simon, 1980) in that it is used to derive descriptions of the processes (both cognitive and physical) that an individual uses to perform a task. It is a method for eliciting expert knowledge and gathering information on the reasoning strategies and cues that guide operator behaviour. It requires an operator to perform a task whilst a colleague, matched as closely in experience as possible, observes, providing verbalisations. The 'narration' is recorded using the same video used to capture task performance.

Domain of Application

Generic.

Procedure and Advice

The approach for capturing data using CONT and the method for analysing the resulting protocols are described below. Though the data collection method is fixed, the method for analysing the verbal data is only a suggestion; as with any data collection method, the resulting information may be analysed in a number of different ways, depending on the needs of the analysis and the nature of the domain.

Step 1: Define the Scenario under Analysis
First, the scenario under analysis should be clearly defined. The analyst must know what he or she is intending to study and the boundaries of the analysis.

Step 2: Instruct/Train the Participant
Once the task to be analysed is clearly defined, the analyst must instruct the individual who is to provide the verbalisations on how to go about narrating the task. The instructions follow those of Ericsson and Simon's (1980) description of VPA; the narrator should verbalise all that he or she sees, should not be concerned with grammar, should not censor his or her thoughts at all and should talk continuously. Depending on the requirements of the analysis, the analyst may ask the narrator to focus on certain aspects of the task, for example, interaction methods, though this should only be done if the analyst is already at least partially familiar with the domain; it is not appropriate for the analyst to judge which parts of the task are pertinent if he or she lacks the domain knowledge to successfully and reliably do so (McIlroy and Stanton, 2011). The operator (the individual performing the task) should also be briefed on the nature of the study, explaining he or she is to undertake the task as would be done in normal circumstances.

Step 3: Introduction to the Physical Context
The analyst should then be introduced to the domain's physical context. This step is important in order to allow the analyst to judge where best to situate the camera and narrating expert such that a balance is reached between being distant enough not to disturb the progression of the task and being close enough to have an adequate view of the system and practitioner.

Step 4: Begin Scenario and Record Data

The recording of the scenario should then commence; the narrator must be sat alongside the camera such that the scenario can be recorded in its entirety, the video capturing the task and the audio capturing the narrations. If the narrator should fall silent for a prolonged period of time, the analyst should prompt him or her to 'continue talking' or to 'try to keep talking'. The length of the silence warranting a prompt will depend on the total task length, though it is typically between 30 and 60 seconds.

Step 5: Transcription and Segmentation of Verbal Data

Upon completion of the session, the recordings should be transcribed verbatim. Once the transcript has been produced, it must be segmented. Though there is no standard scheme for segmentation, most approaches are broadly similar, being based on (for example) pause-bound utterances, content of speech, or single identifiable sentences or reflections. The choice of segmentation scheme relies on the judgment of the analyst and is guided by the purpose of the investigation.

Step 6: Develop a Coding Scheme

The development of an encoding scheme is wholly dependent on the nature and purpose of the investigation. The codes may be developed *a priori*, as in the case of traditional VPA, or *a posteriori*, as in the case of grounded theory (Glaser and Strauss, 1967). If the former option is chosen, the coding scheme should be developed from the literature and subsequently applied to the protocols. If the latter option is chosen, the process is somewhat lengthier as the scheme is developed from the transcripts themselves; an initial coding scheme is based on transcript content, is subsequently applied to different sections of the transcripts and is then added to and amended in order to better fit the content of the transcripts. This process is repeated in an iterative fashion until the coding scheme adequately accounts for all content.

It is also possible to adopt a mixed approach, developing a coding scheme *a priori* and subsequently adapting it iteratively based on repeated application to portions of the transcripts.

Step 7: Encode Verbalisations and Establish Inter-rater Reliability

Once the coding scheme has been finalised, it should be applied to all the transcripts. One or more codes may be assigned to each segment, again based on the needs of the analysis. Once all transcripts have been coded, a randomly selected section (typically around 10 per cent of the whole data set) should be passed to at least one other analyst for encoding. Traditionally, Cohen's Kappa is calculated to statistically assess inter-rater reliability. If the reliability is not sufficient, then the coding scheme may require amendment.

Step 8: Analyse the Structure and Patterns of the Code Frequencies

Finally, the results of the CONT can be analysed. Enumeration of the codes is the first step; the frequency with which each code is present in the transcript must be calculated. Analysis can proceed in a number of ways, which are again dependent on the needs of the analyst – for example, the length of time spent on each code and verbalisation rates.

Advantages

- CONT has significant benefits in situations where the subject-matter expert cannot provide concurrent verbalisations himself or herself, due to, for example, high workload or communication requirements.
- It provides verbalisations that do not require interpretation (the scene has already been interpreted by the narrating subject matter expert), hence it is very useful in complex domains where a layperson would be unlikely to understand the details of the task or subject matter.
- It places minimal strain on resources, in terms of both the time required to gather data and the equipment required to capture that data.

Disadvantages

- Unlike traditional verbalisation techniques (e.g. concurrent and retrospective think aloud), the method requires a colleague to verbalise the work of the operator. As such, it is not possible to say definitively that the narrator's verbalisations match the cognitive processes of the operator. However, the protocols still provide a means for eliciting expert knowledge, information on typical reasoning strategies and cognitive processes, information on the perceptual cues used to guide behaviour and actions, and more general information about what system operators could or should be thinking at a specific moment in time (Seagull and Xiao, 2001). The method thus offers a valuable means of acquiring information.
- Multiple experts are required for the method.

Related Methods

CONT has its theoretical roots in VPA, with the main difference being that an expert provides verbalisation for the work of a colleague rather than verbalising his or her own work. Collegial verbalisation (Erlandsson and Jansson, 2007; Jansson, Olsson and Erlandsson, 2006) also has this requirement; however, in this method the operator is video recorded performing the task and the recording is shown to the expert colleague subsequent to task completion. The individual providing the verbalisations is not situated in the work setting whilst the task is performed.

Tools Needed

A video camera capable of recording both video and audio is required. A word processing package (for example, Microsoft Word) is typically used in the transcription of the recordings; the process can be facilitated by the use of transcription foot-pedals (for example, VEC's Infinity In-USB-2 transcription pedals) and transcription software (for example, NCH© Software's Express Scribe version 5.0.1).

Approximate Application and Training Times

CONT is a simple method to apply and requires minimal training. The application time will be equal to the length of the task under analysis, though coding and analysis of the resulting transcripts will be time-consuming and laborious.

Reliability and Validity

At present, there is no data within the literature regarding the reliability or validity of the method. McIlroy and Stanton (2011) argue, however, that as the task does not rely on the analyst to guide the verbalisations (an inappropriate practice when considering analyst naïveté), validity is ensured in the sense that the protocols can only reflect what the narrator is actually observing; the acquisition of information is not reliant on the probing skills or relevant domain knowledge of the researcher, but on the knowledge of the subject-matter expert and the performance of the operator.

Example

The CONT method was used to gather information about the use of an Instructor Control Console to run and manipulate scenarios in the simulator-based training of Merlin HM Mk1 helicopter rear crew (McIlroy and Stanton, 2011). Across five sessions, each lasting between one and two hours, one subject-matter expert provided verbalisations for the work of a number of colleagues; in session one, the Instructor Control Console (the system under analysis) was manned by two instructors, whilst in the remaining four sessions, only one instructor was required to operate the system (this was due to differences in the nature of the training sessions, not differences in instructor proficiency). The resulting transcripts were subjected to a thematic analysis consistent with the tenets of 'grounded theory' (e.g. Glaser and Strauss, 1967; Xiao et al., 2008), in that the data was the driving factor in the development of codes. However, a 'full-fat' grounded theory approach was not adopted; the primary aim of the research was to gather information regarding the use of the system rather than develop predictive theories of cognition or behaviour. The final coding scheme contained six major themes which were further separated into 19 sub-themes. Codes were enumerated and the relative frequency with which each code was present in each of the transcripts was calculated. Figure 4.4 presents a graphical depiction of the coding categories identified and the frequency of words present in each category.

A wealth of information was obtained regarding the activities and behaviours required from instructors and of the various interaction methods and teaching options available to them. For example, the results provided the researchers with an understanding of how the instructors' time was distributed

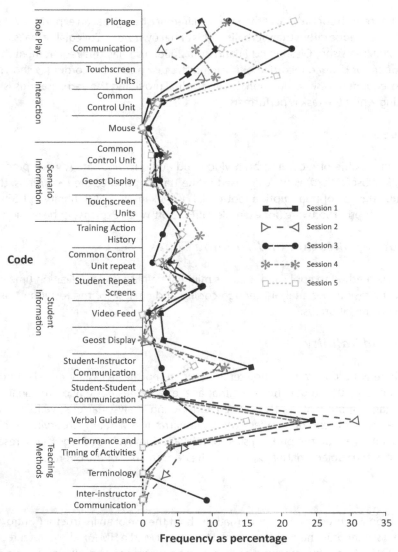

Figure 4.4   Initial output from the CONT method representing coding categories and word frequency

Source: McIlroy and Stanton, 2011.

across the various activities and highlighted system flexibility, with a variety of methods available both for gathering information and for manipulating system variables.

Flowchart

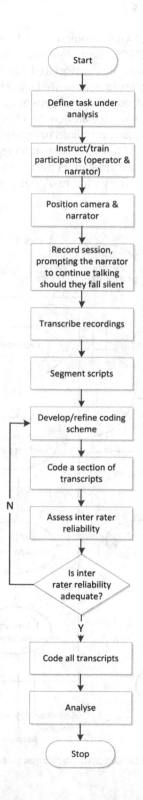

Object-Oriented Cognitive Task Analysis and Design (OOCTAD)

Background and Applications

OOCTAD (Wei and Salvendy, 2006) is a CTA methodology based upon Wickens' (1987b) human information processing theory. Wei and Salvendy (2006) suggested that current CTA methodologies were unable to capture and represent all aspects of cognition associated with task performance. In light of this, Wei and Salvendy (2006) aimed to develop an object oriented model of human information processing that could be used across a wide range of domains and system types. The method focuses on cognition utilised in knowledge-based task performance and enables the categorisation of this cognition into a hierarchy of modules, packages and classes, with links to represent the relationships between them.

The method was originally developed to elicit information regarding cognition in task performance and to provide the ability to utilise this information in the future design of technologies to support task performance (Wei and Salvendy, 2006). The structure of these modules, packages and classes is shown below in Figure 4.5.

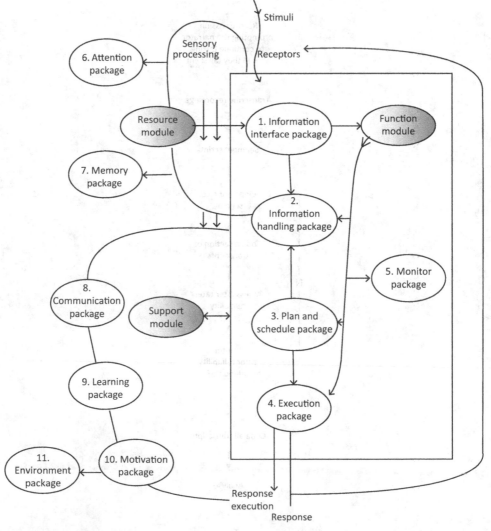

Figure 4.5   OOCTAD modules, classes and packages (adapted from Wei and Salvendy, 2006)

The three modules – support, function and resource – are claimed to be representative of the core functions involved in knowledge-based task performance (Wie and Salvendy, 2006).

The modules and packages are further divided into classes and this breakdown is presented in the box on the right (Wei and Salvendy, 2006):

The OOCTAD classification aims to provide an overview of the core cognitive functions required in task performance (function), the cognitive resources required (resource) and the associated cognitive support requirements (support) (Wei and Salvendy, 2006). In this way, the method enables a breakdown of the cognition involved in task performance. This understanding can be used to inform the design of tools to support these tasks.

Domain of Application

The method is object-oriented and therefore can be applied to analyse the cognitive processes in any knowledge-based task performance.

Procedure and Advice

Step 1: Define the Task under Analysis
The first stage of this methodology involves clearly defining the task under analysis. This definition guides the analysis and ensures that the information collected is relevant and appropriate for the analysis goal.

Function module
1   Information interface
   1.1 Search receive information
   1.2 Identify
2   Information building
   2.1 Handle existing information
   2.2 Handle extra information
3   Plan and schedule
4   Execution
   4.1 Generate ideas
   4.2 Decision-making
   4.3 Problem solving
5   Monitor
   5.1 Sense problem
   5.2 Intervene
Resource Module
6   Attention
   6.1 Visual attention
   6.2 Audio attention
   6.3 Cognitive attention
7   Memory
   7.1 Sensory memory
   7.2 Short-term memory
   7.3 Long-term memory
Support module
8   Communication
   8.1 Personal communication
   8.2 Equipment communication
9   Learning
   9.1 Use learned information
   9.2 Human learning
10 Motivation
11 Environment

Step 2: Recruit Participants
Once the task is clearly defined, the analyst must recruit appropriate users to take part in the analysis as participants. These participants must be a representative sample of the population under analysis.

Step 3: Brief Participants
After participants are recruited, they need to be briefed on the analysis aims and exactly what they will be asked to take part in. Before the analysis proceeds, the analyst should check that all participants are happy with the analysis and what is required of them.

Step 4: Users Perform Task
At this stage, the analyst must carefully observe the participant interacting with the system under analysis.

Step 5: Module Classification
The participants' performance is analysed and cognitive aspects are drawn out. These cognitive aspects are then classified into modules. These modules are linked together based upon the relationships between them.

Step 6: Package Classification
At this stage, the analyst must break down the modules further into packages.

Step 7: Class Classification
Finally, the analyst breaks the cognitive performance down into classes.

Step 8: Graphical Sequencing
Once the analyst has broken the task performance down into a hierarchy of cognitive functions, a graphical representation of the sequence of cognitive activities for the task under analysis should be developed.

Advantages

- OOCTAD provides inputs for the design and redesign of technology and systems.
- It provides a graphical illustration of the cognitive components, and their interrelationships, utilised within knowledge-based task performance.
- All of the modules, packages and classes are generic and so can reused (Wei and Salvendy, 2006).
- It is based upon a well-supported paradigm – human information processing (Wei and Salvendy, 2006).
- It is able to elicit information regarding how the user interacts with a system, which can provide useful information for system evaluation.
- Such a generic toolbox of cognitive modules provides a common language for a range of practitioners involved in system development, from HF professionals to system developers (Wei and Salvedy, 2006).
- Wei and Salvendy (2006) argue that the use of generic categories reduces the time and financial costs associated with system and software design.

Disadvantages

- OOCTAD is object- rather than scenario-based, which could be detrimental in that scenario differences are not fully accounted for in the model.
- It is complicated to implement.

Tools Needed

A pen and paper are required to conduct OOCTAD.

Approximate Application and Training Times

The method is complex to apply and therefore could be time-consuming.

Validity and Reliability

Wei and Salvendy (2006) posit that the method's basis in an accepted theory (Wickens 1987b) provides it with a high level of validity. The method has also been shown to capture a high percentage of the cognitive processes undertaken during task performance illustrating high construct validity (Wei and Salvendy, 2006).

Example

The OOCTAD method was applied to the cognitive task of renting a car online through a website by Wei and Salvendy (2006). The initial stage of analysis involved defining the task that was labelled as *renting a car: determine rental location*. Next, Wei and Salvendy identified a series of appropriate functions, which are presented below in Table 4.10.

Table 4.10   OCCTAD functions identified in the task (adapted from Wei and Salvendy, 2006)

Functions	Actions	Descriptions
Search receive info.	Observe.	The user visualises the rental car's information.
Visual attention.	Unitary resource.	The user identifies the on-screen information relating to the rental car's information.
Identify.	Identify.	The user identifies the on-screen information relating to the rental and return locations.
Handle existing info.	Analyse.	The user analyses possible (nearby) rental and return locations.
Plan schedule.	Set strategy.	The user sets up a strategy to decide the feasible rental and return locations.
Decision-making.	Choose.	The user chooses the most feasible rental and return locations.
Use learnt info.	Use experience.	All the above six operations need to use the learnt related experiences.

The functions, actions and descriptions provide a representation of the way in which the user will interact with the system in order to find the location of the rental centre. Wei and Salvendy (2006) also created a graphical representation of this process in order to summarise the information.

This generic map of the user's interaction with the system was then populated with task-specific data providing a context-specific illustration of the information processing involved.

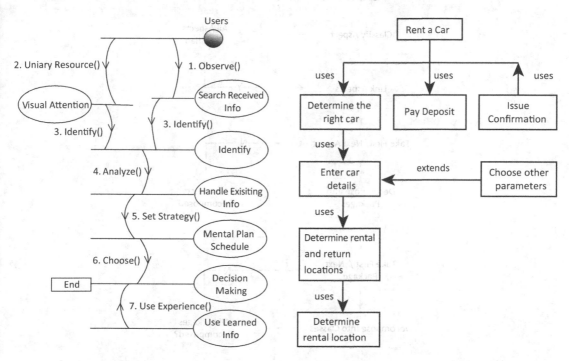

Figure 4.6   OOCTAD model of a user renting a car (adapted from Wei and Salvendy, 2006)

Figure 4.7   User renting a car (adapted from Wei and Salvendy, 2006)

Flowchart

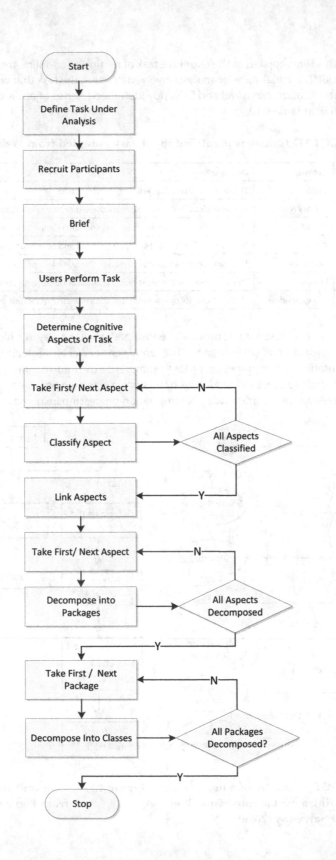

Collegial Verbalisation

Background and Applications

Collegial verbalisation (Erlandsson and Jansson, 2007; Jansson, Olsson and Erlandsson, 2006) is a method of knowledge elicitation grounded in VPA (Ericsson and Simon, 1980). The method was designed to elicit information about participants' everyday work activities in order to understand their mental representations (Erlandsson and Jansson, 2007). Collegial verbalisation employs a pair of experts who are closely matched in terms of experience and role. One expert is asked to perform a task whilst being video recorded. The second expert is then played the recording of task performance and is asked to evaluate the first expert's performance (Jansson, Olsson and Erlandsson, 2006).

Collegial verbalisation represents a hybrid method borne out of think-aloud procedures and retrospective verbalisations (Erlandsson and Jansson, 2007). McIlroy and Stanton (2011) argue that the method is both retrospective and concurrent, since verbalisations occur post-task but whilst the second expert (providing the verbalisations) is experiencing his or her first exposure to the task performance. The hybrid nature of the method enables it to incorporate the benefits of both methods: the minimal interference to task performance afforded by retrospective verbalisations and the reduction in rationalisation afforded by concurrent verbalisation methods (Erlandsson and Jansson, 2007; Jansson, Olsson and Erlandsson, 2006; Stanton and McIlroy, 2011).

The method provides an indepth appreciation of task performance, including resources utilised and decision processes engaged in (Erlandsson and Jansson, 2007), as well as taking into account the impact of contextual, environmental factors on work performance (Jansson, Olsson and Erlandsson, 2007).

Domain of Application

Collegial verbalisation is a generic method that can be used in any context. Previous applications of the method have explored the high-speed ferry domain (Erlandsson and Jansson, 2007) and the rail domain (Jansson, Olsson and Erlandsson, 2006). The method is especially well suited to safety critical domains due to the limited interference with, and disruption to, task performance (Erlandsson and Jansson, 2007).

Procedure and Advice (Erlandsson and Jansson, 2007)

Step 1: Define the Task under Analysis
The first stage of analysis involves clearly defining the scenario and task under analysis. It is recommended that observations of the task are undertaken before the analysis proceeds.

Step 2: Identify and Recruit Participants
Once the task scenario has been clearly defined, the analyst must identify and then recruit expert participants who are closely matched in terms of skill, experience and role. Access to the experts' work domain must also be arranged.

Step 3: Training in Verbalisation
Next, the participants should be briefed regarding what is required of them during analysis. What the second expert should report verbally is clarified here. A small demonstration and, if required, a short practice run should be given at this stage.

Step 4: Understand the System under Analysis
At this stage, the analyst should ensure that he or she has an understanding of the system and task under analysis. Jansson, Olsson and Erlandsson (2006) posit that the analyst should observe the task under analysis before the trial in order to identify appropriate positions for video recording equipment and so forth.

Step 5: Begin Scenario and Record Data
At this stage, the participant should be asked to conduct the task (or tasks) under analysis. The analyst should set up the video recording equipment before the task begins and ensure that this does not interfere with the performance of the task. The participant should then be recorded undertaking the task.

Step 6: Capture Expert Verbalisations
Once the task is complete, the video recordings are shown to a second participant, a work colleague. This second participant is asked to verbalise the cognitive processes undertaken at each stage of the task performance. The analyst should ensure that this stage of the analysis is also recorded, either in a visual or auditory manner.

Step 7: Transcribe Data
All data captured during both the task performance of expert one and the verbalisations of expert two should be transcribed by the analyst.

Step 8: Interview Experts
It is suggested by Erlandsson and Jansson (2007) that a semi-structured interview may be beneficial at this stage to discuss the findings with the experts.

Step 9: Analyse Data
The final stage of the procedure involves collating the transcripts and drawing out key insights concerning the cognitive processes captured.

Advantages

- The method does not interfere with task performance, unlike think-aloud procedures, which removes the additional workload associated with verbalisations and prevents any disruption to task performance (Erlandsson and Jansson, 2007).
- The use of a second expert prevents participants from 'rationalising' their actions (Erlandsson and Jansson, 2007).
- The method allows for the description of processes which may have become automated (Erlandsson and Jansson, 2007).
- It provides data with a higher level of detail and reliability than think-aloud or retrospective methods (Erlandsson and Jansson, 2007).
- It enables the elicitation of implicit knowledge and exploration of mental models of the system (Erlandsson and Jansson, 2007).
- The use of two experts provides a level of validity to the results over those based on a single participant (Erlandsson and Jansson, 2007).
- It removes the need for subjective interpretation by the analyst (Jansson, Olsson and Erlandsson, 2006).
- It provides useful information for system design (Jansson, Olsson and Erlandsson, 2006).

Disadvantages

- The use of video recording equipment could influence task performance (McIlroy and Stanton, 2011); this may be especially true when participants know that colleagues will later evaluate their performance.
- McIlroy and Stanton (2011) argue that the method separates the verbalisations from the task performance through the utilisation of two participants.

- As verbal protocols are not provided by the person conducting the tasks, there is the possibility that the protocols may not exactly match the cognition of the person (Erlandsson and Jansson, 2007).
- Erlandsson and Jansson (2007) argue that further investigation is needed to validate the methodology.
- Due to the naturalistic approach taken by this method, events that are not common may not be captured (Erlandsson and Jansson, 2007).
- It provides a large amount of data to be analysed.

Related Methods

Collegial verbalisation is similar to Elicitation by Critiquing (EBC: Miller, Patterson and Woods, 2006). EBC involves asking participants to think aloud whilst performing a task, and both task performance and the concurrent verbalisations are recorded. Post-task performance, an expert is presented with the recordings and is asked to evaluate the participant's performance. This expert evaluation is also recorded and studied by the analyst. The only difference between EBC and collegial verbalisation is that EBC involves both experts providing verbalisations, rather than only the second expert providing verbalisations.

CONT (McIlroy and Stanton, 2011) is another method similar in approach to collegial verbalisation which utilises work colleagues to provide verbalisations of task performance. Within CONT, the work colleague providing the verbalisations is observing task performance from within the work environment as opposed to the video recordings of task performance utilised in collegial verbalisation.

Video-cued recall procedure (Omodei and McLennan, 1994, cited in Erlandsson and Jansson, 2007) is an additional knowledge elicitation technique which involves a video recording of the participant's task performance, recorded using a head camera. Post-task performance, participants watch the recordings and are asked to provide verbalisations to describe their actions at each stage of performance.

Approximate Training and Application Times

Previous applications of the method have not listed application times; however, it is anticipated that the large amount of data collected by the method would lead to considerable data analysis. The use of two experts will significantly increase the application time of the method.

As with all verbal protocol analyses, time will need to be spent training participants in appropriate verbalisation techniques.

Reliability and Validity

There is no explicit data regarding reliability and validity of the method, although the authors of the method argue that increased validity is afforded by the use of multiple experts over a single participant (Erlandsson and Jansson, 2007). The method also removes reliability constraints associated with post-trial verbalisations, such as not remembering aspects of the trial accurately.

Tools Needed

In order to employ collegial verbalisation, two experts, closely matched on experience and role, are required. In addition, access to the experts' work domain is required, including the ability to employ video and audio recording equipment.

Flowchart

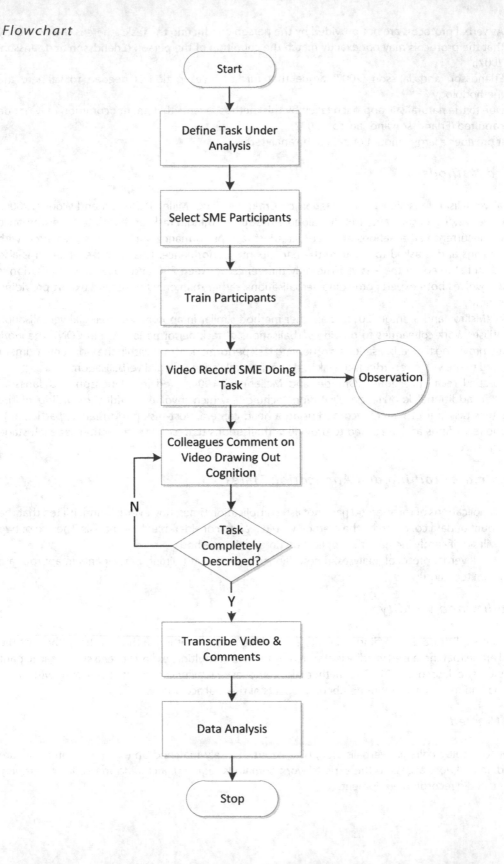

Chapter 5
Process Charting Methods

Process charting techniques are used to represent activity or processes in a graphical format. According to Kirwan and Ainsworth (1992), the first attempt to chart a work process was conducted by Gilbreth and Gilbreth in the 1920s. Process charting techniques have since been used in a number of different domains to provide graphical representations of tasks or sequences of activity. They use standardised symbols to depict task sequences or processes and are employed because they are easier to understand than text descriptions (Kirwan and Ainsworth, 1992). The charting of work processes is also a useful way of highlighting essential task components and requirements. Process chart outputs are extremely useful as they convey a number of different features associated with the activity under analysis, including a breakdown of the component task steps involved, the sequential flow of the tasks, the temporal aspects of the activity, an indication of collaboration between different agents during the tasks, a breakdown of who performs which component task steps and also what technological artefacts are used to perform the activity. Process charting techniques therefore represent both the human and system elements involved in the performance of a certain task or scenario (Kirwan and Ainsworth, 1992). They are particularly useful for representing team-based or distributed tasks, which are often exhibited in command and control systems. A process chart-type analysis allows the specification of which tasks are conducted by which team member or technological component. A number of variations on process charting techniques exist, including techniques used to represent operator decisions (Decision Action Diagrams: DADs) and the causes of hardware and human failures (fault tree analysis, Murphy diagrams). Process charting techniques have been used in a variety of domains in order to understand, evaluate and represent the human and system aspects of a task, including the nuclear, petrochemical, aviation, maritime, railway and air traffic control domains. Sanders and McCormick (1993) suggest that operation sequence diagrams (OSDs) are developed during the design of complex systems in order to advance a detailed understanding of the tasks involved in systems operation and that the process of developing the OSD may be more important than the actual outcome itself. A brief description of the process charting techniques reviewed is given on the following pages.

Process charts are probably the simplest form of charting technique, consisting of a single, vertical flow line which links up the sequence of activities that are performed in order to complete the task under analysis successfully. OSDs are used to graphically describe the interaction between teams of operators and a system. The output of an OSD graphically depicts a task process, including the tasks performed and the interaction between operators over time, using standardised symbols.

Event tree analysis is a task analysis technique that uses tree-like diagrams to represent the various possible outcomes associated with operator task steps in a scenario. Fault trees are used to depict system failures and their causes. A fault tree is a tree-like diagram that defines the failure event and displays the possible causes in terms of hardware failure or human error (Kirwan and Ainsworth, 1992).

DADs are used to depict the process of a scenario through a system in terms of the decisions required and the actions to be performed by the operator in conducting the task or scenario under analysis. Murphy diagrams (Pew, Miller and Feehrer, 1981; cited in Kirwan, 1992a) are also used to graphically describe errors and their causes (both proximal and distal). A summary of the charting techniques reviewed is presented in Table 5.1.

Table 5.1   Summary of charting techniques

Method	Type of method	Domain	Training time	App time	Related methods	Tools needed	Validation studies	Advantages	Disadvantages
Process charts.	Charting technique.	Generic.	Low.	Med.	HTA, observation, interviews.	Pen and paper, Microsoft Visio, video and audio recording equipment.	No.	1) Can be used to graphically depict a task or scenario sequence. 2) Can be used to represent man and machine tasks. 3) Easy to learn and use.	1) For large, complex tasks, the process chart may become too large and unwieldy. Also may be time-consuming to conduct. 2) Some of the process chart symbols are irrelevant to C4i. 3) Only models error-free performance
OSDs.	Charting technique.	Generic.	Low.	Med.	HTA, observation, interviews.	Pen and paper, Microsoft Visio, video and audio recording equipment.	No.	1) Can be used to graphically depict a task or scenario sequence. 2) Can be used to represent man and machine tasks. 3) Seems to be suited for use in analysing C4i or team-based tasks.	1) For large, complex tasks, may become too large and unwieldy. Also may be time-consuming to conduct. 2) Laborious to construct.
Event tree analysis.	Charting technique.	Generic.	Low.	Med.	HTA, observation, interviews.	Microsoft Visio, video and audio recording equipment.	No.	1) Can be used to graphically depict a task or scenario sequence. 2) Can be used to represent man and machine tasks.	1) For large, complex tasks, may become too large and unwieldy. Also may be time-consuming to conduct. 2) Some of the chart symbols are irrelevant to C4i. 3) Only models error-free performance.
Decision Action Diagrams (DADs).	Charting technique.	Generic.	Low.	Med.	HTA, observation, interviews.	Pen and paper, Microsoft Visio, video and audio recording equipment.	No.	1) Can be used to graphically depict a task or scenario sequence. 2) Can be used to represent man and machine tasks. 3) Can be used to analyse decision-making in a task or scenario.	1) For large, complex tasks, may become too large and unwieldy. Also may be time-consuming to conduct.
Fault tree analysis.	Charting technique.	Generic.	Low.	Med.	HTA, observation, interviews.	Pen and paper, Microsoft Visio, video and audio recording equipment.	No.	1) Can be used to graphically depict a task or scenario sequence. 2) Can be used to represent man and machine tasks. 3) Offers an analysis of error events.	1) For large, complex tasks, may become too large and unwieldy. Also may be time-consuming to conduct. 2) Only used retrospectively.
Murphy Diagrams.	Charting technique.	Generic.	Low.	Med.	HTA, observation, interviews.	Pen and paper, Microsoft Visio, video and audio recording equipment.	No.	1) Offers an analysis of task performance and potential errors made. 2) Has a sound theoretical underpinning. 3) Potentially exhaustive.	1) For large, complex tasks, may become too large and unwieldy. Also may be time-consuming to conduct. 2) Only used retrospectively.

Process Charts

Background and Applications

Process charts offer a systematic approach to describing and representing a task or scenario that is easy to follow and understand (Kirwan and Ainsworth, 1992). They are used to graphically represent separate steps or events that occur during the performance of a task. They were originally used to show the path of a product through its manufacturing process, i.e. the construction of a car. Since the original use of process charts, however, there have been many variations in terms of their use. Such variations include operation sequence process charts, which show a chronological sequence of operations and actions that are employed during a particular process, and also various forms of resource charts, which have separate columns for the operator, the equipment used and the material. A set of typical process chart symbols is presented in Figure 5.1 on the right.

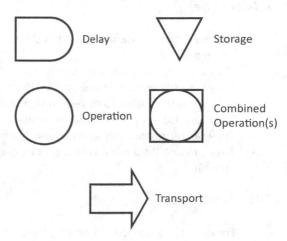

Once completed, a process chart depicts the task in a single, top-down flow line, which represents a sequence of task steps or activities. The time taken for each task step or activity can also be recorded and added to the process chart.

Figure 5.1   Generic process chart symbols
Source: Kirwan and Ainsworth, 1992.

Domain of Application

Generic.

Procedure and Advice

The symbols should be linked together in a vertical chart depicting the key stages of the task or process under analysis.

Step 1: Data Collection
In order to construct a process chart, the analyst must first obtain sufficient data regarding the scenario under analysis. It is recommended that the analyst uses various forms of data collection in this phase, including observations, interviews, questionnaires and walkthrough analyses. The type and amount of data collected in step 1 is dependent upon the analysis requirements.

Step 2: Create Task List
First, the analyst should create a comprehensive list of the task steps involved in the scenario under analysis. These should then be put into chronological order. An HTA for the task or process under analysis may be useful here, as it provides the analyst with a thorough description of the activity under analysis.

Step 3: Task Step Classification
Next, the analyst needs to classify each task step into one of the process chart behaviours: Operation, Transportation, Storage, Inspection, Delay or Combined Operation. To do this, the analyst should take each task step and classify it as one of the process chart symbols employed. This is typically based upon the analyst's subjective judgment, although consultation with appropriate SMEs can also be used.

Step 4: Create the Process Chart
Once all of the task steps are classified into the appropriate symbol categories, the process chart can be constructed. This involves linking each operation, transportation, storage, inspection, delay or combined operation in a vertical chart. Each task step should be placed in the order that it would occur when performing the task. Alongside the task steps symbol, another column should be created, describing the task step fully.

Advantages

- Process charts are useful in that they depict the flow and structure of actions involved in the task under analysis.
- They are simple to learn and construct.
- They have the potential to be applied to any domain.
- They allow the analyst to observe how a task is undertaken.
- They can also display task time information.
- They can represent both operator and system tasks (Kirwan and Ainsworth, 1992).
- They provide the analyst with a simple, graphical representation of the task or scenario under analysis.

Disadvantages

- For large tasks, a process chart may become large and unwieldy.
- When using process charts for complex, large tasks, chart construction will become very time-consuming. Also, complex tasks require complex process charts.
- The process chart symbols are somewhat limited.
- Process charts do not error take into account, modelling only error-free performance.
- Only a very limited amount of information can be represented in a process chart.
- Process charts do not represent the cognitive processes employed during task performance.
- Process charts only offer descriptive information.

Related Methods

The process chart technique belongs to a family of charting or network techniques. Other techniques charting/networking techniques include input-output diagrams, functional flow diagrams, information flow diagrams, Murphy diagrams, critical path analysis, Petri nets and signal flow graphs (Kirwan and Ainsworth, 1992).

Approximate Training and Application Times

The training time for such a technique should be low, representing the amount of time it takes for the analyst to become familiar with the process chart symbols. The application time is dependent upon the size and complexity of the task under analysis. For small, simple tasks, the application time would be very low, while for larger, more complex tasks, it would be high.

Reliability and Validity

No data regarding the reliability and validity of the technique is available in the literature.

Example

The following example is a process chart analysis of the landing task, 'Land aircraft at New Orleans airport using the autoland system' (Marshall et al., 2003). A process chart analysis was conducted in order to assess the feasibility of applying process chart-type analysis in the aviation domain. Initially, an HTA was developed for the landing task, based upon an interview with an aircraft pilot, a video demonstration of the landing task and a walkthrough of the task using Microsoft Flight Simulator 2000. The HTA is presented in list form below. A simplistic process chart was then constructed, using the process chart symbols presented in Figure 5.1.

1.1.1 Check the current speed brake setting

1.1.2 Move the speed brake lever to 'full' position

1.2.1 Check that the auto-pilot is in IAS mode

1.2.2 Check the current airspeed

1.2.3 Dial the speed/Mach knob to enter 210 on the IAS/Mach display

2.1 Check the localiser position on the HSI display

2.2.1 Adjust heading +

2.2.2 Adjust heading -

2.3 Check the glideslope indicator

2.4 Maintain current altitude

2.5 Press 'APP' button to engage the approach system

2.6.1 Check that the 'APP' light is on

2.6.2 Check that the 'HDG' light is on

2.6.3 Check that the 'ALT' light is off

3.1 Check the current distance from runway on the captain's primary flight display

3.2.1 Check the current airspeed

3.2.2 Dial the speed/Mach knob to enter 190 on the IAS/Mach display

3.3.1 Check the current flap setting

3.3.2 Move the flap lever to setting '1'

3.4.1 Check the current airspeed

3.4.2 Dial the speed/Mach knob to enter 150 on the IAS/Mach display

3.5.1 Check the current flap setting

3.5.2 Move the flap lever to setting '2'

3.6.1 Check the current flap setting

3.6.2 Move the flap lever to setting '3'

3.7.1 Check the current airspeed

3.7.2 Dial the speed/Mach knob to enter 140 on the IAS/Mach display

3.8 Put the landing gear down

3.9 Check altitude

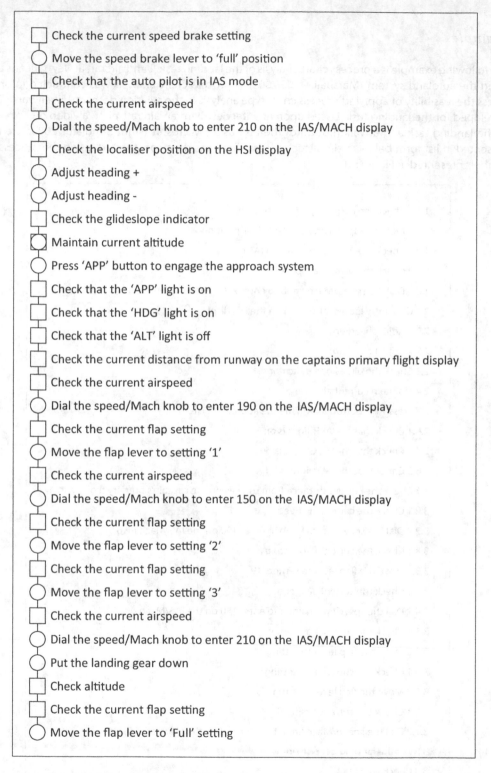

Check the current speed brake setting

Move the speed brake lever to 'full' position

Check that the auto pilot is in IAS mode

Check the current airspeed

Dial the speed/Mach knob to enter 210 on the IAS/MACH display

Check the localiser position on the HSI display

Adjust heading +

Adjust heading -

Check the glideslope indicator

Maintain current altitude

Press 'APP' button to engage the approach system

Check that the 'APP' light is on

Check that the 'HDG' light is on

Check that the 'ALT' light is off

Check the current distance from runway on the captains primary flight display

Check the current airspeed

Dial the speed/Mach knob to enter 190 on the IAS/MACH display

Check the current flap setting

Move the flap lever to setting '1'

Check the current airspeed

Dial the speed/Mach knob to enter 150 on the IAS/MACH display

Check the current flap setting

Move the flap lever to setting '2'

Check the current flap setting

Move the flap lever to setting '3'

Check the current airspeed

Dial the speed/Mach knob to enter 210 on the IAS/MACH display

Put the landing gear down

Check altitude

Check the current flap setting

Move the flap lever to 'Full' setting

Figure 5.2   Extract of process chart for the landing task 'Land at New Orleans Airport using the autoland system' (Marshall et al., 2003)

Flowchart

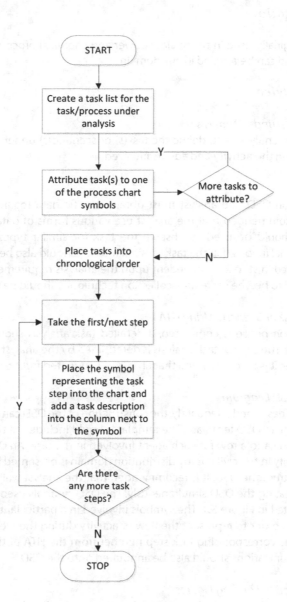

START

Create a task list for the task/process under analysis

Attribute task(s) to one of the process chart symbols

More tasks to attribute?

Y

Place tasks into chronological order

N

Take the first/next step

Place the symbol representing the task step into the chart and add a task description into the column next to the symbol

Y

Are there any more task steps?

N

STOP

Operation Sequence Diagrams (OSDs)

Background and Applications

OSDs are used to graphically describe the activity and interaction between teams of agents within a system. According to Kirwan and Ainsworth (1992), the original purpose of OSD analysis was to represent complex multi-person tasks. The output of an OSD graphically depicts the task process, including the tasks performed and the interaction between operators over time, using standardised symbols. There are various forms of OSD, ranging from a simple flow diagram representing task order to more complex diagrams which account for team interaction and communication. OSDs have recently been used by the authors for the analysis of command and control in a number of domains, including the fire service, naval warfare, aviation, energy distribution, air traffic control and rail domains.

Domain of Application

The technique was originally used in the nuclear power and chemical process industries. However, the technique is generic and can be applied in any domain.

Procedure and Advice

Step 1: Define the Task(s) under Analysis
The first step in an OSD analysis is to define the task(s) or scenario(s) under analysis. These should be defined clearly, including the activity and agents involved.

Step 2: Data Collection
In order to construct an OSD, the analyst must obtain specific data regarding the task or scenario under analysis. It is recommended that the analyst use various forms of data collection in this phase. Observational study should be used to observe the task (or similar types of task) under analysis. Interviews with personnel involved in the task (or similar tasks) should also be conducted. The type and amount of data collected in step 2 is dependent upon the analysis requirements. The more exhaustive the analysis is intended to be, the more data collection techniques should be employed.

Step 3: Describe the Task or Scenario Using HTA
Once the data collection phase is completed, a detailed task analysis should be conducted for the scenario under analysis. The type of task analysis is determined by the analyst and, in some cases, a task list will suffice. However, it is recommended that an HTA is conducted for the task under analysis.

Step 4: Construct the OSD Diagram
Once the task has been described adequately, the construction of the OSD can begin. The process begins with the construction of an OSD template. The template should include the title of the task or scenario under analysis, a timeline and a row for each agent involved in the task. An OSD template used during the analysis of C4i activity in the civil energy distribution domain is presented in Figure 5.3 (Walker et al., 2006). In order to construct the OSD, it is recommended that the analyst walks through the HTA of the task under analysis, creating the OSD simultaneously. The OSD symbols used to analyse C4i activity by the authors are presented in Figure 5.1. The symbols involved in a particular task step should be linked by directional arrows in order to represent the flow of activity during the scenario. Each symbol in the OSD should contain the corresponding task step number from the HTA of the scenario. The artefacts used during the communications should also be annotated onto the OSD.

Step 5: Calculate Operation Loading Figures
Operation loading figures can be calculated for each agent involved in the scenario from the OSD. The figures are calculated for each OSD operator or symbol used, for example, operation, delay and decision. These operation loading figures refer to the extent that each agent was involved in the operation in question during the scenario.

Step 6: Overlay Additional Analyses Results
One of the endearing features of the OSD technique is that additional analysis results can easily be added to the OSD. According to the analysis requirements, additional task features can also be annotated onto the OSD. For example, in the analysis of C4i activity in a variety of domains, the authors annotated coordination values (from a coordination demands analysis: CDA) between team members for each task step onto the OSD.

NGC 16th June 2004		Scenario: Switching Operations
	1. Conduct Switching Operations	
NOC Control Room Operator		
SAP/AP at Barking Substation		
WOK Control Room Operator		
REC Control Room Operator		

Figure 5.3   Example OSD template

Advantages

- OSDs provide an exhaustive analysis of the task in question. The flow of the task is represented in terms of activity and information, the type of activity and the agents involved are specified, while a timeline of the activity, the communications between agents involved in the task, the technology used and also a rating of total coordination for each teamwork activity is also provided. The flexibility of the technique also permits the analyst to add further analysis outputs onto the OSD, adding to its exhaustiveness.
- They are particularly useful for analysing and representing distributed teamwork or collaborated activity.
- They are useful for demonstrating the relationship between tasks, technology and team members.
- They demonstrate high face validity (Kirwan and Ainsworth, 1992).
- They been used extensively in the past and have been applied in a variety of domains.
- A number of different analyses can be overlaid onto an OSD of a particular task. For example, Baber et al. (2004) add the corresponding HTA task step numbers and CDA results to OSDs of C4i activity.
- The OSD technique is very flexible and can be modified to suit the analysis needs.
- The WESTT software package can be used to automate a large proportion of the OSD procedure.
- Despite its exhaustive nature, the OSD technique requires only minimal training.

Disadvantages

- The application time for an OSD analysis is lengthy. Constructing an OSD for large, complex tasks can be extremely time-consuming and the initial data collection stage adds further time to the analysis.
- The construction of large, complex OSDs is also quite a laborious and taxing process.
- OSDs can become cluttered and confusing (Kirwan and Ainsworth, 1992).
- Their output can become large and unwieldy.
- The present OSD symbols are limited for certain applications (e.g. C4i scenarios).
- The reliability of the technique is questionable. Different analysts may interpret the OSD symbols differently.

Related Methods

Various types of OSD exist, including temporal operational sequence diagrams, partitioned operational sequence diagrams and spatial operational sequence diagrams (Kirwan and Ainsworth, 1992). During the OSD data collection phase, traditional data collection procedures such as observational study and interviews are typically employed. Task analysis techniques such as HTA are also used to provide the input for the OSD. Timeline analysis may also be employed in order to construct an appropriate timeline for the task or scenario under analysis. Additional analyses results can also be annotated onto an OSD, such as CDA and comms usage diagrams. The OSD technique has also recently been integrated with a number of other techniques (HTA, observation, CDA, comms usage diagrams, social network analysis and propositional networks) to form the event analysis of systemic teamwork (EAST) methodology (Baber et al., 2004), which has been used by the authors to analyse C4i activity in a number of domains.

Approximate Training and Application Times

No data regarding the training and application time associated with the OSD technique is available in the literature. However, it is apparent that the training time for such a technique would be minimal. The application time for the technique is very high, including the initial data collection phase of interviews and observational analysis, and also the construction of an appropriate HTA for the task under analysis. The construction of the OSD in particular is a very time-consuming process. A typical OSD normally can take up to one week to construct.

Reliability and Validity

According to Kirwan and Ainsworth, OSD techniques possess a high degree of face validity. The intra-analyst reliability of the technique may be suspect, as different analysts may interpret the OSD symbols differently.

Tools Needed

When conducting an OSD analysis, pen and paper may be sufficient. However, to ensure that data collection is comprehensive, it is recommended that video or audio recording devices are used in conjunction with this. For the construction of the OSD, it is recommended that a suitable drawing package such as Microsoft Visio is used. The WESTT software package (Houghton et al., 2008) can also be used to automate a large portion of the OSD procedure. WESTT constructs the OSD based upon an input of observational data for the scenario under analysis.

Example

The OSD technique has recently been used by the authors in the analysis of C4i activity in the fire service (Baber et al., 2004), naval warfare, aviation, energy distribution, air traffic control and rail domains. The following example is an extract of an OSD from an energy distribution scenario (Salmon et al., 2005). The task involved the switching out of three circuits at three substations. Observational data from the substation and the network operations centre (NOC) control room was used to conduct an HTA of the switching scenario. This HTA then acted as the primary input into the OSD diagram. Total coordination values for each teamwork task step (from a CDA – see Chapter 9) were also annotated onto the OSD. The glossary for the OSD is

Agent	Operation	Receive	Transport	Decision	Delay	Total
NOC	98	40				138
SAP	223	21	19		1	264
WOK	40	10				50
REC	15	14				29

Table 5.2   Operational loading results

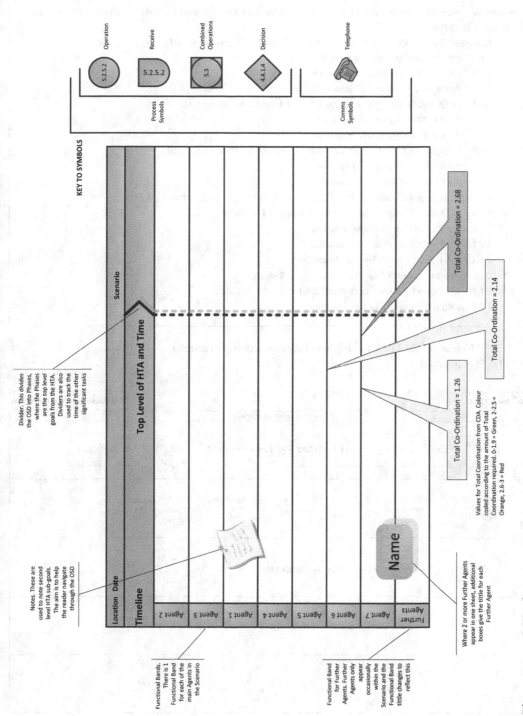

Figure 5.4   OSD glossary

Source: Stanton, Baber and Harris, 2008.

0. Coordinate and carry out switching operations on circuits SGT5. SGT1A and 1B at Bark s/s (Plan 0. Do 1 then 2 then 3, EXIT)
1. Prepare for switching operations (Plan 1. Do 1.1, then 1.2, then 1.3, then 1.4, then 1.5, then 1.6, then 1.7, then 1.8, then 1.9, then 1.10 EXIT)
    1.1.   Agree SSC (Plan 1.1. Do 1.1.1, then 1.1.2, then 1.1.3, then 1.1.4, then 1.1.5, EXIT)
        1.1.1.   (WOK) Use phone to Contact NOC
        1.1.2.   (WOK + NOC) Exchange identities
        1.1.3.   (WOK + NOC) Agree SSC documentation
        1.1.4.   (WOK+NOC) Agree SSC and time (Plan 1.1.4. Do 1.1.4.1, then 1.1.4.2, EXIT)
            1.1.4.1. (NOC) Agree SSC with WOK
            1.1.4.2. (NOC) Agree time with WOK
        1.1.5.   (NOC) Record and enter details (Plan 1.1.5. Do 1.1.5.1, then 1.1.5.2, EXIT)
            1.1.5.1. Record details on log sheet
            1.1.5.2. Enter details into worksafe
    1.2.   (NOC) Request remote isolation (Plan 1.2. Do 1.2.1, then 1.2.2, then 1.2.3, then 1.2.4, EXIT)
        1.1.4.   1.2.1. (NOC) Ask WOK for isolators to be opened remotely
        1.1.4.   1.2.2. (WOK) Perform remote isolation
        1.1.4.   1.2.3. (NOC) Check Barking s/s screen
        1.1.4.   1.2.4. (WOK + NOC) End communications
    1.3.   Gather information on outage at transformer 5 at Bark s/s
    (Plan 1.3. Do 1.3.1, then 1.3.2, then 1.3.3, then 1.3.4, EXIT)
        1.3.1. (NOC) Use phone to contact SAP at Bark

Figure 5.5   Extract of an HTA for high-voltage switching scenario
Source: Salmon et al., 2004.

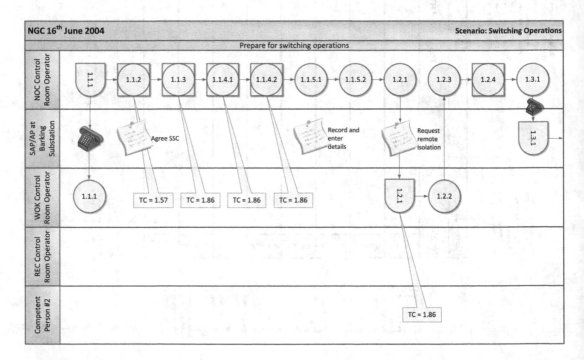

Figure 5.6   Extract of the OSD for the National Grid Transco switching scenario
Source: Stanton, Baber and Harris, 2008.

presented in Figure 5.4. An extract of the HTA for the corresponding energy distribution task is presented in Figure 5.5. The corresponding extract of the OSD is presented in Figure 5.6. Finally, the operational loading figures are presented in Table 5.2.

The operational loading analysis indicates that the senior authorised person (SAP) at the substation has the highest loading in terms of operations, transport and delay, while the network operations centre (NOC) operator has the highest loading in terms of receipt of information. This provides an indication of the nature of the roles involved in the scenario. The NOC operator's role is one of information distribution (giving and receiving), which is indicated by the high receive operator loading, whilst the majority of the work is conducted by the SAP, which is indicated by the high operation and transport loading figures.

Flowchart

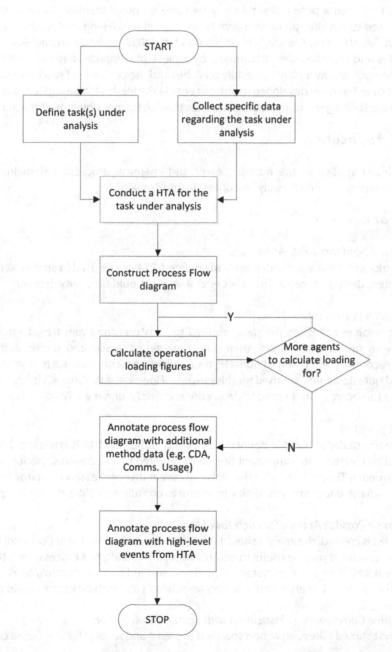

Event Tree Analysis

Background and Applications

As mentioned earlier, event tree analysis (ETA) is a task analysis technique that uses tree-like diagrams to represent possible outcomes associated with operator tasks steps in a scenario. Originally used in system reliability analysis (Kirwan and Ainsworth, 1992), it can also be applied to human operations to investigate possible actions and their consequences. A typical event tree output comprises a tree-like diagram consisting of nodes (representing task steps) and exit lines (representing the possible outcomes). Typically, success and failure outcomes are used, but for more complex analyses, multiple outcomes can be represented (Kirwan and Ainsworth, 1992). ETA can be used to depict task sequences and their possible outcomes, to identify error potential within a system and to model team-based tasks. Event trees have recently been used to identify appropriate underwater tunnel sites (Hong et al., 2009) and nuclear power plants (Mercurio et al., 2009). The study by Mercurio et al. (2009) utilised dynamic event trees in which the probabilities and branches were deciphered by a piece of computer software which simulated the dynamics of the nuclear power plant to identify branches and their chances of success or failure. The use of dynamic event trees has been developed in order to enable the method to be applied to complex systems in which there are such a great number of possibilities that it becomes unfeasible for a human to analyse.

Domain of Application

ETA was originally applied in the nuclear power and chemical processing domains. However, the technique is generic and could feasibly be applied in any domain.

Procedure and Advice

Step 1: Define the Scenario under Analysis
First, the scenario under analysis should be clearly defined. ETA can be used to analyse activity in existing systems or system design concepts. The task under analysis should be clearly defined.

Step 2: Data Collection Phase
The next step involves collecting the data required to construct the event tree diagram. If the ETA is focused upon an operational system, then data regarding the scenario under analysis should be collected. It is recommended that traditional HF data collection techniques, such as observational study, interviews and questionnaires, are used for this purpose. However, if the analysis is based upon a design concept, then storyboards can be used to depict the scenario(s) under analysis.

Step 3: Draw Up Task List
Once the scenario under analysis is defined clearly and sufficient data is collected, a comprehensive task list should be created. The component task steps required for effective task performance should be specified in sequence. This initial task list should be representative of the standard error-free performance of the task or scenario under analysis. It may be useful to consult with SMEs during this process.

Step 4: Determine Possible Actions for Each Task Step
Once the task list is created, the analyst should then describe every possible action associated with each task step in the task list. It may be useful to consult with SMEs during this process. Each task step should be broken down into the human or system operations required and any controls or interface elements used should also be noted. Every possible action associated with each task step should be recorded.

Step 5: Determine Consequences Associated with Each Possible Action
Next, the analyst should take each action specified in step 4 and record the associated consequences.

Step 6: Construct Event Tree
Once steps 4 and 5 are complete, the analyst can begin to construct the event tree diagram. The event tree should depict all possible actions and their associated consequences.

Advantages

- ETA can be used to highlight a sequence of tasks steps and their associated consequences.
- It can be used to highlight error potential and error paths throughout a system.
- It can be used in the early design life-cycle to highlight task steps that may become problematic (multiple associated response options) and also those task steps that have highly critical consequences.
- If used correctly, it can potentially depict anything that could possibly go wrong in a system.
- It is a relatively easy-to-use technique that requires little training.
- It has been used extensively in probabilistic safety assessment (PSA)/human reliability analysis (HRA).
- Hong et al. (2009) argue that ETA can be used in almost all stages of system development: to develop system criteria, to identify possible failures, to test operations and so on.

Disadvantages

- For large, complex tasks, the event tree diagram can become very large and complex.
- ETA can be time-consuming in its application.
- Task steps are often not explained in the output.
- Heavy reliance on the judgments of experts may result in questionable credibility (Ferdous et al., 2009).

Example

An extract of an ETA is presented in Figure 5.7. An event tree was constructed for the landing task 'Land A320 at New Orleans Airport using the autoland system' in order to investigate the use of ETA for predicting design-induced pilot error.

Check current airspeed	Dial in 190Kn using SM knob	Check flap setting	Set flaps to level 3	Lower landing gear
Success	Success	Success	Success	Success
				Fail to lower landing gear
			Set flaps at the wrong time	
			Fail to set flaps	
		Fail to check flap setting		
	Fail to dial in airspeed			
	Dial in airspeed at the wrong time (too early, too late)			
	Dial in wrong airspeed (too much, too little)			
	Dial in airspeed using the heading knob			
Fail to check airspeed				

Figure 5.7   **Extract of an event tree diagram for the flight task 'Land at New Orleans Airport using the autoland system' (Marshall et al., 2003)**

Related Methods

According to Kirwan and Ainsworth (1992), there are a number of variations of the original ETA technique, including operator action event tree analysis (OATS) and human reliability analysis event tree analysis (HRAET). Event trees are also similar to fault tree analysis and OSDs.

Flowchart

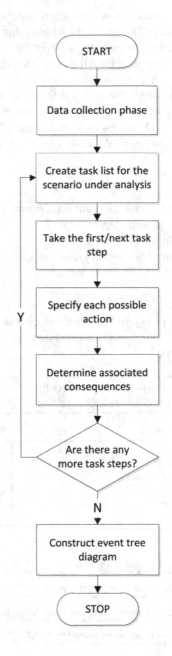

Reliability and Validity

No data regarding the reliability and validity of the event tree technique is available.

Tools Needed

An event tree diagram can be created using pen and paper. If the event tree is based on an existing system, then observational study may be used for data collection purposes, which requires video and audio recording equipment and a computer.

Decision Action Diagrams (DADs)

Background and Applications

DADs, also known as information flow diagrams (Kirwan and Ainsworth, 1992), are used to graphically depict a process in terms of the decisions required and actions to be performed by the operator involved in the activity. Decisions are represented by diamonds and each decision option available to the system operator is represented by exit lines. In their simplest form, the decision options are usually 'Yes' or 'No'; however, depending upon the complexity of the task and system, multiple options can also be represented. The DAD output diagram should display all of the possible outcomes at each task step in a process. DAD analysis can be used to evaluate existing systems or to inform the design of systems and procedures.

Domain of Application

DAD was originally applied in the nuclear power and chemical processing domains. However, the technique is generic and could feasibly be applied in any domain.

Procedure and Advice

Step 1: Define the Task or Scenario under Analysis
First, the scenario(s) under analysis should be clearly defined. DAD analysis can be used to analyse activity in existing systems or system design concepts.

Step 2: Data Collection
In order to construct a DAD, the analyst(s) must obtain sufficient data regarding the task or scenario under analysis. It is recommended that traditional HF data collection techniques, such as observational study, interviews and questionnaires, are used for this purpose. However, if the analysis is based upon a design concept, then storyboards can be used to depict the scenario(s) under analysis.

Step 3: Conduct a Task Analysis
Once the data collection phase is completed, a detailed task analysis should be conducted for the scenario under analysis. The type of task analysis is determined by the analyst and, in some cases, a task list will suffice. However, it is recommended that when constructing a DAD, an HTA for the scenario under analysis is conducted.

Step 4: Construct DAD
Once the task or scenario under analysis is fully understood, the DAD can be constructed. This process should begin with the first decision available to the operator of the system. Each possible outcome or action associated with the decision should be represented with an exit line from the decision diamond. Each resultant action and outcome for each of the possible decision exit lines should then be specified.

This process should be repeated for each task step until all of the possible decision outcomes for each task have been exhausted.

Advantages

- DADs can be used to depict the possible options that an operator faces during each task step in a scenario. This information can be used to inform the design of the system or procedures, i.e. task steps that have multiple options associated with them can be re-designed.
- They are relatively easy to construct and require little training.
- They could potentially be used for error prediction purposes.

Disadvantages

- In their current form, DADs do not cater for the cognitive component of task decisions.
- It would be very difficult to model parallel activity using DADs.
- They do not cater for processes involving teams. Constructing a team DAD would appear to be extremely difficult.
- It appears that an HTA for the task or scenario under analysis would be sufficient. A DAD output is very similar to the plans depicted in an HTA.
- For large, complex tasks, the DAD would be difficult and time-consuming to construct.
- The initial data collection phase involved in the DAD procedure adds a considerable amount of time to the analysis.
- Reliability and validity data for the technique is sparse.

Related Methods

DADs are also known as information flow charts (Kirwan and Ainsworth, 1992). The DAD technique is related to other process chart techniques such as OSDs and also task analysis techniques such as HTA. When conducting a DAD-type analysis, a number of data collection techniques are used, such as observational study and interviews. A task analysis (e.g. HTA) of the task/scenario under analysis may also be required.

Approximate Training and Application Times

No data regarding the training and application times associated with DADs is available in the literature. It is estimated that the training time for DADs would be minimal or low. The application time associated with the DAD technique is dependent upon the task and system under analysis. For complex scenarios with multiple options available to the operator involved, the application time would be high, while for more simple linear tasks, it would be very low. The data collection phase of the DAD procedure adds considerable time, particularly when observational analysis is used.

Reliability and Validity

No data regarding the reliability and validity of the DAD technique is available.

Tools Needed

Once the initial data collection is complete, the DAD technique can be conducted using pen and paper, although it may be more suitable to use a drawing package such as Microsoft Visio. The tools required for the data collection phase are dependent upon the techniques used. Typically, observational study is used, which would require video and audio recording equipment and a computer.

Example

The following example (Figure 5.8) is a DAD taken from Kirwan and Ainsworth (1992).

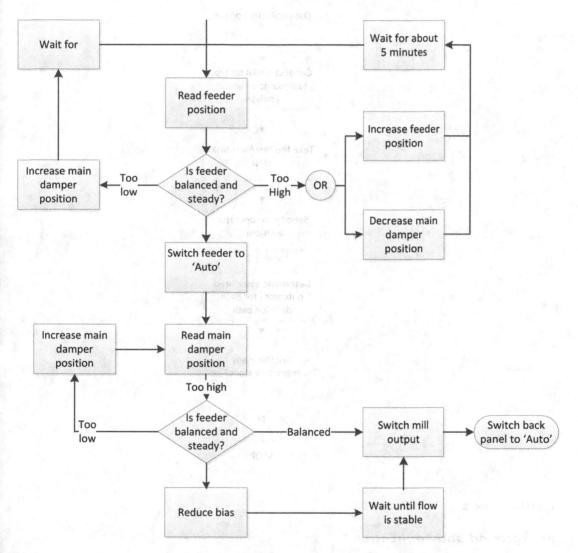

Figure 5.8   Decision Action Diagram (adapted from Kirwan and Ainsworth, 1992)

Flowchart

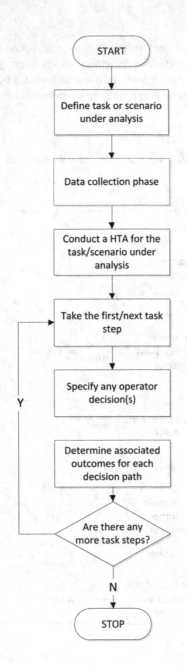

Fault Trees

Background and Application

Fault trees are used to graphically represent system failures and their causes. A fault tree is a tree-like diagram, which defines the failure event and displays the possible causes in terms of hardware failure or human error (Kirwan and Ainsworth, 1992). Fault tree analysis was originally developed for the analysis of complex systems in the aerospace and defence industries and they are now used extensively in PSA. Although typically used to evaluate events retrospectively, fault trees can be used at any stage in the system life-cycle to predict failure events and their causes. Typically, the failure event or top event (Kirwan and Ainsworth, 1992) is placed at the top of the fault tree and the contributing events are placed below. The fault tree is held together by AND and OR gates, which link contributory events together. An AND gate is used when more than one event causes a failure, i.e. when multiple contributory factors are involved. The events placed directly underneath an AND gate must occur together for the failure event above to occur. An OR gate is used when the failure event could be caused by more than one contributory event in isolation, but not together. The event above the OR gate may occur if any one of the events below the OR gate occurs. Fault tree analysis can be used for the retrospective analysis of incidents or for the prediction of failure in a particular scenario.

Recent applications of fault tree analysis span a number of domains, Doytchev and Szwillus (2009) combined fault tree analysis, task analysis and the Human Error Identification in Systems Tool (HEIST) in order to explore the human causality involved in a hydroelectric power plant accident, while Lindhe et al. (2009) employed the method to conduct a risk analysis of the Gothenberg drinking water systems.

Durga Rao et al. (2009) explore an adaptation of the method called dynamic fault tree analysis. Dynamic fault tree analysis consists of traditional fault tree analysis with the addition of the Monte Carlo simulation technique in order to simulate realistic scenarios that could occur in the system. Durga Rao et al. (2009) conducted three case studies to investigate the ability of dynamic fault tree analysis to explore complex systems, with all three case studies providing comparable results with previous analyses. The authors describe the utility of Dynamic Reliability and SIMulation (DRSIM), a software application designed to aid in dynamic fault tree analysis.

Domain of Application

Fault tree analysis was originally applied in the nuclear power and chemical processing domains. However, the technique is generic and could potentially be applied in any domain.

Procedure and Advice

Step 1: Define Failure Event
The failure or event under analysis should be defined first. This may be either an actual event that has occurred (retrospective incident analysis) or an imaginary event (predictive analysis). This event then becomes the top event in the fault tree.

Step 2: Determine Causes of Failure Event
Once the failure event has been defined, the contributory causes associated with the event should be defined. The nature of the causes analysed is dependent upon the focus of the analysis. Typically, human error and hardware failures are considered (Kirwan and Ainsworth, 1992).

Step 3: AND/OR Classification
Once the cause(s) of the failure event are defined, the analysis proceeds with the AND or OR causal classification phase. Each contributory cause identified during step 2 of the analysis should be classified as either an AND or an OR event. If two or more contributory events contribute to the failure event, then they are classified as AND events. If two or more contributory events are responsible for the failure even when they occur separately, then they are classified as OR events.

Steps 2 and 3 should be repeated until each of the initial causal events and associated causes are investigated and described fully.

Step 4: Construct Fault Tree Diagram

Once all events and their causes have been defined fully, they should be put into the fault tree diagram. The fault tree should begin with the main failure or top event at the top of the diagram, with its associated causes linked underneath as AND/OR events. The causes of these events should then be linked underneath as AND/OR events. The diagram should continue until all events and causes are exhausted fully.

Example

The following example (Figure 5.9) is taken from Kirwan (1994) from a brake failure scenario model.

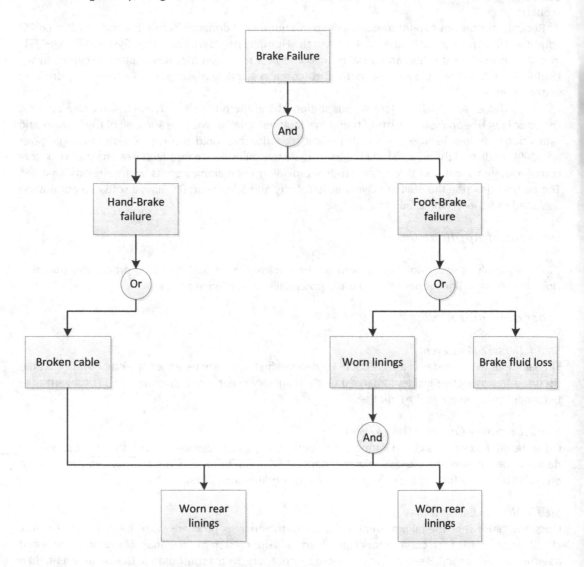

Figure 5.9   Fault tree for a brake failure scenario

Flowchart

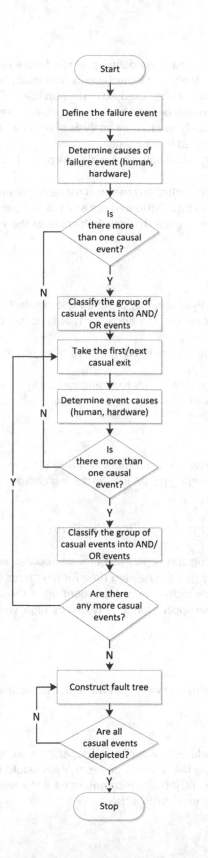

Advantages

- Fault trees are useful in that they define possible failure events and associated causes. This is especially useful when looking at failure events with multiple causes.
- Fault tree-type analysis has been used extensively in PSA.
- Fault trees could potentially be used both predictively and retrospectively.
- Although most commonly used in the analysis of nuclear power plant events, the technique is generic and can be applied in any domain.
- Fault trees can be used to highlight potential weak points in a system design concept (Kirwan and Ainsworth, 1992).
- The technique could be particularly useful in modelling team-based errors, where a failure event is caused by multiple events distributed across a team of personnel.
- It can provide insights into the dynamic behaviour of the system under analysis (Lindhe et al., 2009).

Disadvantages

- When used in the analysis of large, complex systems, fault trees can be complex, difficult and time-consuming to construct. It is apparent that fault tree diagrams can quickly become large and complicated.
- To utilise the technique quantitatively, a high level of training may be required (Kirwan and Ainsworth, 1992).
- The use of fault trees as a predictive tool remains largely unexplored.
- There is little evidence of their use outside of the nuclear power domain.

Related Methods

The fault tree technique is often used in conjunction with ETA (Kirwan and Ainsworth, 1992). Fault trees are similar to many other charting techniques, including cause-consequence charts, DADs and event trees.

Approximate Training and Application Times

No data regarding the training and application times associated with fault tree analysis is available in the literature. It is estimated that the training time for fault trees would be low. The application time associated with the fault tree technique is dependent upon the task and system under analysis. For complex failure scenarios, the application time would be high, while for more simple failure events, it would be very low.

Reliability and Validity

No data regarding the reliability and validity of the fault tree technique is available.

Tools Needed

Fault tree analysis can be conducted using pen and paper. If the analysis is based upon an existing system, an observational study of the failure event under analysis would be useful. This would require video and audio recording equipment. It is also recommended that when constructing fault tree diagrams, a drawing package such as Microsoft Visio is used.

Murphy Diagrams

Background and Applications

Murphy diagrams (Pew, Miller and Feehrer, 1981; cited in Kirwan, 1992a) were originally used for the retrospective examination of errors in process control rooms. They are based on the notion that 'if anything can go wrong, it will go wrong' (Kirwan and Ainsworth, 1992). The technique is very similar to fault tree analysis in that errors or failures are analysed in terms of their potential causes. Murphy diagrams use the following eight behaviour categories:

- activation/detection;
- observation and data collection;
- identification of system state;
- interpretation of the situation;
- task definition/selection of goal state;
- evaluation of alternative strategies;
- procedure selection; and
- procedure execution.

The Murphy diagram begins with the top event being split into success and failure nodes. The analyst begins by describing the failure event under analysis. Next, the 'failure' outcome is specified and the sources of the error that have an immediate effect are defined. These are called the proximal sources of error. The analyst then takes each proximal error source and breaks it down further so that the causes of the proximal error sources are defined. These proximal error causes are termed the distal causes. For example, if the failure was 'Procedure incorrectly executed', the proximal sources could be 'wrong switches chosen', 'switches incorrectly operated' or 'switches not operated'. The distal sources for 'wrong switches chosen' could then be further broken down into 'deficiencies in placement of switches', 'inherent confusability in switch design' or 'training deficiency' (Kirwan and Ainsworth, 1992). The Murphy diagram technique is typically used for the retrospective analysis of failure events.

Domain of Application

Nuclear power and chemical process industries.

Procedure and Advice

Step 1: Define the Task or Scenario under Analysis
The first step in a Murphy diagram analysis is to define the task or scenario under analysis. Although typically used in the retrospective analysis of incidents, it is feasible that the technique could be used proactively to predict potential failure events and their causes.

Step 2: Data Collection
If the analysis is retrospective, then data regarding the incident under analysis should be collected. This may involve the interviews with the actors involved in the scenario or a walkthrough of the event. If the analysis is proactive and concerns an event that has not yet happened, then walkthroughs of the events should be used.

Step 3: Define Error Events
Once sufficient data regarding the event under analysis is collected, the analysis begins with the definition of the first error. The analyst should define the error as clearly as possible.

Step 4: Classify Error Activity into Decision-Making Category
Once the error event under analysis is described, the activity leading up to the error should be classified into one of the eight decision-making process categories.

Step 5: Determine Error Consequence and Causes
Once the error is described and classified, the analyst should determine the consequences of the error event and also determine possible consequences associated with the error. The error causes should be explored fully, with proximal and distal sources described.

Step 6: Construct Murphy Diagram
Once the consequences, proximal and distal sources have been explored fully, the Murphy diagram for the error in question should be constructed.

Step 7: Propose Design Remedies
For the purpose of error prediction in the design of systems, it is recommended that the Murphy diagram be extended to include an error or design remedy column. The analyst should use this column to propose design remedies for the identified errors based upon the causes identified.

Advantages

- The technique is easy to use and learn, requiring little training.
- Murphy diagrams present a useful way for the analyst to identify a number of different possible causes for a specific error or event.
- It has high documentability.
- Each task step failure is exhaustively described, including proximal and distal sources.
- The technique has the potential to be applied to team-based tasks, depicting teamwork and failures with multiple team-based causes.
- Murphy diagrams use very little resources (low cost, time spent, etc.).
- Although developed for the retrospective analysis of error, it is feasible that the technique could be used proactively.

Disadvantages

- The use of the technique as a predictive tool remains largely unexplored.
- It could become large and unwieldy for large, complex tasks.
- There is little guidance on the technique for the potential analyst.
- The consistency of the method is questionable.
- Design remedies are based entirely upon the analyst's subjective judgment.
- It is a dated technique that is not commonly used.

Example

A Murphy diagram analysis was conducted for the flight task 'Land aircraft X at New Orleans Airport using the autoland system'. An extract of the analysis is presented in Figure 5.10.

Related Methods

Murphy diagrams are very similar to fault tree diagrams and ETA in that they depict failure events and their causes.

| ACTIVITY | OUTCOME | PROXIMAL SOURCES | DISTAL SOURCES |

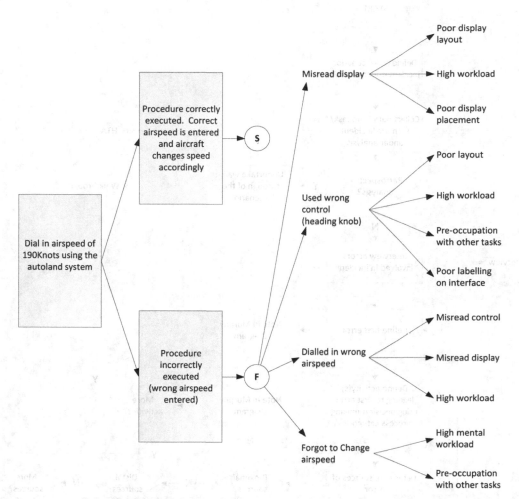

Figure 5.10   Murphy diagram for the flight task 'Land aircraft X at New Orleans Airport using the autoland system'

Approximate Training and Application Times

The training time for the technique would be minimal. The application time would depend upon the task or scenario under analysis. For error incidences with multiple causes and consequences, the application time would be high.

Reliability and Validity

No data regarding the reliability and validity of Murphy diagrams is available in the literature.

Tools Needed

The technique can be conducted using pen and paper. It is recommended that a drawing package such as Microsoft Visio be used to construct the Murphy diagram outputs.

Flowchart

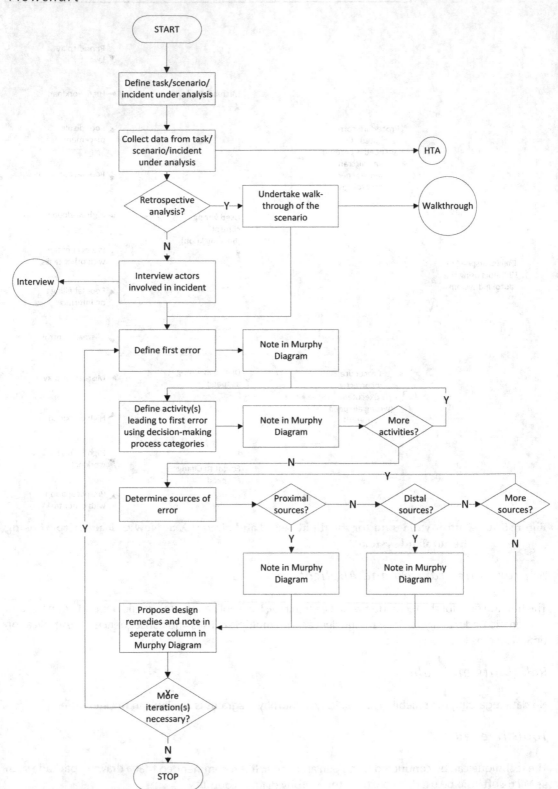

Chapter 6

Human Error Identification and Accident Analysis Methods

Human error is a complex construct that has received considerable attention from the HF community. It has been consistently identified as a contributory factor in a high proportion of incidents in complex, dynamic systems. For example, within the civil aviation domain, recent research indicates that human or pilot error is the major cause of all commercial aviation incidents (Harris and Li, 2011; McFadden and Towell, 1999). Within the rail transport field, human error was identified as a contributory cause of almost half of all collisions occurring on the UK's rail network between 2002 and 2003 (Lawton and Ward, 2005). In the health-care domain, the US Institute of Medicine estimates that between 44,000 and 88,000 people die as a result of medical errors (Helmreich, 2000) and it has also been estimated that human or driver error contributes to as much as 75 per cent of road accidents (Medina et al., 2004).

Although human error has been investigated since the dawn of the discipline, research into the construct only increased around the late 1970s and early 1980s in response to a number of high-profile catastrophes in which human error was implicated. Major incidents such as the Three Mile Island, Chernobyl and Bhopal disasters, and the Tenerife and Papa India air disasters (to name but a few) were all attributed, in part, to human error. As a result, the construct began to receive considerable attention from the HF community and also the general public, and has been investigated in a number of different domains ever since, including the military and civil aviation (Shappell and Wiegmann, 2000, Marshall et al., 2003; Griffin, Young and Stanton, 2010; Li, Harris and Yu, 2008), road transport (Reason et al., 1990), shipping (Celik and Cebi, 2009), nuclear power and petrochemical reprocessing (Kirwan, 1992a, 1992b, 1996b, 1998a, 1998b), military, rail (Reinach and Viale, 2006; Baysari, McIntosh and Wilson, 2008), mining (Patterson and Shappell, 2010), medicine, air traffic control (Shorrock and Kirwan, 1999) and even the space travel domains (Nelson et al., 1998).

Human error is formally defined as: 'All those occasions in which a planned sequence of mental or physical activities fails to achieve its intended outcome, and when these failures cannot be attributed to the intervention of some chance agency' (Reason, 1990: 9). Further classifications of human error have also been proposed, such as the slips (and lapses), mistakes and violations taxonomy proposed by Reason (1990), which also includes a complete description of error classifications and error theory.

The prediction of human error in complex systems was widely investigated in response to the Three Mile Island, Chernobyl and Bhopal disasters. HEI or error prediction techniques are used to identify potential errors that may arise as a result of man–machine interactions in complex systems. The prediction of human error is used within complex, dynamic systems in order to identify the nature of potential human or operator errors and the causal factors, recovery strategies and consequences associated with them. Information derived from HEI analyses is then typically used to propose remedial measures designed to eradicate the potential errors identified. HEI works on the premise that an understanding of an individual's work task and the characteristics of the technology being used allows us to indicate potential errors that may arise from the resulting interaction (Stanton and Baber, 1996a). HEI techniques can be used either during the design process to highlight potential design-induced error or to evaluate error potential in existing systems. These are typically conducted on a task analysis of the activity under investigation. The output of HEI techniques usually describes potential errors, their consequences, recovery potential, probability and criticality, and offers associated design remedies or error reduction strategies. HEI approaches can be broadly categorised into two groups: qualitative and quantitative. Qualitative approaches are used to determine the nature of errors that might occur within a particular system, whilst quantitative approaches are used to provide a numerical probability of error occurrence within a particular system. There is a broad range of HEI approaches available to the HEI practitioner, ranging from simplistic external error mode taxonomy-based approaches to more sophisticated human performance simulation techniques. The techniques reviewed can be further categorised into the following types:

- taxonomy-based techniques;
- error identifier techniques;
- error quantification techniques.

In order to familiarise the reader with the different HEI techniques available, a brief overview of the different HEI techniques is presented below.

Taxonomy-based HEI techniques use external error mode taxonomies and typically involve the application of these error modes to a task analysis of the activity under investigation. Techniques such as SHERPA (Embrey, 1986), HET (Marshall et al., 2003), TRACEr (Shorrock and Kirwan, 2002) and CREAM (Hollnagel, 1998) all use domain-specific external error mode taxonomies designed to aid the analyst in identifying potential errors. Taxonomic approaches to HEI are typically the most successful in terms of sensitivity and are the quickest and simplest to apply with only limited resource usage. However, these techniques also place a great amount of dependence upon the judgment of the analyst and as a result there are concerns associated with the reliability of the error predictions made. Different analysts often make different predictions for the same task using the same technique (intra-analyst reliability). Similarly, the same analyst may make different judgments on different occasions (inter-analyst reliability). A brief description of the taxonomy-based HEI techniques considered here is provided below.

SHERPA (Embrey, 1986) uses a behavioural classification linked to an external error mode taxonomy (action, retrieval, check, selection and information communication errors) to identify potential errors associated with human activity. The SHERPA technique works by indicating which error modes are credible for each bottom-level task step in an HTA. The analyst classifies a task step into a behaviour and then determines whether any of the associated error modes are credible. For each credible error, the analyst describes the error and determines the consequences, error recovery, probability and criticality. Finally, design remedies are proposed for each error identified.

The human error template (HET: Marshall et al., 2003) technique was developed for the certification of civil flight-deck technology and is a checklist approach that is applied to each bottom-level task step in an HTA of the task under analysis. The HET technique works by indicating which of the HET error modes are credible for each task step, based upon the subjective judgment of the analyst. The analyst applies each of the HET error modes to the task step in question and determines whether any of the modes produce any credible errors or not. The HET error taxonomy consists of 12 error modes that were selected based upon a study of actual pilot error incidence and existing error modes used in contemporary HEI methods. For each credible error (i.e. those judged by the analyst to be possible), the analyst should give a description of the form that the error would take, such as 'pilot dials in the airspeed value using the wrong knob'. The associated error consequences, likelihood of occurrence and criticality in relation to the task under analysis are then specified. Finally, a pass or fail rating is assigned to the interface element in question.

HAZOP (Kletz, 1974) is a well-established engineering approach that was developed in the late 1960s by ICI (Swann and Preston, 1995) for use in process design audit and engineering risk assessment (Kirwan, 1992a). It involves a team of analysts applying guidewords, such as 'Not done', 'More than' or 'Later than', to each step in a process in order to identify potential problems that may occur. Human Error HAZOP uses a set of human error guidewords (Whalley, 1988) to identify potential human error. These guidewords are applied to each step in an HTA to determine any credible errors. For each credible error, a description of the error is offered and the associated causes, consequences and recovery steps are specified. Finally, design remedies for each of the errors identified are proposed.

The technique for the retrospective analysis of cognitive errors (TRACEr: Shorrock and Kirwan, 2002) was developed specifically for use in the air traffic control (ATC) domain and can be used either proactively to predict error or retrospectively to analyse errors that have occurred. It uses a series of decision flow diagrams and comprises the following eight error classification schemes: task error, information, performance shaping factors (PSFs), external error modes (EEMs), internal error modes (IEMs), psychological error mechanisms (PEMs), error detection and error correction.

The system for predictive error analysis and reduction (SPEAR: CCPS, 1994) is another taxonomic approach to HEI that is similar to the SHERPA approach described above. SPEAR uses an error taxonomy consisting of action, checking, retrieval, transmission, selection and planning errors, and operates on an HTA of the task under analysis. The analyst considers a series of performance-shaping factors for each bottom-level task step and determines whether or not any credible errors can occur. For each credible error, a description of the error, its consequences and any error reduction measures are provided.

The Cognitive Reliability and Error Analysis Method (CREAM: Hollnagel, 1998) is a recently developed human reliability analysis technique that can be used either predictively or retrospectively. CREAM uses a model of cognition, the Contextual Control Model (COCOM), which focuses on how actions are chosen and assumes that the degree of control that an operator has over his or her actions is variable and determines the reliability of his or her performance. It also uses an error taxonomy containing phenotypes (error modes) and genotypes (error causes), as well as common performance conditions (CPCs) to account for context.

Error identifier HEI techniques, such as HEIST and the technique for human error assessment (THEA), use a series of error identifier prompts or questions linked to external error modes to aid the analyst in identifying potential human error. Examples of typical error identifier prompts include 'could the operator fail to carry out the act in time?', 'could the operator carry out the task too early?' and 'could the operator carry out the task inadequately?' (Kirwan, 1994). The error identifier prompts are typically linked to external error modes and remedial measures. Whilst these techniques attempt to remove the reliability problems associated with taxonomic-based approaches, they add considerable time to the analysis, as each error identifier prompt must be considered. A brief description of the error identifier-based techniques considered in this review is presented below.

HEIST (Kirwan, 1994) uses a set of error identifier prompts designed to aid the analyst in the identification of potential errors. There are eight sets of error identifier prompts, including activation/detection, observation/data collection, identification of system state, interpretation, evaluation, goal selection/task definition, procedure selection and procedure execution. The analyst applies each error identifier prompt to each task step in an HTA and determines whether any of the errors are credible or not. Each error identifier prompt has a set of linked error modes. For each credible error, the analyst records the system causes, the psychological error mechanism and any error reduction guidelines.

THEA (Pocock et al., 2001a, 2001b) is a highly structured approach that employs cognitive error analysis based upon Norman's (1988) model of action execution. THEA uses a scenario analysis to consider context and then employs a series of questions in a checklist-style approach based upon goals, plans, performing actions and perception/evaluation/interpretation.

Error quantification techniques are used to determine the numerical probability of error occurrence. Identified errors are assigned a numerical probability value that represents their associated probability of occurrence. Performance shaping factors (PSFs) are typically used to aid the analyst in the identification of potential errors. Error quantification techniques, such as JHEDI and the human error assessment and reduction technique (HEART), are typically used in probabilistic safety assessments (PSAs) of nuclear processing plants. For example, Kirwan (1996b) reports the use of JHEDI in an HRA risk assessment for the BNFL Thermal Oxide Reprocessing Plant at Sellafield, and also the use of HEART in a HRA risk assessment of the Sizewell B pressurised water reactor. The main advantage of error quantification approaches lies in the numerical probability of error occurrence that they offer. However, error quantification approaches are typically difficult to use and may require some knowledge of PSA and mathematical procedures. Doubts also remain over the consistency of such approaches.

HEART (Williams, 1986) attempts to predict and quantify the likelihood of human error or failure within complex systems. The analyst begins by classifying the task under investigation into one of the HEART generic categories, such as category a which states 'Totally familiar, performed at speed with no real idea of the likely consequences'. Each HEART generic category has a human error probability associated with it. The analyst then identifies any error-producing conditions (EPCs) associated with the task. Each EPC has an associated HEART effect. Examples of HEART EPCs include 'Shortage of time available for error detection and correction' and 'No obvious means of reversing an unintended action'. Once EPCs have been assigned, the analyst calculates the assessed proportion of effect of each EPC (between 0 and 1). Finally an error probability value is derived and remedial measures are proposed.

A more recent development within HEI is to use a toolkit of different HEI techniques in order to maximise the comprehensiveness of the error analysis. The HERA framework is a prototype multiple method or 'toolkit' approach to HEI that was developed by Kirwan (1998a, 1998b). In response to a review of HEI methods, Kirwan (1998b) suggested that the best approach would be for practitioners to utilise a framework-type approach to HEI, whereby a mixture of independent HRA/HEI tools would be used under one framework. Consequently, Kirwan proposed the Human Error and Recovery Assessment (HERA) framework, which was developed for the UK nuclear power and reprocessing industry. Whilst the technique has yet to be applied, it is offered in this review as a representation of the form that a HEI 'toolkit' or framework approach may take.

Task Analysis for Error Identification (TAFEI: Baber and Stanton, 1996a) combines HTA with state-space diagrams (SSDs) in order to predict illegal actions associated with the operation of a system or device. In conducting a TAFEI analysis, plans from an HTA of the task under analysis are mapped onto SSDs for the device in question and a TAFEI diagram is produced. This diagram is then used to highlight any illegal transitions or errors that might arise as a result of the interaction between the two. Remedial measures or strategies are then proposed for each of the illegal transitions identified.

In terms of performance, the literature consistently suggests that SHERPA is the most promising of the HEI techniques available to the HF practitioner. Kirwan (1992b) conducted a comparative study of six HEI techniques and reported that SHERPA achieved the highest overall rankings in terms of performance. In conclusion, Kirwan recommended that a combination of expert judgment together with SHERPA would be the best approach to HEI. Other studies have also produced encouraging reliability and validity data for SHERPA (Baber and Stanton, 1996b, 2001; Stanton and Stevenage, 1998). In a more recent comparative study of HEI techniques, Kirwan (1998b) used 14 criteria to evaluate 38 HEI techniques. In conclusion, it was reported that, of the 38 techniques, only nine are available in the public domain and are of practical use (Kirwan, 1998b).

In general, the main problem surrounding the application of HEI techniques is related to their validation. There have only been a limited number of HEI validation studies reported in the literature (Williams, 1989; Whalley and Kirwan, 1989; Kirwan 1992a, 1992b, 1998a, 1998b; Kennedy, 1995; Baber and Stanton, 1996a, 2002; Stanton and Stevenage, 1998). Considering the number of HEI techniques available and the importance of their use, this represents a very limited set of validation studies. Problems such as resource usage (e.g. financial and time costs) and also access to systems under analysis often affect attempts to validate HEI techniques. As a result, the validation of approaches is often assumed rather than tested.

Stanton (2002) suggests that HEI techniques suffer from two further problems. The first of these relates to the lack of representation of the external environment or objects. Typically, HEI techniques do not represent the activity of the device and material that the human interacts with in more than a passing sense. A large proportion of the HEI techniques available do not consider the context in which the activity under analysis occurs. Second, HEI techniques place a great amount of dependence upon the judgment of the analyst. This subjectivity of analysis often weakens the confidence that can be placed in the error predictions made. A summary of the HEI techniques reviewed is presented in Table 6.1.

Table 6.1   Summary of HEI techniques

Training time	Type of method	Domain	Training time	App time	Related methods	Tools needed	Validation studies	Advantages	Disadvantages
Systematic human error reduction and prediction approach (SHERPA).	HEI.	Nuclear power, generic.	Low.	Med.	HTA.	Pen and paper, system diagrams.	Yes.	1) Encouraging reliability and validity data. 2) Probably the best HEI technique available. 3) Has been used extensively in a number of domains and is quick to learn and easy to use.	1) Can be tedious and time-consuming for large, complex tasks. 2) Extra work may be required in conducting an appropriate HTA.
Human error template (HET).	HEI.	Aviation, generic.	Low.	Med.	HTA.	Pen and paper, system diagrams.	Yes.	1) Very easy to use, requiring very little training. 2) Taxonomy is based upon an analysis of pilot error occurrence. 3) Taxonomy is generic.	1) Can be tedious and time-consuming for large, complex tasks. 2) Extra work may be required in conducting an appropriate HTA.
Technique for the retrospective analysis of cognitive errors (TRACEr).	HEI. HRA.	ATC.	Med.	High.	HTA.	Pen and paper, system diagrams.	No.	1) Appears to be a very comprehensive approach to error prediction and error analysis, including IEM, PEM, EEM and PSF analysis. 2) Based upon sound scientific theory, integrating Wickens' (1992) model of information processing into its model of ATC. 3) Can be used predictively and retrospectively.	1) Appears to be complex for a taxonomic error identification tool. 2) No validation evidence.
Task Analysis for Error Identification (TAFEI).	HEI.	Generic.	Med.	Med.	HTA. SSD.	Pen and paper, system diagrams.	Yes.	1) Uses HTA and SSDs to highlight illegal interactions. 2) Structured and thorough procedure. 3) Sound theoretical underpinning.	1) Can be tedious and time-consuming for large, complex tasks. 2) Extra work may be required in conducting an appropriate HTA. 3) It may be difficult to get hold of SSDs for the system under analysis.
Human Error HAZOP.	HEI.	Nuclear power.	Low.	Med.	HAZOP. HTA.	Pen and paper, system diagrams.	Yes.	1) Very easy to use, requiring very little training. 2) Generic error taxonomy.	1) Can be tedious and time-consuming for large, complex tasks. 2) Extra work may be required in conducting an appropriate HTA.
Technique for Human Error Assessment (THEA).	HEI.	Design, generic.	Low.	Med.	HTA.	Pen and paper, system diagrams.	No.	1) Uses error identifier prompts to aid the analyst in the identification of error 2) Highly structured procedure 3) Each error question has associated consequences and design remedies	1) High resource usage. 2) No error modes are used, making it difficult to interpret which errors could occur. 3) Limited usage.
Human Error Identification in Systems Tool (HEIST).	HEI.	Nuclear power.	Low.	Med.	HTA.	Pen and paper, system diagrams.	No.	1) Uses error identifier prompts to aid the analyst in the identification of errors. 2) Each error question has associated consequences and design remedies.	1) High resource usage. 2) Limited usage.

Table 6.1   Continued

Training time	Type of method	Domain	Training time	App time	Related methods	Tools needed	Validation studies	Advantages	Disadvantages
Human Error and Recovery Assessment (HERA) framework.	HEI. HRA.	Generic.	High.	High.	HTA. HEIST. JHEDI.	Pen and paper, system diagrams.	No.	1) Exhaustive technique that covers all aspects of error. 2) Employs a toolkit methods approach, ensuring comprehensiveness.	1) Time-consuming in its application. 2) No evidence of usage available. 3) High training and application times.
System for Predictive Error Analysis and Reduction (SPEAR).	HEI.	Nuclear power.	Low.	Med.	SHERPA. HTA.	Pen and paper, system diagrams.	No.	1) Easy to use and learn 2) Analyst can choose specific taxonomy	1) Almost exactly the same as SHERPA. 2) Limited use. 3) No validation evidence available.
Human Error Assessment and Reduction Technique (HEART).	HEI quantification.	Nuclear power.	Low.	Med.	HTA.	Pen and paper, system diagrams.	Yes.	1) Offers a quantitative analysis of potential error. 2) Considers PSFs. 3) Quick and easy to use.	1) Doubts over consistency of the technique. 2) Limited guidance given to the analyst. 3) Further validation required.
Cognitive Reliability Analysis Method (CREAM).	HEI. HRA.	Generic.	High.	High.	HTA.	Pen and paper, system diagrams.	Yes.	1) Potentially very comprehensive. 2) Has been used both predictively and retrospectively.	1) Time-consuming in terms of both training and application. 2) Limited use. 3) Over-complicated.
Systems Theory Accident Modelling and Process (STAMP).	Accident analysis.	Generic.	High.	High.	Data collection methods.	Pen and paper.	No.	1) Has been used in multiple domains. 2) Provides comprehensive analysis.	1) Resource-intensive. 2) Complex.
AcciMap.	Accident analysis.	Generic.	Low.	Med.	Data collection methods. HFACS. STAMP.	Pen and paper.	No.	1) Focus on interactions between failures. 2) Based on a sound theoretical model. 3) Used in multiple domains.	1) Can be time-consuming. 2) No taxonomy of error types.
Human Factors Analysis and Classification System (HFACS).	Accident analysis.	Aviation.	Low.	Low.	SOAM, HFIX, HFACS-ME, HFACS RR, HFACS ATC	Pen and paper.	No.	1) Simple to use. 2) Popular method applied in multiple domains. 3) Strong theoretical grounding.	1) Needs tighter category definitions. 2) Doesn't explore the context of error.
Safety Occurrence Analysis Methodology (SOAM).	Accident analysis.	Generic.	Low.	Low.	HFACS.	Pen and paper.	No.	1) Simple and quick to use. 2) Provides a clear overview of errors.	1) Doesn't explore interactions between failures. 2) Doesn't explore evolution of failure.
Functional Resonance Accident Model (FRAM).	Accident analysis.	Generic.	Low.	High.	HTA, STAMP, AcciMap	Pen and paper.	No.	1) Has a strong theoretical base. 2) Explores interactions. 3) Simple to learn and use.	1) Requires large amounts of data. 2) Can be time-consuming.
Why-Because Analysis (WBA).	Accident analysis.	Generic.	Low.	Low.	Data collection methods.	Pen and paper.	No.	1) Explores interactions between factors. 2) Based on formal logic.	1) May not identify all failures. 2) Can become complex.

Although both HEI and human error analysis methods can be used to analyse accidents, a sub-section of methods designed specifically to focus on accident analysis has arisen as a growing field in HF. Accident analysis methods are employed in order to derive an accident aetiology and identify contributory factors in the deviation from safe performance. Salmon et al. (2010b) identified over 30 accident analysis-related methods, illustrating the prominence of accident analysis methods in contemporary HF. Six accident analysis methods are introduced in this chapter with succinct discussions of example applications. For a more indepth review of a number of accident analysis methods, the reader is referred to Salmon et al. (in press). A brief introduction to the six accident analysis methods is provided below.

The STAMP method combines a taxonomy and a systemic structure of control to aid in the identification or errors, or causal factors, that were present in accidents within complex systems. The method is based upon the supposition that accidents occur due to inappropriate energy transfers which are allowed to evolve due to ineffective safety barriers. As such, the main focus of the method is on the identification of barriers within the system. The human factors analysis and classification system (HFACS) is a taxonomic method designed to aid in the analysis of accidents and incidents within complex systems. It consists of a number of taxonomies to guide error identification across a series of systemic levels based on Reason's (1990) model of accident causation, ranging from unsafe acts to organisational influences. AcciMap is a method that uses a systemic control structure (based upon Rasmussen's risk framework) to guide the analyst in the identification of errors across six systemic levels. No taxonomy of error types is included, but the physical activities of the accident scenario are used as a basis to link out to contributory factors elsewhere in the system. The FRAM technique focuses upon the way in which deviations from standard operating procedures can combine, amalgamate and escalate into system failures (Hollnagel et al., 2008). The technique uses a series of categories to aid the analyst in identifying and linking such deviations, enabling a depiction of accident aetiology to be built. SOAM is another accident analysis technique which utilises Reason's (1990) model of error. The method has been successfully applied in the analysis of air traffic incidents for a number of years. It provides the analyst with a structured taxonomy of failures across four systemic levels, enabling a standardised process of error identification and resulting in a summarised graphical illustration of aetiology. WBA is a method built developed for accident analysis based upon theories of formal logic. It aids the analyst in the identification of causal factors which are then linked together using mathematical equations of causality. The output of the method is a series of interrelated causal factors, which is represented in a graphical summary.

The Systematic Human Error Reduction and Prediction Approach (SHERPA)

Background and Applications

SHERPA (Embrey, 1986) was originally developed for use in the nuclear reprocessing industry and is probably the most commonly used HEI approach, with further applications in a number of domains, including aviation (Salmon et al., 2002), health care (Smith et al., 2011; Phipps et al., 2008), public technology (Baber and Stanton, 1996a; Stanton and Stevenage, 1998) and even in-car radio-cassette machines (Stanton and Young, 1999a). SHERPA comprises of an error mode taxonomy linked to a behavioural taxonomy and is applied to an HTA of the task or scenario under analysis in order to predict potential human or design-induced error. As well as being the most commonly used of the various HEI techniques available, according to the literature, it is also the most successful in terms of accuracy of error predictions.

Domain of Application

Despite being originally developed for use in the process industries, the SHERPA behaviour and error taxonomy is generic and can be applied in any domain involving human activity.

Procedure and Advice

Step 1: Conduct an HTA
The first step in a SHERPA analysis involves describing the task or scenario under investigation. For this purpose, an HTA of the task or scenario under analysis is normally conducted. The SHERPA technique works by indicating which of the errors from the SHERPA error taxonomy are credible at each bottom-level task step in an HTA of the task under analysis. A number of data collection techniques may be used in order to gather the information required for the HTA, such as interviews with SMEs and observations of the task under analysis.

Step 2: Task Classification
Next, the analyst should take the first (or next) bottom-level task step in the HTA and classify it according to the SHERPA behaviour taxonomy, which is presented below (Stanton, 2005a):

- Action (e.g.,pressing a button, pulling a switch, opening a door).
- Retrieval (e.g. getting information from a screen or manual).
- Checking (e.g. conducting a procedural check).
- Selection (e.g. choosing one alternative over another).
- Information communication (e.g. talking to another party).

Action Errors
A1 - Operation too long/short
A2 – Operation mistimed
A3 – Operation in wrong direction
A4 – Operation too little/much
A5 – Misalign
A6 – Right operation on wrong object
A7 – Wrong operation on right object
A8 – Operation omitted
A9 – Operation incomplete
A10 – Wrong operation on wrong object

Checking Errors
C1 – Check omitted
C2 – Check incomplete
C3 – Right check on wrong object
C4 – Wrong check on right object
C5 – Check mistimed
C6 – Wrong check on wrong object

Retrieval Errors
R1 – Information not obtained
R2 – Wrong information obtained
R3 – Information retrieval incomplete

Communication Errors
I1 – Information not communicated
I2 – Wrong information communicated
I3 – Information communication

Selection Errors
S1 – Selection omitted
S2 – Wrong selection made

Figure 6.1   SHERPA external error mode taxonomy

Step 3: HEI
The analyst then uses the associated error mode taxonomy and domain expertise to determine any credible error modes for the task in question. For each credible error (i.e. those judged by the analyst to be possible), the analyst should give a description of the form that the error would take, such as 'pilot dials in wrong airspeed'. The SHERPA error mode taxonomy is presented in Figure 6.1.

Step 4: Consequence Analysis
The next step involves determining and describing the consequences associated with the errors identified in step 3. The analyst should consider the consequences associated with each credible error and provide clear descriptions of the consequences in relation to the task under analysis.

Step 5: Recovery Analysis
Next, the analyst should determine the recovery potential of the identified error. If there is a later task step in the HTA at which the error could be recovered, it is entered here. If there is no recovery step, then 'None' is entered.

Step 6: Ordinal Probability Analysis
Once the consequence and recovery potential of the error have been identified, the analyst should rate the probability of the error occurring. An ordinal probability scale of low, medium or high is typically used. If the error has not occurred previously, a low (L) probability is assigned. If the error has occurred on previous occasions,

a medium (M) probability is assigned. Finally, if the error has occurred on frequent occasions, a high (H) probability is assigned.

Step 7: Criticality Analysis
Next, the analyst rates the criticality of the error in question. A scale of low, medium and high is also used to rate error criticality. Normally, if the error would lead to a critical incident (in relation to the task in question), it is rated as a highly critical error.

Step 8: Remedy Analysis
The final stage in the process is to propose error reduction strategies. Normally, remedial measures comprise suggested changes to the design of the process or system. According to Stanton (2005a), remedial measures are normally proposed under the following four categories:

1. equipment (e.g. redesign or modification of existing equipment);
2. training (e.g. changes in training provided);
3. procedures (e.g. provision of new, or redesign of old, procedures); and
4. organisational (e.g. changes in organisational policy or culture).

Advantages

- The SHERPA technique offers a structured and comprehensive approach to the prediction of human error.
- The SHERPA taxonomy prompts the analyst for potential errors.
- According to the HF literature, SHERPA is the most promising HEI technique available. It has been applied in a number of domains with considerable success. There is also a wealth of encouraging validity and reliability data available.
- It is quick to apply compared to other HEI techniques.
- It is also easy to learn and apply, requiring minimal training.
- It is exhaustive, offering error reduction strategies in addition to predicted errors, associated consequences, probability of occurrence, criticality and potential recovery steps.
- The SHERPA error taxonomy is generic, allowing the technique to be used in a number of different domains.

Disadvantages

- SHERPA can be tedious and time-consuming for large, complex tasks.
- The initial HTA adds additional time to the analysis.
- It only considers errors at the 'sharp end' of system operation and does not consider system or organisational errors.
- It does not model cognitive components of error mechanisms.
- Some predicted errors and remedies are unlikely or lack credibility, thus posing a false economy (Stanton, 2005a).
- Its current taxonomy lacks generalisability (Stanton, 2005a).
- It is a subjective method based on analysts ability (Phipps et al., 2008).
- It is unable to explore contextual factors (Phipps et al., 2008).

Example

The following example is a SHERPA analysis of the measurement of an individual's blood glucose level and the administration of insulin. The high-level task network of the insulin task is presented in Figure

6.2. An extract of the HTA for the insulin task is presented below this, followed by a corresponding extract of the SHERPA output for the insulin task in Table 6.2.

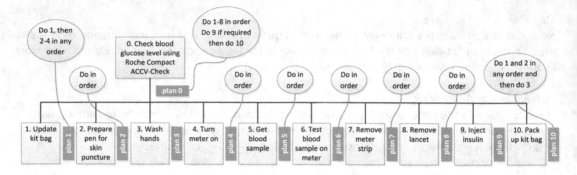

Figure 6.2   HTA of the insulin task

Table 6.2   SHERPA output for the insulin task

Task step	Error mode	Error description	Consequence	Recovery	P	C	Remedial strategy
1. Unpack kit bag.							
Plan 1. Do 1, then do 2 to 4 in any order.							
1.1. Unzip kit bag.							
1.2. Remove meter from kit bag.	A8	Fail to unpack the meter.	Cannot test blood.	Immediate.	L	L	Remove reliance on a kit bag.
1.3 Remove pen from kit bag.	A8	Fail to unpack the pen.	Cannot get blood.	Immediate.	L	L	Remove reliance on a kit bag.
1.4. Remove a lancet from the kit bag.							
Plan 1.4. Do in order.							
1.4.1. Unzip the lancet pouch.	A8	Fail to unzip the pouch.	Cannot get lancet.	Immediate.	L	L	Remove reliance on a kit bag. Integrate lancet into pen
1.4.2. lift a lancet out of the pouch.	A8	Fail to unpack the lancet.	Cannot pierce skin or have to reuse old lancet.	Immediate.	L	M	Remove reliance on a kit bag. Integrate lancet into pen.
1.4.3. Zip up the lancet pouch.	A8 A9	Fail to zip up the pouch. Pouch partially unzipped.	May lose lancets.	?	M	M	Remove reliance on a kit bag. Integrate lancet into pen.

Related Methods

The initial data collection for SHERPA might involve a number of data collection techniques, including interviews, observation and walkthroughs. An HTA of the task or scenario under analysis is typically used as the input to a SHERPA analysis. The taxonomic approach to error prediction employed by the SHERPA technique is similar to a number of other HEI approaches, such as HET (Marshall et al., 2003), Human Error HAZOP (Kirwan and Ainsworth, 1992) and TRACEr (Shorrock and Kirwan, 2002).

Approximate Training and Application Times

In order to evaluate the reliability, validity and trainability of various techniques, Stanton and Young (1998) compared SHERPA to 11 other HF techniques. Based on the application of the technique to the operation of an in-car radio-cassette machine, the authors reported training times of around three hours (this is doubled if training in HTA is included). It took an average of two hours and 40 minutes for people to evaluate the radio-cassette machine using SHERPA. In a study comparing the performance of SHERPA, Human Error HAZOP, HEIST and HET when used to predict design-induced pilot error, Salmon et al. (2002) reported that participants achieved acceptable performance with the SHERPA technique after only two hours of training.

Harvey and Stanton (2012) applied SHERPA to the evaluation of in-vehicle computer interfaces and suggested that 2–4 hours were required to collect data for the method and a further 8–10 hours were needed to undertake the analysis.

Flowchart

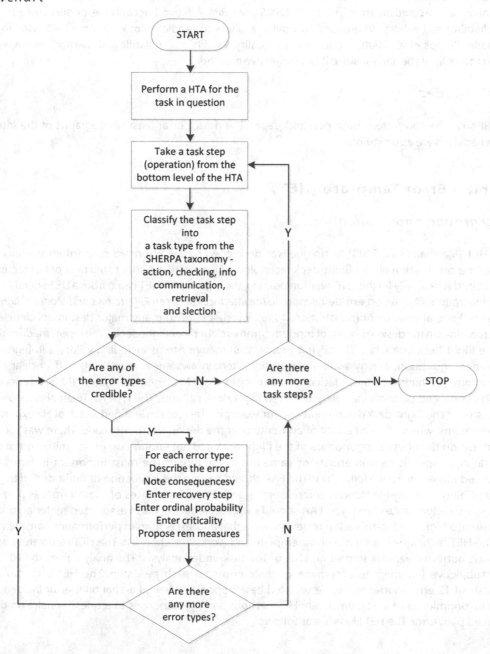

Reliability and Validity

There is a wealth of promising validation data associated with the SHERPA technique. Kirwan (1992b) reported that SHERPA was the most highly rated of the five human error prediction techniques by expert users. Baber and Stanton (1996a) reported a concurrent validity statistic of 0.8 and a reliability statistic of 0.9 in the application of SHERPA by two expert users to prediction of errors on a ticket vending machine. Stanton and Stevenage (1998) reported a concurrent validity statistic of 0.74 and a reliability statistic of 0.65 in the application of SHERPA by 25 novice users to prediction of errors on a confectionery vending machine. According to Stanton and Young (1999a), SHERPA achieved a concurrent validity statistic of 0.2 and a reliability statistic of 0.4 when used by eight novices to predict errors on an in-car radio-cassette machine task. According to Harris et al. (2005), SHERPA achieved acceptable performance in terms of reliability and validity when used by novice analysts to predict pilot error on a civil aviation flight scenario. Phipps et al. (2008) argue that the quality, validity and reliability of method are dependent upon the skills of the analyst since it is a subjective method.

Tools Needed

SHERPA can be conducted using pen and paper. The device or at least photographs of the interface under analysis are also required.

Human Error Template (HET)

Background and Applications

The HET (Marshall et al., 2003) technique was developed by the ErrorPred consortium specifically for use in the certification of civil flight-deck technology. Along with a distinct shortage of HEI techniques developed specifically for the civil aviation domain, the impetus for HET came from a US Federal Aviation Administration (FAA) report entitled *Report on the Interfaces between Flightcrews and Modern Flight Deck Systems* (Federal Aviation Administration, 1996), which identified many major design deficiencies and shortcomings in the design process of modern commercial airliner flight decks. The report made criticisms of the flight deck interfaces, identifying problems in many systems including: pilots' autoflight mode awareness/indication; energy awareness; position/terrain awareness; confusing and unclear display symbology and nomenclature; a lack of consistency in Flight Management System (FMS) interfaces and conventions; and poor compatibility between flight deck systems. The FAA HF Team also made many criticisms of the flight-deck design process. For example, the report identified a lack of HF expertise on design teams, which also had a lack of authority over the design decisions made. There was too much emphasis on the physical ergonomics of the flight deck and not enough on the cognitive ergonomics. A total of 51 specific recommendations came out of the report. The most important in terms of the ErrorPred project were the following: 'The FAA should require the evaluation of flight deck designs for susceptibility to design-induced flightcrew errors and the consequences of those errors as part of the type certification process' and 'The FAA should establish regulatory and associated material to require the use of a flight deck certification review process that addresses human performance considerations'.

The HET technique is a simple error template and works as a checklist. The HET template is applied to each bottom-level task step in an HTA of the task under analysis. The analyst uses the HET EEMs and subjective judgment to determine credible errors for each task step. The HET error taxonomy consists of 12 error modes that were selected based upon a review of actual pilot error incidence, the EEM taxonomies used in contemporary HEI methods and the responses to a questionnaire on design-induced pilot error. The HET EEMs are as follows:

- Fail to execute.
- Task execution incomplete.
- Task executed in the wrong direction.
- Wrong task executed.
- Task repeated.
- Task executed on the wrong interface element.
- Task executed too early.
- Task executed too late.
- Task executed too much.
- Task executed too little.
- Misread Information.
- Other.

For each credible error (i.e. those judged by the analyst to be possible), the analyst should give a description of the form that the error would take, such as 'pilot dials in the airspeed value using the wrong knob'. Next, the analyst has to determine the outcome or consequence associated with the error, e.g. 'Aircraft stays at current speed and does not slow down for approach'. Finally, the analyst then has to determine the likelihood of the error (low, medium or high) and the criticality of the error (low, medium or high). If the error is assigned a high rating for both likelihood and criticality, the aspect of the interface involved in the task step is then rated as a 'fail', meaning that it is not suitable for certification.

Domain of Application

The HET technique was developed specifically for the aviation domain and is intended for use in the certification of flight deck technology. However, the HET EEM taxonomy is generic, allowing the technique to be applied in any domain.

Procedure and Advice

Step 1: Conduct an HTA
The first step in a HET analysis is to conduct an HTA of the task or scenario under investigation. The HET technique works by indicating which of the errors from the HET error taxonomy are credible at each bottom-level task step in an HTA of the task under analysis. A number of data collection techniques may be used in order to gather the information required for the HTA, such as interviews with SMEs and observations of the task under analysis.

Step 2: HEI
In order to identify potential errors, the analyst takes each bottom-level task step from the HTA and considers the credibility of each of the HET EEMs. Any EEMs that are deemed to be credible by the analyst are recorded and analysed further. At this stage, the analyst ticks each credible EEM and provides a description of the form that the error will take.

Step 3: Consequence Analysis
Once a credible error is identified and described, the analyst should then consider and describe the consequence(s) of the error. The analyst should consider the consequences associated with each credible error and provide clear descriptions of the consequences in relation to the task under investigation.

Step 4: Ordinal Probability Analysis
Next, the analyst should provide an estimate of the probability of the error occurring, based upon his or her subjective judgment. An ordinal probability value is entered as low, medium or high. If the analyst

feels that chances of the error occurring are very small, a low (L) probability is assigned. If the analyst thinks that the error may occur and has knowledge of the error occurring on previous occasions, a medium (M) probability is assigned. Finally, if the analyst thinks that the error would occur frequently, a high (H) probability is assigned.

Step 5: Criticality Analysis

Next, the criticality of the error is rated. Error criticality is rated as low, medium or high. If the error would lead to a serious incident (this would have to be defined clearly before the analysis), it is labelled as high. Typically, a high criticality would be associated with error consequences that would lead to substantial damage to the aircraft, injury to crew and passengers, or complete failure of the flight task under analysis. If the error has consequences that still have a distinct effect on the task, such as heading the wrong way or losing a large amount of height or speed, the criticality is labelled as medium. If the error would have minimal consequences that are easily recoverable, such as a small loss of speed or height, the criticality is labelled as low.

Step 6: Interface Analysis

The final step in an HET analysis involves determining whether or not the interface under analysis passes the certification procedure. The analyst assigns a 'pass' or 'fail' rating to the interface under investigation (dependent upon the task step) based upon the associated error probability and criticality ratings. If a high probability and a high criticality were assigned previously, the interface in question is classed as a 'fail'. Any other combination of probability and criticality and the interface in question is classed as a 'pass'.

Advantages

- The HET methodology is quick, simple to learn and use and requires very little training.
- It utilises a comprehensive error mode taxonomy based upon existing HEI EEM taxonomies, actual pilot error incidence data and pilot error case studies.
- It is easily auditable as it comes in the form of an error pro-forma.
- The HET taxonomy prompts the analyst for potential errors.
- The HET methodology has encouraging reliability and validity data (Marshall et al., 2003; Salmon et al., 2002; Stanton et al., 2006a).
- Although the error modes in the HET EEM taxonomy were developed specifically for the aviation domain, they are generic, ensuring that the HET technique can potentially be used in a wide range of different domains, such as command and control, ATC and nuclear reprocessing.
- It is a useful tool for HF certification (Stanton et al., 2006b).
- Li et al. (2009) argue that the method enables the design of a user-friendly interface that can improve task performance and increase task effectiveness.

Disadvantages

- For large, complex tasks, an HET analysis may become tedious and time-consuming.
- Extra work is involved if an HTA is not already available.
- It does not deal with the cognitive component of errors.
- It only considers errors at the 'sharp end' of system operation and does not consider system or organisational errors.
- Stanton et al. (2006b) point out that the method was originally developed for use in aviation and application outside of this domain would require adaptation of the method.
- It is best applied by an analyst with domain relevant knowledge (Stanton et al., 2006b).

Flowchart

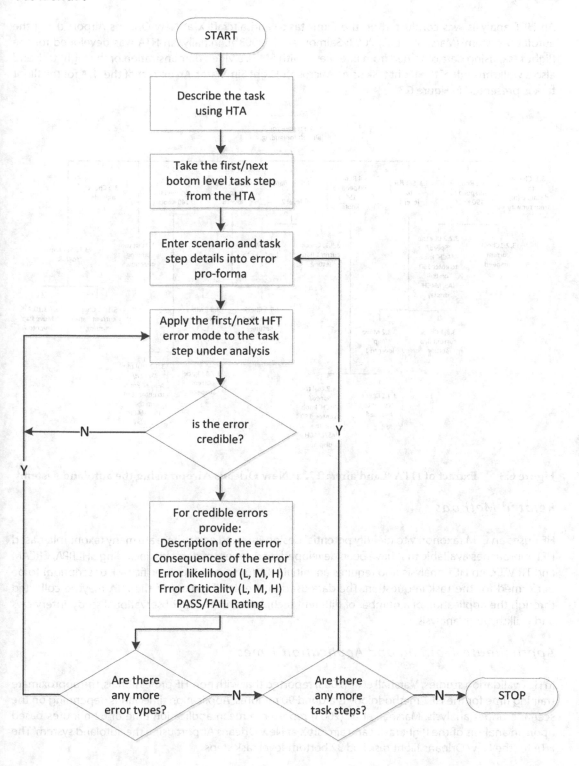

Example

An HET analysis was conducted on the flight task 'Land aircraft X at New Orleans Airport using the autoland system' (Marshall et al., 2003; Salmon et al., 2003b). Initially, an HTA was developed for the flight task, using data obtained from interviews with SMEs, a video demonstration of the flight task and also a walkthrough of the flight task using Microsoft Flight Simulator. An extract of the HTA for the flight task is presented in Figure 6.3.

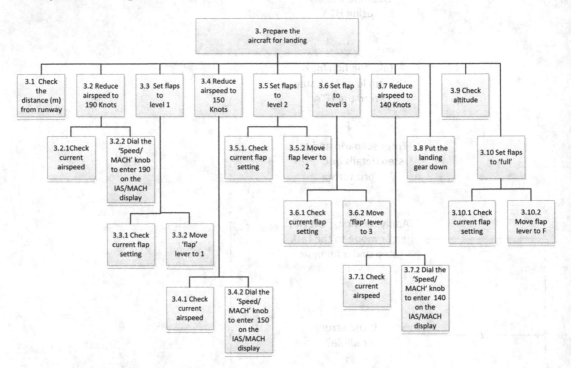

Figure 6.3   Extract of HTA 'Land aircraft X at New Orleans Airport using the autoland system'

Related Methods

HET uses an EEM taxonomy to identify potential design-induced error. There are many taxonomic-based HEI approaches available that have been developed for a variety of domains, including SHERPA, CREAM and TRACEr. An HET analysis also requires an initial HTA (or some other specific task description) to be performed for the task in question. The data used in the development of the HTA may be collected through the application of a number of different techniques, including observational study, interviews and walkthrough analysis.

Approximate Training and Application Times

In HET validation studies, Marshall et al. (2003) reported that with non-HF professionals, the approximate training time for the HET methodology is around 90 minutes. Application time varies depending on the scenario under analysis. Marshall et al. (2003) reported a mean application time of 62 minutes based upon an analysis of the flight task 'Land aircraft X at New Orleans Airport using the autoland system'. The HTA for the New Orleans flight task had 32 bottom-level task steps.

Table 6.3   Example of HET output

Scenario: Land A320 at New Orleans using the Autoland system			Task step: 3.4.2 Dial the 'Speed/MACH' knob to enter 150 on IAS/MACH display								
Error Mode		Description	Outcome	Likelihood			Criticality			Pass	Fail
				H	M	L	H	M	L		
Fail to execute											
Task execution incomplete											
Task executed in wrong direction	✓	Pilot turns the Speed/MACH knob the wrong way	Plane speeds up instead of slowing down	✓			✓			✓	
Wrong task executed											
Task repeated											
Task executed on wrong interface element	✓	Pilot dials using the HDG knob instead	Plane changes course and not speed	✓			✓				✓
Task executed too early											
Task executed too late											
Task executed too much	✓	Pilot turns the Speed/MACH knob too much	Plane slows down too much	✓				✓		✓	
Task executed too little	✓	Pilot turns the Speed/MACH knob too little	Plane does not slow down enough/Too fast for approach	✓				✓		✓	
Misread information											
Other											

Reliability and Validity

Salmon et al. (2003b) reported sensitivity index ratings between 0.7 and 0.8 for subjects using the HET methodology to predict potential design-induced pilot errors for the task 'Land aircraft X at New Orleans Airport using the autoland system'. Furthermore, it was reported that subjects using the HET method were more successful in their error predictions than subjects using SHERPA, Human Error Hazard and Operability (HAZOP) and HEIST.

Stanton et al. (2006a) compare HET to SHERPA, Human Error HAZOP and HEIST in their ability to predict errors during the landing of a commercial aircraft. The analysts used all methods on two separate occasions (separated by a month) and the results of these were compared to an independent collection of all errors possible during task performance (derived from interviews with, and questionnaires completed by, pilots) using a Signal Detection paradigm. The results of the study revealed that the HET analysis was statistically more accurate than the other methods, each of which were statistically comparable to one another. Stanton et al. (2006a) conclude that HET is able to predict a higher proportion of actual errors than HEIST, SHERPA or Human Error HAZOP.

Tools Needed

HET can be carried out using the HET error pro-forma, an HTA of the task under analysis, functional diagrams of the interface under analysis and a pen and paper. In the example HET analysis described above, subjects were provided with an error pro-forma, an HTA of the flight task, diagrams of the autopilot panel, the captain's primary flight display, the flap lever, the landing gear lever, the speed brake, the attitude indicator and an overview of the A320 cockpit (Marshall et al., 2003).

Technique for the Retrospective and Predictive Analysis of Cognitive Errors (TRACEr)

Background and Applications

TRACEr (Shorrock and Kirwan, 2002) is an HEI technique that was developed specifically for use in the ATC domain as part of the human error in European air traffic management project (Isaac, Shorrock and Kirwan, 2002). Under the European project, Isaac, Shorrock and Kirwan were required to develop a human error incidence analysis technique that conformed to the following criteria (Isaac, Shorrock and Kirwan, 2002):

- the technique should be flowchart-based for ease of use;
- it should utilise a set of interrelated taxonomies (EEMs, IEMs, PEMs, PSFs, tasks and information);
- it must be able to deal with chains of events and errors;
- the PSF taxonomy should be hierarchical and may need a deeper set of organisational causal factor descriptors;
- the technique must be comprehensive, accounting for situation awareness, signal detection theory and control theory; and
- it must be able to account for maintenance errors, latent errors, violations and errors of commission.

TRACEr can be used both predictively and retrospectively and is based upon a literature review of a number of domains, including experimental and applied psychology, HF and communication theory (Isaac, Shorrock and Kirwan, 2002).

Different versions of TRACEr can be found in the literature, including an adaptation for use within the rail domain (TRACEr-Rail: Baysari et al., 2009) and for retrospective rail application in Australia (Baysari, Cappnecchia and McIntosh, 2011). Baysari et al. (2009) compared TRACEr-Rail with HFACS in the categorisation of Australian rail accidents spanning an eight-year time period (1998–2006). Their comparison concluded that TRACEr-Rail failed to identify all organisational factors and all 'extraneous' factors, and they argue that additional categories need to be incorporated as it is currently not comprehensive, suggesting that greater coverage of environmental, safety culture and infrastructure factors should be taken into consideration. They also argue that the method's error detection and error reduction phases are confusing, are not comprehensive and are unable to take account of a number of common strategies. The Baysari et al. (2009) study concluded that TRACEr-Rail provided a more indepth description of the context of error occurrence than HFACS, but that HFACS provided a more indepth description of the organisational context.

Domain of Application

TRACEr was originally developed for the ATC domain. However, the technique has since been applied in the rail domain and it is feasible that the technique could be applied in any domain.

Procedure and Advice (Predictive Analysis)

Step 1: Conduct an HTA
The first step in a TRACEr analysis involves conducting an HTA in order to describe the task or scenario under investigation. A number of data collection techniques may be used in order to gather the information required for the HTA, such as interviews with SMEs and observations of the task under analysis.

Step 2: PSF and EEM Consideration

The analyst takes the first bottom-level task step from the HTA and considers each of the TRACEr PSFs for the task step in question. The purpose of this is to identify any environmental or situational factors that could influence the controllers' performance during the task step in question. Once the analyst has considered all of the relevant PSFs, the EEMs are considered for the task step under analysis. Based upon subjective judgment, the analyst determines whether any of the TRACEr EEMs are credible for the task step in question. The TRACEr EEM taxonomy is presented in Table 6.4. If there are any credible errors, the analyst proceeds to step 3. If there are no errors deemed to be credible, the analyst goes back to the HTA and takes the next task step.

Selection and quality	Timing and sequence	Communication
Omission.	Action too long.	Unclear info transmitted.
Action too much.	Action too short.	Unclear info recorded.
Action too little.	Action too early.	Info not sought/ obtained.
Action in wrong direction.	Action too late.	Info not transmitted.
Wrong action on right object.	Action repeated.	Info not recorded.
Right action on wrong object.	Misordering.	Incomplete info transmitted.
Wrong action on wrong object.		Incomplete info recorded.
Extraneous act.		Incorrect info transmitted.
		Incorrect info recorded.

Table 6.4   TRACEr's EEM taxonomy

Step 3: IEM and Information Classification

For any credible errors, the analyst then determines which of the IEMs are associated with the error. IEMs describe which cognitive function failed or could fail (Shorrock and Kirwan, 2002). Examples of TRACEr IEMs include late detection, misidentification, hearback error, forget previous actions, prospective memory failure, misrecall stored information and misprojection. The topic of the error is also determined using the information error modes (Shorrock and Kirwan, 2002).

Step 4: PEM Classification

Next, the analyst has to determine the psychological cause or PEM behind the error. Examples of TRACEr PEMs include insufficient learning, expectation bias, false assumption, perceptual confusion, memory block, vigilance failure and distraction.

Step 5: Error Recovery

Finally, once the error analyst has described the error and has determined the EEM, IEMs and PEMs, error recovery steps for each error should be offered. This is based upon the analyst's subjective judgment.

Procedure and Advice (Retrospective Analysis)

Step 1: Classify Incident into 'Error Events'

First, the analyst has to classify the task steps into error events, i.e. the task steps in which an error was produced. This is based upon the analyst's subjective judgment.

Step 2: Task Error Classification

The analyst then takes the first/next error from the error events list and classifies it into a task error from the task error taxonomy, which contains 13 categories describing controller errors. Examples of task error categories include 'radar monitoring error', 'coordination error' and 'flight progress strip use error' (Shorrock and Kirwan, 2002).

Step 3: IEM and Information Classification

Next, the analyst has to determine the IEM associated with the error. IEMs describe which cognitive function failed or could fail (Shorrock and Kirwan, 2002). Examples of TRACEr IEMs include late

Table 6.5   Extract from TRACEr's PSF taxonomy

PSF category	Example PSF keyword
Traffic and airspace.	Traffic complexity.
Pilot/controller communications.	RT workload.
Procedures.	Accuracy.
Training and experience	Task familiarity.
Workplace design, HMI and equipment factors.	Radar display.
Ambient environment.	Noise.
Personal factors.	Alertness/fatigue.
Social and team factors.	Handover/takeover.
Organisational factors.	Conditions of work.

detection, misidentification, hearback error, forget previous actions, prospective memory failure, misrecall stored information and misprojection. The analyst also has to use the information taxonomy to describe the 'subject matter' of the error, i.e. what information the controller misperceived. The information terms used are related directly to the IEMs in the IEM taxonomy. The information taxonomy is important as it forms the basis of error reduction within the TRACEr technique.

Step 4: PEM Classification
The analyst then has to determine the 'psychological cause' or PEM behind the error. Example TRACEr PEMs include insufficient learning, expectation bias, false assumption, perceptual confusion, memory block, vigilance failure and distraction.

Step 5: PSF Classification
PSFs are factors that influenced or have the potential to have influenced the operator's performance. The analyst uses the PSF taxonomy to select any PSFs that were evident in the production of the error under analysis. TRACEr's PSF taxonomy contains both PSF categories and keywords. Examples of TRACEr PSF categories and associated keywords are presented in Table 6.5.

Step 6: Error Detection and Correction
Unique to retrospective TRACEr applications, the error detection and correction stage provides the analyst with a set of error detection keywords. Four questions are used to prompt the analyst in the identification and selection of error detection keywords (Shorrock and Kirwan, 2002):

- How did the controller become aware of the error (e.g. action feedback, inner feedback, outcome feedback)?
- What was the feedback medium (e.g. radio, radar display)?
- Did any factors, internal or external to the controller, improve or degrade the detection of the error?
- What was the separation status at the time of error detection?

Once the analyst has identified the error detection features, the error correction or reduction should also be determined. TRACEr uses the following questions to prompt the analyst in error correction/reduction classification (Shorrock and Kirwan, 2002):

- What did the controller do to correct the error (e.g. reversal or direct correction, automated correction)?
- How did the controller correct the error (e.g. turn or climb)?
- Did any factors, internal or external to the controller, improve or degrade the detection of the error?
- What was the separation status at the time of the error correction?

Once the analyst has completed step 6, the next error should be analysed. Alternatively, if there are no more 'error events', then the analysis is complete.

Flowchart (Predictive Analysis)

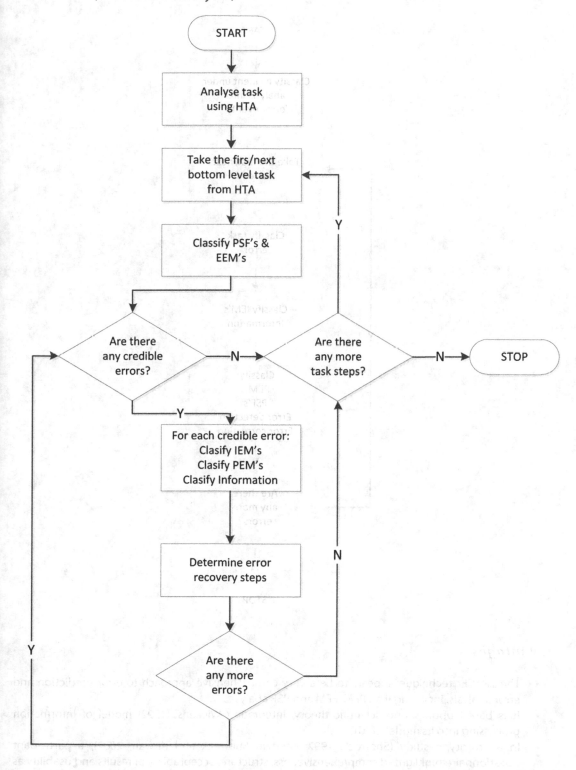

Flowchart (Retrospective Analysis)

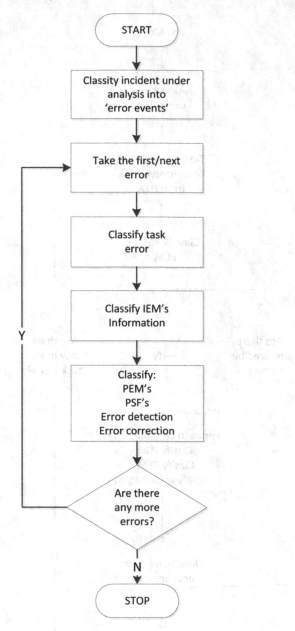

Advantages

- The TRACEr technique appears to be a very comprehensive approach to error prediction and error analysis, including IEM, PEM, EEM and PSF analysis.
- It is based upon sound scientific theory, integrating Wickens' (1992) model of information processing into its model of ATC.
- In a prototype study (Shorrock, 1997; cited in Shorrock and Kirwan, 2002), a participant questionnaire highlighted comprehensiveness, structure, acceptability of results and usability as strong points of the technique.

- It has proved successful in analysing errors from AIRPROX (air proximity) reports and providing error reduction strategies.
- It has been developed specifically for ATC, based upon previous ATC incidents and interviews with ATC controllers.
- It considers PSFs within the system that may have contributed to the errors identified.

Disadvantages

- The TRACEr technique appears unnecessarily overcomplicated. A prototype study (Shorrock, 1997; cited in Shorrock and Kirwan, 2002) highlighted a number of areas of confusion in participant use of the different categories. Much more simple error analysis techniques exist, such as SHERPA and HET.
- There is no validation evidence or studies using TRACEr.
- For complex tasks, a TRACEr analysis may become laborious and unwieldy.
- A TRACEr analysis typically incurs high resource usage. In a participant questionnaire used in the prototype study, resource usage (time and expertise) was the most commonly reported area of concern (Shorrock and Kirwan, 2002).
- Training time would be extremely high for such a technique and a sound understanding of psychology would be required in order to use the technique effectively.
- Extra work is involved if an HTA not already available.
- Existing techniques using similar EEM taxonomies appear to be far simpler and much quicker to apply (SHERPA, HET, etc.).
- It requires greater levels of guidance on error identification and additional detail in category descriptions is required to avoid misinterpretation (Baysari et al., 2009).

Example

For an example TRACEr analysis, the reader is referred to Shorrock and Kirwan (2002).

Related Methods

TRACEr is a taxonomy-based approach to HEI. A number of error taxonomy techniques exist, such as SHERPA, CREAM and HET. When applying TRACEr (both predictively and retrospectively), an HTA for the task/scenario under analysis is required.

Approximate Training and Application Times

No data regarding training and application times for the TRACEr technique is presented in the literature. It is estimated that both the training and application times for TRACEr would be high.

Reliability and Validity

There is no data available regarding the reliability and validity of the TRACEr technique. According to Shorrock and Kirwan (2002), such a study is being planned. In a small study analysing error incidences from AIRPROX reports (Shorrock and Kirwan, 2002), it was reported, via participant questionnaire, that the TRACEr technique's strengths are its comprehensiveness, structure, acceptability of results and usability.

Baysari et al. (2009) argue that TRACEr is prone to misinterpretation and inconsistent applications by analysts, leading to poor levels of inter-rater reliability, although their analysis of the data revealed a moderate level of inter-rater reliability.

Tools Needed

TRACEr analyses can be carried out using pen and paper. PEM, EEM, IEM and PSF taxonomy lists are also required, as is an HTA for the task under investigation.

Task Analysis For Error Identification (TAFEI)

Background and Applications

TAFEI is a method that enables people to predict errors with device use by modelling the interaction between the user and the device under analysis. It assumes that people use devices in a purposeful manner, such that the interaction may be described as a 'cooperative endeavour', and it is through this cooperative process that problems arise. Furthermore, the technique makes the assumption that actions are constrained by the state of the product at any particular point in the interaction and that the device offers information to the user about its functionality. Thus, the interaction between users and devices progresses through a sequence of states. At each state, the user selects the action most relevant to their goal, based on the System Image.

The foundation for the approach is based on general systems theory. This theory is potentially useful in addressing the interaction between subcomponents in systems (that is, the human and the device). It also assumes a hierarchical order of system components, i.e. all structures and functions are ordered by their relation to other structures and functions, and any particular object or event is comprised of lesser objects and events. Information regarding the status of the machine is received by the human part of the system through sensory and perceptual processes and is converted to physical activity in the form of input to the machine. The input modifies the internal state of the machine and feedback is provided to the human in the form of output. Of particular interest here is the boundary between humans and machines, as this is where errors become apparent. It is believed that it is essential for a method of error prediction to examine explicitly the nature of this interaction.

The theory draws upon the ideas of scripts and schema. It can be imagined that a person approaching a ticket-vending machine might draw upon a 'vending machine' or a 'ticket kiosk' script when using a ticket machine. From one script, the user might expect the first action to be 'Insert Money', but from the other script, the user might expect the first action to be 'Select Item'. The success, or failure, of the interaction would depend on how closely he or she was able to determine a match between the script and the actual operation of the machine. The role of the comparator is vital in this interaction. If it detects differences from the expected states, then it is able to modify the routines. Failure to detect any differences is likely to result in errors. Following Bartlett's lead, the notion of schema is assumed to reflect a person's 'effort after meaning' (Bartlett, 1932), arising from the active processing (by the person) of a given stimulus. This active processing involves combining prior knowledge with information contained in the stimulus. While schema theory is not without its critics (see Brewer, 2000 for a review), the notion of an active processing of stimuli has resonance with a proposal for rewritable routines. The reader might feel that there are similarities between the notion of rewritable routines and some of the research on mental models that was popular in the 1980s. Recent developments in the theory underpinning TAFEI by the authors have distinguished between global prototypical routines (i.e. a repertoire of stereotypical responses that allow people to perform repetitive and mundane activities with little or no conscious effort) and local, state-specific routines (i.e. responses that are developed only for a specific state of the system). The interesting part of the theory is the proposed relationship between global and local routines. It is posited that these routines are analogous to global and local variables in computer programming code. In the same manner as a local variable in programming code, a local routine is overwritten (or rewritable in TAFEI terms) once the user have moved beyond the specific state for which it was developed. See Baber and Stanton (2002) for a more detailed discussion of the theory.

Examples of applications of TAFEI include prediction of errors in boiling kettles (Baber and Stanton, 1994, 1999), comparison of word processing packages (Stanton and Baber, 1996b, 1998), withdrawing cash from automatic cash machines (Burford, 1993), medical applications (Baber and Stanton, 1999; Yamaoka and Baber, 2000), recording on tape-to-tape machines (Baber and Stanton, 1994), programming a menu on cookers (Crawford, Taylor and Po, 2001), programming video-cassette recorders (Baber and Stanton, 1994; Stanton and Baber, 1998), operating radio-cassette machines (Stanton and Young, 1999a), recalling a phone number on mobile phones (Baber and Stanton, 2002), buying a train ticket on the ticket machines on the London Underground (Baber and Stanton, 1996a) and operating high-voltage switchgear in substations (Glendon and McKenna, 1995).

Domain of Application

Public technology and product design.

Procedure and Advice

Step 1: Construct an HTA
First, HTA is performed to model the human side of the interaction. Of course, one could employ any technique to describe human activity. However, HTA suits this purpose for the following reasons:

- it is related to goals and tasks;
- it is directed at a specific goal;
- it allows consideration of task sequences (through 'plans').

As will become apparent, TAFEI focuses on a sequence of tasks aimed at reaching a specific goal.

For illustrative purposes of how to conduct the method, a simple, manually-operated electric kettle is used. The first step in a TAFEI analysis is to obtain an appropriate HTA for the device, as shown in Figure 6.4. As TAFEI is best suited to scenario analyses, it is wise to consider just one specific goal, as described

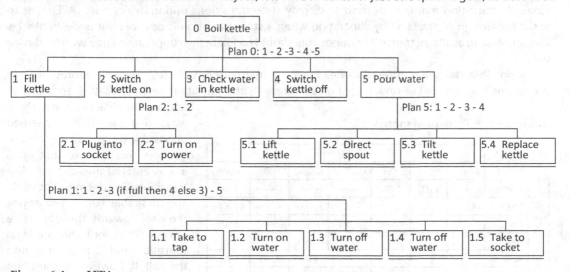

Figure 6.4   HTA

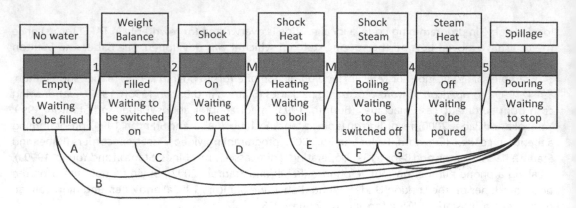

Figure 6.5   State-space TAFEI diagram

by the HTA (e.g. a specific, closed-loop task of interest) rather than the whole design. Once this goal has been selected, the analysis proceeds to constructing SSDs for device operation.

Step 2: Construct SSDs
Next, SSDs are constructed to represent the behaviour of the artefact. An SSD essentially consists of a series of states that the device passes through, from a starting state to the goal state. For each series of states, there will be a current state and a set of possible exits to other states. At a basic level, the current state might be 'off', with the exit condition 'switch on' taking the device to the state 'on'. Thus, when the device is 'off', it is 'waiting to …' an action (or set of actions) that will take it to the state 'on'. On completing the SSD, it is very important to have an exhaustive set of states for the device under analysis. Numbered plans from the HTA are then mapped onto the SSD, indicating which human actions take the device from one state to another. Thus, the plans are mapped onto the state transitions (if a transition is activated by the machine, this is also indicated on the SSD, using the letter 'M' on the TAFEI diagram). This results in a TAFEI diagram, as shown in Figure 6.4. Potential state-dependent hazards have also been identified.

Step 3: Create Transition Matrix
Finally, a transition matrix is devised to display state transitions during device use. TAFEI aims to assist the design of artefacts by illustrating when a state transition is possible but undesirable (i.e. illegal). Making all illegal transitions impossible should facilitate the cooperative endeavour of device use.

All possible states are entered as headers on a matrix (see Table 6.6). The cells represent state transitions (e.g. the cell at row 1, column 2 represents the transition between state 1 and state 2) and are then filled in one of three ways. If a transition is deemed impossible (i.e. you simply cannot go from this state to that one), a '-' is entered into the cell. If a transition is deemed possible and desirable (i.e. it progresses the user towards the goal state – a correct action), this is a legal transition and 'L' is entered into the cell. If, however, a transition is both possible but undesirable (a deviation from the intended

Table 6.6   Transition matrix

		TO STATE						
		Empty	Filled	On	Heating	Boiling	Off	Pouring
FROM STATE	Empty	---------	L (1)	I (A)	---------	---------	---------	I (B)
	Filled		---------	L (2)	---------	---------	---------	I (C)
	On			---------	L (M)	---------	---------	I (D)
	Heating				---------	L (M)	---------	I (E)
	Boiling					I (F)	L (4)	I (G)
	Off						---------	L (5)
	Pouring							---------

path – an error), this is deemed to be illegal and the cell is filled with an 'I'. The idea behind TAFEI is that usability may be improved by making all illegal transitions (errors) impossible, thereby limiting the user to only performing desirable actions. It is up to the analyst to conceive of design solutions to achieve this.

The states are normally numbered, but in this example the text description is used. The character 'L' denotes all of the error-free transitions and the character 'I' denotes all of the errors. Each error has an associated character (i.e. A to G) for the purposes of this example and so that it can be described in Table 6.7. Obviously, the design solutions in Table 6.7 are just illustrative and would need to be formally assessed in terms of their feasibility and cost.

Table 6.7   Error descriptions and design solutions

Error	Transition	Error description	Design solution
A	1 to 3	Switch empty kettle on.	Transparent kettle walls and/or link to water supply.
B	1 to 7	Pour empty kettle.	Transparent kettle walls and/or link to water supply.
C	2 to 7	Pour cold water.	Constant hot water or auto heat when kettle placed on base after filling.
D	3 to 7	Pour kettle before boiled	Kettle status indicator showing water temperature.
E	4 to 7	Pour kettle before boiled.	Kettle status indicator showing water temperature.
F	5 to 5	Fail to turn off boiling kettle.	Auto cut-off switch when kettle boiling.
G	5 to 7	Pour boiling water before turning kettle off.	Auto cut-off switch when kettle boiling.

What TAFEI does best is to enable the analysis to model the interaction between human action and system states. This can be used to identify potential errors and consider the task flow in a goal-oriented scenario. Potential conflicts and contradictions in task flow should come to light. For example, in a study of medical imaging equipment design, Baber and Stanton (1999) identified disruptions in task flow that made the device difficult to use. TAFEI enabled the design to be modified and led to the development of a better task flow. This process of analytical prototyping is key to the use of TAFEI in designing new systems. Obviously, TAFEI can also be used to evaluate existing systems. There is a potential problem that the number of states that a device can be in could overwhelm the analyst. Our experience suggests that there are two possible approaches. First, only analyse goal-oriented task scenarios – the process is pointless without a goal and HTA can help focus the analysis. Second, the analysis can be nested at various levels in the task hierarchy, revealing more and more detail. This can make each level of analysis relatively self-contained and not overwhelming. The final piece of advice is to start with a small project and build up from that position.

Example

The following example of TAFEI was used to analyse the task of programming a DVD recorder. The task analysis, SSDs and transition matrix are all presented. First of all, the task analysis is performed to describe human activity, as shown in Figure 6.6 (on the following page).

Next, the SSDs are drawn as shown in Figure 6.7 (on the following page).

From the TAFEI diagram, a transition matrix is compiled and each transition is scrutinised, as presented in Figure 6.8 (on the following page).

Thirteen of the transitions are defined as 'illegal'; these can be reduced to a subset of six basic error types:

- Switch DVD recorder off inadvertently.
- Insert DVD into machine when switched off.
- Programme without DVD inserted.
- Fail to select programme number.

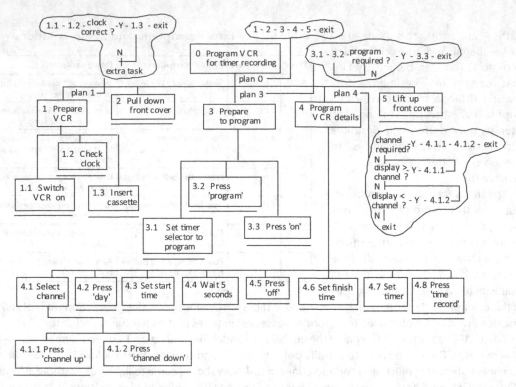

Figure 6.6   HTA of DVD recorder programming task

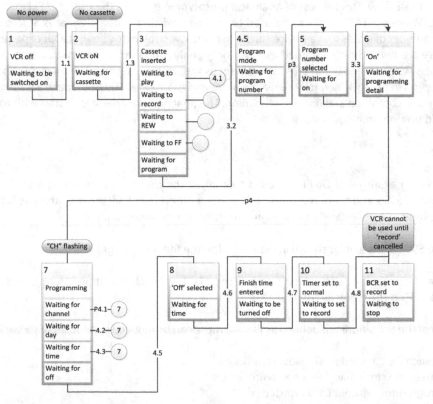

Figure 6.7   The TAFEI description

- Fail to wait for 'on' light.
- Fail to enter programming information.

In addition, one legal transition has been highlighted because it requires a recursive activity to be performed. These activities seem to be particularly prone to errors of omission. These predictions then serve as a basis for the designer to address the re-design of the DVD recorder. A number of illegal transitions could be dealt with fairly easily by considering the use of modes in the operation of the device, such as switching off the DVD recorder without stopping the DVD and pressing play without inserting the DVD.

	1	2	3	4.5	5	6	7	8	9	10	11
1		L	I	-	-	-	-	-	-	-	-
2	L	-	L	-	-	-	-	-	-	-	-
3	L	-	-	L	-	-	-	-	-	-	-
4.5	I	-	-	-	L	I	-	-	-	-	-
5	I	-	-	-	-	L	I	I	-	-	-
6	I	-	-	-	-	-	L	I	-	-	-
7	I	-	-	-	-	-	L	L	-	-	-
8	I	-	-	-	-	-	-	-	L	-	-
9	I	-	-	-	-	-	-	-	-	L	-
10	I	-	-	-	-	-	-	-	-	-	L
11	-	-	-	-	-	-	-	-	-	-	-

Figure 6.8   The transition matrix

Related Methods

TAFEI is related to HTA for a description of human activity. Like SHERPA, it is used to predict human error with artefacts. Kirwan and colleagues recommend that multiple HEI methods can be used to improve the predictive validity of the techniques. This is based on the premise that one method may identify an error that another one misses. Therefore, using SHERPA and TAFEI may be better than using either by itself. It has been found that multiple analysts similarly improve the performance of a method. This is based on the premise that one analyst may identify an error that another one misses. As such, using SHERPA or TAFEI with multiple analysts may perform better than one analyst with SHERPA or TAFEI.

Advantages

- TAFEI is a structured and thorough procedure.
- It has sound theoretical underpinning.
- It has a flexible, generic methodology.
- It can include error reduction proposals.
- It appears to be relatively simple to apply.
- TAFEI can be applied to a diverse range of systems (Baber and Stanton, 1994).

Disadvantages

- TAFEI is not a rapid technique, as HTA and SSD are prerequisites. Kirwan (1998a) suggested that it is a resource-intensive technique and that the transition matrix and SSDs may rapidly become unwieldy for even moderately complex systems.
- It requires some skill to perform effectively.
- It is limited to goal-directed behaviour.
- It may be difficult to learn and also time-consuming to train.
- It may also be difficult to acquire or construct the SSDs required for a TAFEI analysis. A recent study investigated the use of TAFEI for evaluating design-induced pilot error and found that SSDs do not exist for Boeing and Airbus aircraft.

Approximate Training and Application Times

Stanton and Young (1998, 1999a) report that observational techniques are relatively quick to train and apply. For example, in their study of radio-cassette machines, training in the TAFEI method took approximately three hours, while application of the method by recently trained people took approximately three hours in the radio-cassette study to predict the errors.

Reliability and Validity

There are some studies that report on the reliability and validity of TAFEI for both expert and novice analysts.

Table 6.8   Reliability and validity data for TAFEI

	Novices*	Experts**
Reliability	r = 0.67	r = 0.9
Validity	SI = 0.79	SI = 0.9

Note: * taken from Stanton and Baber (2002); ** taken from Baber and Stanton (1996a).

Tools Needed

TAFEI is a pen-and-paper-based tool. There is currently no software available to undertake TAFEI, although there are software packages to support HTA.

Flowchart

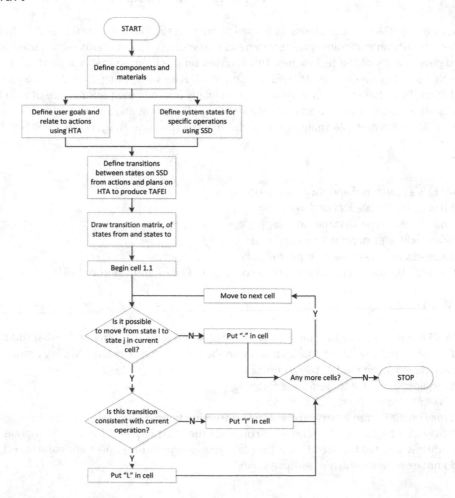

Human Error HAZOP

Background and Applications

The HAZOP technique was first developed by ICI in the late 1960s in order to investigate the safety or operability of a plant or operation (Swann and Preston, 1995) and has been used extensively in the nuclear power and chemical process industries. HAZOP (Kletz, 1974) is a well-established engineering approach that was developed for use in process design audit and engineering risk assessment (Kirwan, 1992a). Originally applied to engineering diagrams (Kirwan and Ainsworth, 1992), the HAZOP technique involves the analyst applying guidewords, such as 'Not done', 'More than' or 'Later than', to each step in a process in order to identify potential problems that may occur. When conducting a HAZOP-type analysis, a HAZOP team is assembled, usually consisting of operators, design staff, HF specialists and engineers. The HAZOP leader (who should be extensively experienced in HAZOP-type analyses) guides the team through an investigation of the system design using the HAZOP 'deviation' guidewords. The HAZOP team consider guidewords for each step in a process to identify what may go wrong. The guidewords are proposed and the leader then asks the team to consider the problem in the following fashion (Swann and Preston, 1995):

- Which section of the plant is being considered?
- What is the deviation and what does it mean?
- How can it happen and what is the cause of the deviation?
- If it cannot happen, move on to the next deviation.
- If it can happen, are there any significant consequences?
- If there are not, move on to the next guideword.
- If there are any consequences, what features are included in the plant to deal with these consequences?

If the HAZOP team believes that the consequences have not been adequately covered by the proposed design, then solutions and actions are considered.

Applying guidewords like this in a systematic way ensures that all of the possible deviations are considered. Typically, the efficiency of the actual HAZOP analysis is largely dependent upon the HAZOP team. There are a number of different variations of HAZOP-style approaches, such as CHAZOP (Swann and Preston, 1995) and SCHAZOP (Kennedy and Kirwan, 1998). An HEI-based approach emerged in the form of the Human Error HAZOP technique, which was developed for the analysis of human error issues (Kirwan and Ainsworth, 1992). In the development of another HEI tool (PHECA), Whalley (1988) also created a set of HF-based guidewords, which are more applicable to human error. These Human Error guidewords are presented in Table 6.9. The error guidewords are applied to each bottom-level task step in the HTA to determine any credible errors (i.e. those judged by the subject matter expert to be possible). Once the analyst has recorded a description of the error, the consequences, cause and recovery path of the error are also recorded. Finally, the analyst then identifies any design improvements that could potentially be used to remedy the error.

Dunjo et al. (2010) recently presented a review of HAZOP, describing its cross-domain application in fields such as medicine, road safety, steel works and liquid hydrogen filling stations to name but a few. Their review examined over 160 papers discussing HAZOP over a 35 year time period. They explore the numerous adaptations and variations of HAZOP, discussing the common combinations of HAZOP and failure modes effect analysis (FMEA), HAZOP and fault tree analysis, as well as the mixture of HAZOP, FMEA and layer of protection analysis (LOPA). The review identifies a plethora of work attempting to develop automated systems to aid in HAZOP analysis, including HAZOPEX (Heino et al., cited in Dunjo et al., 2010), OptHAZOP (Khan and Abbasi, cited in Dunjo et al., 2010) and HAZOPExpert (Vaidhyanathan, cited in Dunjo et al., 2010). Specifically focusing on the Human Error HAZOP, Dunjo et al. (2010) discuss a number of authors who have created variants of the human error guidewords, such as Aspinall, Baybutt,

Table 6.9   Human Error HAZOP guidewords

Not Done	Repeated
Less Than	Sooner Than
More Than	Later Than
As Well As	Misordered
Other Than	Part Of

Shurman and Fleger, Kennedy and Kirwan, and Kletz. Kennedy and Kirwan in particular are emphasised as they empirically validated their Human Error HAZOP through a comparison with management oversight and risk tree (MORT) and FMEA.

Domain of Application

HAZOP was originally developed for the nuclear power and chemical processing industries. However, it is feasible that it could be applied in any domain involving human activity.

Procedure and Advice (Human Error HAZOP)

Step 1: Assembly of the HAZOP Team
The most important part of any HAZOP analysis is assembling the correct HAZOP team (Swann and Preston, 1995). The HAZOP team needs to possess the right combination of skills and experience in order to make the analysis efficient. The HAZOP team leader should be experienced in HAZOP-type analysis so that the team can be guided effectively. For a Human Error HAZOP analysis of a nuclear petrochemical plant, it is recommended that the team be comprised of the following personnel.

- HAZOP team leader.
- HF specialist(s).
- Human Reliability Analysis (HRA)/HEI specialist.
- Project engineer.
- Process engineer.
- Operating team leader.
- Control room operator(s).
- Data recorder.

Step 2: Conduct an HTA
Next, an exhaustive description of the task and system under analysis should be created. There are a number of task analysis techniques that can be used for this purpose. It is recommended that an HTA of the task under analysis is conducted. The Human Error HAZOP technique works by indicating which of the errors from the HAZOP EEM taxonomy are credible at each bottom-level task step in an HTA of the task under analysis. A number of data collection techniques may be used in order to gather the information required for the HTA, such as interviews with SMEs and observations of the task under investigation.

Step 3: Guideword Consideration
The HAZOP team takes the first/next bottom-level task step from the HTA and considers each of the associated HAZOP guidewords for the task step under analysis. This involves discussing whether the guideword could have any effect on the task step or not and also what type of error would result. If any of the guidewords are deemed to be credible by the HAZOP team, then they move on to step 4.

Step 4: Error Description
For any credible guidewords, the HAZOP team should provide a description of the form that the resultant error would take, such as 'operator fails to check current steam pressure setting'. The error description should be clear and concise.

Step 5: Consequence Analysis
Once the HAZOP team have described the potential error, its consequence should be determined. The consequence of the error should be described clearly, such as 'operator fails to comprehend high steam pressure setting'.

Step 6: Cause Analysis
Next, the HAZOP team should determine the cause(s) of the potential error. The cause analysis is crucial to the remedy or error reduction part of the HAZOP analysis. Any causes associated with the identified error should be described clearly.

Step 7: Recovery Path Analysis
In the recovery path analysis, any recovery paths that the operator might potentially take after the described error has occurred to avoid the associated consequences are recorded. The recovery path for an error will typically be another task step in the HTA or a description of a recovery step.

Step 8: Error Remedy
Finally, the HAZOP team propose any design or operational remedies that could be implemented in order to reduce the chances of the error occurring. This is based upon subjective analyst judgment and domain expertise.

Advantages

- A correctly conducted HAZOP analysis has the potential to highlight all of the possible errors that could occur in the system.
- It has been used emphatically in many domains. HAZOP-style techniques have received wide acceptance by both the process industries and the regulatory authorities (Andrews and Moss, 1993).
- Since a team of experts is used, the technique should be more accurate and comprehensive than other 'single analyst' techniques. Using a team of analysts should ensure that no potential errors are missed and should also remove the occurrence of non-credible errors.
- It is easy to learn and use.
- Whalley's (1988) guidewords are generic, allowing the technique to be applied to a number of different domains.
- It utilises a hybrid blend of deductive and inductive approaches (Hoepffner, cited in Dunjo et al., 2010).
- It is a structured and systematic method (Dunjo et al., 2010).

Disadvantages

- HAZOP can be extremely time-consuming in its application. Typical HAZOP analyses can take up to several weeks to be completed.
- It requires a mixed team made up of operators, HF specialists, designers, engineers, etc. Building such a team and ensuring that they can all be brought together at the same time is often a difficult task.
- It generates huge amounts of information that has to be recorded and analysed.
- It is laborious.
- Disagreement and personality clashes within the HAZOP team may be a problem.
- The guidewords used are either limited or specific to the nuclear petrochemical industry.
- The Human Error HAZOP guidewords lack comprehensiveness (Salmon et al., 2002).
- It focuses on physical causes and 'ignores' management and organisational descriptions/factors (Soukas and Rouhiainen, cited in Dunjo et al., 2010).

- Not all of the HAZOP questions are relevant in all situations, which can waste time (Grossman and Fromm, cited in Dunjo et al., 2010).
- It is dependent on subjective judgment and so is affected by analyst bias and impacted by personal factors such as experience (Dunjo et al., 2010).
- Dunjo et al. (2010) argue that the integration of the human error aspect of HAZOP is still incomplete.

Flowchart

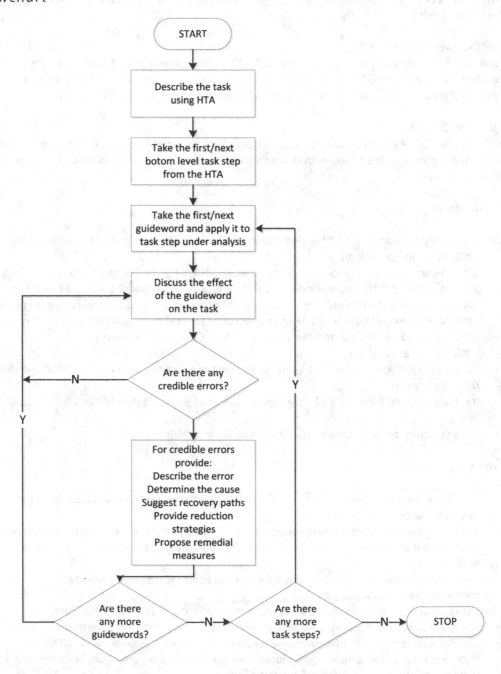

Example

A Human Error HAZOP analysis was conducted for the flight task 'Land aircraft X at New Orleans Airport using the autoland system' (Marshall et al., 2003). An extract of the HTA for the flight task is presented in Figure 6.9, while an extract of the Human Error HAZOP analysis for the flight task is presented in Table 6.10.

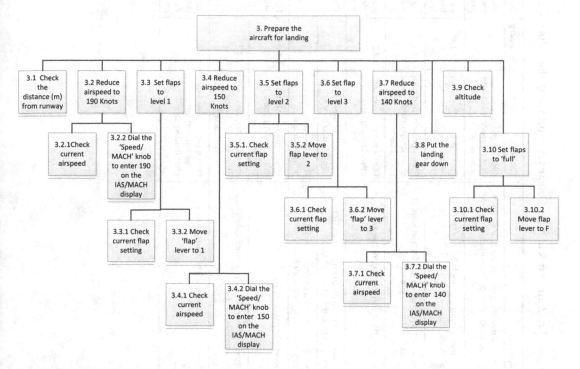

Figure 6.9   Extract of HTA of task 'Land aircraft X at New Orleans Airport using the autoland system'

Related Methods

A number of variations of the HAZOP technique exist, such as Human Error HAZOP (Kirwan and Ainsworth, 1992), CHAZOP (Computer HAZOP: Swann and Preston, 1995) and SCHAZOP (Safety Culture HAZOP: Kennedy and Kirwan, 1998). HAZOP-type analyses are typically conducted on an HTA of the task under analysis. Engineering diagrams, flow-sheets, operating instructions and plant layouts are also typically required (Kirwan and Ainsworth, 1992). Human Error HAZOP is a taxonomy-based HEI technique, of which there are many, including SHERPA, CREAM and HET.

Approximate Training and Application Times

Whilst the HAZOP technique appears to be one in which it is quick to train users, Swann and Preston (1995) report that studies on the duration of the HAZOP analysis process have been conducted, with the conclusion that a thorough HAZOP analysis carried out correctly would take over five years for a typical processing plant. This is clearly a worst-case scenario and is impractical. More realistically, Swann and Preston report that ICI benchmarking shows that a typical HAZOP analysis would require about 40 meetings lasting approximately three hours each.

Table 6.10   Extract of Human Error HAZOP analysis of the task 'Land aircraft X at New Orleans Airport using the autoland system'

Task step	Guide-word	Error	Consequence	Cause	Recovery	Design improvements
3.1 Check the distance from runway.	Later than.	Pilot checks the distance from the runway later than he should.	Plane may be travelling too fast for that stage of the approach and also may have the wrong level of flap.	Pilot inadequacy. Pilot is preoccupied with another landing task.	3.9	Auditory distance countdown inside 25N miles
3.2.1 Check current airspeed.	Not done.	Pilot fails to check current airspeed.	Pilot changes airspeed wrongly, i.e. may actually increase airspeed.	Pilot is preoccupied with other landing tasks.	3.4.1	Auditory speed updates. bigger, more apparent speedo
	Misordered.	Pilot checks the current airspeed after he has altered the flaps.	Plane may be travelling too fast for that level of flap or that leg of the approach.	Pilot is preoccupied with other landing tasks, pilot inadequacy.	3.4.1	Design flaps so that each level can only be set within certain speed level windows
3.2.2 Dial the speed/ mach knob to enter 190.	Not done.	Pilot fails to enter new airspeed.	Plane may be travelling too fast for the approach.	Pilot is preoccupied with other landing tasks.	3.4.2	Auditory reminder that the plane is travelling too fast, e.g. overspeed display
	Less than.	Pilot does not turn the speed/ MACH knob enough.	Plane's speed is not reduced enough and it may be travelling too fast for the approach.	Poor control design, pilot inadequacy.	3.4.2	One full turn for 1 knot. Improved control feedback
	More than.	Pilot turns the speed/MACH knob too much.	Plane's speed is reduced too much and so it is travelling too slow for the approach.	Poor control design, pilot inadequacy.	3.4.2	Improved control feedback
	Sooner than.	Pilot reduces the planes speed too early.	Plane slows down too early.	Pilot is preoccupied with other landing tasks, pilot inadequacy.	3.4.2	Plane is travelling too slow auditory warning
	Other than.	Pilot reduces the planes using the wrong knob e.g. HDG knob.	Plane does not slow down to desired speed and takes on a heading of 190.	Pilot is preoccupied with other landing tasks, pilot inadequacy.	3.4.2	Clearer labelling of controls. Overspeed auditory warning
3.3.1 Check the current flap setting.	Not done.	Pilot fails to check the current flap setting.	Pilot does not comprehend the current flap setting.	Pilot is preoccupied with other landing tasks, pilot inadequacy.	3.4.2	Bigger/improved flap display/control. Auditory flap setting reminders

Reliability and Validity

The HAZOP-type approach has been used emphatically over the last four decades in process control environments. However Kennedy reports that it has not been subjected to rigorous academic scrutiny (Kennedy and Kirwan, 1998). Stanton et al. (2006a) reported that in a comparison of four HEI methods (HET, Human Error HAZOP, HEIST and SHERPA) when used to predict potential design-induced pilot error, subjects using the Human Error HAZOP method achieved acceptable sensitivity in their error predictions (mean sensitivity index 0.62). Furthermore, only those subjects using the HET methodology performed better.

Tools Needed

HAZOP analyses can be carried out using pen and paper. Engineering diagrams are also normally required. The EEM taxonomy is also required for the Human Error HAZOP variation, as is an HTA for the task under analysis.

The Technique for Human Error Assessment (THEA)

Background and Applications

THEA (Pocock et al., 2001b) was developed to aid designers and engineers in the identification of potential user interaction problems in the early stages of interface design. The impetus for the development of THEA was the requirement for a HEI tool that could be used effectively and easily by non-HF specialists. Although THEA has its roots firmly in HRA methodology, it is suggested by the authors that the technique contains a high level of prompts for the analyst and is also much easier to apply than typical HRA methods (Pocock et al., 2001a). The technique itself is a structured approach to HEI and is based upon Norman's model of action execution (Norman, 1988). Like HEIST (Kirwan, 1994), THEA uses a series of questions in a checklist-style approach based upon goals, plans, performing actions and perception/evaluation/interpretation. THEA also utilises a scenario-based analysis, whereby the analyst exhaustively describes the scenario under investigation before any error analysis is performed.

Domain of Application

Generic.

Procedure and Advice

Step 1: System Description
Initially, a THEA analysis requires a formal description of the system and task or scenario under investigation. Pocock et al. (2001a: 9) state that this should include details regarding the specification of the 'system's functionality and interface, and how it interacts with other systems'.

Step 2: Scenario Description
Next, the analyst should provide a description of the type of scenario under investigation. The authors have developed a scenario template that assists the analyst in developing the scenario description. The scenario description is conducted in order to give the analyst a thorough description of the scenario under investigation, including information such as actions and any contextual factors which may provide error potential. The scenario description template is presented in Table 6.11.

Table 6.11   A template for describing scenarios

AGENTS
The human agents involved and their organisations.
The roles played by the humans, together with their goals and responsibilities.
RATIONALE
Why is this scenario an interesting or useful one to have picked?
SITUATION AND ENVIRONMENT
The physical situation in which the scenario takes place.
External and environmental triggers, problems and events that occur in this scenario.
TASK CONTEXT
What tasks are performed?
Which procedures exist and will they be followed as prescribed?
SYSTEM CONTEXT
What devices and technology are involved?
What usability problems might participants have?
What effects can users have?
ACTION
How are the tasks carried out in context?
How do the activities overlap?
Which goals do actions correspond to?
EXCEPTIONAL CIRCUMSTANCES
How might the scenario evolve differently, either as a result of uncertainty in the environment or because of variations in agents, situation, design options, system and task context?
ASSUMPTIONS
What, if any, assumptions have been made that will affect this scenario?

Source: Pocock et al., 2001a.

Step 3: Task Description
A description of the tasks that the operator or user would perform in the scenario is also required. This should describe goals, plans and intended actions. It is recommended that an HTA of the task under analysis is conducted for this purpose.

Step 4: Goal Decomposition
The HTA developed for step 3 of the THEA analysis should be used for step 4, which involves breaking down the task goals into operations.

Step 5: Error Analysis
Next, the analyst has to identify and explain any human error that may arise during task performance. THEA provides a structured questionnaire or checklist-style approach in order to aid the analyst in identifying any possible errors. The analyst simply asks questions (from THEA) about the scenario under analysis in order to identify potentially errors. For any credible errors, the analyst should record the error, its causes and its consequences. Then questions are normally asked about each goal or task in the HTA, or, alternatively, the analyst can select parts of the HTA where problems are anticipated. The THEA error analysis questions are comprised of the following four categories:

- goals;
- plans;
- performing actions; and
- perception, interpretation and evaluation.

Examples of the THEA error analysis questions for each of the four categories are presented in Table 6.12 on the next page.

Step 6: Design Implications/Recommendations
Once the analyst has identified any potential errors, the final step of the THEA analysis is to offer any design remedies for each error identified. This is based primarily upon the analyst's subjective judgment. However, the design issues section of the THEA questions also prompts the analyst for design remedies.

Advantages

- THEA offers a structured approach to HEI.
- It is easy to learn and use, and can be used by non-HF professionals.
- As it is recommended that It be used very early in the system life-cycle, potential interface problems can be identified and eradicated very early in the design process.
- Its error prompt questions have a solid theoretical underpinning (Norman's action execution model).
- It contains a high level of prompts to guide the analyst, making it easier to apply than typical HRA methods (Pocock et al., 2001a).

Table 6.12   Example THEA error analysis questions

Questions	Consequences	Design issues
Goals		
G1 – Are items triggered by stimuli in the interface, environment or task?	If not, goals (and the tasks that achieve them) may be lost, forgotten or not activated, resulting in omission errors.	Are triggers clear and meaningful? Does the user need to remember all of the goals?
G2 – Does the user interface 'evoke' or 'suggest' goals?	If not, goals may not be activated, resulting in omission errors. If the interface does 'suggest' goals, they may not always be the right ones, resulting in the wrong goal being addressed.	For example, graphical display of flight plan shows pre-determined goals as well as current progress.
Plans		
P1 – Can actions be selected *in situ* or is pre-planning required?	If the correct action can only be taken by planning in advance, then the cognitive work may be harder. However, when possible, planning ahead often leads to less error-prone behaviour and fewer blind alleys.	
P2 – Are there well practised and pre-determined plans?	If a plan isn't well known or practised, then it may be prone to being forgotten or remembered incorrectly. If plans are not pre-determined and must be constructed by the user, their success depends heavily on the user possessing enough knowledge about their goals and the interface to construct a plan. If pre-determined plans do exist and are familiar, then they might be followed inappropriately, not taking account of the peculiarities of the current context.	
Performing actions		
A1 – Is there physical or mental difficulty in executing the actions?	Difficult, complex or fiddly actions are prone to being carried out incorrectly.	
A2 – Are some actions made unavailable at certain times?		
Perception, interpretation and evaluation.		
I1 – Are changes in the system resulting from user action clearly perceivable?	If there is no feedback that an action has been taken, the user may repeat actions, with potentially undesirable effects.	
I2 – Are the effects of user actions perceivable immediately?	If feedback is delayed, the user may become confused about the system state, potentially leading to a supplemental (perhaps inappropriate) action being taken.	

Source: Pocock et al., 2001a.

- Each error question has associated consequences and design issues to aid the analyst.
- It appears to be a generic technique, allowing it to be applied in any domain.
- It enables insights into user difficulties with interface design in a more cost-effective way than user studies (Maxiom and Reeder, 2005).

Disadvantages

- Although error questions prompt the analyst for potential errors, THEA does not use any error modes and so the analyst may be unclear on the types of errors that may occur. However, HEIST (Kirwan, 1994) uses error prompt questions linked with an error mode taxonomy, which seems to be a much sounder approach.
- It is very resource-intensive, particularly with respect to time taken to complete an analysis.
- Error consequences and design issues provided by THEA are generic and limited.
- At the moment, there appears to be no validation evidence associated with it.
- HTA, task decomposition and scenario description create additional work for the analyst.

- For a technique that is supposed to be usable by non-HF professionals, the terminology used in the error analysis questions section is confusing and hard to decipher. This could cause problems for non-HF professionals.

Flowchart

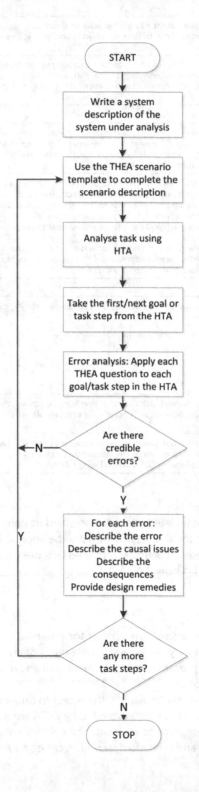

Example

The following example (Table 6.13, Figure 6.10 and Table 6.14) is a THEA analysis of a video recorder programming task (Pocock et al., 2001).

Table 6.13   Scenario details

SCENARIO NAME: Programming a video recorder to make a weekly recording.
ROOT GOAL: Record a weekly TV programme.
SCENARIO SUB-GOAL: Setting the recording date.
ANALYST NAME(S) AND DATE:
AGENTS: A single user interfacing with a domestic video cassette recorder (VCR) via a remote control unit (RCU).
RATIONALE: The goal of programming this particular VCR is quite challenging. Successful programming is not certain.
SITUATION AND ENVIRONMENT: A domestic user wishes to make a recording of a television programme that is broadcast on a particular channel at the same time each week. The user is not very technologically aware and has not programmed this VCR previously. A reference handbook is not available, but there is no time pressure to set the machine – recording is not due to commence until tomorrow.
TASK CONTEXT: The user must perform the correct tasks to set the VCR to record a television programme on three consecutive Monday evenings from 6pm to 7pm on Channel 3. Today is Sunday.
SYSTEM CONTEXT: The user has a RCU containing navigation keys used in conjunction with programming the VCR as well as normal VCR playback operation. The RCU has four scrolling buttons, indicating left, right, up and down. Other buttons relevant to programming are labelled OK and I.
ACTIONS: The user is required to enter a recording date into the VCR via the RCU using the buttons listed above. The actions appear in the order specified by the task decomposition.
EXCEPTIONAL CIRCUMSTANCES: None.
ASSUMPTIONS: None.

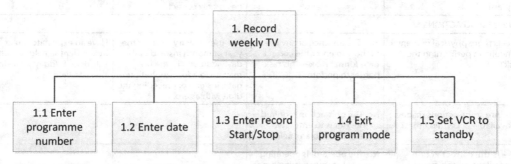

Figure 6.10   Video recorder HTA (adapted from Pocock et al., 2001a)

Table 6.14   Error Analysis Questionnaire

SCENARIO NAME: Programming a video recorder to make a weekly recording			
TASK BEING ANALYSED: Setting the recording date			
ANALYST NAME(S) AND DATE:			
QUESTION	**CAUSAL ISSUES**	**CONSEQUENCES**	**DESIGN ISSUES**
GOALS, TRIGGERING, INITIATION			
G1 – Is the task triggered by stimuli in the interface, environment or the task itself?	Yes. The presence of an 'enter date' prompt is likely to trigger the user to input the date at this point.		
G2 – Does the UI 'evoke' or 'suggest' goals?	N/A. The UI does not per se, strictly evoke or suggest the goal of entering the date.		

Table 6.14   Continued

QUESTION	CAUSAL ISSUES	CONSEQUENCES	DESIGN ISSUES
GOALS, TRIGGERING, INITIATION (Continued)			
G3 – Do goals come into conflict?	There are no discernible goal conflicts.		
G4 – Can the goal be satisfied without all its sub-goals being achieved?	NO. The associated sub-goal on this page of setting the DAILY/WEEKLY function may be overlooked. Once the date is entered, pressing the right cursor key on the RCU will enter the next 'ENTER HOUR' setting.	Failure to set the DAILY/ WEEKLY option. Once the ENTER HOUR screen is entered, the DAILY/WEEKLY option is no longer available.	Suggest addition of an interlock so that the DAILY/ WEEKLY option cannot be bypassed.
PLANS			
P1 – Can actions be selected *in situ* or is pre-planning required?	True. Entering the date can be done 'on-the-fly'. No planning is required.		
P2 – Are there well-practised and pre-determined plans?	N/A. A pre-determined plan, as such, does not exist, but the user should possess enough knowledge to know what to do at this stage.		
P3 – Are there plans or actions that are similar? Are some used more often than others?	There are no similar or more frequently used plans or actions associated with this task.		
P4 – Is there feedback to allow the user to determine that the task is proceeding successfully towards the goal and according to plan?	Yes. As the user enters digits into the date field via the RCU, they are echoed back on screen.	Task is proceeding satisfactorily towards the goal of setting the date, although the date being entered is not necessarily correct.	See A1.
PERFORMING ACTIONS			
A1 – Is there physical or mental difficulty in performing the task?	Yes. The absence of any cues for how to enter the correct date format makes this task harder to perform.	The user may try to enter the year or month instead of the day. Additionally, the user may try to add a single-figure date, instead of preceding the digit with a zero.	Have an explanatory text box under the field or, better still, default today's date in the date field.
A2 – Are some actions made unavailable at certain times?	No. The only action required of the user is to enter two digits into the blank field.		
A3 – Is the correct action dependent on the current mode?	No. The operator is operating in a single programming mode		
A4 – Are additional actions required to make the right controls and information available at the right time?	Yes. The date field is presented blank. If the user does not know the date for recording (or today's date), the user must know to press the 'down' cursor key on the RCU to make today's date visible.	The user may be unable to enter the date or the date must be obtained from an external source. Also, if the user presses either the left or right cursor key, the 'enter date' screen is exited.	Default current date into field. Prevent user from exiting 'enter date' screen before an entry is made (e.g. software lock-in).
PERCEPTION, INTERPRETATION AND EVALUATION			
I1 – Are changes to the system resulting from user action clearly perceivable?	Yes. Via on-screen changes to the date field.		
I2 – Are effects of such user actions perceivable immediately?	Yes. Digit echoing of RCU key presses is immediate.		

Table 6.14   Continued

QUESTION	CAUSAL ISSUES	CONSEQUENCES	DESIGN ISSUES
PERCEPTION, INTERPRETATION AND EVALUATION (Continued)			
I3 – Are changes to the system resulting from autonomous system actions clearly perceivable?	N/A. The VCR performs no autonomous actions.		
I4 – Are the effects of such autonomous system actions perceivable immediately?	N/A		
I5 – Does the task involve monitoring, vigilance or spells of continuous attention?	No. There is no monitoring or continuous attention requirements on the user.		
I6 – Can the user determine relevant information about the state of the system from the total information provided?	No. The user cannot determine the current date without knowing about the 'down' cursor key. Also, if the date of recording is known, the user may not know about the need to enter two digits.	If the user doesn't know today's date and only knows that, say, Wednesday, is when he or she wants the recordings to commence, then the user is stuck.	See A1.
I7 – Is complex reasoning, calculation or decision-making involved?	No.		
I8 – If the user is interfacing with a moded system, is the correct interpretation dependent on the current mode?	N/A	It is not considered likely that the date field will be confused with another entry field, e.g. hour.	

Source: Pocock et al., 2001a.

Related Methods

THEA is one of a number of HEI techniques. It is very similar to HEIST (Kirwan, 1994) in that it uses error prompt questions to aid the analysis. A THEA analysis should be conducted on an initial HTA of the task under analysis. THEA can be used with numerous other HF techniques – for example, Maxiom and Reeder (2005) applied THEA alongside VPA and HTA, using HTA to develop a description of correct actions on two security-sensitive interfaces. Comparison with these HTAs enabled identification of user variation. This variation was then classified using THEA to uncover the cognitive basis of these erroneous actions.

Approximate Training and Application Times

Although no training and application time is offered in the literature, it is apparent that the amount of training time would be minimal. The application time, however, would be high, especially for large, complex tasks.

Reliability and Validity

No data regarding reliability and validity is offered by the authors.

Tools Needed

To conduct a THEA analysis, pen and paper is required. The analyst would also require functional diagrams of the system/interface under analysis and the THEA error analysis questions.

Human Error Identification in Systems Tool (HEIST)

Background and Applications

HEIST (Kirwan, 1994) is based upon a series of tables containing questions or 'error identifier prompts' surrounding EEMs, PSFs and PEMs. When using HEIST, the analyst identifies errors through applying the error identifier prompt questions to all of the tasks involved in the task or scenario under analysis. The questions link EEMs (type of error) to relevant PSFs. All EEMs are then linked to PEMs. The method comprises eight HEIST tables, each containing a series of pre-defined error-identifier questions linked to EEMs, associated causes (system cause or PEM) and error reduction guidelines. The HEIST tables and questions are based upon the Skill, Rule and Knowledge (SRK) framework (Rasmussen at al., 1981), i.e. activation/detection, observation/data collection, identification of system state, interpretation, evaluation, goal selection/task definition, procedure selection and procedure execution. These error prompt questions are designed to prompt the analyst for potential errors. All of the error identifying prompts are PSF-based questions which are coded to indicate one of six PSFs. These PSFs are Time (T), Interface (I), Training/Experience (E), Procedures (P), Task Organisation (O) and Task Complexity (C). The analyst classifies the task step under analysis into one of the HEIST behaviours and then applies the associated error prompts to the task step and determines whether any of the proposed errors are credible or not. For each credible error, the analyst then records the system cause or PEM and error reduction guidelines (both of which are provided in the HEIST tables) and also the error consequence. Although it can be used as a stand-alone method, HEIST is also used as part of the HERA 'toolkit' methodology (Kirwan, 1998b) as a back-up check for any of the errors identified.

Domain of Application

The nuclear power and chemical process industries. However, it is feasible that the HEIST technique can be applied in any domain.

Procedure and Advice

Step 1: Conduct an HTA
The HEIST procedure begins with the development of an HTA of the task or scenario under analysis. A number of data collection techniques may be used in order to gather the information required for the HTA, such as interviews with SMEs and observations of the task under analysis.

Step 2: Task Step Classification
The analyst takes the first task step from the HTA and classifies it into one or more of the eight HEIST behaviours listed above. For example, the task step 'Pilot dials in airspeed of 190 using the speed/Mach selector knob' would be classified as procedure execution. This part of the HEIST analysis is based entirely upon the subjective judgment of the analyst.

Step 3: Error Analysis
Next, the analyst takes the appropriate HEIST table and applies each of the error identifier prompts to the task step under analysis. Based upon his or her subjective judgment, the analyst should determine whether or not any of the associated errors could occur during the task step under analysis. If the analyst deems an error to be credible, then the error should be described and the EEM, system cause and PEM should be determined from the HEIST table.

Step 4: Error Reduction Analysis

For each credible error, the analyst should select the appropriate error reduction guidelines from the HEIST table. Each HEIST error prompt has an associated set of error reduction guidelines. Whilst it is recommended that the analyst should use these, it is also possible for analysts to propose their own design remedies based upon domain knowledge.

Advantages

- As HEIST uses error identifier prompts, the technique has the potential to be very exhaustive.
- Error identifier prompts aid the analyst in error identification.
- Once a credible error has been identified, the HEIST tables provide the EEMs, PEMs and error reduction guidelines.
- The technique is easy to use and learn, and requires only minimal training.
- It offers a structured approach to error identification.
- It considers PSFs and PEMs.

Disadvantages

- The use of error identifier prompts means that HEIST is time-consuming in its application.
- The need for an initial HTA creates further work for HEIST analysts.
- Although the HEIST tables provide error reduction guidelines, these are generic and do not offer specific design remedies, such as ergonomic design of equipment and good system feedback.
- A HEIST analysis requires HF/psychology professionals.
- No validation evidence is available for HEIST.
- There is only limited evidence of HEIST applications in the literature.
- Many of the error identifier prompts used by HEIST are repetitive.
- Salmon et al. (2002) reported that HEIST performed poorly when used to predict potential design-induced error on the flight task 'Land aircraft at New Orleans Airport using the autoland system'. Out of the four techniques (HET, SHERPA, Human Error HAZOP and HEIST), subjects using HEIST achieved the lowest error prediction accuracy.

Example

A HEIST analysis was conducted on the flight task 'Land aircraft at New Orleans Airport using the autoland system' in order to investigate the potential use of the HEIST approach for predicting design-induced pilot error on civil flight decks (Salmon et al., 2002, 2003b). An extract of the HEIST analysis is presented in Table 6.15 below, while an extract of the HTA for the flight task is presented in Figure 6.11.

Table 6.15   Extract of HEIST analysis of the task 'Land aircraft X at New Orleans Airport using the autoland system' (Salmon et al., 2003b)

Task step	Error code	EEM	Description	PEM System cause	Consequence	Error reduction guidelines
3.2.2	PEP3	Action on wrong object.	Pilot alters the airspeed using the wrong knob, e.g. heading knob.	Topographic misorientation. Mistakes alternatives. Similarity matching.	The airspeed is not altered and the heading will change to the value entered.	Ergonomic design of controls and displays. Training. Clear labelling.
3.2.2	PEP4	Wrong action.	Pilot enters the wrong airspeed.	Similarity matching. Recognition failure. Stereotype takeover. Misperception. Intrusion.	Airspeed will change to the wrong airspeed.	Training. Ergonomic procedures with checking facilities. Prompt system feedback.

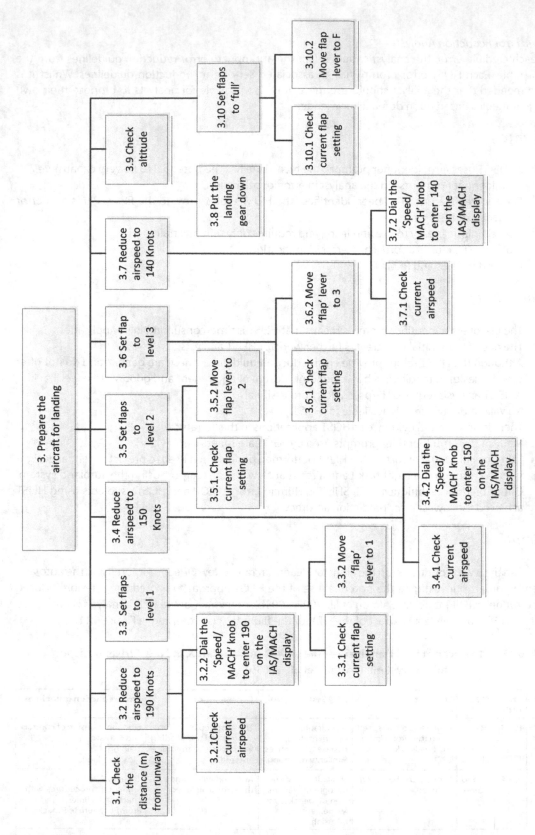

Figure 6.11   Extract of HTA 'Land aircraft X at New Orleans Airport using the autoland system' (Marshall et al., 2003)

Flowchart

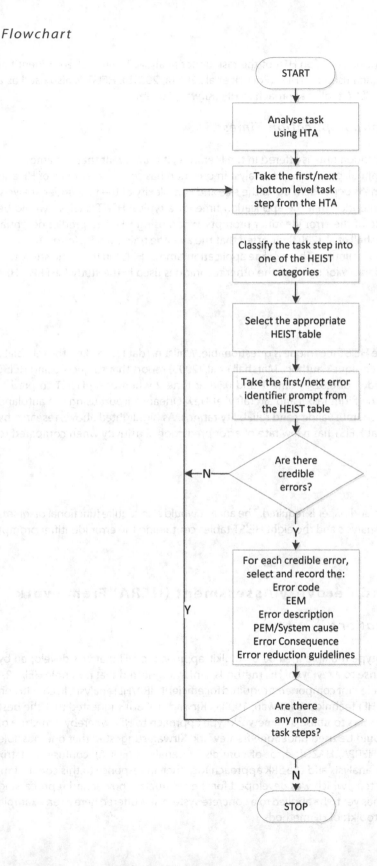

Related Methods

A HEIST analysis is typically conducted on an HTA of the task under analysis. The use of error identifier prompts is similar to the approach used by THEA (Pocock et al., 2001a, 2001b). HEIST is also used as a back-up check when using the HERA toolkit approach to HEI (Kirwan, 1998b).

Approximate Training and Application Times

Although no training and application time is offered in the literature, it is apparent that the amount of training required would be minimal, providing the analyst in question has some experience of HF and psychology. The application time is dependent upon the size and complexity of the task under analysis. However, it is generally recommended that the application time for a typical HEIST analysis would be medium to high, due to the use of the error identifier prompts. When using HEIST to predict potential design-induced pilot error, Marshall et al. (2003) reported that the average training time for participants using the HEIST technique was 90 minutes. The average application time of HEIST in the same study was 110 minutes, which was considerably longer than the other techniques used in the study (SHERPA, HET and Human Error HAZOP).

Reliability and Validity

The reliability and validity of the HEIST technique is questionable. Whilst no data regarding the reliability and validity is offered by the technique's authors, Marshall et al. (2003) report that subjects using HEIST achieved a mean sensitivity index of 0.62 at time 1 and 0.58 at time 2 when using HEIST to predict design-induced pilot error on the flight task 'Land aircraft X at New Orleans Airport using the autoland system'. This represents only moderate validity and reliability ratings. As highlighted above, research by Salmon et al. (2003b) found that HEIST has a low rate of error prediction sensitivity when compared to other methods.

Tools Needed

To conduct a HEIST analysis, pen and paper is required. The analyst would also require functional diagrams of the system/interface under analysis and the eight HEIST tables containing the error identifier prompt questions.

The Human Error and Recovery Assessment (HERA) Framework

Background and Applications

The HERA framework is a prototype multiple method or 'toolkit' approach to HEI that was developed by Kirwan (1998a, 1998b) in response to a review of HEI methods, which suggested that no single HEI/HRA technique possessed all of the relevant components required for efficient HRA/HEI analysis. In conclusion to a review of 38 existing HRA/HEI techniques (Kirwan, 1998a), Kirwan (1998b) suggested that the best approach would be for practitioners to utilise a framework-type approach to HEI, whereby a mixture of independent HRA/HEI tools would be used under one framework. Kirwan suggested that one possible framework would be to use SHERPA, HAZOP, error of commission analysis (EOCA), confusion matrix analyses, fault symptom matrix analysis and the SRK approach together. In response to this conclusion, Kirwan proposed the HERA system, which was developed for the UK nuclear power and reprocessing industry. Whilst the technique has yet to be applied to a concrete system, it is offered here as an example of an integrated framework or toolkit of HF methods.

Domain of Application

The nuclear power and chemical process industries.

Procedure and Advice

Step 1: Critical Task Identification
Before a HERA analysis is undertaken, the HERA team should determine the depth of analysis required and also which tasks are to be analysed. Kirwan (1998b) suggests that the following factors should be taken into account: the nature of the plant being assessed and the cost of failure, the criticality of human operator roles in the plant, the novelty of the plant's design, the system life-cycle, the extent to which the analysis is PSA-driven and the resources available for the analysis. A new plant that is classed as highly hazardous with critical operator roles would require an exhaustive HERA analysis, whilst an older plant that has no previous accident record and in which operators only take minor roles would require a scaled-down, less exhaustive analysis. Once the depth of the analysis is determined, the HERA assessment team must then determine which operational stages are to be the focus of the analysis, e.g. normal operation, abnormal operation and emergency operation.

Step 2: Task Analysis
Once the scope of the analysis is determined and the scenarios under analysis are defined, the next stage of the HERA analysis is to describe the tasks or scenarios under analysis. It is recommended that task analysis is used for this purpose. According to Kirwan (1998b), two forms of task analysis are used during the HERA process. These are initial task analysis (Kirwan, 1994) and HTA (Annett et al., 1971; Kirwan and Ainsworth, 1992; Shepherd, 1989). Initial task analysis involves describing the scenario under analysis, including the following key aspects:

- the scenario starting condition;
- the goal of the task;
- the number and type of tasks involved;
- the time available;
- the personnel available;
- any adverse conditions;
- the availability of equipment;
- the availability of written procedures;
- training; and
- the frequency and severity of the event.

Once the initial task analysis is completed, HTAs for the scenarios under analysis should be developed. A number of data collection techniques may be used in order to gather the information required for the HTAs, such as interviews with SMEs and observations of the scenario(s) under analysis

Step 3: Error Analysis
The error analysis part of the HERA framework comprises nine overlapping error identification modules. A brief description of these is presented in the paragraphs below.

Mission analysis   The mission analysis part of the HERA analysis involves determining the scope for failure that exists for the task or scenario under analysis. The mission analysis module uses the following questions to identify the scope for failure:

- Could the task fail to be achieved in time?

- Could the task be omitted entirely?
- Could the wrong task be carried out?
- Could only part of the task be carried out unsuccessfully?
- Could the task be prevented or hampered by a latent or coincident failure?

For the HERA analysis to proceed further, one of the answers to the mission analysis questions must be yes.

Operations-level analysis   The operations-level analysis involves the identification of the mode of failure for the task or scenario under analysis.

Goals analysis   Goals analysis involves focusing on the goals identified in the HTA and determining if any goal-related errors can occur. To do this, the HERA team use 12 goal analysis questions designed to highlight any potential goal errors. An example of a goals analysis question used in HERA is 'Could the operators have no goal, e.g. due to a flood of conflicting information; the sudden onset of an unanticipated situation; a rapidly evolving and worsening situation; or due to a disagreement or other decision-making failure to develop a goal?'. The goal error taxonomy used in the HERA analysis is presented below:

- no goal;
- wrong goal;
- outside procedures;
- goal conflict;
- goal delayed;
- too many goals;
- goal inadequate.

Plans analysis   Plans analysis involves focusing on the plans identified in the HTA to determine whether any plan-related errors could occur. The HERA team uses 12 plans analysis questions to identify any potential 'plan errors'. HERA plans analysis questions include 'Could the operators fail to derive a plan, due to workload, or decision-making failure?' or 'Could the plan not be understood or communicated to all parties?'. The HERA plan error taxonomy is presented below:

- no plan;
- wrong plan;
- incomplete plan;
- plan communication failure;
- plan co-ordination failure;
- plan initiation failure;
- plan execution failure;
- plan sequence error;
- inadequate plan;
- plan termination failure.

Error analysis   The HERA approach employs an EEM taxonomy derived from the SHERPA (Embrey, 1986) and the Technique for Human Error Rate Prediction (THERP: Swain and Guttman, 1983) HEI approaches. This EEM taxonomy is used to identify potential errors that may occur during the task or scenario under analysis. This involves applying the EEMs to each bottom-level task step in the HTA. Any credible errors are identified based upon the subjective judgment of the analyst. The HERA EEM taxonomy is presented in Table 6.16.

PSF analysis   The HERA approach also considers the effect of PSFs on potential error. Explicit questions regarding environmental influences on performance are applied to each of the task steps in the HTA. This allows the HERA team to identify any errors that might be caused by situational or environmental factors. The HERA approach uses the following PSF categories: time, interface, training and experience, procedures, organisation, stress and complexity. Each PSF question has an EEM associated with it. Examples of HERA PSF questions from each category are provided below:

- Time: is there more than enough time available (Too late)?
- Interface: is the onset of the scenario clearly alarmed or cued, and is this alarm or cue compelling (Omission or detection failure)?
- Training and experience: have operators been trained to deal with this task in the past 12 months (Omission, too late, too early)?

Table 6.16   HERA EEM taxonomy

Omission.	Action too little.
Omits entire task step.	Action in the wrong direction.
Omits step in the task.	Misalignment error.
Timing.	Other quality or precision error.
Action too late.	Selection error.
Action too early.	Right action on wrong object.
Accidental timing with other event.	Wrong action on right object.
Action too short.	Wrong action on wrong object.
Action too long.	Substitution error.
Sequence.	Information transmission error.
Action in the wrong sequence.	Information not communicated.
Action repeated.	Wrong information communicated.
Latent error prevents execution.	Rule violation.
Quality.	
Action too much.	Other.

- Procedures: are procedures required (Rule violation, wrong sequence, omission, quality error):
- Organisation: are there sufficient personnel to carry out the task and to check for errors (Action too late, wrong sequence, omission, error of quality)?
- Stress: will the task be stressful and are there significant consequences of task failure (Omission, error of quality, rule violation)?
- Complexity: is the task complex or novel (Omission, substitution error, other)?

PEM analysis   The PEM analysis part of the HERA approach is used to identify potential errors based upon the associated PEMs. The HERA approach uses 14 PEM questions which are applied to each task step in the HTA. Each PEM question is linked to a set of associated EEMs.

HEIST analysis   The HEIST approach is then used by the HERA team as a back-up check to ensure analysis comprehensiveness (i.e. that no potential errors have been missed). It is also used to provide error reduction guidelines.

Human Error HAZOP analysis   Finally, to ensure maximum comprehensiveness, a Human Error HAZOP-style analysis should be performed.

Advantages

- The multi-method HERA framework ensures that it is highly exhaustive and comprehensive.
- The HERA team are provided with maximum guidance when conducting the analysis. Each of the questions used during the approach prompt the analyst for potential errors and are also linked to the relevant EEMs.
- The framework approach offers the analyst more than one chance to identify potential errors. This should ensure that no potential errors are missed.
- The HERA framework allows analysis teams to see the scenario from a number of different perspectives.

- It uses existing, proven HEI techniques, such as the Human Error HAZOP, THERP and SHERPA techniques.

Disadvantages

- A HERA analysis would require a huge amount of time and resources.
- The technique could potentially become very repetitive, with many errors being identified over and over again by the different techniques employed within the HERA framework.
- Domain expertise would be required for a number of the modules.
- Due to the many different techniques employed within the HERA framework, the training time for such an approach would be extremely high.
- A HERA team would have to be constructed. Such a team requires a mixed group made up of operators, HF specialists, designers, engineers, etc. Building such a team and making sure they can all be brought together at the same time would be a difficult thing to do.
- Although the HERA technique is vast and contains a number of different modules, it is difficult to see how such an approach (using traditional EEM taxonomies) would perform better than far simpler and quicker approaches to HEI such as SHERPA and HET.
- There is only limited evidence of the application of the HERA framework available in the literature.

Example

HERA has yet to be applied in a concrete analysis. The following examples are extracts of a hypothetical analysis described by Kirwan (1992b). As the output is so large, only a small extract is presented in Table 6.17. For a more comprehensive example, the reader is referred to Kirwan (1992b).

Table 6.17   Extract of mission analysis output

Identifier	Task step	Error identified	Consequence	Recovery	Comments
1) Fail to achieve in time.	Goal 0: Restore power and cooling.	Fail to achieve in time.	Reactor core degradation.	Grid re-connection.	This is at the highest level of task-based failure description.
2) Omit entire task.	Goal 0: Restore power and cooling. Goal A: Ensure reactor trip.	Fail to restore power and cooling.	Reactor core degradation. Reactor core melt (ATWS).	Grid re-connection. None.	This is the anticipated transient without SCRAM (ATWS) scenario. It is not considered here but may be considered in another part of the risk assessment.

Source: Kirwan, 1992b.

Related Methods

The HERA framework employs a number of different techniques, including initial task analysis, HTA, HEIST and Human Error HAZOP.

Approximate Training and Application Times

Although no training and application time is offered in the literature, it is apparent that the amount of time in both cases would be high. The training time would be considerable as analysts would have to be trained in the different techniques employed within the HERA framework, such as initial task analysis, Human Error HAZOP and HEIST. The application time would also be extremely high, due to the various different analyses that are conducted as part of a HERA analysis.

Reliability and Validity

No data regarding reliability and validity is offered by the authors. The technique was proposed as an example of the form that such an approach would take. At the present time, there are no reported applications of the HERA framework in the literature.

Tools Needed

The HERA technique comes in the form of a software package, although HERA analysis can be performed without using the software. This would require pen and paper and the goals, plans, PEM and PSF analysis questions. Functional diagrams for the system under analysis would also be required as a minimum.

System for Predictive Error Analysis and Reduction (SPEAR)

Background and Applications

SPEAR was developed by the Center for Chemical Process Safety for use in the American chemical processing industry's HRA programme. It is a systematic taxonomy-based approach to HEI that is very similar to the SHERPA technique (Embrey, 1986). In addition to an external error mode taxonomy, the SPEAR technique also uses a PSF taxonomy to aid the identification of environmental or situational factors that may enhance the possibility of error. It is typically applied to the bottom-level tasks (or operations) of an HTA of the task under analysis. Using subjective judgment, the analyst uses the SPEAR human error taxonomy to classify each task step into one of the five following behaviour types:

- action;
- retrieval;
- check;
- selection; and
- transmission.

Each behaviour has an associated set of EEMs, such as action incomplete, action omitted and right action on wrong object. The analyst then uses the taxonomy and domain expertise to determine any credible error modes for the task in question. For each credible error (i.e. those judged by the analyst to be possible), the analyst provides a description of the form that the error would take, such as 'pilot dials in wrong airspeed'. Next, the analyst has to determine how the operator can recover the error and also any consequences associated with the error. Finally, error reduction measures are proposed, under the categories of procedures, training and equipment.

Domain of Application

The SPEAR technique was developed for the chemical process industry. However, the technique employs a generic external error mode taxonomy and can be applied in any domain.

Procedure and Advice

Step 1: Conduct an HTA
The first step in a SPEAR analysis is to conduct an HTA of the task or scenario under analysis. The SPEAR technique works by indicating which of the errors from the SPEAR EEM taxonomy are credible at each

bottom-level task step in an HTA of the task under analysis. A number of data collection techniques may be used in order to gather the information required for the HTA, such as interviews with SMEs and observations of the task under analysis.

Step 2: PSF Analysis
The analyst should take the first/next bottom-level task step from the HTA and consider each of the PSFs for that task step. This allows the analyst to determine whether any of the PSFs are relevant for the task step in question. The SPEAR technique provides the analyst with a specific PSF taxonomy, and in the past, the PSF taxonomy from the THERP technique (Swain and Guttman, 1983) has been used in conjunction with SPEAR.

Step 3: Task Classification
Next, the analyst should classify the task step under analysis into one of the behaviour categories from the SPEAR behaviour taxonomy. The analyst should select appropriate behaviour and EEM taxonomies based upon the task under analysis. The analyst has to classify the task step into one of the behaviour categories; action, retrieval, check, selection and transmission.

Step 4: Error Analysis
Taking the PSFs from step 2 into consideration, the analyst next considers each of the associated EEMs for the task step under analysis. The analyst uses his or her subjective judgment to identify any credible errors associated with the task step in question. Each credible error should be recorded and a description of the error should be provided.

Step 5: Consequence Analysis
For each credible error, the analyst should record the associated consequences.

Step 6: Error Reduction Analysis
For each credible error, the analyst should offer any potential error remedies. The SPEAR technique uses three categories of error reduction guideline: procedures, training and equipment. It is normally expected that a SPEAR analysis should provide at least one remedy for each of the three categories.

Advantages

- SPEAR provides a structured approach to HEI.
- It is simple to learn and use, and requires minimal training.
- The taxonomy prompts the analyst for potential errors.
- Unlike SHERPA, it also considers PSFs.
- It is quicker than most HEI techniques.
- It is generic, allowing the technique to be applied in any domain.

Disadvantages

- For large, complex tasks, the technique may become laborious and time-consuming to apply.
- The initial HTA adds additional time to the analysis.
- The consistency of such techniques is questionable.
- It appears to be an almost exact replica of SHERPA.
- For large, complex tasks, the analysis may become time-consuming and unwieldy.
- It does not consider the cognitive component of error.

Related Methods

The SPEAR technique is a taxonomy-based approach to HEI. There are a number of similar HEI techniques available, such as SHERPA (Embrey, 1986) and HET (Marshall et al., 2003). A SPEAR analysis also requires an initial HTA to be performed for the task under analysis. The development of the HTA may involve the use of a number of data collection procedures, including interviews with SMEs and observational study of the task or scenario under analysis.

Flowchart

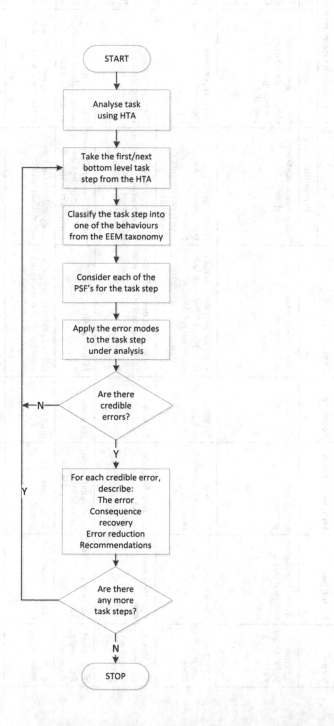

Table 6.18   Example SPEAR output

Step	Error type	Error description	Recovery	Consequences	Error reduction recommendations		
					Procedures	Training	Equipment
2.3 Enter tanker target weight.	Wrong information obtained (R2).	Wrong weight entered.	On check.	Alarm does not sound before tanker overfills.	Independent validation of target weight	Ensure operator double checks entered date. Recording of values in checklist.	Automatic setting of weight alarms from unladen weight. Computerise logging system and build in checks on tanker reg. no. and unladen weight linked to warning system. Display differences.
3.2.2 Check tanker while filling.	Check omitted (C1).	Tanker not monitored while filling.	On initial weight alarm.	Alarm will alert the operator if correctly set. Equipment fault (e.g. leaks) not detected early and remedial action delayed.	Provide secondary task involving other personnel. Supervisor periodically checks operation	Stress importance of regular checks for safety.	Provide automatic log in procedure.
3.2.3 Attend tanker during filling of last 2–3 tons.	Operation omitted (O8).	Operator fails to attend.	On step 3.2.5.	If alarm is not detected within 10 minutes, tanker will overfill.	Ensure work schedule allows operator to do this without pressure	Illustrate consequences of not attending.	Repeat alarm in secondary area. Automatic interlock to terminate loading if alarm not acknowledged. Visual indication of alarm.
3.2.5 Cancel final weight alarm.	Operation omitted (O8).	Final weight alarm taken as initial weight alarm.	No recovery.	Tanker overfills.	Note differences between the sound of the two alarms in checklist	Alert operators during training about differences in sounds of alarms.	Use completely different tones for initial and final weight alarms.
4.1.3 Close tanker valve.	Operation omitted (O8).	Tanker valve not closed.	4.2.1	Failure to close tanker valve would result in pressure not being detected during the pressure check in 4.2.1.	Independent check on action. Use checklist	Ensure operator is aware of consequences of failure.	Valve position indicator would reduce probability of error.
4.2.1 Vent and purge lines.	Operation omitted (O8).	Lines not fully purged.	4.2.4	Failure of operator to detect pressure in lines could lead to leak when tanker connections broken.	Procedure to indicate how to check if fully purged.	Ensure training covers symptoms of pressure in line.	Line pressure indicators at controls. Interlock device on line pressure.
4.4.2 Secure locking nuts.	Operation omitted (O8).	Locking nuts left unsecured.	No recovery.	Failure to secure locking nuts could result in leakage during transportation.	Use checklist.	Stress safety implications of training.	Locking nuts to give tactile feedback when secure.

Approximate Training and Application Times

It is estimated that the training time associated with the SPEAR technique is low. The SPEAR technique is very similar to the SHERPA technique, which typically takes around two to three hours to train novice analysts. The application time is based on the size and complexity of the task under analysis. In general, the application time associated with the SPEAR technique would be low. However, for large, complex scenarios, the application time may increase considerably.

Reliability and Validity

No data regarding the reliability and validity of the SPEAR technique are available in the literature. Since the technique is very similar to the SHERPA technique, it estimated that the reliability and validity of the SPEAR technique would be high.

Tools Needed

A SPEAR analysis can be conducted using pen and paper. The analyst would also require functional diagrams of the system/interface under analysis and an appropriate EEM taxonomy, such as the SHERPA (Embrey, 1986) error mode taxonomy. A PSF taxonomy is also required, such as the one employed by the THERP technique (Swain and Guttman, 1983).

Example

The example output presented in Table 6.18 on the previous page is an extract from a SPEAR analysis of a chlorine tanker-filling problem (CCPS, 1994).

The Human Error Assessment and Reduction Technique (HEART)

Background and Applications

HEART (Williams, 1986) offers an approach for deriving numerical probabilities associated with error occurrence. It was designed as a quick, easy to use and understand HEI technique, and is a highly structured approach that allows the analyst to quantify human error potential. One of the features of the HEART approach is that, in order to reduce resource usage, it only deals with those errors that will have a significant effect on the system in question (Kirwan, 1994). The method uses its own values of reliability and also 'factors of effect' for a number of EPCs. The HEART approach has been used in the UK for the risk assessment of Sizewell B station and also the risk assessments for UK Magnox and Advanced Gas-Cooled Reactor stations.

Domain of Application

HEART was developed for the nuclear power and chemical process industries.

Procedure and Advice

Step 1: Determine the Task or Scenario under Analysis
The first step in a HEART analysis is to select an appropriate set of tasks for the system under analysis. In order to ensure that the analysis is as exhaustive as possible, it is recommended that the analyst selects a set of tasks that are as representative of the system under analysis as possible.

Step 2: Conduct an HTA for the Task or Scenario under Analysis

Once the tasks or scenarios under analysis are defined clearly, the next step involves describing the tasks or scenarios. It is recommended that an HTA is used for this purpose. A number of data collection techniques may be used in order to gather the information required for the HTA, such as interviews with SMEs and observational study of the task under analysis.

Step 3: Conduct HEART Screening Process

The HEART technique uses a screening process, in the form of a set of guidelines that allow the analyst to identify the likely classes, sources and strengths of human error for the scenario under analysis (Kirwan, 1994).

Step 4: Task Unreliability Classification

Once the screening process has been conducted, the analyst must define the proposed nominal level of human unreliability associated with the task under analysis. To do this, the analyst uses the HEART generic categories to assign a human error probability for the task in question. For example, if the analysis was focused upon a non-routine, emergency situation in the control room, this would be classed as 'Category A: Totally unfamiliar, performed at speed with no real idea of likely consequences'. The probability associated with this would be 0.55. The generic HEART categories are presented in Table 6.19.

Step 5: Identification of EPCs

The next stage of a HEART analysis is the identification of EPCs associated with the task under analysis. To do this, the analyst uses the associated HEART EPCs to identify any EPCs that are applicable to the task under analysis. These are presented in Table 6.20.

Step 6: Assessed Proportion of Effect

Once the analyst has identified any EPCs associated with the task under analysis, the next step involves determining the assessed proportion of effect of each of the EPCs identified. This involves providing a rating between 0 and 1 (0 = low, 1 = high) for each EPC. The ratings offered are based upon the subjective judgment of the analyst involved.

Table 6.19   Generic HEART categories

Generic task	Proposed nominal human unreliability (5th–95th percentile bounds)
Totally unfamiliar, performed at speed with no real idea of the likely consequences.	0.55 (0.35–0.97)
Shift or restore system to a new or original state on a single attempt without supervision or procedures.	0.26 (0.14–0.42)
Complex task requiring a high level of comprehension and skill.	0.16 (0.12–0.28)
Fairly simple task performed rapidly or given scant attention.	0.09 (0.06–0.13)
Routine, highly practised, rapid task involving a relatively low level of skill.	0.02 (0.007–0.045)
Restore or shift a system to original or new state following procedures, with some checking.	0.003 (0.0008–0.007)
Completely familiar, well designed, highly practised, routine task occurring several times per hour, performed at the highest possible standards by highly motivated, highly trained and experienced person, totally aware of the implications of failure, with time to correct potential error, but without the benefit of significant job aids.	0.0004 (0.00008–0.009)
Respond correctly to system command even when there is an augmented or automated supervisory system providing accurate interpretation of system stage.	0.00002 (0.000006–0.009)

Table 6.20   HEART EPCs

Error-producing condition (EPC)	Maximum predicted amount by which unreliability might change, going from good conditions to bad
Unfamiliarity with a situation which is potentially important but which only occurs infrequently or which is novel.	X17
A shortage of time available for error detection and correction.	X11
A low signal-to-noise ratio.	X10
A means of suppressing or overriding information or features which is too easily accessible.	X9
No means of conveying spatial and functional information to operators in a form which they can readily assimilate.	X8
A mismatch between an operator's model of the world and that imagined by a designer.	X8
No obvious means of reversing an unintended action.	X8
A channel capacity overload, particularly one caused by the simultaneous presentation of non-redundant information.	X6
A need to unlearn a technique and apply one which requires the application of an opposing philosophy.	X6
The need to transfer specific knowledge from task to task without loss.	X5.5
Ambiguity in the required performance standards.	X5
A mismatch between perceived and real risk.	X4
Poor, ambiguous or ill-matched system feedback.	X4
No clear, direct and timely confirmation of an intended action from the portion of the system over which control is exerted.	X4
Operator inexperience.	X3
An impoverished quality of information conveyed procedures and person–person interaction.	X3
Little or no independent checking or testing of output.	X3
A conflict between immediate and long-term objectives.	X2.5
No diversity of information input for veracity checks.	X2
A mismatch between the educational achievement level of an individual and the requirements of the task.	X2
An incentive to use other more dangerous procedures.	X2
Little opportunity to exercise mind and body outside the immediate confines of the job.	X1.8
Unreliable instrumentation.	X1.6
A need for absolute judgments which are beyond the capabilities or experience of an operator.	X1.6
Unclear allocation of function and responsibility.	X1.6
No obvious way to keep track of progress during an activity.	X1.2
A danger that finite physical capabilities will be exceeded.	X1.4
Little or no intrinsic task meaning.	X1.4
High level of emotional stress.	X1.3
Evidence of ill health amongst operatives.	X1.2
Low workforce morale.	X1.2
Inconsistency of meaning of displays and procedures.	X1.2
A poor or hostile environment.	X1.15
Prolonged inactivity or highly repetitious cycling of mental workload (for initial half hour).	X1.1
(for every following half hour).	X1.05

Table 6.20   Continued

Error-producing condition (EPC)	Maximum predicted amount by which unreliability might change, going from good conditions to bad
Disruption of normal work sleep cycles.	X1.1
Task pacing caused by intervention of others.	X1.06
Additional team members over and above those necessary to perform the task normally and satisfactorily.	X1.03
Age of personnel performing the perceptual task.	X1.02

Source: Kirwan, 1994.

Step 7: Remedial Measures

The next step involves identifying and proposing possible remedial measures for the errors identified. Although the HEART technique does provide some generic remedial measures, the analyst may also be required to provide more specific measures, depending upon the nature of the error and the system under analysis. The remedial measures provided by the HEART methodology are generic and are not system-specific.

Step 8: Documentation Stage

It is recommended that the HEART analysis is fully documented by the analyst. Throughout the analysis, every detail should be recorded by the analyst. Once the analysis is complete, the HEART analysis should be converted into a suitable presentation format.

Example

An example of a HEART analysis output is presented in Table 6.21.

Table 6.21   HEART output

Type of task – F		Nominal human reliability – 0.003	
Error-producing conditions	Total HEART effect	Engineers POA	Assessed effect
Inexperience	X3	0.4	((3 – 1) x 0.4) + 1 = 1.8
Opp technique	X6	1.0	((6 – 1) x 1.0) + 1 = 6.0
Risk misperception	X4	0.8	((4 – 1) x 0.8 + 1 = 3.4
Conflict of objectives	X2.5	0.8	((2.5 – 1) x 0.8) + 1 = 2.2
Low morale	X1.2	0.6	((1.2 – 1) x 0.6 + 1 = 1.12

Source: Kirwan, 1994.

Assessed, nominal likelihood of failure = 0.27 (0.003 x 1.8 x 6 x 3.4 x 2.2 x 1.12).

For the example presented above, a nominal likelihood of failure of 0.27 was identified. According to Kirwan (1994), this represents a high predicted error probability and would warrant error reduction measures. In this instance, technique unlearning is the biggest contributory factor and so if error reduction measures were required, retraining or re-designing could be offered. Table 6.22 presents the remedial measures offered for each EPC in this example.

Table 6.22   Remedial measures

Technique unlearning (x6)	The greatest possible care should be exercised when a number of new techniques are being considered that all set out to achieve the same outcome. They should not involve the adoption of opposing philosophies.
Misperception of risk (x4)	It must not be assumed that the perceived level of risk, on the part of the user, is the same as the actual level. If necessary, a check should be made to ascertain where any mismatch might exist, and what its extent is.
Objectives conflict (x2.5)	Objectives should be tested by management for mutual compatibility, and where potential conflicts are identified, these should either be resolved, so as to make them harmonious, or made prominent so that a comprehensive management-control programme can be created to reconcile such conflicts, as they arise, in a rational fashion.
Inexperience (x3)	Personnel criteria should contain experience parameters specified in a way relevant to the task. Chances must not be taken for the sake of expediency.
Low morale (x1.2)	Apart from the more obvious ways of attempting to secure high morale – by way of financial rewards, for example – other methods, involving participation, trust and mutual respect, often hold out at least as much promise. Building up morale is a painstaking process, which involves a little luck and great sensitivity.

Source: Kirwan, 1994.

Advantages

- The HEART approach is a simplistic one requiring only minimal training.
- It is quick and simple to use.
- Each EPC has a remedial measure associated with it.
- It gives the analyst a quantitative output.
- It uses fewer resources than other techniques such as SHERPA.
- A number of validation studies have produced encouraging results for the HEART approach, e.g. Kirwan (1996b, 1997a, 1997b).

Disadvantages

- Little guidance is offered to the analyst in a number of the key HEART stages, such as the assignment of EPCs. As a result, there are doubts over the reliability of the HEART approach.
- Although HEART has been subject to a number of validation studies, the methodology still requires further validation.
- Neither dependence nor EPC interaction is accounted for by HEART (Kirwan, 1994).
- It is very subjective, reducing its reliability and consistency.
- It was developed specifically for the nuclear power domain and would require considerable development to be applied in other domains.

Related Methods

Normally, a HEART analysis requires a task analysis description of the task or scenario under analysis. HTA (Annett et al., 1971; Kirwan and Ainsworth, 1992; Shepherd, 1989) is normally used for this purpose. The HEART technique is an HRA technique, of which there are many, such as THERP (Swain and Guttman, 1983) and Justification of Human Error Data Information (JHEDI: Kirwan, 1994).

Approximate Training and Application Times

According to Kirwan (1994), the HEART technique is quick to apply and requires minimal training. The technique is certainly simple in its application and so the associated training and application times are estimated to be low.

Reliability and Validity

Kirwan (1992a, 1992b) describes a validation of nine HRA techniques and reports that, of the nine techniques, HEART, THERP, absolute probability judgment (APJ) and JHEDI performed moderately well. A moderate level of validity for HEART was reported. In a second validation study (Kirwan 1997a), HEART, THERP and JHEDI were subject to a validation study. The highest precision rating associated with the HEART technique was 76.67 per cent. Of 30 assessors using the HEART approach, 23 displayed a significant correlation between their error estimates and the real HEPs. According to Kirwan (1997a, 1997b), the results demonstrate a level of empirical validity for the three techniques.

Tools Needed

The HEART approach can be applied using pen and paper. The associated HEART documentation is also required (HEART generic categories, HEART EPCs, etc.).

Flowchart

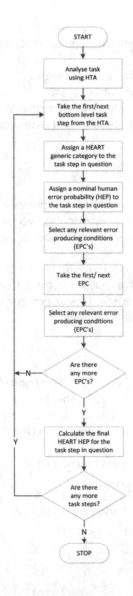

The Cognitive Reliability and Error Analysis Method (CREAM)

Background and Applications

CREAM (Hollnagel, 1998) is an HEI/HRA method that was developed in response to an analysis of existing HRA approaches. It can be used both predictively, to predict potential human error, and retrospectively, to analyse and quantify error. According to Hollnagel (1998), CREAM enables the analyst to:

- identify those parts of the work, tasks or actions that require or depend upon human cognition, and which therefore may be affected by variations in cognitive reliability;
- determine the conditions under which the reliability of cognition may be reduced, and where therefore the actions may constitute a source of risk;
- provide an appraisal of the consequences of human performance on system safety, which can be used in probabilistic risk assessment (PRA)/PSA; and
- develop and specify modifications that improve these conditions and hence serve to increase the reliability of cognition and reduce the risk.

CREAM uses a model of cognition, the COCOM, which focuses on how actions are chosen and assumes that the degree of control that an operator has over his or her actions is variable, and determines the reliability of his or her performance. The COCOM describes four modes of control: scrambled control, opportunistic control, tactical control and strategic control. According to Hollnagel (1998), when the level of operator control rises, so does the operators' performance reliability. The CREAM technique uses a classification scheme consisting of a number of groups that describe the phenotypes (error modes) and genotypes (causes) of the erroneous actions. The CREAM classification scheme is used by the analyst to predict and describe how errors could potentially occur. It allows the analyst to define the links between the causes and consequences of the error under analysis. Within the CREAM classification scheme, there are three categories of causes (genotypes): individual, technological and organisational causes. A brief description of each genotype category is provided below:

- Individual-related genotypes: specific cognitive functions, general person-related functions (temporary) and general person-related functions (permanent).
- Technology-related genotypes: equipment, procedures, interface (temporary) and interface (permanent).
- Organisation-related genotypes: communication, organisation, training, ambient conditions and working conditions.

The CREAM technique uses a number of linked classification groups. The first classification group describes the CREAM error modes, which are presented below:

- Timing – too early, too late, omission.
- Duration – too long, too short.
- Sequence – reversal, repetition, commission, intrusion.
- Object – wrong action, wrong object.
- Force – too much, too little.
- Direction – Wrong direction.
- Distance – too short, too far.
- Speed – too fast, too slow.

These eight different error mode classification groups are then divided further into the four sub-groups:

- Action at the wrong time – includes the error mode's timing and duration.
- Action of the wrong type – includes the error mode's force, distance, speed and direction.
- Action at the wrong object – includes the error mode 'object'.
- Action in the wrong place – includes the error mode 'sequence'.

In addition to the exploration of error modes (phenotypes) and causes of error (genotypes), the CREAM technique also uses a set of CPCs that are used by the analyst to describe the context in the scenario/task under analysis. These are similar to the PSFs used by other HEI/HRA techniques. The CREAM common performance conditions are presented in Table 6.23.

CREAM is highly successful and as such a number of variations on the method have been developed. A recent adaptation of the CREAM technique is the CEAM (Communication Error Analysis Method) (Lee, Ha and Seong, 2011) which focuses upon errors of communication. Kennedy et al. (2007) also identified a simplified version of THERP called the Accident Sequence Evaluation Programme (ASEP), which they suggest is beneficial for the identification of high-level hazards. A frequently utilised version of CREAM has also been developed called DREAM.

DREAM is an adaptation of the CREAM method designed for use within the driving domain. It was developed in order to provide a series of codes capable of describing both genotype and phenotype causality in car crashes (Ljung, 2010).

Domain of Application

Although the technique was developed for the nuclear power industry, it is a generic approach and can be applied in any of domain involving the operation of complex, dynamic systems.

Table 6.23   Cream CPCs

CPC name	Level/descriptors
Adequacy of organisation.	The quality of the roles and responsibilities of team members, additional support, communication systems, safety management systems, instructions and guidelines for externally orientated activities, etc.: very efficient/efficient/inefficient/deficient.
Working conditions.	The nature of the physical working conditions such as ambient lighting, glare on screens, noise from alarms, task interruptions, etc.: advantageous/compatible/incompatible.
Adequacy of MMI and operational support.	The man–machine interface in general, including the information available on control panels, computerised workstations and operational support provided by specifically designed decision aids: supportive/adequate/tolerable/inappropriate.
Availability of procedures/plans.	Procedures and plans, including operating and emergency procedures, familiar patterns of response heuristics, routines, etc.: appropriate/acceptable/inappropriate..
Number of simultaneous goals	The number of tasks a person is required to pursue or attend to at the same time: fewer than capacity/matching current capacity/more than capacity.
Available time	The time available to carry out the task: adequate/temporarily inadequate/continuously inadequate..
Time of day (Circadian rhythm).	Time at which the task is carried out, in particular whether or not the person is adjusted to the current time: daytime (adjusted)/night-time (unadjusted).
Adequacy of training and experience.	The level and quality of training provided to operators, such as familiarisation with new technology, refreshing old skills, etc. Also refers to operational experience: adequate, high experience/adequate, limited experience/inadequate..
Crew collaboration quality.	The quality of collaboration between the crew members, including the overlap between the official and unofficial structure, level of trust and the general social climate among crew members: very efficient/efficient/inefficient/deficient.

Procedure and Advice (Prospective Analysis)

Step 1: Task Analysis
The first step in a CREAM analysis involves describing the task or scenario under analysis. It is recommended that an HTA of the task or scenario under analysis is developed for this purpose. A number of data collection procedures may be used to collect the data required for the HTA, including interviews with SMEs and observational study of the task or scenario under analysis.

Step 2: Context Description
Once the task or scenario under analysis is described, the analyst should begin by first describing the context in which the scenario under analysis takes place. This involves describing the context using the CREAM CPCs (Table 6.23). To do this, the analyst uses his or her subjective judgment to rate each CPC regarding the task under analysis. For example, if the analyst assumes that the operator has little experience or training for the task under analysis, then the CPC 'Adequacy of training and experience' should be rated 'limited experience/inadequate'.

Step 3: Specification of the Initiating Events
The analyst then needs to specify the initiating events that will be subject to the error predictions. Hollnagel (1998) suggests that PSA event trees can be used for this step. However, since a task analysis has already been conducted in step 1 of the procedure, it is recommended that this be used. The analyst(s) should specify the tasks or task steps that are to be subject to further analysis.

Step 4: Error Prediction
Once the CPC analysis has been conducted and the initiating events are specified, the analyst should then determine and describe how an initiating event could potentially into an error occurrence. To predict errors, the analyst constructs a modified consequent/antecedent matrix. The rows on the matrix show the possible consequents, while the columns show the possible antecedents. The analyst starts by finding the classification group in the column headings that correspond to the initiating event (e.g. for missing information, it would be communication). The next step is to find all the rows that have been marked for this column. Each row should point to a possible consequent, which in turn may be found amongst the possible antecedents. Hollnagel (1998) suggests that in this way, the prediction can continue in a straightforward way until there are no further paths left. Each error should be recorded along with the associated causes (antecedents) and consequences (consequents).

Step 5: Selection of Task Steps for Quantification
Depending upon the analysis requirements, a quantitative analysis may be required. If so, the analyst should select the error cases that require quantification. It is recommended that if quantification is required, all of the errors identified should be selected for quantification.

Step 6: Quantitative Performance Prediction
CREAM has a basic and extended method for quantification purposes. Since this review is based upon the predictive use of CREAM, the error quantification procedure is not presented. For a description of the quantification procedure, the reader is referred to Hollnagel (1998).

Advantages

- CREAM has the potential to be extremely exhaustive.
- Context is considered when using CREAM.

- It is a clear, structured and systematic approach to error identification and quantification.
- It can be used both proactively to predict potential errors and retrospectively to analyse error occurrence.
- It is not domain-specific and the potential for application in different domains is apparent.
- Its classification scheme is detailed and exhaustive, even taking into account system and environmental (socio-technical) causes of error.
- Sandin (2009) argues that the links in a DREAM analysis (and correspondingly a CREAM analysis) illustrate causal relationships.

Disadvantages

- To the novice analyst, CREAM appears complicated and daunting.
- The exhaustiveness of the classification scheme serves to make it larger and more resource-intensive than other methods.
- It has not been used extensively.
- It is apparent that the training and application time for the CREAM technique would be considerable.
- It does not offer remedial measures, i.e. ways to recover erroneous human actions are not provided or considered.
- It appears to be very complicated in its application.
- It would presumably require analysts with knowledge of HF and cognitive ergonomics.
- Its application time would be high, even for very basic scenarios.

Related Methods

CREAM analyses are typically conducted on an HTA of the task or scenario under analysis. A number of data collection procedures may be used during the development of the HTA, including interviews with SMEs and observational study of the task or scenario in question. CREAM is a taxonomy-based approach to HEI. Other taxonomic approaches include SHERPA (Embrey, 1986), HET (Marshall et al., 2003) and TRACEr (Shorrock and Kirwan, 2002).

Approximate Training and Application Times

Although there is no data regarding training and application times presented in the literature, it is estimated that the associated times will be high in both cases.

Reliability and Validity

Validation data for the CREAM technique is limited. Hollnagel, Kaarstad and Lee (1999) report a 68.6 per cent match between errors predicted and actual error occurrences and outcomes when using the CREAM error taxonomy.

Tools Needed

At its simplest, CREAM can be applied using pen and paper only. A prototype software package has also been developed to aid analysts (Hollnagel, 1998). Kennedy et al. (2007) identify a recent software development to aid in THEA analysis: ProtoTHEA.

Flowchart – Prospective Analysis

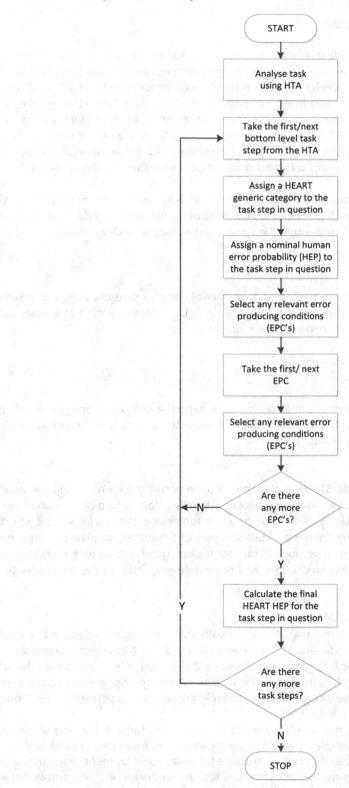

```
              ( START )
                 │
                 ▼
        ┌──────────────────┐
        │   Analyse task    │
        │   using HTA       │
        └──────────────────┘
                 │
                 ▼
        ┌──────────────────┐
    ┌──│  Take the first/next │
    │   │  bottom level task   │
    │   │  step from the HTA   │
    │   └──────────────────┘
    │            │
    │            ▼
    │   ┌──────────────────┐
    │   │  Assign a HEART   │
    │   │  generic category to the │
    │   │  task step in question │
    │   └──────────────────┘
    │            │
    │            ▼
    │   ┌──────────────────┐
    │   │ Assign a nominal human │
    │   │ error probability (HEP) to │
    │   │ the task step in question │
    │   └──────────────────┘
    │            │
    │            ▼
    │   ┌──────────────────┐
    │   │ Select any relevant error │
    │   │ producing conditions │
    │   │ (EPC's) │
    │   └──────────────────┘
    │            │
    │            ▼
    │   ┌──────────────────┐
    │   │ Take the first/ next │
    │   │ EPC │
    │   └──────────────────┘
    │            │
    │            ▼
    │   ┌──────────────────┐
    │   │ Select any relevant error │
    │   │ producing conditions │
    │   │ (EPC's) │
    │   └──────────────────┘
    │            │
    │            ▼
    │         ◇ Are there
    │  N ◄──  any more
    │         EPC's? ◇
    │            │ Y
    │            ▼
    │   ┌──────────────────┐
    │   │ Calculate the final │
    │ Y │ HEART HEP for the │
    │   │ task step in question │
    │   └──────────────────┘
    │            │
    │            ▼
    │         ◇ Are there
    └────────  any more
              task steps? ◇
                 │ N
                 ▼
              ( STOP )
```

The Systems Theory Accident Modelling and Process (STAMP)

Background and Applications

STAMP (Leveson, 2004) is an accident analysis technique developed to explore accident causation from a systems theory perspective. Leveson argues against traditional event chain models of accident causation and instead posits that accident causation should account for processes such as adaptation and emergence that are present within complex systems. The method is based upon the hypothesis that accidents occur due to 'external disturbances, component failures, or dysfunctional interactions among system components' (Leveson, 2004: 250) which are not sufficiently constrained or controlled by the system. Leveson posits that a system is made up of multiple levels which interact in unpredictable ways; in order to prevent an accident occurring, sufficient control should be enforced upon the system in order to prevent unsafe evolution.

The method provides an organisational hierarchy, control structure and taxonomy of causation, the classification of flawed control, which is applied to each level of the hierarchy in order to identify where the dysfunctional interactions occurred within the system under analysis (Leveson, 2004).

Domain of Application

The method is generic and has been applied to a number of complex systems, including aerospace systems (Leveson, 2004), friendly fire incidents (Leveson, 2002), the contamination of a water supply (Leveson, 2004) and and air traffic management (Arnold, 2009).

Procedure and Advice (Leveson, 2004)

Step 1: Define the Task under Analysis
The initial stage of analysis involves the analyst clearly defining the task under analysis. A definition of the core goals and boundaries should be developed in order to guide analysis and ensure that the investigation is appropriate and relevant.

Step 2: Data Collection
Like all accident analysis methods, STAMP is dependent upon accurate data regarding the accident in question. The next step therefore involves collecting detailed data regarding the accident and about the domain and organisation in which the accident took place. Data collection tools such as reviewing accident reports, inquiry reports and task analyses of the system in question, interviewing personnel involved in the accident, reviewing documents regarding the domain in question (e.g. rules and regulations, standard operating procedures) and/or interviewing SMEs for the domain/system in question should all be utilised.

Step 3: Construct Hierarchical Control Structure
Using the data collected in step 2, the analyst must identify the key people involved in the accident scenario. These people will be spread across the different levels of the hierarchical control structure and will include those responsible for producing guidelines, developing policies and so on. The actors should be plotted onto a graphical illustration of the structure at the appropriate hierarchical level. In addition to plotting actors onto the control structure, the relevant constraints between levels should be annotated onto the structure as well.

STAMP provides a generic control model to guide the analyst in the construction of his or her specific control structure, an example of which is presented on the next page in Figure 6.12. The left-hand side of the diagram shows the control structure for system development, whereas the right-hand side of the diagram shows the control structure for system operations. The arrows between

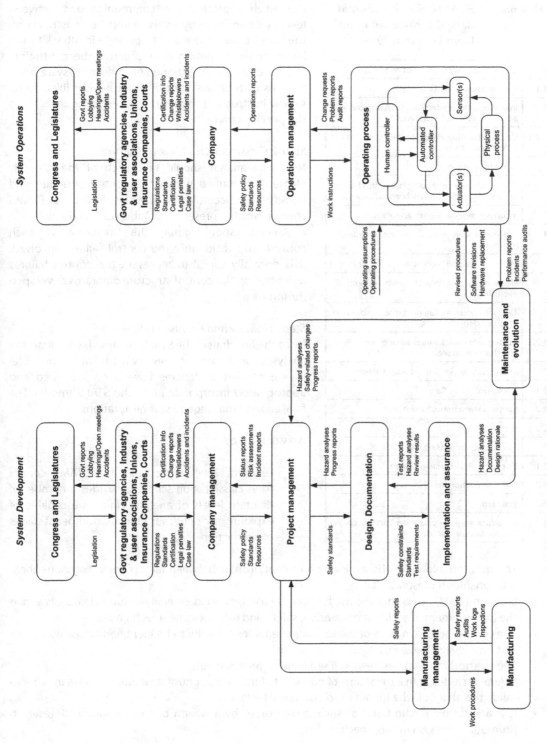

Figure 6.12   Generic control structure model (adapted from Leveson, 2004)

Table 6.24   STAMP classification of flawed control (adapted from Leveson, 2004)

1	Inadequate enforcement of constraints.
1.1	Unidentified hazards.
1.2	Inappropriate, ineffective or missing control actions for identified hazards.
1.2.1	Design of control algorithm (process) does not enforce constraints.
	Flaws in creation process.
	Process changes without appropriate change in control algorithm (asynchronous evolution).
	Incorrect modification or adaptation.
1.2.2	Process models inconsistent, incomplete or incorrect.
	Flaws in creation process.
	Flaws in updating process (asynchronous evolution).
	Time lags and measurement inaccuracies not accounted for.
1.2.3	Inadequate coordination among controllers and decision makers.
2	Inadequate execution of control action.
2.1	Communication flaw.
2.2	Inadequate actuator operation.
2.3	Time lag.
3	Inadequate or missing feedback.
3.1	Not provided in system design.
3.2	Communication flow.
3.3	Time lag.
3.4	Inadequate sensor operation (incorrect or no information provided).

the levels represent the communications between levels that are used by levels to impose constraints on the levels below them and to provide feedback to the levels above regarding how effective the constraints are (Leveson, 2004). Although each domain/system will have its own unique control structure, it is likely to be similar in structure to that presented in Figure 6.12 on the previous page.

Step 4: Classification of Flawed Control
Next, the analyst should take each of the systemic elements identified within the control structure and classify each according to the classification of flawed control, which is presented in Table 6.24 on the right.

Analysts should apply the taxonomy to each control loop to identify the control failures involved. It is normally useful to represent the control failures identified on the control structure diagrams developed during step 3.

Step 5: Review and Finalise Analysis
Once the first draft of the STAMP analysis is complete, the analyst should conduct a review (with SMEs if possible) to ensure that all failures have been identified and appropriately incorporated into the STAMP model. The final analysis may require several iterations.

Advantages

- Leveson (2004) argues that the STAMP control flaws classification scheme provides a number of different levels of analysis which are capable of exploring accident causation at a number of stages of abstraction.
- The method allows for the exploration of relationships between factors, including non-linear relationships (Leveson, 2004).
- Leveson (2004) posits that the method can be used for accident analysis, hazard analysis and in the development of accident prevention, safety and risk assessment techniques.
- The comprehensive nature of the technique enables causality to be identified across numerous systemic levels (Leveson, 2004).
- The method can, and has been, utilised in numerous domains.
- STAMP includes both a taxonomy of possible failures and a control structure template in order to guide the analyst in the identification of causal factors.
- It is a systemic method and as such is supported by a wealth of contemporary HF research promoting the systems approach.

Disadvantages

- The analysis is resource-intensive, especially with respect to time (Braband, Evers and Stefano, 2003).
- A significant amount of detailed data is required in order to conduct the comprehensive method.
- Previous research has highlighted the need to increase the level of guidance within the STAMP method (Almedia and Johnson, 2005; Qureshi, 2007).
- Previous research by Almedia and Johnson (2005) has also highlighted the inability of STAMP to fully explore the reasoning behind actions within the accident scenario.

Related Methods

Conducting a STAMP analysis requires the utilisation of data collection techniques such as observations and interviews.

Approximate Training and Application Times

STAMP analysis is a time-consuming procedure (Braband, Evers and Stefano, 2003) and involves high training times.

Reliability and Validity

There is currently no data available on the reliability or validity of STAMP.

Tools Needed

A STAMP analysis can be conducted using a pen and paper, but it benefits from software drawing packages such as Microsoft Office Visio to create high-quality control structure diagrams.

Example

STAMP has been applied to numerous incidents occurring within complex socio-technical systems, for which the reader is referred to Leveson (2002). Here a short extract of the analysis conducted into an incident of friendly fire is presented to summarise the methodology. In 1994 the USA was one of the countries engaged in a multi-national effort to ensure the safety of northern Iraq. One aspect of the mission involved preventing Iraqi forces from flying through specific areas, known as no-fly zones, where they might attempt to engage refugees. During the enforcement of one such no-fly zone, two US fighter jets wrongly identified two US helicopters as enemy aircraft and proceeded to destroy them. The devastation caused by this misidentification has been subject to numerous analyses (USAF Accident Investigation Board, 1994; United States General Accounting Office, 1997; Leveson, 2002, 2004; Leveson, Allen and Storey, 2002; Snook, 2000).

Leveson, Allen and Storey (2002) begin their analysis of the incident by developing a control structure of the military system in operation at the time of the incident. Using the generic framework presented earlier, agents and teams were plotted onto the appropriate systemic levels with communication channels annotated onto them. The agents identified by Leveson, Allen and Storey are presented in a hierarchical fashion on the next page in Figure 6.13, which illustrates the first stage in the construction of the STAMP control structure.

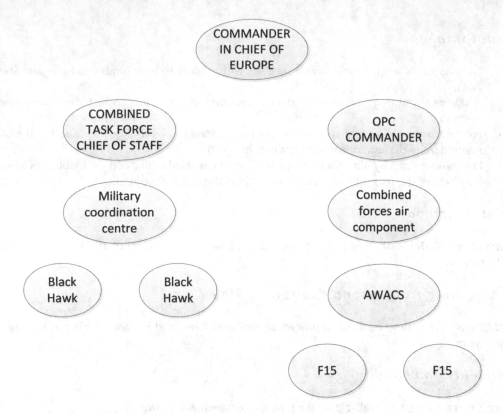

Figure 6.13   Control structure for the friendly-fire incident (adapted from Leveson, Allen and Storey, 2002)

Once the control structure was created, Leveson, Allen and Storey (2002) applied the taxonomy to each aspect of this control structure to identify potential causal factors. For each agent, contextual descriptions are provided that discuss the role of each level of the classification scheme on that agent during the accident scenario. The output produced by Leveson, Allen and Storey's analysis is extensive and so is not presented in full here. Example factors are presented below for a number of levels of the classification scheme in order to illustrate the manner in which it was applied. Focusing on the systemic level in charge of the helicopters which were destroyed, the Military Coordination Center (MCC), the authors argue that multiple errors occurred within this systemic level:

1.2.3   Inadequate coordination among controllers and decision makers: helicopter schedules were not sufficiently comprehensive, failing to include adequate flight details about helicopter missions.

1.2   Inappropriate, ineffective or missing control actions for identified hazards: the helicopters were tasked to enter the no fly zone before the daily morning safety search flight had been conducted by fighter jets to ensure the area was free from enemy aircraft.

Leveson, Allen and Storey (2002) provide indepth descriptions of how accident factors were able to arise and evolve within the military system converging on the occurrence of this incident of friendly fire. For more information, see also Leveson (2004).

Flowchart

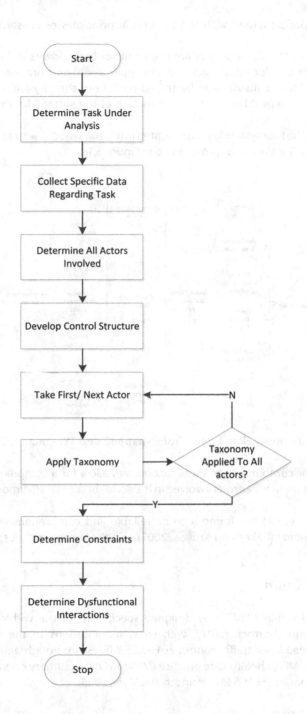

Human Factors Analysis Classification System (HFACS)

Background and Applications

HFACS (Shappell and Wiegmann, 2000, 2003) is a taxonomic error method designed to provide the US military with a tool for the analysis of aviation accidents. The method is the result of a comprehensive review of accident reports merged with the theoretical principles of Reason's Swiss Cheese model (Reason, 1990).

According to Reason (1990), accidents are the result of breakdowns in four core system layers: unsafe acts; pre-conditions for unsafe acts; unsafe supervision; and organisational influences. The theory posits that accident causality can be traced back from the physical acts explicitly involved in the accident scenario to problems in the organisation of the system, for example, problems with resource allocation.

The method utilises the four system layers presented in Reason's model, with an appropriate taxonomy for each of these layers. The taxonomy is presented in Figure 6.14.

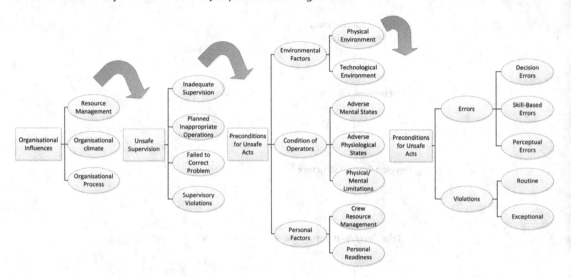

Figure 6.14   HFACS framework (adapted from Shappell and Wiegmann, 2000)

In accident analysis, the combination of these taxonomies allow for a comprehensive classification of both the 'latent failures' and active errors involved in the accident scenario (Salmon, Regan and Johnston, 2005).

Research into the method has found it to be reliable and comprehensive, as well as the most developed error taxonomy (Baker and Krokos, 2007) and the most widely used (Chin Li, Harris and San You, 2008).

Domain of Application

The taxonomy utilised within HFACS was designed specifically for use within the aviation domain. According to Olsen and Shorrock (2010), with small modifications to the taxonomy, HFACS has been successfully applied in air traffic control as HFACS-ATC (Scarborough and Pounds, 2001) and to maintenance as HFACS-ME, to health-care practice (Milligan, 2007), surgery operations (El Bardissi et al., 2007) and to the rail domain (HFACS RR – Reinach and Viale, 2006).

Procedure and Advice

Step 1: Define the Task under Analysis
The initial stage of analysis involves the development of a clear definition of the task under analysis. This is an important stage of the analysis as it ensures that the analysis is focused on collecting appropriate and relevant information

Step 2: Data Collection
The next stage of analysis involves comprehensive data collection to ensure the analyst has a sound understanding of the accident scenario. The analyst should review relevant documents, such as the accident investigation report, and interview appropriate personnel.

Step 3: Identification of Unsafe Acts
At this stage, the analyst, or team of analysts as is normally the case, begins to identify unsafe acts based upon the data collected in step 2. This initial stage focuses on the identification and classification of errors made by front-line actors during the accident scenario.

Step 4: Identify Failures at the Pre-conditions for Unsafe Acts Level
Once all unsafe acts have been identified, the analyst should categorise all pre-conditions to unsafe acts. Pre-conditions to unsafe acts should be categorised into one of three areas: personnel factors; environmental factors; and conditions of operators (as illustrated in Table 6.25). Each of these is further sub-divided into more detailed categories. Each pre-condition to unsafe acts should be mapped onto the corresponding unsafe act identified in step 3. This mapping of hierarchical categories onto one another should continue at each stage of analysis for each level of the hierarchy.

Step 5: Identify Failures at the Unsafe Supervision Level
Next, the analyst should identify any erroneous actions at the unsafe supervision level, including inadequate, inappropriate, violated and inactive supervision. Again, each of these categories is further sub-divided.

Step 6: Identify Failures at the Organisational Influences Level
After exhausting the analysis of supervision, the analyst should move on to examining any organisational influences that impacted the accident scenario. The HFACS taxonomy divides organisational influences into three core categories: resource management; organisational climate; and organisational process.

Step 7: Produce a Short Narrative Discussing Each Error
Once all errors have been identified, the analyst produces a short description of each, outlining the specific context and characteristics.

Step 8: Review and Refine Analysis
As with other complex error methods, the analysis typically takes multiple iterations. Once the first draft is complete, the analyst should review (with SMEs if possible) and amend the categorisations accordingly. Particular attention should be paid to ensuring that all contributory factors have been identified and that their categorisation is appropriate. A useful procedure is to review the HFACS taxonomy as a whole, with each error considered individually and included or excluded in the analysis.

Step 9: Data Analysis
The different categories of HFACS and different hierarchical levels enable data to be represented in various summation formats.

Step 10: Analyse Associations between Failures Across the Different HFACS Levels
Once the data has been appropriately organised, statistical analyses can be conducted, enabling the examination of the relationships between categories and hierarchical levels. Research by Li, Harris and Yu argues that HFACS explores 'both latent and active failures and their inter-relationships' (2008: 427). These authors were able to find statistical significance in the relationships between errors across the systemic levels, e.g. organisational influences to violations. One way of examining the associations between failures across the different levels is through the use of Fisher's exact test for contingency

tables, where odds ratios (ORs) are calculated to assess the strength of association. Odds are calculated for lower-level factors – the odds are the ratio of the probability that a (lower-level) factor is present to the probability that it is absent. The odds can be calculated under two conditions: one for when a higher-level factor is present and another for when a higher-level factor is absent. An odds ratio is calculated by dividing these two odds.

Advantages

- HFACS has been applied to multiple domains and is discussed in numerous publications.
- It provides the analyst with a taxonomy for each level of failure (Salmon, Regan and Johnston, 2005).
- It is simple to use (Baysari et al., 2009).
- It provides a consistent structure for accident analysis.
- The taxonomy is comprehensive enough to consider all levels of the system under analysis.
- It is based upon a well-regarded model of human error.

Disadvantages

- HFACS, as with most accident analysis methods, is limited by the data available. In many cases, accident reports do not contain sufficient detail for a comprehensive analysis (Salmon, Regan and Johnston, 2005).
- Critiques of taxonomic methods such as HFACS discuss the 'fitting' of data into the categories provided by the analyst which would equate to low levels of validity in the method (Salmon, Regan and Johnston, 2005).
- HFACS as a taxonomic method can be criticised for renaming error rather than exploring causality (Dekker, 2002a, 2002b).
- There has been confusion over the taxonomy, with critics suggesting that tighter definitions for the categories are required to avoid misinterpretation (Baysari, McIntosh and Wilson, 2008).
- It is unable to describe the error context – factors such as the task being performed or the way in which the error is responded to (Baysari et al., 2009).

Related Methods

HFACS has also been extended to include an intervention matrix to guide analysts in choosing an intervention strategy for the system under analysis. This matrix is called the Human Factors Intervention Matrix (HFIX). For each of the error types identified within the HFACS analysis, the matrix can be used to identify the corresponding intervention required. HFIX is not only used for retrospective analysis but can also be used in helping to design prospective safety programmes, identifying gaps in these programs (Shappell and Wiegmann, 2009).

An extension of HFACS has been created in order to take account of social psychological phenomena. The extension contains an increased taxonomy which includes items such as *mental states: progressive commitment and self-motives* (Paletz et al., 2009).

Approximate Training and Application Times

HFACS is a simple method with low training times. Application time is dependent upon the size and complexity analysed, but should be relatively low.

Reliability and Validity

The reliability of HFACS has been proven by its ability to accommodate all human causal factors associated with the commercial accidents examined during 1990–1996 (Wiegmann and Shappell, 2001a). The HFACS taxonomy has been criticised for poor levels of inter-rater reliability; a study by Baysari, McIntosh and Wilson found that the percentage agreement between raters was sometimes as low as 40 per cent. The reason behind the poor percentage agreement was reported to be a lack of confidence in the coding. The authors concluded that a comprehensive training programme is required for HFACS, along with improvements to clarifications between error types to prevent confusion.

Tools Needed

HFACS requires pen and paper.

Example

HFACS was applied to an incident of fratricide in order to represent the processes involved in the methodology. The incident under analysis occurred in 1994 and has already been described above.

The first stage of the HFACS procedure involved developing an understanding of the events that took place and classifying any unsafe acts using the first level of the HFACS taxonomy. Once this initial level of analysis was conducted, the unsafe acts were traced back through the military system to identify causal factors in the subsequent four levels of the HFACS taxonomy. For each factor identified, a short narrative description was developed. A summarised version of the results is given below in Table 6.25.

HFACS analysis allows for the results to be summarised into five core error categories, relating to the five core sections of the taxonomy. From the summary, we can see that HFACS identified unsafe supervision and organisational issues as being the highest categories of error. This mirrors the idea present in many analyses of this incident that the F15 pilots were not solely to blame (USAF Accident Investigation Board, 1994; United States General Accounting Office, 1997; Leveson, 2002, 2004; Leveson, Allen and Storey, 2002; Snook, 2000).

HFACS analysis also allows us to look at the categories in more depth. If we focus on organisational issues, we can see that this category is predominantly made up of resource and acquisition management errors. Similarly, focusing on unsafe supervision illustrates the dominance of erroneous actions in planned inappropriate actions. HFACS allows for the identification of key causal areas, but also allows the analyst to identify lower-level errors.

Table 6.25   Summarised HFACS taxonomy

1	Unsafe acts = 18
1.1	Errors (AE xxx) = 13
1.2	Violations (AV xxx) = 5
2	Preconditions for unsafe acts = 4
2.1	Environmental factors (PE xxx) = 4
3	Condition of operators = 12
3.2	Personnel factors (PP xxx) = 12
4	Unsafe supervision = 19
4.1	Inadequate supervision (SI xxx) = 7
4.2	Planned inappropriate operations (SP xxx) = 8
4.3	Failure to correct known problem (SF xxx) = 1
4.4	Supervisory violations (SV xxx) = 3
5	Organisational influences = 19
5.1	Resource/acquisition management (OR xxx) = 11
5.2	Organisational climate (OC xxx) = 3
5.3	Organisational processes (OP xxx) = 5

Flowchart

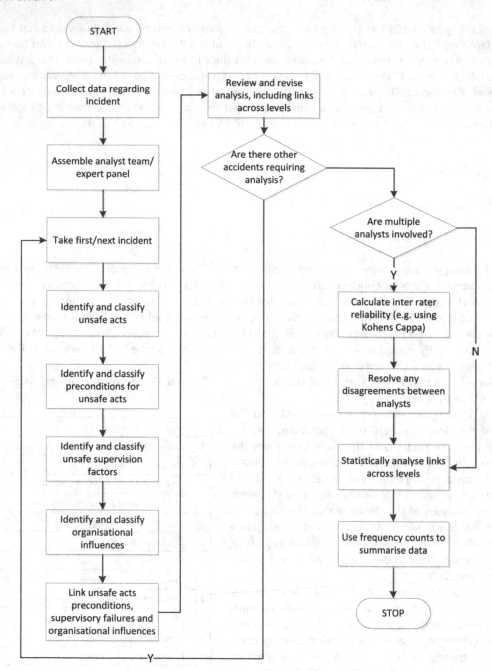

AcciMaps

Background and Applications

The AcciMap method (Rasmussen, 1997; Svedung and Rasmussen, 2002) is an accident analysis technique based upon the notion that there are multiple levels of causality involved in accidents, as put forward in Rasmussen's (1997) model of risk management. The method presents a graphical portrayal of factors within the system that impacted the occurrence of an accident. Typically, the following six levels of complex socio-technical systems are considered (although these can be modified to suit analysis needs): government policy and budgeting; regulatory bodies and associations; local area government planning and budgeting (including company management); technical and operational management; physical processes and actor activities; and equipment and surroundings. Levels 1–4 represent all of the decision-makers that may have been involved in the accident scenario, described by levels 5 and 6, through the decisions they make in their normal work context (Woo and Vicente, 2003). AcciMap analysis focuses upon the causal relationships between these levels, which allows for a vertical analysis across the levels of a system rather than the horizontal generalisation within individual levels that is normally found in accident analysis methods (Svedung and Rasmussen, 2002).

Domain of Application

AcciMap analysis is a generic approach which has been utilised in multiple domains, including aviation accidents (Royal Australian Aviation Force, 2001), police incidents (Jenkins et al., 2010a) and both rail and road accidents (Hopkins, 2005; Svendung and Rasmussen, 2002).

Procedure and Advice

Step 1: Data Collection
The first stage of the analysis involves the analyst reviewing the accident investigation and any other data associated with the system under analysis. The analyst should conduct interviews with SMEs or actors involved, observe recordings of the incident and review all relevant documentation. This data collection phase is the basis of the analysis and should be comprehensive.

Step 2: Construct an Actor Map
Once the data collection is complete, the analyst should identify all actors involved in the scenario and annotate these onto an actor map. Actors should be linked to one another to reflect the communication structure of the system.

Step 3: Identify Physical Process/Actor Activities Failures
After the key actors have been identified, the analyst can begin to develop the AcciMap. The first stage of the AcciMap involves identifying the errors involved and identifying the links between these errors. This step is concerned with the identification of errors which occurred at the level of physical process and actor activities.

Step 4: Identify Causal Factors
The analyst now needs to identify causal factors for each of the physical and actor failures identified in step 3. Each failure is taken in turn and the analyst identifies all related failures at the remaining five levels of the AcciMap: government policy and budgeting; regulatory bodies and associations; local area government planning and budgeting; physical processes and actor activities; and equipment and surroundings.

Step 5: Identify Failures at Other Levels
Once step 4 has been completed, the analyst should review the six systemic levels to ensure that all relevant failures have been identified. He or she should also take each level in turn and, whilst reviewing the data collected in step 1, ensure that no failures have been missed on the AcciMap.

Step 6: Finalise and Review AcciMap Diagram
The AcciMap should be constructed whilst the analyst steps through these stages. At this stage, the analyst should review the AcciMap and ensure that all links between causal factors have been identified and that all annotated links are appropriate. SMEs should be asked to review the AcciMap to ensure its validity.

This review and revise stage of the AcciMap process normally requires several iterations.

Advantages

- AcciMaps enable the identification of system-wide errors that led to the occurrence of the accident at the sharp end. The complete accident aetiology is exposed.
- The method is simple to learn and use.
- It is based upon a sound theoretical model.
- It considers causal factors across systemic levels.
- Its output offers an exhaustive analysis of accidents and incidents.
- It provides a clear visual interpretation of the accident aetiology.
- It is a generic approach which has been applied across many domains.
- Considering the different levels involved in the accident enables an extended timeline of causality to be established.
- It focuses on systematic improvements rather than focusing on blaming individuals.

Disadvantages

- The method can be time-consuming.
- The quality of the analysis produced is entirely dependent upon the quality of the accident report.
- AcciMap analysis does not provide a method to develop corrective measures; these are based on the judgment of the analyst.
- It does not provide a structured taxonomy for error classification.
- AcciMap analysis can only be used retrospectively.
- Its graphical output can become extensive and hard to decipher when used in the analysis of complex accidents.
- Almedia and Johnson (2005) argue that it is unable to adequately explore the local rationality of those involved in the accident scenario.

Related Methods

In order to conduct AcciMap analysis, data collection methods such as interviews and document reviews must be utilised first.

AcciMap analysis is based on the model proposed by Reason (1990) and as such is related to methods such as the Incident Cause Analysis Method (ICAM) and HFACS (Wiegmann and Shappell, 2003) which are also developed from this model.

Approximate Training and Application Times

AcciMaps is a simple method to learn and apply, but can become time-consuming when applied to complex systems. Estimated timescales for the method are expected to be around one to two weeks for data collection and a further week for the initial construction of the AcciMap. However, the final procedural stage of review can take additional time.

Reliability and Validity

There is no reliability and validity data available, but the reliability of the approach is questionable due to the low level of guidance provided and the consequent reliance on the subjective judgment of the analyst to classify causal factors.

Tools Needed

Pen and paper are all that are required for AcciMaps analysis; however, Microsoft Office tools such as Visio can be used to create more professional AcciMaps.

Example

Jenkins et al. (2010a) employed the AcciMap methodology in order to analyse the shooting of Charles de Menezes in 2005 at a London Underground station.

Charles de Menezes was an electrician who had moved to the UK from Brazil. On 22 July 2005, he was shot dead by the UK's Metropolitan Police Service specialist firearms department. Due to a case of mistaken identity, he had been targeted as a suspected suicide bomber. The analysis enabled the identification of failures across all six systemic levels and the ability to trace errors as the errors moved through these levels. An extract is presented below that illustrates the causal factors across the systemic levels which contributed to the misidentification of Charles de Menezes. The example illustrates only one factor from each systemic level; for the complete AcciMap, the reader is referred to Jenkins et al. (2010a).

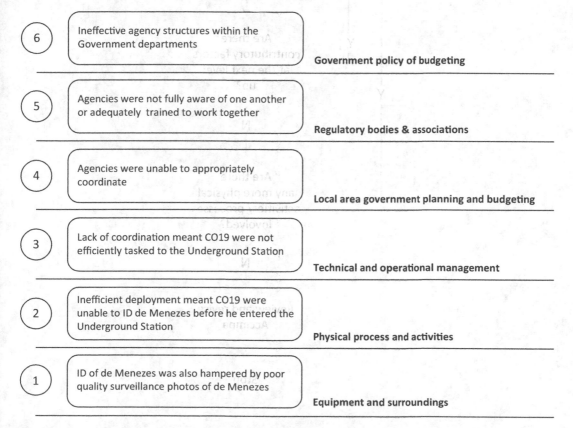

Figure 6.15   AcciMap analysis of the London Underground shooting

Flowchart

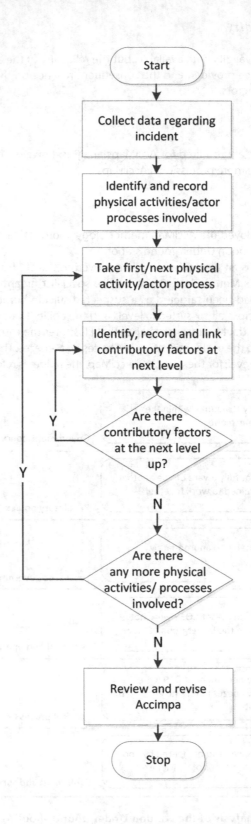

The Functional Resonance Accident Model (FRAM)

Background and Applications

FRAM (Hollnagel, 2004) is an accident analysis methodology designed to elicit information about how variations in the performance of individuals, technology and organisations impact accident aetiology with complex socio-technical systems (Hollnagel et al., 2008). It emphasises the importance of the relationships between causal factors, arguing that 'accidents can be the outcome of unexpected combinations of normal performance variability, i.e., arise from normal actions' (Hollnagel et al., 2008: 1). It is based upon the core supposition that task performance is in a state of constant variation in line with changing demands and that humans must adapt their performance in order to maintain effective task completion. Hollnagel et al. (2008) argue that although this variation is normally beneficial, it can also be inappropriate. Although each act of variability alone may have little effect on the system, the variability within the system that results from multiple adaptations can interact in unanticipated ways and produce accident scenarios (Hollnagel et al., 2008).

FRAM has its own dedicated website (see https://sites.google.com/site/erikhollnagel2/fram), which contains detailed information on the background, methodological procedure, publications and teaching material for the method.

The aim of FRAM is to identify potential variability within the functions of a system and depict the manner in which such variability could have interacted to cause an accident (Hollnagel et al., 2008). Once potential variability has been identified, FRAM defines a series of barriers, or recommendations, that can prevent the variability from evolving into an accident in the future.

In order to identify variability, FRAM uses six parameters, as illustrated below in Table 6.26, in addition to a series of context dependent performance indicators, as presented in Table 6.27.

Table 6.26   FRAM parameters (adapted from Hollnagel et al., 2008)

Input.	The input into the function process or the trigger of the function process.
Precondition.	The preconditions necessary for the function to occur.
Resource.	The resources required for the function to create the output.
Output.	The result of the function.
Time.	The time constraints which impact upon the function.
Control.	The method through which the function is controlled.

Table 6.27   FRAM common performance conditions (adapted from Hollnagel, 2004)

Personnel and material.	The quality of personnel and materials.
Training and experience.	The quality and level of experience and training.
Quality of communication.	The efficiency and adequacy of communications.
Human–Machine Interface (HMI) and operational support.	The quality of HMI design.
Access to procedures and methods.	The ability to access procedures, such as emergency procedure guides.
Conditions of work.	The physical aspects of the work environment.
Number of goals and conflict resolution.	The average number of tasks assigned to each team member and their strategy for resolving task conflicts.
Available time.	The time available to perform each task.
Circadian rhythm.	The adaptation to the current time zone (e.g. night shift adaptation).
Crew collaboration quality.	The level of formal and informal cohesion and collaboration within a team.
Quality and support of organisation.	The quality of team roles and of the organisation's safety culture.

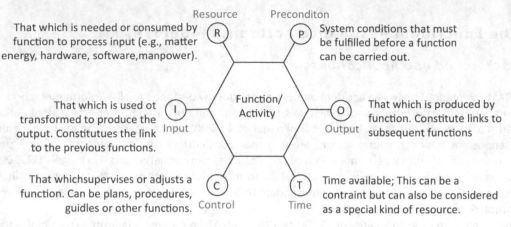

Figure 6.16   An example of the FRAM unit

Domain of Application

The method is generic and can be applied to other complex systems.

Procedure and Advice (Hollnagel, 2004; Hollnagel et al., 2008)

Step 1: Define the Task under Analysis
The first stage of FRAM involves producing a clear definition of the task under analysis. This definition should be comprehensive in order to guide the analyst in conducting a relevant investigation.

Step 2: Understand the System under Analysis
The next stage of the FRAM analysis involves collecting and reviewing data on the system under analysis.

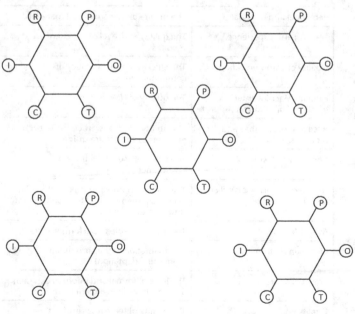

Step 3: Define Core System Functions
Using the six function categories (input, pre-condition, resource, output, time and control, presented in Table 6.26), the analyst should identify all relevant functions of the system under analysis.

Hollnagel et al. (2008) suggest two formats for presenting this data: a tabular format and a graphical format (using hexagonal templates). The blank hexagonal functions shown on the right would include text and shading to represent the essential functions of the system.

It is important for the analyst to remember that at this stage, the focus is on normal, or non-

Figure 6.17   Hexagonal functions (adapted from Hollnagel
et al., 2008)

accident, system functioning. The aim of this stage is to create a baseline against which to compare the accident scenario.

Step 4: Identify Variation of System Functions
Once the baseline of system functions is complete, the analyst should move on to identify all variations from this that occurred within the accident scenario. The context of each function is explored and then the variability of each function is characterised using the 11 common performance conditions presented in Table 6.27.

The analyst should take each function in turn and provide a rating for each of the common performance conditions. FRAM provides three ratings: stable or variable but adequate; stable or variable but inadequate; and unpredictable.

This stage of analysis should continue until all aspects of the accident scenario have been described.

Step 5: Functional Resonance Description
Hollnagel et al. (2008) describe how at this stage, the analyst should have a list of functions for the non-accident system (step 3) and a list of functions for the accident system (step 4). Next, the analyst must attempt to define the relationships between the functions, which is conducted by identifying functions that input into one another, constrain certain activities and so on. The analyst should take each function and compare it to every other function in order to define which functions are related to which others. Hollnagel et al. propose that, once complete, this stage of analysis enables a depiction of the accident aetiology.

Hollnagel et al. (2008) recommend that the analysis conducted within this stage be formatted into a graphical display illustrating a summary of the interrelationships between functions using the hexagonal templates. Nouvel, Travadel and Hollnagel (2007) provide a description of FRAM visualiser: a software application designed to aid in the analysis and development of these graphical illustrations.

Step 6: Development of Counter-measures
The final stage of a FRAM analysis involves the identification of counter-measures or barriers within the FRAM terminology. Hollnagel et al. (2008) divide barrier measures into two main categories: systems (the construct type of the barrier) and functions (the way in which the barrier works). Guidance on the development of these barriers is provided, with four barrier types described:

- physical barriers – barriers constraining movement;
- functional barriers – pre-conditions that must be met;
- symbolic barriers – physical constraints;
- incorporeal barriers – non-physical constraints.

Hollnagel et al. (2008) also suggest that the FRAM analysis enables the analyst to identify appropriate techniques and performance indicators to gauge system safety and identify when variability may become inappropriate in the future.

Tools Needed

The method can be conducted with paper and pen, although Nouvel, Travadel and Hollnagel (2007) and Bergqvist et al. (2007) also describe a software application to aid in the development of the graphical summary illustrations: FRAM visualiser.

Advantages

- FRAM can be applied both retrospectively to accident analysis and predictively to risk assessment.

- It is a systemic method which explores the interactions between causal factors in a system (Hollnagel et al., 2008).
- It can identify causal factors that were not explored in the accident report (Hollnagel et al, 2008).
- It provides guidance on the development of counter-measures (barriers)
- It aligns with popular notions of systems theory.
- It has a strong theoretical basis (Hollnagel et al., 2008).
- It provides a graphical illustration of the accident (Hollnagel et al., 2008).
- It is simple to learn and use (Hollnagel et al., 2008).

Disadvantages

- Detailed and explicit performance indicators are required to identify unsafe performance (Dijkstra, 2011).
- It requires the analyst to have an indepth knowledge of the system under analysis.
- It is dependent upon the quality and degree of data gathered.
- It can be time-consuming (Hollnagel et al., 2008).

Related Methods

FRAM analysis requires the use of multiple data collection methods such as interviews and generally makes use of HTA.

As an accident analysis method based on the systems perspective, FRAM is related to methods such as STAMP and AcciMap which also aim to explore the system as a whole and the role of interrelationships between causal factors.

Approximate Training and Application Times

Hollnagel et al. (2008) propose that the method is easy to learn and apply, although the data collection and analysis stages may be time-consuming.

Reliability and Validity

Currently, there is no empirical data regarding the reliability or validity of FRAM, although several studies have subjectively evaluated the method. Dijkstra (2011) compared FRAM to other techniques and concluded that the common performance conditions within FRAM provided the most comprehensive and appropriate set of conditions. Woltjer (2010) examined the ability of FRAM to evaluate core factors of safety, including flexibility, tolerance and safety margins. The evaluation concluded that FRAM was able to provide an indepth assessment of each of the safety components suggested.

Example

Hollnagel et al (2008) applied the FRAM method to the crash of Comair Airlines Flight 5191 in Lexington, KY on 27 August 2006. The analysis explored why a Comair flight followed an incorrect taxi path and attempted take-off on a shorter runway, which resulted in the aircraft engaging in an unsuccessful take-off, colliding with trees, crashing and catching fire. Hollnagel et al. (2008) describe analysis beginning with the identification of the main functions associated with the plane taking off, such as the function *Taxi to runway*; in order to complete its function, the plane must taxi to the runway. According to Hollnagel et al. (2008), once a large proportion of the core functions had been identified, the analyst attempted to understand them in more detail, deriving the associated contextual conditions.

The separate functions of the FRAM analysis (each having been explored in detail) were then collated into a complete FRAM network representing the system, all of its essential functions and the way in which these functions were interrelated. An example of this is too large to be shown here, but a template extract of the composition is presented below in Figure 6.18. For a complete, data-populated illustration, the reader is directed to Hollnagel et al. (2008). Hollnagel et al.'s analysis of the incident using the FRAM technique enabled insights into the causality associated with the incident, including numerous causal factors that were not identified by the official inquiry.

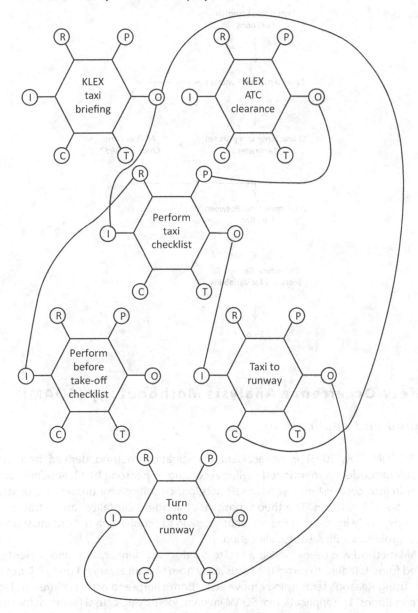

Figure 6.18   Example of an FRAM network representing the system with essential and interrelated functions

Flowchart

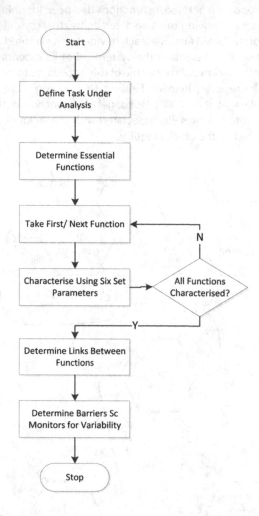

The Safety Occurrence Analysis Methodology (SOAM)

Background and Applications

SOAM (EUROCONTROL, 2005) is an accident investigation method derived from Reason's 'Swiss Cheese' accident model. As mentioned earlier, according to Reason (1990), accidents are the result of breakdowns in four core system layers: unsafe acts; pre-conditions for unsafe acts; unsafe supervision; and organisational influences. The theory posits that accident causality can be traced back from the physical acts explicitly involved in the accident scenario to problems in the organisation of the system, for example, problems with resource allocation.

The SOAM method was developed as a tool to conduct accident analysis and present the results in a standardised format. It aims to move the focus away from human error and on to failures deeper within the system, using Reason's terminology move away from 'sharp-end active failures' and focus on 'blunt end latent conditions'. The application of SOAM involves identifying causal factors at four systemic levels:

- failed barriers;
- human involvement;

- contextual conditions;
- organisational factors.

Once all causal factors have been identified, the analyst develops a series of safety recommendations in order to increase the resilience of the system and prevent the future occurrence of similar accidents.

Domain of Application

SOAM was used by EUROCONTROL for the analysis of ATM occurrences for a period of over five years.

Procedure and advice (EUROCONTROL, 2005)

Step 1: Understand the System under Analysis

The first stage of the analysis should involve the analyst gaining a comprehensive knowledge of the system and the accident scenario under investigation. The accident investigation report should be studied along with any other relevant documentation, and interviews with subject matter experts and those involved should be conducted where possible.

Step 2: Sort Data

The analyst must now sort the data gathered in step 1. Each piece of information is classified through the use of two tests:

(a) Test 1: does the fact represent a condition or event that contributed to the eventual occurrence?

Each piece of information that is not a contributory factor is removed from the analysis at this stage, although a record of the information is maintained.

Each piece of information that is a contributory factor is then assessed using test 2.

(b) Test 2: under which category can the fact be classified?

This test allows for the contributory factors to be divided into one of the four SOAM categories or system levels: barrier; human involvement; contextual condition; or organisational factor. Questions are provided at each system level in order guide the analyst further in this categorisation.

Step 3: Identify Barriers

In order to identify and classify the failed barriers within the system under analysis, a check question is applied: does the item describe a work procedure, aspect of human awareness, physical obstacle, warning or control system, or protection measure designed to prevent an occurrence or lessen its consequences?

Based upon the answer to this question, the analyst can place the barrier into one of the six categories used to classify the ineffective barriers (see Figure 6.19).

Step 4: Identify Human Involvement

At this stage of the analysis, the analyst must identify and classify the causal factors that represent human involvement. Again, a check question is provided to aid the analyst: does the item describe an action or non-action taking place immediately prior to and contributing to the occurrence?

Step 5: Identify Contextual Conditions

The analyst should now identify the pre-existing conditions within the system that may have affected the accident aetiology. Again, a check question is provided: does the item describe an aspect of the workplace, local organisational climate or a person's attitudes, personality, performance limitations, physiological or emotional state that helps explain their actions?

This question allows the analyst to classify the contextual condition into one of the five categories (see Figure 6.20).

Step 6: Identify Organisational Factors

The final systemic level is organisational factors. Each organisational factor is classified into one of 12 categories (see Figure 6.21) through the use of a check question: does the item describe an aspect of the organisation's culture, systems, processes, or decision-making that existed before the occurrence and which resulted in the contextual conditions or allowed those conditions to continue?

Organisational factors are then described in detail based on three concepts: definition; indicators and consequences.

Step 7: Production of the SOAM Chart

Once all causal factors have been identified, the analyst should collate the data into a simple graphical format – the SOAM chart. The chart should contain all causal factors identified in the analysis grouped into the appropriate system levels. Links should be annotated onto the chart in order to ensure that relationships between factors are clearly represented. The links between factors are identified by ascertaining which factors influenced which others.

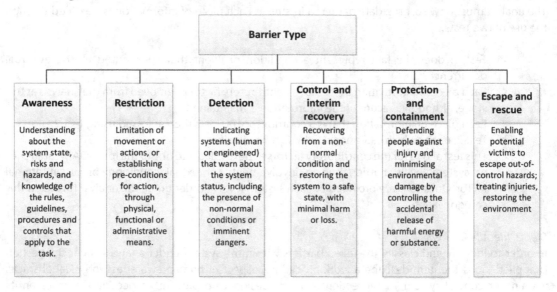

Figure 6.19   Barrier-type definitions (EUROCONTROL, 2005)

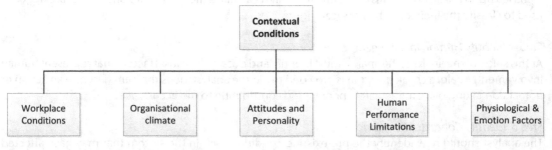

Figure 6.20   Contextual conditions categories (EUROCONTROL, 2005)

Figure 6.21   Organisational factor categories (EUROCONTROL, 2005)

Step 8: Identify Recommendations
The final part of the SOAM analysis is the identification of recommendations to reduce the risk of accidents evolving. Recommendations are focused on two areas: deficient barriers and organisational factors. For each factor identified in the analysis, the analyst must derive a corresponding recommendation.

Advantages

- According to Arnold (2009), the method is a 'useful heuristic and a powerful communication device'.
- SOAM is quick and easy to apply.
- It provides a clear summary of the analysis.
- It provides a standardised process and terminology.
- It offers an overview of casual factors within a system.
- It creates a strong framework which allows reliability and removes a great deal of subjective judgment from the analysis.

Disadvantages

- Arnold (2009) posits that the method fails to comprehensively explore 'emergent phenomena' and 'non-linear interactions'.
- It does not fully explore the manner in which the causal factors occur (Arnold, 2009).

Reliability and Validity

Licu et al. (2007) posit that the formal structure of the method and detailed descriptions required for each factor suggest that SOAM could have high levels of reliability.

Approximate Training and Application Times

SOAM is a simple method which can be conducted with little training in a short time period.

Tools Needed

SOAM analysis requires pen and paper. Microsoft Visio, or similar drawing tools, can be used to create well-presented SOAM charts.

Related Methods

SOAM is similar to other methods based upon the Swiss Cheese model such as HFACS.

Example

EUROCONTROL (2005) presents a number of examples of the SOAM technique being applied, and one such example is reviewed here. EUROCONTROL employed SOAM in order to investigate the causal factors associated with a near-mid-air collision between two planes over Sydney in 1991. Both planes were attempting to land on intersecting runways at Sydney Airport and came within 33 metres of colliding. Plane A was given permission to land on runway 1 and was told to stop before the intersection with a second runway (runway 2). Plane B was given permission to land on runway 2 and was told to stop before the intersection with runway 1. When plane A landed, it became clear that it would not stop before the intersection – this was because of incomplete dispersion of the order, inaccurate knowledge of the intersection location and an automatic brake malfunction. The crash was averted as plane B elected to conduct a go-around and not land immediately.

EUROCONTROL began its analysis by identifying the causal factors or conditions surrounding the accident. Once identified, the causal factors and conditions were categorised into the appropriate SOAM category: barrier; human involvement; contextual condition; or organisational factor. An example of a causal factor from each category is presented below:

- Barrier: lack of positive sequencing of aircraft landing on runways 34 and 25.
- Human involvement: the captain of plane A did not devote sufficient attention the task of landing.
- Contextual condition: the captain of plane A was under additional strain due to his responsibility to train an inexperienced co-pilot.
- Organisational factor: the inherent risk in simultaneous procedures which were fundamentally weak and relied on near-perfect human performance.

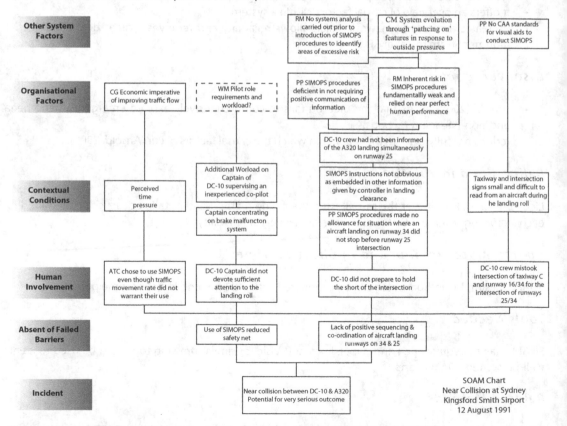

Figure 6.22   Extract of SOAM chart (adapted from EUROCONTROL, 2005)

Each of these errors is further classified into type, for example, the lack of sufficient attention provided by pilot A was classified as human involvement – incorrect execution.

The result of the EUROCONTROL (2005) analysis was a set of classified causal factors which were then linked to create a representation of the accident aetiology. EUROCONTROL created a graphical representation of this aetiology, an extract of which is presented in Figure 6.22 on the previous page.

Flowchart

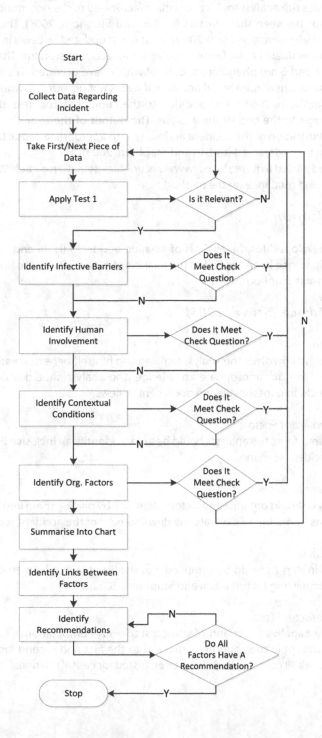

Why-Because Analysis (WBA)

Background and Applications

WBA (Ladkin and Loer, 1998) is an accident analysis methodology which utilises formal mathematical equations to identify causal factors in complex systems. The technique is based upon the assumption that formal logic enables the analyst to discover which factors led to the occurrence of an accident and the causal relationships between these factors (Ladkin and Stuphorn, 2003). The method utilises the counter-factual test initially developed over 200 years ago and updated by Lewis in 1973 (cited in Ladkin and Stuphorn, 2003) to evaluate causal factors involved in an accident scenario. The counter-factual test posits that 'Suppose A and B are phenomena, both of which have occurred. Then phenomenon A is a necessary causal factor of phenomenon B if and only if had A not occurred, B would not have occurred either' (Ladkin and Stuphorn, 2003: 5). In addition to the counter-factual test, the method provides a formal validation stage at the end of the analysis. The output of the analysis is a WB graph, which presents a graphical illustration of the accident aetiology with appropriate causal factors and the causal relationships between them depicted (Ladkin and Stuphorn, 2003).

WBA has a website devoted to it (see http://www.rvs.uni-bielefeld.de/research/WBA/), which contains multiple publications and guidance on the method.

Domain of Application

WBA was originally developed for the analysis of aviation accidents (Höhl and Ladkin, 1997), but is a generic method and has since been applied to the analysis of rail accidents (Ladkin, 2005) and incidents of friendly fire (Ladkin and Stuphorn, 2003).

Procedure and Advice (Stuve, 2005)

Step 1: Data Collection
The initial stage of analysis involves the analyst familiarising himself or herself with the scenario under investigation. In order to gain appropriate knowledge, the analyst should utilise a variety of data collection methods, including interviews and document reviews.

Step 2: Determine Significant Factors
From the data collection stage, the analyst should be able to identify an inclusive list of factors that may have impacted the accident scenario.

Step 3: Develop Incident Timeline
At this stage, the analyst should organise the factors identified by placing them into a temporal sequence and displaying them as a timeline to illustrate the development of the accident scenario.

Step 4: Derive List of Facts
The factors identified in step 2 should be compiled into the list of facts. The method includes guidance on the identification of missing facts (Ladkin and Stuphorn, 2003).

Step 5: Apply Counter-factual Test
At this stage, the analyst applies the counter-factual test to every combination of facts. Starting with the first fact, the analyst will apply the counter-factual test to the first and second facts, then the first and third facts and so on until all combinations have been tested for causal relationships.

Step 6: Produce Why-Because Graph
The analyst now annotates this information onto a WB graph. The analyst should place the accident as the top event and plot all necessary causal factors underneath. The factors are linked to one another in order to represent their causal relationships and provide an overview of the accident scenario.

Step 7: Formal Verification
Once complete, the analyst can use WBA formal verification criteria to ensure the applicability of the WB graph. Many analyses successfully apply WBA without completing step 7 due to its time-intensive nature (Ladkin and Stuphorn, 2003).

Advantages

- The method enables the reliable identification of causal factors and the relationships between them (Ladkin and Stuphorn, 2003).
- It provides guidance to the analyst regarding the factors which require further investigation (Ladkin and Stuphorn, 2003).
- It provides a clear graphical illustration of root causes and causal relationships (Ladkin, 2005; Braband, Evers and Stefano, 2003).
- According to Qureshi (2007), formal methods have been successfully applied to the design and certification of safety-critical systems.
- It utilises a formal definition of causality (Braband, Evers and Stefano, 2003).
- It can discover causal factors that were dismissed by the accident analysis report as having no effect (Qureshi, 2007).
- It can be used both reactively and proactively.

Disadvantages

- The method requires experts in formal logic (Qureshi, 2007).
- The formal verification stage is time-consuming (Ladkin and Stuphorn, 2003).
- It is based on an accident analysis report and therefore is dependent on the quality of the accident analysis conducted.
- WBA graphs can become too big and cumbersome (Qureshi, 2007).
- The strict definition of causality may prevent the identification of human or organisational factors (Braband, Evers and Stefano, 2003).
- Qureshi (2007) states that although formal methods have been successfully applied, they 'need to be extended to capture the many factors and aspects that are found in accidents'.
- An indepth knowledge of the system under analysis is required to link factors to one another.

Training and Application

The method is easy to understand and to apply, with little training required. The original terminology was found to be confusing, leading to the development of a simplified procedure for non-experts (Braband and Brehmke, 2002).

Reliability and Validity

There is no data regarding the reliability or validity of the method.

Tools Needed

The analysis can be conducted using pen and paper. For increased levels of professionalism, Braband and Brehmke (2002) suggest the use of Microsoft Office and its hyperlink facility to link elements of the WB graph to documents supporting the graph.

Ladkin (2005) highlights numerous software tools to aid in WBA, including YBEdit, VDAS, YBFactor and YBTimeliner. YBEdit is a graphical editor assisting in the construction of the WB graphs; VDAS is an archiving system to allow analysts to collaborate on WB Graphs; YBFactor is a tool to aid in the development of the list of causal factors; and YBTimeliner aids the analyst in constructing the timeline.

Related Methods

WBA utilises data collection tools such as interviews and observation. The method is similar to event and fault tree analysis, although WBA adds statistical inferences to the tree mapping.

Example

For multiple examples of the application of WBA to accidents within the domains of rail, air, maritime and computer security, the reader is referred to the WBA website (see http://www.rvs.uni-bielefeld.de/research/WBA/#wba-examples).

Flowchart

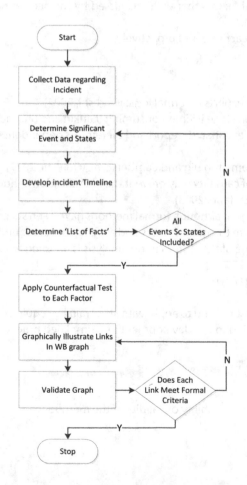

Chapter 7
Situation Awareness Assessment Methods

Introduction

Situation awareness (SA) is the term that is used within HF circles to describe the level of awareness that operators have of the situation that they are engaged in; it focuses on how operators develop and maintain a sufficient understanding of 'what is going on' (Endsley, 1995a) in order to achieve success in task performance. A critical commodity in the safety critical domains, SA is now recognised as a key consideration in system design and evaluation (e.g. Endsley, Bolte and Jones, 2003; Salmon et al., 2009a). Reflecting this, SA has been explored in a range of domains over the past 30 years, ranging from the military (e.g. Salmon et al., 2009a) and transportation domains (e.g. Ma and Kaber, 2005, 2007; Golightly et al., 2010) to sport (James and Patrick, 2004), health care and medicine (Hazlehurst, McCullen and Gorman, 2007), and the emergency services (e.g. Blandford and Wong, 2004). A contentious concept, various models of SA have been postulated, focusing either on the awareness held by individuals (e.g. Endsley, 1995a; Smith and Hancock, 1995), teams (e.g. Endsley and Robertson, 2000; Salas et al., 1995) or socio-technical systems (e.g. Artman and Garbis, 1998; Salmon et al., 2009a). As a corollary, various different approaches for assessing SA have been developed and applied in a range of domains.

Situation Awareness Theory

Various definitions of SA are presented in the academic literature (e.g. Adams, Tenney and Pew, 1995; Bedny and Meister, 1999; Billings, 1995; Dominguez, 1994; Fracker, 1991; Sarter and Woods, 1991; Smith and Hancock, 1995; Stanton et al., 2006b; Taylor, 1990). Still by far the most prominent is that offered by Endsley, who defines SA as a cognitive product (resulting from a separate process labelled situation assessment) comprising 'the perception of the elements in the environment within a volume of time and space, the comprehension of their meaning, and the projection of their status in the near future' (Endsley, 1995a: 36). More appropriate, given the increased presence of complex socio-technical systems in which teams of humans together collaboratively, is Stanton et al.'s (2006b) systems theory-oriented definition, which asserts that SA represents 'activated knowledge for a specific task, at a specific time within a system' and echoes Bell and Lyon's (2000) presumption that SA can be defined as knowledge in working memory about elements in the environment.

Models Accounting for SA in the Head

Inaugural SA models were, in the main, focused on how individual operators develop and maintain SA whilst undertaking activity within complex systems (e.g. Adams, Tenney and Pew, 1995; Endsley, 1995a; Smith and Hancock, 1995). As such, these models primarily focus on the awareness held in the minds of individual operators, that is, SA as experienced in the mind of the person (Stanton et al., 2010b). Two of these models in particular stand out from the literature, one being by far the most popular, the other being, in the author's view at least, the most appropriate for describing the concept, albeit at an individual level. Endsley's three-level model (Endsley, 1995a) has undoubtedly received the most attention. The information processing-based model describes SA as an internally held cognitive product comprising three levels (see Figure 7.1) that follows perception and leads to decision-making and action execution.

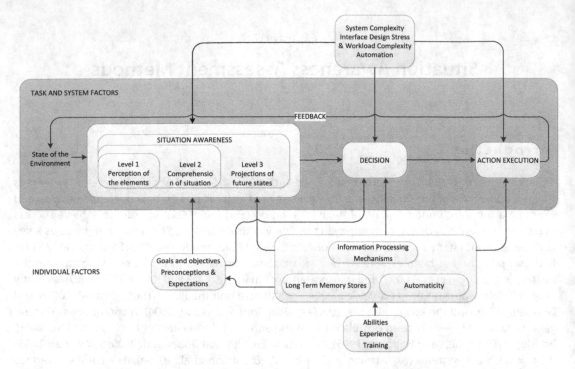

Figure 7.1   Endsley's three-level model of situation awareness (adapted from Endsley, 1995a)

The first level involves perceiving the status, attributes and dynamics of task-related elements in the surrounding environment. A range of factors influence the data perceived, including the task being performed, the individual's goals, experience, expectations and also systemic factors such as interface design, level of complexity and automation. To achieve level 2 SA, the individual interprets the level 1 data in such a way that he or she can then comprehend its relevance to his or her task and goals. The interpretation and comprehension of SA-related data is influenced by an individual's goals, expectations, experience in the form of mental models and pre-conceptions regarding the situation. Level 3 SA involves prognosticating future system states. Using a combination of level 1 and 2 SA-related knowledge and experience in the form of mental models, individuals can forecast likely future states in the situation. For example, a land warfare commander can forecast, based on level 1 and 2-related information, that an enemy unit might attack in a given manner. The commander can do this through perceiving elements such as the location of an enemy unit, its formation, the number of troops, comprehending what the elements mean and then comparing this to experience (in the form of mental models) to forecast what might happen next. Mental models are therefore used to facilitate the achievement of SA by directing attention to critical elements in the environment (level 1), integrating the elements to aid understanding of their meaning (level 2) and generating possible future states and events (level 3). According to the model, SA acquisition and maintenance is influenced by other factors, including individual factors (e.g. training, workload), task factors (e.g. complexity) and systemic factors (e.g. interface design) (Endsley, 1995a).

Smith and Hancock's ecological approach, based on Neisser's (1976) perceptual cycle model, takes a more holistic stance, viewing SA as a 'generative process of knowledge creation and informed action taking' (Smith and Hancock, 1995: 138). According to the perceptual cycle model, our interaction with the world (termed explorations) is directed by internally held schemata. The outcome of interaction modifies the original schemata, which in turn directs further exploration. This process of directed interaction and modification continues in an infinite cyclical nature. Using this model, Smith and Hancock (1995) suggest that SA is neither resident in the world nor in the person, but resides through

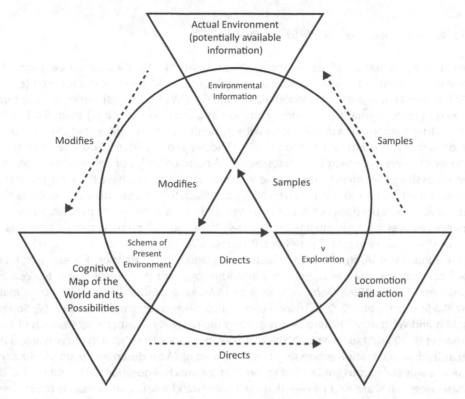

Figure 7.2   Smith and Hancock's perceptual cycle model of situation awareness (adapted from Smith and Hancock, 1995)

the interaction of the person with the world. They describe SA as 'externally, directed consciousness' that is an 'invariant component in an adaptive cycle of knowledge, action and information' (1995: 138). In addition, they argue that the process of achieving and maintaining SA revolves around internally held schema, which contain information regarding certain situations. These schema facilitate the anticipation of situational events, directing an individual's attention to cues in the environment and directing his or her eventual course of action. An individual then conducts checks to confirm that the evolving situation conforms to his or her expectations. Any unexpected events serve to prompt further search and explanation, which in turn modifies the operator's existing model. The perceptual cycle model of SA is presented in Figure 7.2.

Smith and Hancock identify SA as a sub-set of the content of working memory in the mind of the individual (in one sense it is a product). However, they emphasise that attention is externally directed rather than introspective (and thus is contextually linked and dynamic). Unlike the three-level model, which depicts SA as a product separate from the processes used to achieve it, SA is therefore viewed as both process and product. Smith and Hancock's complete model therefore views SA as more of a holistic process that influences the generation of situational representations. The model has sound underpinning theory (Neisser, 1976) and is complete in that it refers to the continuous cycle of SA acquisition and maintenance, including both the process (the continuous sampling of the environment) and the product (the continually updated schema) of SA. Their description also caters for the dynamic nature of SA and more clearly describes an individual's interaction with the world in order to achieve and maintain SA, whereas Endsley's model seems to place the individual as a passive information receiver. The model therefore considers the individual, the situation and the interactions between the two.

Models Accounting for SA Held by Teams

Of course, due to the nature of teamwork and the systems in which it takes place, team SA has to be more complex than individual SA. Various models of team SA have been proposed (e.g. Salas et al., 1995; Shu and Furuta, 2005; Stanton et al., 2006b; 2009; Wellens, 1993). Reflecting the three-level model's widespread popularity, the most common approach to describing team SA has involved applying Endsley's model to team SA, along with the addition of the related but distinct concepts of 'team' and 'shared' SA (e.g. Endsley and Jones, 1997; Endsley and Robertson, 2000). Team SA reflects 'the degree to which every team member possesses the SA required for his or her responsibilities' (Endsley, 1989), whereas shared SA refers to 'the degree to which team members have the same SA on shared SA requirements' (Endsley and Jones, 1997). Endsley's approach to team SA therefore suggests that team members have distinct portions of SA, but also overlapping or 'shared' portions of SA. Successful team performance requires that individual team members have good SA on their specific elements and the same SA for shared SA elements (Endsley and Robertson, 2000).

Much like individual SA, the concept of team SA is plagued by contention. For example, many have expressed concern over the use of Endsley's individual operator three-level model to describe team SA (Artman and Garbis, 1998; Gorman, Cooke and Winner, 2006; Patrick et al., 2006; Salmon et al., 2006, 2009a; Stanton et al., 2009; Shu and Furuta, 2005; Siemieniuch and Sinclair, 2006; Sonnenwald, Maglaughlin and Whitton, 2004) and also regarding the relatively blunt characterisation of shared SA (e.g. Salmon et al., 2009a; Stanton et al., 2009, 2010b). Putting aside the obvious concerns associated with using an individual information processing-based account of SA for describing team SA, the concept of shared SA remains ambiguous, and recent research in the area has questioned the notion that different human operators can 'share' SA in a way that they understand a situation in exactly the same manner (e.g. Salmon et al, 2009a; Stanton et al., 2010b).

Models Accounting for SA Held by Systems

Recent advances in the area have led to the description of SA as a social phenomenon that is held by systems comprising human and technological agents. Known as 'distributed situation awareness' (DSA; Salmon et al., 2009a; Stanton et al., 2006b), this approach uses as its basis distributed cognition-based accounts of system performance (e.g. Hutchins, 1995), which move the focus on cognition out of the heads of individual operators and on to the overall system consisting of human and technological agents; here cognition transcends the boundaries of individual actors and 'systemic' cognition is achieved by the transmission of representational states throughout the system (Hutchins, 1995).

SA was first discussed in this context by Artman and Garbis (1998), who, due to the flaws evident when applying individualistic models to complex socio-technical systems, called for a systems perspective model on SA. They subsequently defined team SA as, "the active construction of a model of a situation partly shared and partly distributed between two or more agents, from which one can anticipate important future states in the near future" (Artman and Garbis, 1998: 2). Following this, the foundations for a theory of DSA in complex systems, laid by Stanton et al. (2006) were built upon by Salmon et al. (2009a) who outlined a model of DSA, developed based on applied research in a range of military and civilian command and control environments.

Briefly, Stanton et al.'s model is underpinned by four theoretical concepts: schema theory (e.g. Bartlett, 1932), genotype and phenotype schema, Neisser's (1976) perceptual cycle model of cognition and, of course, Hutchin's (1995) distributed cognition approach. SA is viewed as an emergent property of collaborative systems, arising from the interactions between agents, both human and technological. According to Stanton et al. (2006b, 2009), a system's awareness comprises a network of information on which different components of the system have distinct views and ownership of information. Scaling the model down to individual team members, it is suggested that team member SA represents the state of their perceptual cycle (Neisser, 1976); individuals possess genotype schema that are triggered by the task-relevant nature of task performance, and during task performance, the phenotype schema comes to the fore. It is this task and schema-driven content of team member SA that brings the shared SA (e.g. Endsley and Robertson, 2000) notion into question. Rather than possess shared SA (which suggests that team members understand a situation or elements of a situation in the same manner), the model instead suggests that team members possess unique, but compatible, portions of awareness. Team members experience a situation in different ways, as defined by their own personal experience, goals, roles, tasks, training, skills, schema and so on. Compatible awareness is therefore the phenomenon that holds distributed systems together (Stanton et al., 2006b, 2009). Each team member has his or her own awareness related to the goals that he or she is working towards. This is not the same as other team members, but is such that it enables him or her to work with adjacent team members. Although different team members may have access to the same information, differences in goals, roles, the tasks being performed, experience and their schema mean that their resultant awareness of it is not shared; instead, the situation is viewed differently based on these factors. However, each team member's SA is compatible since it is different in content but is collectively required for the system to perform collaborative tasks optimally.

The compatible SA view does not discount the sharing of information, nor does it discount the notion that different team members have access to the same information; this is where the concept of SA 'transactions' applies. Transactive SA describes the notion that DSA is acquired and maintained through transactions in awareness that arise from communications and the sharing of information. A transaction in this case represents an exchange of SA between one agent and another (where 'agent' refers to humans and artefacts). Agents receive information, which is integrated with other information and acted on, and then is passed on to other agents. The interpretation of that information changes depending on the team member. The exchange of information between team members leads to transactions in SA; for example, when a soccer coach provides instruction to a soccer team during the game, the resultant transaction in SA for each player is different, depending on his or her role in the team. Each player is using the information for his or her own ends, integrated into his or her own schemata and used to reach an individual interpretation. Thus, the transaction is an exchange rather than a sharing of awareness. Each agent's SA (and so the overall DSA) is therefore updated via SA transactions. Transactive SA elements from one model of a situation can form an interacting part of another without any necessary requirement for parity of meaning or purpose; it is the systemic transformation of situational elements as they cross the system boundary from one team member to another that bestows upon team SA an emergent behaviour.

Measuring SA

The popularity of the concept is such that various methods of measuring SA have been proposed and applied (for detailed reviews, see Salmon et al., 2006, 2009a). These SA measures can be broadly categorised into the following methodological groups: SA requirements analysis, freeze probe recall methods, real-time probe methods, post-trial subjective rating methods, observer rating methods, process indices and team SA measures. SA requirements analysis forms the first step in an SA assessment effort and is used to identify what exactly it is that comprises SA in the scenario and environment in question. Endsley defines SA requirements as 'those dynamic information needs associated with the major goals or sub-goals of the operator in performing his or her job (2001: 8). According to Endsley, they concern not only the data that operators need but also how the data is integrated to address decisions. Matthews, Strater and Endsley (2004) highlight the importance of conducting SA requirements analysis when developing reliable and valid SA metrics. Freeze probe methods, popular in this area initially due to their relationship with Endsley's three-level model, involve the administration of SA-related queries during 'freezes' in a simulation of the task under

Table 7.1   SA assessment methods summary table

Method	Type of method	Domain	Training time	Application time	Related methods	Tools needed	Validation studies	Advantages	Disadvantages
SAGAT.	Freeze online probe technique.	Aviation (military).	Low.	Med.	SACRI. SALSA.	Simulator, computer.	Yes.	1) Widely used in a number of domains. 2) Subject to numerous validation studies. 3) Removes problems associated with collecting SA data post-trial.	1) Requires expensive simulators. 2) Intrusive to primary task. 3) Substantial work is required to develop appropriate queries.
Propositional networks.	Modelling technique.	Generic.	Low.	High.	Semantic networks. Concept maps.	Pen and paper.	No.	1) Explores DSA. 2) Explores SA at multiple systemic levels. 3) Considers mapping between elements of information.	1) Can be time-consuming and laborious. 2) Can become unwieldy for complex systems. 3) More focused on SA modelling than measurement.
SART.	Self-rating technique.	Aviation (military).	Low.	Low.	CARS. MARS. SARS.	Pen and paper.	Yes.	1) Quick and easy to administer. Also low cost. 2) Generic – can be used in other domains. 3) Widely used in a number of domains.	1) Correlation between performance and reported SA. 2) Participants are not aware of their low SA. 3) Construct validity is questionable.
SA-SWORD.	Paired comparison technique.	Aviation.	Low.	Low.	SWORD. Pro-SWORD.	Pen and paper.	Yes.	1) Easy to learn and use. Also low cost. 2) Generic – can be used in other domains. 3) Useful when comparing two designs.	1) Post-trial administration – correlation with performance, forgetting, etc. 2) Limited use and validation evidence. 3) Does not provide a measure of SA.
SPAM.	Real-time probe technique.	ATC.	High.	Low.	SASHA_L.	Simulator, computer, telephone.	Yes.	1) No freeze required.	1) Low construct validity. 2) Limited use and validation. 3) Participants may be unable to verbalise spatial representations.
SA requirements analysis.	N/A	Aviation, generic.	High.	High.	Interview. Task analysis. Obs. Quest.	Pen and paper, recording equipment.	No.	1) Specifies the elements that comprise SA in the task environment under analysis. 2) Can be used to generate SA queries/probes. 3) Has been used extensively in a number of domains.	1) A huge amount of resources are required. 2) Analysts may require training in a number of different HF techniques, such as interviews, task analysis and observations.

analysis. Participant responses are compared to the state of the system at the point of the freeze and an overall SA score is calculated. Due to criticisms regarding the high level of intrusion on task performance imposed by simulation freezes, real-probe techniques were developed, whereby SA queries are administered during task performance but with no freeze of the task. Self-rating methods involve participants providing subjective ratings of their own SA post-trial using pre-defined rating scales. Observer rating methods involve SMEs observing participants during task performance and then providing a rating of SA based upon pre-defined observable SA-related behaviours exhibited by those being assessed. Using performance measures to assess SA involves measuring the relevant aspects of participant performance during task performance; depending upon the task, certain aspects of performance are recorded in order to determine an indirect measure of SA. Process indices involve analysing the processes that participants use in order to develop SA during task performance. Examples of SA-related process indices include the use of eye-tracking devices to measure participant eye movements during task performance (e.g. Smolensky, 1993), the results of which are used to determine how the participant's attention was allocated during task performance, and concurrent VPA, which involves creating a written transcript of operator behaviour as he or she performs the task under analysis. Finally, team SA measures represent a general class of measures that have been developed and applied specifically for the assessment of team SA, or SA in collaborative systems. Whilst the classes of SA measures above all represent methods designed to assess the level of SA held by individual operators, team SA methods are used to assess or describe the level of SA possessed by teams or systems. A range of methods exist within this category, including Coordinated Awareness of Situations by Teams (CAST: Gorman, Cooke and Winner, 2006), process tracing (Patrick and Morgan, 2010) and propositional networks (Salmon et al., 2009a).

This chapter focuses on the following SA assessment methods: the SA requirements analysis method (Endsley, 1993); the situation awareness global assessment technique (SAGAT: Endsley, 1995b); the situation present assessment method (SPAM: Durso et al., 1998); the situation awareness rating technique (SART: Taylor, 1990), the SA-subjective workload dominance method (SA-SWORD: Vidulich and Hughes, 1991) and the propositional network method (Salmon et al., 2009a; Stanton et al., 2006b, 2009). VPA, process indices that have also previously been used to assess SA, are described in Chapter 3.

The SA requirements analysis method uses SME interviews and goal-directed task analysis to identify the SA requirements for a particular task or scenario. SAGAT is the most popular freeze probe method and was developed to assess pilot SA-based on the three levels of SA postulated in Endsley's three-level model. The SPAM method is a real-time probe method that was developed for use in the assessment of the SA of air traffic controllers and uses online real-time probes to assess participant SA. SART is a subjective rating method that uses 10 dimensions to measure operator SA. Participant ratings on each dimension are combined in order to calculate a measure of participant SA. SA-SWORD is a subjective rating method that is used to compare the levels of SA afforded by two different displays, devices or interfaces. The propositional network method (Salmon et al., 2009a) is used to model DSA in collaborative environments and represents SA as a network of information elements on which different team members have differing views. A summary of the SA assessment methods described is presented in Table 7.1.

SA Requirements Analysis

Background and Applications

SA requirements analysis is used to identify exactly what it is that comprises SA in the scenario and environment under analysis. According to Endsley (2001), SA requirements concern not only the data that operators need but also how the data is integrated to address decisions. Matthews, Strater and Endsley (2004) suggest that a fundamental step in developing reliable and valid SA metrics is to identify the SA requirements of a given task. Further, the authors point out that knowing what the SA requirements are for a given domain provides engineers and technology developers with a basis to develop optimal system designs that maximise human performance rather than overloading workers and degrading their performance.

Endsley (1993) and Matthews, Strater and Endsley (2004) describe a generic procedure for conducting SA requirements analyses that uses unstructured interviews with SMEs, goal-directed task analysis and questionnaires in order to determine the SA requirements for a particular task or system. Endsley's methodology focuses on SA requirements across the three levels of SA specified in her information processing-based model of SA (level 1: perception of elements, level 2: comprehension of meaning, level 3: projection of future states).

Domain of Application

The SA requirements analysis procedure is generic and has been applied in various domains, including the military (Bolstad et al., 2002; Matthews, Strater and Endsley, 2004) and ATC (Endsley, 1993).

Procedure and Advice

Step 1: Define the Task or Scenario under Analysis
The first step in an SA requirements analysis is to clearly define the task or scenario under investigation. It is recommended that the task be described clearly, including the different actors involved, the task goals and the environment within which the task is to take place. An SA requirements analysis requires that the task be defined explicitly in order to ensure that the appropriate SA requirements are comprehensively assessed.

Step 2: Select Appropriate SMEs
The SA requirements analysis procedure is based upon eliciting SA-related knowledge from SMEs. Therefore, the analyst should next select a set of appropriate SMEs. The more experienced the SMEs are in the task environment under analysis, the better, and the analyst should strive to use as many SMEs as possible to ensure comprehensiveness.

Step 3: Conduct SME Interviews
Once the task under analysis is defined clearly and appropriate SMEs are identified, a series of unstructured interviews with the SMEs should be conducted. First, participants should be briefed on the topic of SA and the concept of SA requirements analysis. Following this, Endsley (1993) suggests that the SME should be asked to describe, in their own words, what they feel comprises 'good' SA for the task in question. They should then be asked what they would want to know in order to achieve perfect SA. Finally, the SMEs should be asked to describe what each of the SA elements identified is used for during the task under analysis, e.g. decision-making, planning and actions. Endsley (1993) also suggests that once the interviewer has exhausted the SME's knowledge, he or she should offer his or her own suggestions regarding SA requirements and should discuss their relevance. It is recommended that each interview is recorded using either video or audio recording equipment. Following completion of the interviews, all data should be transcribed.

Step 4: Conduct Goal-Directed Task Analysis
Once the interview phase is complete, a goal-directed task analysis should be conducted for the task or scenario under investigation. Endsley (1993) prescribes her own goal-directed task analysis method; however, it is also possible to use HTA (Annett et al., 1971) for this purpose, since it focuses on goals and their decomposition. For this purpose, the HTA procedure presented in Chapter 3 should be used. Once the HTA is complete, the SA elements required for the completion of each step in the HTA should be

added. This step is intended to ensure that the list of SA requirements identified during the interview phase is comprehensive. Upon completion, the task analysis output should be reviewed and refined using the SMEs utilised during the interview phase.

Step 5: Compile List of SA Requirements Identified
The outputs from the SME interview and goal-directed task analysis phases should then be used to compile a list of SA requirements for the different actors involved in the task under analysis.

Step 6: Rate SA Requirements
Endsley's method uses a rating system to sort the SA requirements identified based on their importance. These should be compiled into a rating-type questionnaire, along with any other elements that the analyst feels are pertinent. Appropriate SMEs should then be asked to rate the criticality of each of the SA elements identified in relation to the task under analysis. Items should be rated as not important (1), somewhat important (2) or very important (3). The ratings provided should then be averaged across subjects for each item.

Step 7: Determine SA Requirements
Once the questionnaires have been collected and scored, the analyst should use them to determine the SA elements for the task or scenario under analysis. How this is done is dependent upon the analyst's judgment. It may be that the elements specified in the questionnaire are presented as SA requirements, along with a classification in terms of importance (e.g. not important, somewhat important or very important).

Step 8: Create SA Requirements Specification
The final stage involves creating an SA requirements specification that can be used by other practitioners (e.g. system designers or methods developers). The SA requirements should be listed for each actor involved in the task or scenario under analysis. Endsley (1993) and Matthews, Strater and Endsley (2004) demonstrate how the SA requirements can be categorised across the three levels of SA, as outlined by the three-level model; however, this may not be necessary, depending on the specification requirements. It is recommended that SA requirements should be listed in terms of what it is that needs to be known, what information is required, how this information is used (i.e. what the linked goals and decisions are) and what the relationships between the different pieces of information actually are, i.e. how they are integrated and used by different actors. Once the SA requirements are identified for each actor in question, a list should be compiled, including tasks, SA elements, the relationships between them and the goals and decisions associated with them.

Advantages

- SA requirements analysis provides a structured approach for identifying the SA requirements associated with a particular task or scenario.
- The output tells us exactly what it is that needs to be known by different actors during task performance.
- The output has many uses, including for developing SA measures or to inform the design of coaching and training interventions, procedures or new technology.
- If conducted properly, it has the potential to be exhaustive.
- It is generic and can be used to identify the SA requirements associated with any task in any domain.
- It has great potential to be used as a method for identifying the SA requirements associated with different sports and also with different positions in sporting teams, the outputs of which can be used to inform the design of training interventions, tactics, performance aids and sports technology.

- It has been used to identify the SA requirements associated with various roles, including infantry officers (Matthews, Strater and Endsley, 2004), pilots (Endsley, 1993), aircraft maintenance teams (Endsley and Robertson, 2000) and air traffic controllers (Endsley and Rogers, 1994).

Disadvantages

- Due to the use of interviews and task analysis methods, the method is very time-consuming to apply.
- It requires a high level of access to multiple SMEs for the task under analysis.
- Identifying SA elements and the relationships between them requires significant skill on the part of the analyst involved.
- Analyses may become large and unwieldy for complex collaborative systems.
- Analysts require an indepth understanding of the SA concept.
- It does not directly inform design.

Related Methods

The SA requirements analysis procedure outlined by Endsley (1993) was originally conceived as a way of identifying the SA elements to be tested using the SAGAT freeze probe recall method. The SA requirements analysis method itself uses interviews with SMEs and also goal-directed task analysis, which is similar to the HTA approach.

Approximate Training and Application Times

Provided that analysts have significant experience of the SA concept, interviews and task analysis methods, the training time for the SA requirements analysis method is low; however, for novice analysts new to the area and without experience in interview and task analysis methods, the time required is high. The application time for the SA requirements analysis method is high, including the conduct of interviews, transcription of interview data, conduct of task analysis for the task in question, the identification and rating of SA elements and finally the compilation of SA requirements and the relationships between them.

Reliability and Validity

The reliability and validity of the SA requirements method is difficult to assess. As long as appropriate SMEs are used throughout the process, the validity should be high; however, the method's reliability may be questionable.

Tools Needed

At its most basic, the SA requirements analysis procedure can be conducted using pen and paper; however, in order to make the analysis as simple and as comprehensive as possible, it is recommended that video and audio recording equipment is used to record the interviews and that a computer with a word processing package (such as Microsoft Word) and the Statistical Package for the Social Sciences (SPSS) are used during the design and analysis of the questionnaire. A drawing package such as Microsoft Visio is also useful when producing the task analysis and SA requirements analysis outputs.

Example

Salmon et al. (2009b) recently compared two different SA measurement approaches – a SAGAT-style probe recall measure and Taylor's (1990) SART questionnaire – to assess SA during a military planning task. In order to develop appropriate SAGAT probes for the task under analysis, an initial SA requirements analysis was undertaken so as to identify the different SA requirements associated with the planning task. This involved developing an HTA (Stanton, 2006) for the planning task (the goal-directed task analysis approach produces similar outputs to HTA) and then identifying SA requirements from the HTA. An extract of the HTA developed for the military planning task is presented in Figure 7.3. Six sets of multi-question SA probes were subsequently developed. Examples of the SA elements included in the probes are presented in Table 7.2.

In conclusion to the study, Salmon et al. (2009b) reported that only the overall and level 2 SAGAT scores produced a statistically significant correlation with performance. They concluded that, out of the methods tested, the SAGAT approach was the most accurate at measuring participant SA during the study. For more information, see Endsley, Bolte and Jones (2003) and Matthews, Strater and Endsley (2004).

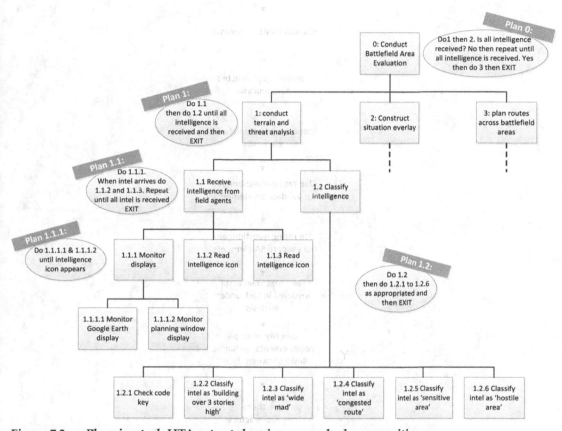

Figure 7.3   Planning task HTA extract showing example decomposition

Table 7.2   Example SA probes derived from SA requirements analysis

Level 1 SA elements	Level 2 SA probes	Level 3 SA probes
Location and number of: Schools. Helicopters. Buildings. Fires. Wide roads. Sensitive areas. Building heights (above or below five storeys). Armoured vehicles. Foot platoons. Enemy.	Areas on the battlefield where it would be the most difficult to evacuate civilians. Most vulnerable locations.	Armoured vehicles. Foot platoons. Potential routes to target areas (based on resources available).

Flowchart

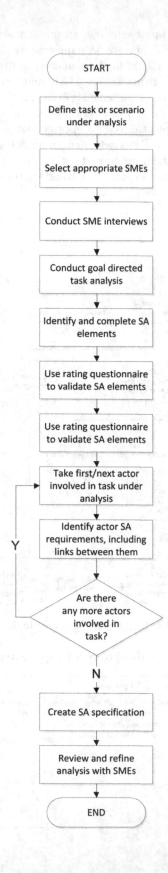

The Situation Awareness Global Assessment Technique (SAGAT)

Background and Applications

SAGAT (Endsley, 1995b) is a freeze probe recall method that was developed to assess pilot SA based on the three levels of SA postulated by Endsley's information processing model. It is a simulator-based measure and involves querying participants regarding their knowledge of SA elements during random freezes in a simulation of the task under analysis. During the freezes, all simulation screens and displays are blanked, and relevant SA queries for that point of the task or scenario are administered.

Domain of Application

SAGAT was originally developed for use in the military aviation domain; however, numerous variations of the method have since been applied in other domains, including an air-to-air tactical aircraft version (Endsley, 1990), an advanced bomber aircraft version (Endsley, 1989) and an ATC version (Endsley and Kiris, 1995). SAGAT-style approaches can be applied in any domain provided the queries are developed based on an SA requirements analysis for the domain and activities under investigation.

Procedure and Advice

Step 1: Define the Aims of the Analysis
First, the aims of the analysis should be clearly defined, since this affects the scenarios used and the types of SAGAT queries administered. For example, the aims of the analysis may be to evaluate the impact that a new performance aid or technological device has on SA during task performance, or it may be to compare novice and expert performer SA during a particular task.

Step 2: Define the Task or Scenario under Analysis
The next step involves clearly defining the task or scenario under analysis. It is recommended that the task be described clearly, including the different actors involved, the task goals and the environment within which the task is to take place.

Step 3: Conduct SA Requirements Analysis and Generate SAGAT Queries
To support query development, an SA requirements analysis is required for the activity or system under investigation. The SA requirements analysis output is then used to inform the development of appropriate SAGAT queries. Jones and Kaber (2005) highlight the importance of this phase, suggesting that the foundation of successful SAGAT data collection efforts rests solely on the efficacy of the queries used. The queries generated should cover the three levels of SA as prescribed by Endsley's model (i.e. perception, comprehension and projection). Jones and Kaber stress that the wording of the queries should be compatible with the operator's frame of reference and appropriate to the language typically used in the domain under analysis.

Step 4: Brief Participants
Once appropriate participants have been recruited based on the analysis requirements, the data collection phase can begin. First, however, it is important to brief the participants involved. This should include an introduction to the area of SA and a description and demonstration of the SAGAT methodology. At this stage, participants should also be briefed on what the aims of the study are and what is required of them as participants.

Step 5: Conduct Pilot Run(s)
Before the data collection process proper begins, it is recommended that pilot runs of the SAGAT data collection procedure are undertaken. A number of small test scenarios, incorporating multiple SAGAT freezes and query administrations, should be used to iron out any problems with the data collection procedure, and the participants should be encouraged to ask any questions. Once the participant is familiar with the procedure and is comfortable with his or her role, the 'real' data collection process can begin.

Step 6: Begin SAGAT Data Collection
Next, the SAGAT data collection phase can begin. The experimenter should initiate this by instructing the participant(s) to undertake the task under analysis.

Step 7: Freeze the Simulation
SAGAT works by temporarily freezing the simulation at pre-determined random points and blanking all displays or interfaces. Jones and Kaber (2005) offer the following guidelines for task freezes:

- the timing of freezes should be randomly determined;
- SAGAT freezes should not occur within the first three to five minutes of the trial;
- SAGAT freezes should not occur within one minute of each other; and
- multiple SAGAT freezes should be used.

Step 8: Administer SAGAT Queries
Once the simulation is frozen at the appropriate point, the analyst should probe the participant's SA using the pre-defined SA queries. These queries are designed to allow the analyst to gain a measure of the participant's knowledge of the situation at that exact point in time and should be directly related to the participant's SA at the point of the freeze. A computer programmed with the SA queries is normally used to administer the queries; however, queries can also be administered using pen and paper. To stop any overloading of the participants, not all SA queries are administrated in any one freeze and only a randomly selected portion of the SA queries is administrated at any one time. Jones and Kaber (2005) recommend that no outside information should be available to the participants during query administration. For evaluation purposes, the correct answers to the queries should also be recorded; this can be done automatically by sophisticated computers/simulators or manually by an analyst. Once all queries are completed, the simulation should resume from the exact point at which it was frozen (Jones and Kaber, 2005). Steps 7 and 8 are repeated throughout the task until sufficient data is obtained.

Step 9: Query Response Evaluation and SAGAT Score Calculation
Upon completion of the simulator trial, participant query responses are compared to what was actually happening in the situation at the time of the query administration. To achieve this, query responses are compared to the data recorded by the simulation computers or analysts involved. Endsley (1995b) suggests that this comparison of the real and perceived situation provides an objective measure of participant SA. Typically, responses are scored as either correct (1) or incorrect (0), and a SAGAT score is calculated for each participant, including an overall score and scores for each of the three SA levels. Additional measures or variations on the SAGAT score can be taken depending on study requirements, such as time taken to answer queries.

Step 10: Analyse SAGAT Data
SAGAT data is typically analysed across conditions (e.g. trial 1 versus trial 2) and the three SA levels specified by Endsley's three-level model. This allows query responses to be compared across conditions and also levels of SA to be compared across participants.

Advantages

- SAGAT provides an online measure of SA, removing the problems associated with collecting subjective SA data (e.g. a correlation between SA ratings and task performance).
- Online data collection avoids the problems associated with collecting SA data post-task, such as memory degradation and forgetting low SA periods of the task.
- SA scores can be viewed in total and also across the three levels specified in Endsley's model. Further, the specification of SA scores across Endsley's three levels is useful for designers and easy to understand.
- SAGAT is the most popular approach for measuring SA and has the most validation evidence associated with it (Jones and Endsley, 2000, Durso et al., 1998, Endsley and Garland, 2000).
- Evidence suggests that SAGAT is a valid metric of SA (Jones and Kaber, 2005).
- The method is generic and can be applied in any domain.

Disadvantages

- Various preparatory activities are required, including the conduct of SA requirements analysis and the generation of numerous SAGAT queries.
- The total application time for the whole procedure (i.e. including SA requirements analysis and query development) can be high.
- Using the SAGAT method typically requires expensive high-fidelity simulators and computers.
- The use of task freezes and online queries is highly intrusive to performance on the primary task.
- It cannot be applied during real-world and/or collaborative tasks.
- It does not account for distributed cognition or distributed situation awareness theory (Salmon et al., 2009a). For example, in a joint cognitive system, it may be that operators do not need to be aware of certain elements as they are held by displays and devices. In this case, SAGAT would score participant SA as low even though this may not in fact be the case.
- It is based upon the three-level model of SA (Endsley, 1995a), which has various flaws (Salmon et al., 2008a).
- Participants may be directed to elements of the task that they are unaware of.
- In order to use the approach, one has to be able to determine what SA consists of *a priori*. This might be particularly difficult, if not impossible, for some scenarios.

Related Methods

SAGAT queries are generated based on an initial SA requirements analysis conducted for the task in question. Various versions of SAGAT have been applied, including an air-to-air tactical aircraft version (Endsley, 1990), an advanced bomber aircraft version (Endsley, 1989) and an ATC version (Endsley and Kiris, 1995). Further, many freeze probe methods based on the SAGAT approach have been developed for use in other domains. SALSA (Hauss and Eyferth, 2003), for example, was developed specifically for use in ATC, whereas the Situation Awareness Control Room Inventory method (SACRI: Hogg et al., 1995) was developed for use in nuclear power control rooms.

Approximate Training and Application Times

The training time for the SAGAT approach is low; however, if analysts require training in the SA requirements analysis procedure, then the training time incurred will increase significantly. The application time for the overall SAGAT procedure, including the conduct of an SA requirements analysis and the development of SAGAT queries, is typically high. The actual data collection process of administering queries and gathering responses requires relatively little time, although this is dependent upon the task under analysis.

Reliability and Validity

There is considerable validation evidence for the SAGAT approach presented in the literature. Jones and Kaber (2005) point out that numerous studies have been undertaken to assess the validity of the SAGAT and the evidence suggests that the method is a valid metric of SA. Endsley (2000) reports that SAGAT has been shown to have a high degree of validity and reliability for measuring SA. Collier and Folleso (1995) also reported good reliability for SAGAT when measuring nuclear power plant operator SA. Fracker (1991), however, reported low reliability for SAGAT when measuring participant knowledge of aircraft location. Regarding validity, Endsley et al. (2000) reported a good level of sensitivity for SAGAT, but not for real-time probes (online queries with no freeze) and subjective SA measures. Endsley (1995b) also reported that SAGAT showed a degree of predictive validity when measuring pilot SA, with SAGAT scores indicative of pilot performance in a combat simulation. The study found that pilots who were able to report on enemy aircraft via SAGAT were three times more likely to later kill that target in the simulation. However, it is certainly questionable whether good performance is directly correlated with good or high SA. Presumably, within the three-level model of SA, a pilot could theoretically have very high SA and still fail to kill the enemy target, thus achieving low performance.

Tools Needed

Typically, a high-fidelity simulation of the task under analysis and computers with the ability to generate and score SAGAT queries are required. The simulation and computer used should possess the ability to randomly blank all operator displays and 'window' displays, randomly administer relevant SA queries and calculate participant SA scores.

Example

Walker et al. (2009b) employed SAGAT as part of their evaluation of distributed military planning teams. Thirty-six planning teams (each comprised of three people) were tasked with completing a game of chess utilising one of a number of different technological communication mediums. The technology-mediated performance of the planning teams was compared to their performance in a collocated (face-to-face) condition. Within the technology-mediated condition, there were four separate levels of mediation:

1. voice (telephone), video and live updateable shared tools;
2. voice and video (no live updateable tools);
3. voice and live updateable tools;
4. voice (telephone) only.

One aspect of this analysis involved the comparison of participants' recall performance across the different conditions. The surprise recall task utilised the SAGAT methodology and involved presenting the participants with a piece of paper on which a chessboard template was drawn. This template was presented to participants at the end of each condition, out of view of the chessboard. Participants were asked to complete the template with the board's current piece positions. This remembered version was then compared with the actual position of pieces on the board, allowing for a percentage-based score of accuracy to be derived. Figure 7.4 below summarises the results of this analysis based on collocation versus the technology medium.

Figure 7.4 indicates that there was a higher mean accuracy rate in the distributed condition compared to the collocated condition. Analysis of this data using a paired t-test confirmed that this difference was significant at the one per cent level (T ¼ 8.73, df ¼ 83, p < 0.01).

From these results, it could be suggested that SA within the distributed condition, as measured by the surprise recall task, was more accurate. In order to explore these results further, Walker et al. (2009b) compared the results of the four levels of technology mediation as summarised below in Figure 7.5.

Analysis of the distributed data in terms of the type of technology medium revealed significant differences between the technology mediums. The condition in which only voice communication was allowed resulted in the highest levels of recall accuracy.

From the analysis, Walker et al. (2009b) concluded that the face-to-face conditions involved a low level of 'knowledge in the head' due to the utilisation of external artefacts such as acetate overlays, and that situation awareness became distributed over these planning materials. These external artefacts meant that participants did not need to remember planning moves, as this information was externally stored for them. The same conclusion was drawn for those distributed conditions which employed the shared data workspace, whereas in the voice-only distributed condition, there was no way to externalise SA, resulting in a higher level of information in the head. For more information on this, see Endsley (1995b) and Jones and Kacer (2005).

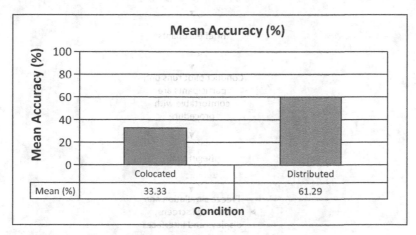

Figure 7.4   Mean percentage of accuracy for collocated and distributed teams

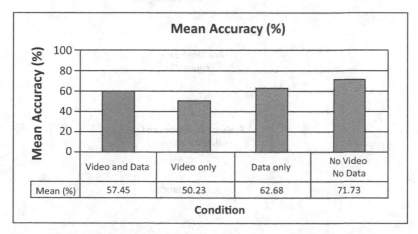

Figure 7.5   Mean percentage of accuracy for each communication medium

Flowchart

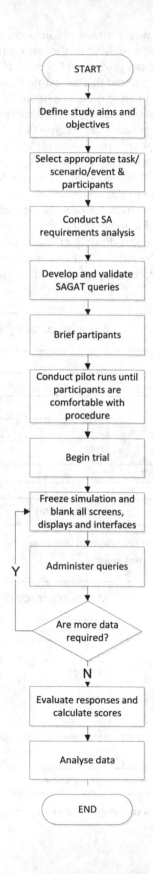

The Situation Present Assessment Method (SPAM)

Background and Applications

SPAM (Durso et al., 1998) is an online, real-time probe method that was developed for use in the assessment of ATC SA. The idea behind real-time online probe methods is that they retain the objectivity of online freeze probe approaches but reduce the level of intrusion on task performance by not using task freezes. SPAM focuses upon operator ability to locate information in the environment as an indicator of SA, rather than the recall of specific information regarding the current situation. The analyst probes the operator for SA using task-related online SA queries based on pertinent information in the environment (e.g. which of the two aircraft A and B has the highest altitude?) via a telephone landline. Query response accuracy and query response time (for those responses that are correct) is taken as an indicator of the operator's SA.

Domain of Application

The SPAM method was developed for measuring ATC SA. However, the principles behind the approach (assessing participant SA using real-time probes) could be applied in any domain.

Procedure and Advice

Step 1: Define the Aims of the Analysis
First, the aims of the analysis should be clearly defined, since this affects the scenarios used and the SPAM queries administered. For example, the aims of the analysis may be to evaluate the impact that a new performance aid or technological device has on SA during task performance, or it may be to compare novice and expert performer SA during a particular task.

Step 2: Define The Task or Scenario under Analysis
The next step involves clearly defining the task or scenario under analysis. It is recommended that the task be described clearly, including the different actors involved, the task goals and the environment within which the task is to take place.

Step 3: Conduct SA Requirements Analysis and Generate Queries
Once the task or scenario under analysis is determined, an SA requirements analysis procedure should be undertaken and a series of appropriate queries should be developed based on the SA requirements analysis outputs. For SPAM analyses, the queries developed should be generic so that they can be applied throughout the task under investigation. Rather than concentrate on information regarding single aircraft (like the SAGAT method), SPAM queries normally ask for 'gist-type' information (Jeannott, Kelly and Thompson, 2003).

Step 4: Brief Participants
Once appropriate participants have been recruited based on the analysis requirements, the data collection phase can begin. It is important first to brief the participants involved. This should include an introduction to the field of SA and a description and demonstration of the SPAM methodology. Participants should also be briefed on what the aims of the study are and what is required of them as participants.

Step 5: Conduct Pilot Runs
Before the data collection process begins, it is recommended that a pilot run of the SPAM data collection procedure be undertaken in order to iron out any problems with the procedure. The pilot run should include multiple SPAM query administrations and participants should also be encouraged to ask any questions. Once the participant is familiar with the procedure and is comfortable with his or her role, the data collection process can begin.

Step 6: Undertake Task Performance
The task is normally performed using a simulation of the system and task under analysis; however, it may also be performed in the field (i.e. real-world task performance). Participants should be instructed to undertake the task as normal.

Step 7: Administer SPAM Queries
SPAM queries should be administered at random points during the task. This involves asking the participant a question or series of questions regarding the current situation. Once the analyst has asked the question, a stopwatch should be started in order to measure participant response time. The query answer, query response time and time to answer the landline should be recorded for each query administered. Step 7 should be repeated until the required amount of data is collected.

Step 8: Calculate Participant SA/Workload Scores
Once the task is complete, the analyst should calculate participant SA based upon the correct query responses and response times (only correct responses are taken into account). A measure of workload can also be derived from the landline response times recorded.

Advantages

- SPAM is quick and easy to use, requiring minimal training.
- There is no need for task freezes.
- It offers an objective measure of SA.
- Online administration removes the various problems associated with collecting SA data post-trial.

Disadvantages

- Even without task freezes, SPAM remains highly intrusive to task performance.
- Various preparatory activities are required, including the conduct of SA requirements analysis and the generation of queries.
- SPAM probes could potentially alert participants to aspects of the task that they are unaware of.
- The total application time for the whole procedure (i.e. including SA requirements analysis and query development) could be high.
- There is only limited evidence of its application.
- Often it is required that the SA queries are developed online during task performance, which places a considerable burden on the analyst involved.
- It is based upon the three-level model of SA, which has various flaws (Salmon et al., 2008a).
- In order to use the approach, one has to be able to determine what SA should comprise *a priori*.

Related Methods

SPAM is essentially the SAGAT approach, but without the use of task freezes. An SA requirements analysis should be used when developing SPAM queries. Other real-time probe methods also exist, including the situational awareness for SHAPE (Solutions for Human-Automation Partnerships in European ATM) (SASHA) method (Jeannott, Kelly and Thompson, 2003), which is a modified version of the SPAM method.

Approximate Training and Application Times

Training time for the SPAM approach is low; however, if analysts require training in the SA requirements analysis procedure, then the training time incurred will increase significantly. The application time for the overall SPAM procedure, including the conduct of an SA requirements analysis and the development of queries, can be high. The actual data collection process of administering queries and gathering responses requires relatively little time, although this is dependent on the task under analysis.

Reliability and Validity

There is only limited data regarding the reliability and validity of the SPAM method available in the open literature. Jones and Endsley (2000) conducted a study to assess the validity of real-time probes as a measure of SA. They reported that the real-time probe measure demonstrated a level of sensitivity to SA in two different scenarios and that the method was measuring participant SA and not simply participant response time.

Tools Needed

SPAM could be applied using pen and paper; however, Durso et al. (1998) used a landline telephone located in close proximity to the participant's workstation in order to administer SPAM queries. A simulation of the task and system under analysis may also be required.

Flowchart

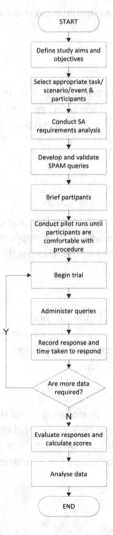

START

Define study aims and objectives

Select appropriate task/scenario/event & participants

Conduct SA requirements analysis

Develop and validate SPAM queries

Brief partipants

Conduct pilot runs until participants are comfortable with procedure

Begin trial

Administer queries

Record response and time taken to respond

Are more data required?

Y

N

Evaluate responses and calculate scores

Analyse data

END

The Situation Awareness Rating Technique (SART)

Background and Applications

SART (Taylor, 1990: Figure 7.6) is a post-trial subjective rating method, which uses the a rating scale comprising the following 10 dimensions to derive a measure operator SA: familiarity of the situation; focusing of attention; information quantity; information quality; instability of the situation; concentration of attention; complexity of the situation; variability of the situation; arousal; and spare mental capacity. SART is typically administered post-trial and involves participants subjectively rating each dimension on a seven-point rating scale (1 = low, 7 = high) based on their own perceived performance during the task under analysis. The ratings are then combined in order to calculate a measure of participant SA. A quicker version of the SART approach also exists, known as the 3D SART. The 3D SART uses the 10 dimensions described above grouped into the following three dimensions:

1.   demands on attentional resources – a combination of complexity, variability and instability of the situation;
2.   supply of attentional resources – a combination of arousal, focusing of attention, spare mental capacity and concentration of attention; and
3.   understanding of the situation – a combination of information quantity, information quality and familiarity of the situation.

Domain of Application

The SART approach was originally developed for use in the military aviation domain; however, the dimensions used are generic and it has since been applied in various domains to assess operator SA, including command and control (Salmon et al., 2009a) and road transport (Walker, Stanton and Young, 2008).

Procedure and Advice

Step 1: Define the Tasks under Analysis
The first step involves defining the tasks that are to be subjected to analysis. The type of tasks analysed are dependent upon the focus of the analysis. For example, when assessing the effects on operator SA caused by a novel design or training programme, it is useful to analyse as representative a set of tasks as possible.

Step 2: Select Participants
Once the task (or tasks) under analysis is clearly defined, it is useful to select the participants that are to be involved in the analysis. This may not always be necessary and it may suffice to select participants randomly on the day; however, if SA is being compared across rank or experience levels, then effort is required to select the appropriate participants.

Step 3: Brief Participants
Before the task (or tasks) under analysis is performed, all of the participants involved should be briefed regarding the purpose of the study, the concept of SA, and the SART method. It may useful at this stage to take the participants through an example SART analysis so that they understand how the method works and what is required of them as participants. Explanation of the different SART dimensions should also be provided.

Step 4: Conduct Pilot Run
It is recommended that participants take part in a pilot run of the SART data collection procedure. A number of small test scenarios should be used to iron out any problems with the data collection procedure, and the participants should be encouraged to ask any questions. Once the participants are familiar with the procedure and are comfortable with their role, the data collection process can begin.

Step 5: Performance of Task
The next step involves the performance of the task or scenario under analysis. Participants should be asked to perform the task as normal.

Step 6: Complete SART Questionnaires
Once the task is completed, participants should be given a SART pro-forma and asked to provide ratings for each dimension based on how they felt during task performance. The participants are permitted to ask questions in order to clarify the dimensions; however, the participants' ratings should not be influenced in any way by external sources. In order to reduce the correlation between SA ratings and performance, no performance feedback should be given until after the participants have completed the self-rating process.

Step 7: Calculate Participant SART Scores
The final step in the SART analysis involves calculating each participant's SA score. When using SART, participant SA is calculated using the following formula:

$$SA = U-(D-S)$$

Where: U = summed understanding
             D = summed demand
             S = summed supply

Typically, for each participant, an overall SART score is calculated along with total scores for the following three dimensions: understanding, demand and supply.

Advantages

- SART is quick and simple to apply.
- It requires little training, both for analysts and participants.
- The data obtained is easily and quickly analysed.
- It provides a low-cost approach to assessing participant SA.
- Its dimensions are generic and so it can be applied in any domain.
- It is non-intrusive to task performance.
- It is a widely used method and has been applied in a range of different domains.
- It provides a quantitative assessment of SA.

Disadvantages

- SART suffers from a host of problems associated with collective subjective SA ratings, including a correlation between performance and SA ratings, and questions regarding whether or not participants can accurately rate their own awareness (e.g. how can one be aware that they are not aware?).

- It suffers from a host of problems associated with collecting SA data post-task, including memory degradation and poor recall, a correlation of SA ratings with performance and also participants forgetting low SA portions of the task.
- Its dimensions are not representative of SA. Upon closer inspection, the dimensions are more representative of workload than anything else.
- It has performed poorly in a number of SA methodology comparison studies (e.g. Endsley, Sollenberger and Stein, 2000; Endsley et al., 1998; Salmon et al., 2009b).
- The method is dated.

Related Methods

SART is one of many questionnaire-based subjective rating SA measurement approaches. Other such methods include the situation awareness rating scale (SARS: Waag and Houck, 1994), the crew awareness rating scale (CARS: McGuinness and Foy, 2000) and the SA-SWORD method (Vidulich and Hughes, 1991).

Approximate Training and Application Times

The training and application times associated with the SART method are very low. As it is a self-rating questionnaire, there is very little training involved. In our experience, the SART questionnaire takes no longer than five minutes to complete and it is possible to set up programs that auto-calculate SART scores based on raw data entry.

Reliability and Validity

Along with SAGAT, SART is the most widely used and tested measure of SA (Endsley and Garland, 2000); however, it has performed poorly in a number of validation and methods comparison studies (e.g. Endsley, Sollenberger and Stein, 2000; Endsley et al., 1998; Salmon et al., 2009b). In particular, it has been found to be insensitive to display manipulations. Further, the construct validity of SART is limited and many have raised concerns regarding the degree to which its dimensions are actually representative of SA (e.g. Endsley, 1995b; Salmon et al., 2009b; Uhlarik and Comerford, 2002).

Tools Needed

SART is applied using pen and paper only; however, it can also be administered using Microsoft Excel, which can also be used to automate the SART score calculation process.

Example

Salmon et al. (2009b) recently compared SAGAT and SART when used to measure SA during a military mission planning scenario. Twenty participants (13 female and 7 male, mean age 30 years), acting in the role of mission commander, completed question one of the Combat Estimate (CE) in Brunel University's command and control experimental test bed environment (see Green et al., 2005 for a full description). The CE is the current planning process adopted by UK armed forces during land warfare activities. Question one involves analysing the battlefield area and the likely threat imposed by the enemy by undertaking a Battlefield Area Evaluation (BAE) and a threat evaluation. This involves using maps of the battlefield area and incoming intelligence to analyse the terrain in terms of its effects on friendly and enemy actions and the identification of likely mobility corridors, avenues of approach and manoeuvre areas. Following this, the threat evaluation involves identifying the enemy's likely modus operandi by looking at the ground and the enemy's strengths and weaknesses. The SA requirements during question one are therefore primarily related to the battlefield area itself and the enemy. The output is an understanding of the battlefield and its effects on how the enemy (and friendly forces) are likely to operate. This is expressed using a situational overlay, which depicts the ground's key features and the enemy's likely movements.

Each participant was asked to undertake question one of the CE process for an experimental Military Operations in Urban Terrain (MOUT) warfare scenario. Based on incoming intelligence from simulated sensors on the battlefield, participants were required to perform a BAE and threat evaluation for a MOUT scenario. The information sent to the participant by the sensors was pre-programmed to enable information to appear on the commander's map display at pre-determined intervals. The participant's task was to use the incoming information from the field agents in order to construct a situation overlay that represented the BAE. In particular, participants were asked to highlight on the map display the following types of key terrain:

1. buildings three storeys or higher;
2. routes and roads;
3. sensitive areas; and
4. hostile areas.

In addition, participants were also asked to highlight various types of events on the map, including the locations of a sniper, barricade, road traffic accident, bomb and fire.

SAGAT probes were administered at random points during the experiment to assess participant SA. This involved the task being frozen, the command wall screens being blanked and the administration (via Microsoft PowerPoint on a laptop computer) of SAGAT probes. Upon completion of the probes, the task continued. The time taken to complete the entire experiment and to respond to each SAGAT probe was recorded by the experimenter. Upon completion of the experimental trial, participants were asked to complete a SART SA questionnaire and a NASA TLX workload questionnaire.

Salmon et al. (2009b) reported that only the overall and level 2 SAGAT scores produced a statistically significant correlation with performance. This indicates that the higher the participant's overall and level 2 SAGAT scores were, the more accurate they were in the situation overlay construction task. It was concluded from this that, out of the methods tested, the SAGAT approach was the most accurate at measuring participant SA during the study. No significant correlation was found between the participant SA scores provided by the two SA measures used. Salmon et al. (2009b) concluded from this that the SAGAT and SART measures were measuring different things during the study, suggesting that the two methods view the construct differently and were measuring different elements of the participant's awareness (or different things entirely?). According to Salmon et al. (2009b), SAGAT measured the extent to which participants were aware of pre-defined elements in the environment, their understanding of the properties of these elements in relation to the task they were performing, and also what the potential future states of these elements might be. SART, on the other hand, provided a measure of how aware participants perceived themselves to be during task performance (based on ratings of understanding, supply and demand) and did not refer to the different elements within the environment. They concluded that each method took a different view on SA in terms of what it is and what it comprises and, as the lack of a correlation between the measures demonstrated, the methods were measuring different things when assessing participant SA. For more information, see also Taylor (1990).

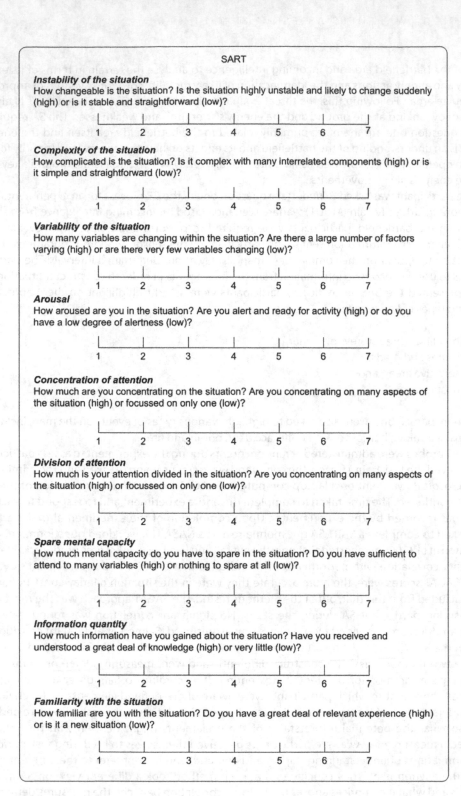

SART

Instability of the situation
How changeable is the situation? Is the situation highly unstable and likely to change suddenly (high) or is it stable and straightforward (low)?

1   2   3   4   5   6   7

Complexity of the situation
How complicated is the situation? Is it complex with many interrelated components (high) or is it simple and straightforward (low)?

1   2   3   4   5   6   7

Variability of the situation
How many variables are changing within the situation? Are there a large number of factors varying (high) or are there very few variables changing (low)?

1   2   3   4   5   6   7

Arousal
How aroused are you in the situation? Are you alert and ready for activity (high) or do you have a low degree of alertness (low)?

1   2   3   4   5   6   7

Concentration of attention
How much are you concentrating on the situation? Are you concentrating on many aspects of the situation (high) or focussed on only one (low)?

1   2   3   4   5   6   7

Division of attention
How much is your attention divided in the situation? Are you concentrating on many aspects of the situation (high) or focussed on only one (low)?

1   2   3   4   5   6   7

Spare mental capacity
How much mental capacity do you have to spare in the situation? Do you have sufficient to attend to many variables (high) or nothing to spare at all (low)?

1   2   3   4   5   6   7

Information quantity
How much information have you gained about the situation? Have you received and understood a great deal of knowledge (high) or very little (low)?

1   2   3   4   5   6   7

Familiarity with the situation
How familiar are you with the situation? Do you have a great deal of relevant experience (high) or is it a new situation (low)?

1   2   3   4   5   6   7

Figure 7.6   SART rating scale

Flowchart

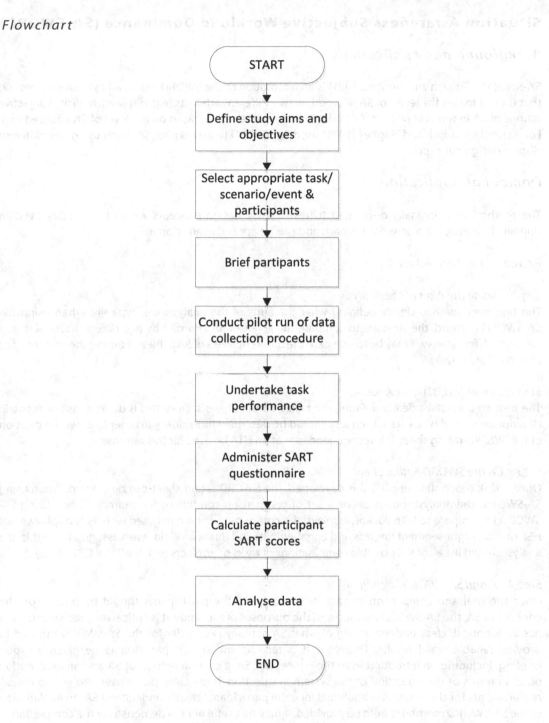

Situation Awareness Subjective Workload Dominance (SA-SWORD)

Background and Applications

SA-SWORD (Vidulich and Hughes, 1991) is an adaptation of the SWORD workload assessment method that is used to test the levels of SA afforded by two different artefacts (e.g. displays). It elicits subjective ratings of SA in terms of one artefact's dominance over another based on the level of SA afforded to it. For example, Vidulich and Hughes (1991) used SA-SWORD to assess pilot SA when using two different display design concepts.

Domain of Application

The method was originally developed to test display design concepts within the military aviation domain; however, it is a generic approach and can be applied in any domain.

Procedure and Advice

Step 1: Define the Aims of the Analysis
The first step involves clearly defining what the aims of the analysis are. Typically, when using the SA-SWORD method, the aims are to compare the levels of SA afforded by two different artefacts (e.g. displays). Alternatively, it may be to compare the different levels of SA achieved during the performance of two different tasks.

Step 2: Define Task(s) under Analysis
The next step involves clearly defining the task(s) under analysis. Once this is done, a task or scenario description should be created. Each task should be described individually in order to allow the creation of the SWORD rating sheet. It is recommended that an HTA be used for this purpose.

Step 3: Create SWORD Rating Sheet
Once a task description (e.g. HTA) is developed, the SWORD rating sheet can be created. When using SA-SWORD, the analyst should define a set of comparison conditions. For example, when using SA-SWORD to compare to F-16 cockpit displays, the comparison conditions used were FCR display versus HSF display, flight segment (ingress and engagement) and threat level (low versus high). To do this, the analyst should list all of the possible combinations of tasks or artefacts (e.g. A v B, A v C, B v C).

Step 4: SA and SA-SWORD Briefing
Once the trial and comparison conditions are defined, the participants should be briefed on the concept of SA, the SA-SWORD method and the purposes of the study. It is critical that each participant has an identical, clear understanding of what SA actually is in order for the SA-SWORD method to provide reliable, valid results. Therefore, it is recommended that participants be given a group briefing, including an introduction to the concept of SA, a clear definition of SA and an explanation of SA in terms of the operation of the system in question. It may also prove useful to define the SA requirements for the task under analysis. Once the participants clearly understand SA, an explanation of the SA-SWORD method should be provided. It may be useful here to demonstrate the completion of an example SA-SWORD questionnaire. Finally, the participants should then be briefed on the purpose of the study.

Step 5: Conduct Pilot Run
Next, a pilot run of the data collection process should be conducted. Participants should perform a small task and then complete a SA-SWORD rating sheet. The participants should be taken step by step through the SA-SWORD rating sheet and should be encouraged to ask any questions regarding any aspects of the data collection procedure that they are not sure about.

Step 6: Undertake Trial
SA-SWORD is administered post-trial. Therefore, the task(s) under analysis should be performed next.

Step 7: Administer SA-SWORD Rating Sheet
Once task performance is complete, the SA-SWORD rating procedure can begin. This involves the administration of the SA-SWORD rating sheet. The participant should be presented with the SWORD rating sheet immediately after task performance has ended. The SWORD rating sheet lists all possible SA paired comparisons of the task conducted in the scenario under analysis e.g. display A versus display B, condition A versus condition B. The analyst has to rate the two variables (e.g. display A versus display B) in terms of the level of SA that they provided during task performance. For example, if the participant feels that the two displays provided a similar level of SA, he or she should mark the 'EQUAL' point on the rating sheet. However, if the participant feels that display A provided a slightly higher level of SA than display B did, he or she would move towards task A on the sheet and mark the 'weak' point on the rating sheet. If the participant felt that display A imposed a much greater level of SA than display B, he or she would move towards display A on the sheet and mark the 'absolute' point on the rating sheet. This allows the participant to provide a subjective rating of one display's dominance in terms of the SA level afforded over the over display. This procedure should continue until all of the possible combinations of SA variables in the task under analysis are rated.

Step 8: Constructing the Judgment Matrix
Once all ratings have been elicited, the SWORD judgment matrix should be conducted. Each cell in the matrix should represent the comparison of the variables in the row with the variable in the associated column. The analyst should fill each cell with the participant's dominance rating. For example, if a participant rated displays A and B as equal, a '1' is entered into the appropriate cell. If display A is rated as dominant, then the analyst simply counts from the 'Equal' point to the marked point on the sheet and enters the number in the appropriate cell. The rating for each variable (e.g. display) is calculated by determining the mean for each row of the matrix and then normalising the means (Vidulich, Ward and Schueren, 1991).

Step 9: Matrix Consistency Evaluation
Once the SWORD matrix is complete, the consistency of the matrix can be evaluated by ensuring that there are transitive trends amongst the related judgments in the matrix.

Advantages

- SA-SWORD is quick and easy to use, and requires only minimal training.
- It is generic can be applied in any domain.
- It has a high level of face validity and user acceptance (Vidulich and Hughes, 1991).
- It is useful when comparing two different interface concepts and their effect upon operator SA.

• Intrusiveness is reduced, as SA-SWORD is administered post-trial.

Disadvantages

• What is good SA for one operator may be poor SA for another, and a very clear definition of SA would need to be developed in order for the method to work. For example, each participant may have different ideas as to what SA actually is and, as a result, the data obtained would be incorrect. In a study testing the SA-SWORD method, it was reported that the participants had very different views on what SA actually was (Vidulich and Hughes, 1991).
• The method does not provide a direct measure of SA; the analyst is merely given an assessment of the conditions in which SA is highest.
• The reporting of SA post-trial has a number of problems associated with it, such as a correlation between SA rating and task performance, and participants forgetting low SA periods during task performance.
• There is limited evidence of the use of the SA-SWORD method in the literature.

Related Methods

The SA-SWORD method is an adaptation of the SWORD workload assessment method. It appears to be unique in its use of paired comparisons to measure SA.

Approximate Training and Application Times

The SA-SWORD method is both easy to learn and apply, and thus has a low training and application time associated with it.

Reliability and Validity

Administered in its current form, the SA-SWORD method suffers from a poor level of construct validity, i.e. the extent to which it is actually measuring SA. Vidulich and Hughes (1991) encountered this problem and found that half of the participants understood SA to represent the amount of information that they were attempting to track, whilst the other half understood SA to represent the amount of information that they may be missing. This problem could potentially be eradicated by incorporating an SA briefing session or a clear definition of what constitutes SA on the SA-SWORD rating sheet. In a study comparing two different cockpit displays, the SA-SWORD method demonstrated a strong sensitivity to display manipulation (Vidulich and Hughes, 1991). Vidulich and Hughes (1991) also calculated inter-rater reliability statistics for the SA-SWORD method, reporting a grand inter-rater correlation of 0.705. According to them, this indicates that participant SA-SWORD ratings were reliably related to the conditions apparent during the trials.

Tools Needed

SA-SWORD can be administered using pen and paper. The system under analysis or a simulation of this is also required for the task performance component of the data collection procedure. For more information, see Vidulich and Hughes (1991).

Flowchart

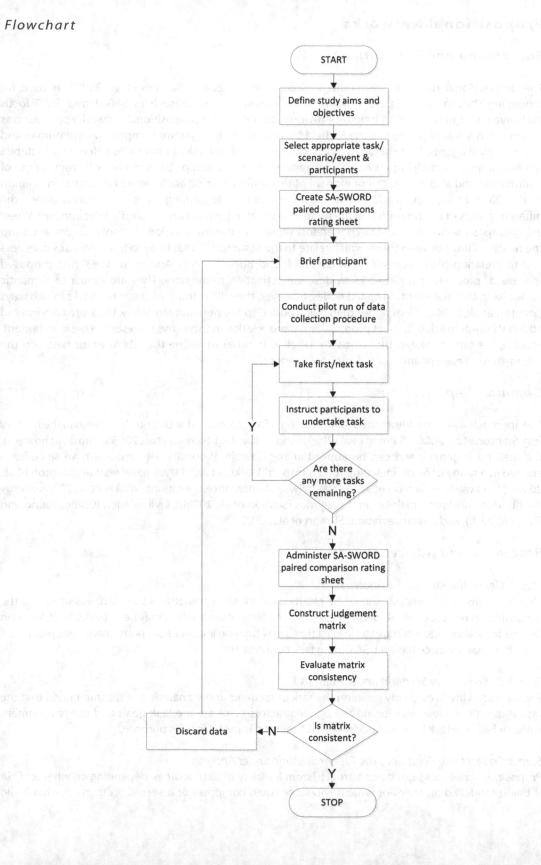

Propositional Networks

Background and Applications

The propositional network methodology (Stanton et al., 2006b; Salmon et al., 2009a) is used for modelling DSA in collaborative systems. Since existing measures such as SAGAT and SART focus exclusively on the levels of SA held by individual operators, the propositional network approach was proposed as a way of modelling the SA held by socio-technical systems comprising both human and technological agents. Propositional networks use networks of linked information elements to depict the information underlying a system's awareness, the relationships between the different pieces of information and also how each component of the system is using each piece of information. Salmon et al. (2009a) argue that, in addition to the information underpinning SA, it is the links between the different pieces of information that are more important in terms of understanding the concept. When using propositional networks, DSA is therefore represented as information elements (or concepts) and the relationships between them, which relate to the assumption that knowledge comprises concepts and the relationships between them (Shadbolt and Burton, 1995). Anderson (1983) first proposed the use of propositional networks to describe activation in memory. They are similar to semantic networks in that they contain linked nodes; however, they differ from semantic networks in two ways (Stanton et al., 2006b). First, rather than being added to the network randomly, the words are instead added through the definition of propositions. A proposition in this sense represents a basic statement. Second, the links between the words are labelled in order to define the relationships between the propositions, i.e. elephant 'has' tail, mouse 'is' rodent.

Domain of Application

The approach was originally applied for modelling DSA in command and control scenarios in the military (e.g. Stanton et al., 2006b; Salmon et al., 2009a) and civilian (e.g. Salmon et al., 2008b) domains; however, the method is generic and can be applied in any domain. Propositional networks have since been applied in a range of fields, including road transport (Walker et al., 2009a), naval warfare (Stanton et al., 2006b), land warfare (Salmon et al., 2009a), railway maintenance operations (Walker et al., 2006), energy distribution substation maintenance scenarios (Salmon et al., 2008b), civil aviation (Griffin, Young and Stanton, 2010), and military aviation (Stewart et al., 2008).

Procedure and Advice

Step 1: Define the Aims of the Analysis
First, the aims of the analysis should be clearly defined, since this affects the scenarios used and the propositional networks developed. For example, the aims of the analysis may be to evaluate DSA during task performance or to evaluate the impact that a new training intervention, performance aid, procedure or technological device has on DSA during task performance.

Step 2: Define Task or Scenario under Analysis
The next step involves clearly defining the task or scenario under analysis. It is recommended that the task is described clearly, including the different actors involved, the task goals and the environment within which the task is to take place. The HTA method is useful for this purpose.

Step 3: Collect Data Regarding the Task or Scenario under Analysis
Propositional networks can be constructed from a variety of data sources, depending on whether DSA is being modelled (in terms of what it should or could comprise) or assessed (in terms of what it did

comprise). These include observational study and/or verbal transcript data, CDM data, HTA data or data derived from work-related artefacts such as standard operating instructions (SOIs), user manuals, standard operating procedures and training manuals. Data should be collected regarding the task based on the opportunities available, although it is recommended that, as a minimum, the task in question is observed and verbal transcript recordings are made.

Step 4: Define Concepts and Relationships between Them
It is normally useful to identify distinct task phases. This allows propositional networks to be developed for each phase, which is useful for depicting the dynamic and changing nature of DSA throughout a task or scenario. For example, in an evaluation of DSA during land warfare mission planning activities (Salmon et al., 2009a), each question planning phase of the seven-question planning process was used as a distinct phase. In order to construct propositional networks, first the concepts need to be defined, followed by the relationships between them. For the purposes of DSA assessments, the term 'information elements' is used to refer to concepts. To identify the information elements related to the task under analysis, a simple content analysis is performed on the input data and keywords are extracted. These keywords represent the information elements, which are then linked based on their causal links during the activities in question (e.g. enemy 'has' plan, commander 'has' intent, terrain 'restricts' routes, etc.). Links are represented by directional arrows and should be overlaid with the linking proposition.

The output of this process is a network of linked information elements; the network contains all of the information that is used by the different actors and artefacts during task performance and thus represents the system's awareness, or what the system 'needed to know' in order to successfully undertake task performance.

Step 5: Define Information Element Usage
Information element usage is normally represented via shading of the different nodes within the network based on their usage by different actors during task performance. During this step, the analyst identifies which information elements the different agents, including both human and technological agents, used during task performance. This can be done in a variety of ways, including by further analysing input data (e.g. observational transcripts, verbal transcripts, HTA) and by holding discussions with those involved or relevant SMEs.

Step 6: Review and Refine Network
Constructing propositional networks is a highly iterative process that normally requires numerous reviews and reiterations. It is recommended that once a draft network is created, it is subject to at least three reviews. It is normally useful to involve domain SMEs or the participants who performed the task in this process. The review normally involves checking the information elements and the links between them and also the usage classification. Reiterations to the networks normally include the addition of new information elements and links, the revision of existing information elements and links and also the modification of the information element usage based on SME opinion.

Step 7: Analyse Networks Mathematically
Depending on the aims and requirements of the analysis, it may also be pertinent to analyse the propositional networks mathematically using social network statistics. For example, in the past we have used sociometric status and centrality calculations to identify the key information elements within propositional networks. Sociometric status provides a measure of how 'busy' a node is relative to the total number of nodes present within the network under analysis (Houghton et al., 2006). In this case, sociometric status gives an indication of the relative prominence of information elements based on their

links to other information elements in the network. Centrality is also a metric of the standing of a node within a network (Houghton et al., 2006), but here this standing is in terms of its 'distance' from all other nodes in the network. A central node is one that is close to all other nodes in the network and a message conveyed from that node to an arbitrarily selected other node in the network would, on average, arrive via the least number of relaying hops (Houghton et al., 2006). Key information elements are defined as those that have salience for each scenario phase, salience being defined as those information elements that act as hubs to other knowledge elements. Those information elements with a sociometric status value above the mean sociometric status value and a centrality score above the mean centrality value are identified as key information elements.

Advantages

- Propositional networks depict the information elements underlying a system's DSA and the relationships between them.
- In addition to modelling the system's awareness, they also depict the awareness of individuals and sub-teams working within the system.
- The networks can be analysed mathematically in order to identify the key pieces of information underlying a system's awareness.
- Unlike other SA measurement methods, they consider the mapping between the information elements underlying SA.
- The propositional network procedure avoids some of the flaws typically associated with SA measurement methods, including intrusiveness, high levels of preparatory work (e.g. SA requirements analysis, development of probes) and the problems associated with collecting subjective SA data and SA data post-trial.
- The outputs can be used to inform training, system, device and interface design and evaluation.
- They are easy to learn and use.
- Software support is available via the WESTT software tool (see Houghton et al., 2008).

Disadvantages

- Constructing propositional networks for complex tasks can be very time-consuming and laborious.
- It is difficult to present larger networks within articles, reports and/or presentations.
- No numerical value is assigned to the level of SA achieved by the system in question.
- They offer more of a modelling approach than a measure, although SA failures can be represented.
- The initial data collection phase may involve a series of activities and often adds considerable time to the analysis.
- Many find the departure from viewing SA in the heads of individual operators (i.e. what operators know) to viewing SA as a systemic property that resides in the interactions between actors and between actors and artefacts (i.e. what the system knows) a difficult one to grasp.
- The reliability of the method is questionable, particularly when being used by inexperienced analysts.

Related Methods

Propositional networks are similar to other network-based knowledge representation methods such as semantic networks (Eysenck and Keane, 1990) and concept maps (Crandall, Klein and Hoffman., 2006). The data collection phase typically utilises a range of approaches, including observational study, CDM interviews, verbal protocol analysis and HTA. The networks can also be analysed using metrics derived from social network analysis methods.

Approximate Training and Application Times

Providing the analysts involved have some understanding of DSA theory, the training time required for the propositional network method is low; our experiences in suggest that around one or two hours of training is required. Following training, however, considerable practice is required before analysts become proficient in the method. The application time is typically high, although it can be low if the task is simplistic and short.

Reliability and Validity

The content analysis procedure should ease some reliability concerns; however, the links between concepts are made on the basis of the analyst's subjective judgment and so the reliability of the method may be limited, particularly when being used by inexperienced analysts. The validity of the method is difficult to assess, although our experiences suggest that validity is high, particularly when appropriate SMEs are involved in the process.

Tools Needed

On a simple level, propositional networks can be conducted using pen and paper; however, video and audio recording devices are typically used during the data collection phase and a drawing package such as Microsoft Visio is used to construct the propositional networks. Houghton et al. (2008) describe the WESTT software tool, which contains a propositional network construction module that auto-builds propositional networks based on text data entry.

Example

As part of an evaluation of a new digitised battle management system, Salmon et al. (2009a) used the propositional network approach to model DSA during land warfare mission planning and battle execution activities. A total of six analysts located within the Brigade and Battle Group headquarters undertook direct observation of the planning and battle execution activities over the course of a three-week operational field trial involving a fully functional Division, Brigade and Battle Group. The trial was set up specifically in order to test the new system and closely represented a real-world operational situation. Propositional networks were developed for each question of the CE planning process observed during the field trial. Various planning cycles were observed over different missions. The seven questions propositional networks for 'mission 1' are presented in Figures 7.7–7.13 on the following pages.

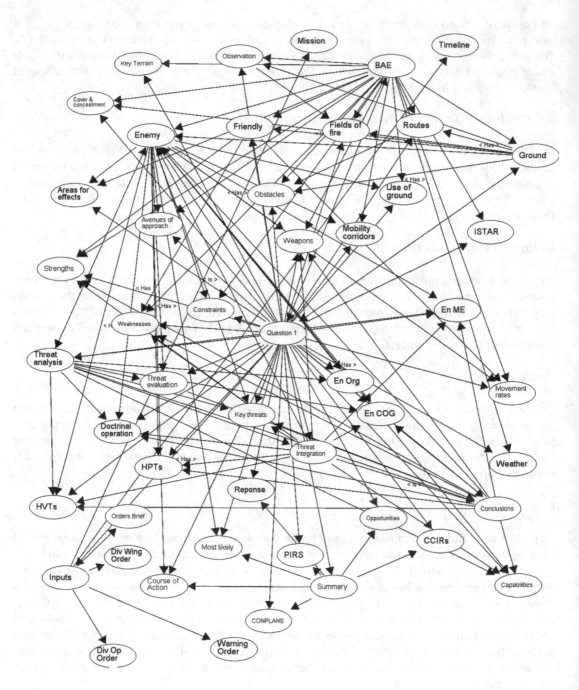

Figure 7.7   Question one propositional network

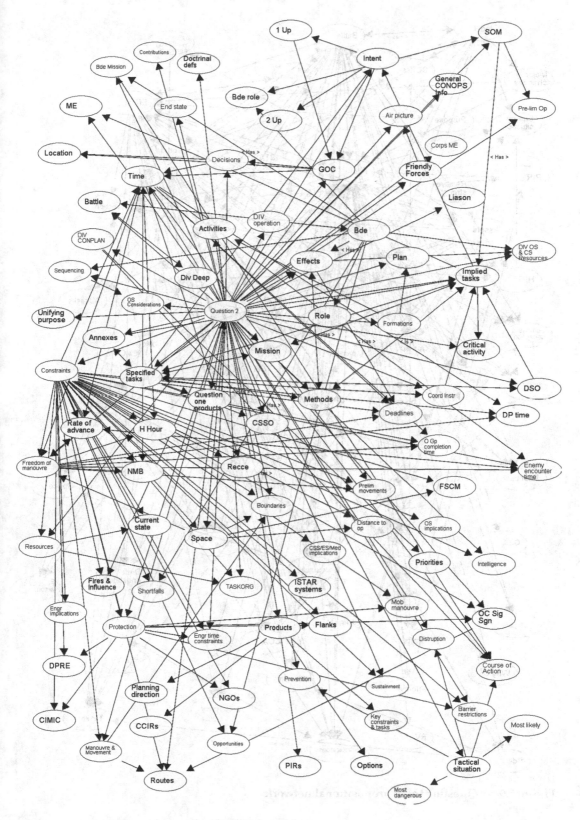

Figure 7.8   Question two propositional network

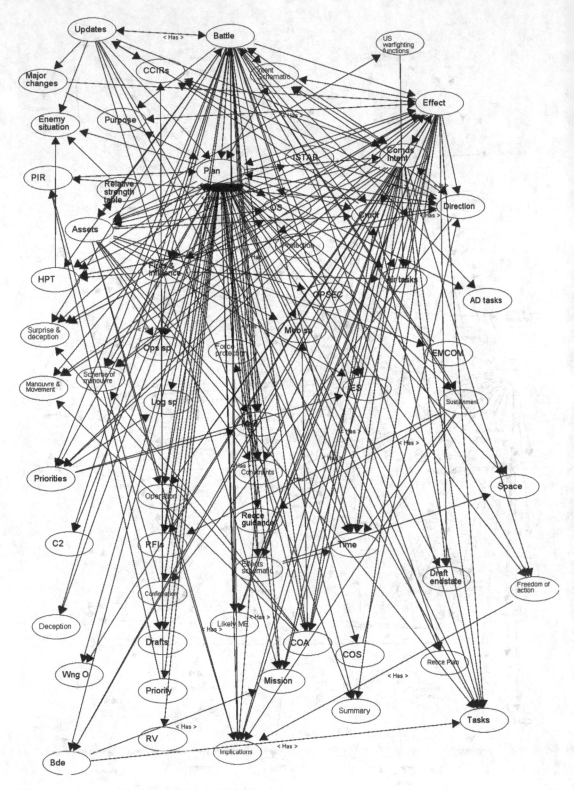

Figure 7.9   Question three propositional network

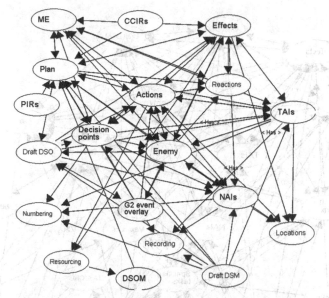

Figure 7.10   Question four propositional network

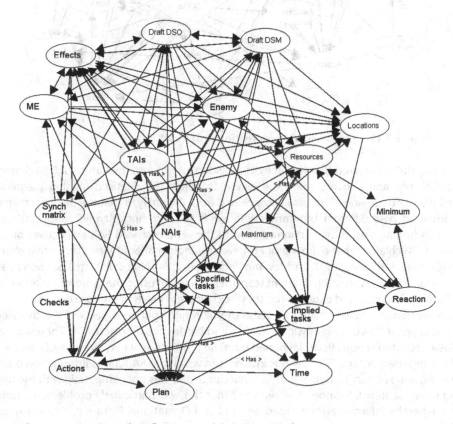

Figure 7.11   Question five propositional network

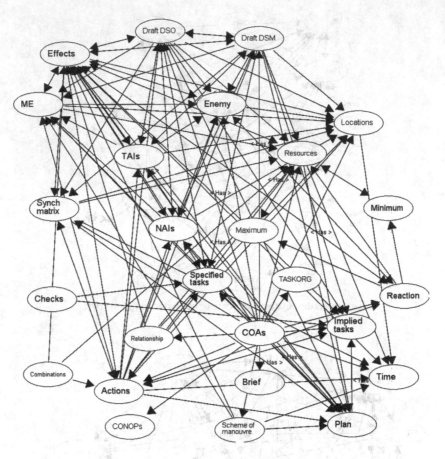

Figure 7.12   Question six propositional network

Salmon et al. (2009a) reported that the analysis provided insight into the nature of DSA during mission planning activities and also regarding the extent to which the digitised mission planning system supported planning activities. First, it was concluded from the analysis that the different teams (known as cells) involved held different but compatible levels of SA during planning and battle execution activities. This finding was in direct contradiction to the various shared SA perspectives prominent in the literature, which argue that team members have common or 'shared' views on certain elements of a situation (Endsley and Jones, 1997; Endsley and Robertson, 2000; Nofi, 2000, etc.). Salmon et al. (2009a) further concluded that, even when different team members were using the same information, due to the nature of their roles, goals and experiences, they held different levels of SA from one another.

The propositional network analysis was also useful in highlighting aspects of the digitised system that did not support DSA during mission planning and battle execution activities. For example, Salmon et al. (2009a) reported various flaws, identified from the propositional network analysis, that limited the utility of the digitised system. First, various instances in which the SA-related information presented by the digital mission support system was in fact inaccurate and was not compatible with the real state of the world were identified. Salmon et al. reported that this was particularly problematic during battle execution, whereby information presented on the Local Operational Picture (LOP) display was either out of date or spurious, which often led to the Brigade and Battle Group's SA of enemy and friendly force locations, movements, numbers and capabilities being inaccurate. Second, the timeliness of the SA-related information presented by the digital system was also found to be problematic; due to data bandwidth limitations, voice transmission was given precedence over global positioning data regarding the locations and movements of entities on the battlefield. Because of this, contact reports

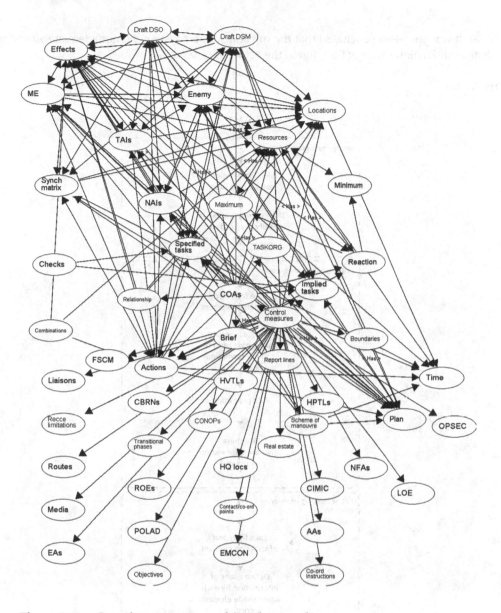

Figure 7.13   Question seven propositional network

and positional information presented on the LOP were often out of date (on occasions being presented up to 20 minutes late). As a corollary, the system's DSA at times was often 'out of date' or at least lagging behind the real state of the world. The problem of 'delayed' SA information presented by the system was a found to be a bandwidth issue. Specifically, because of the amount of data being transmitted and the limited bandwidth of the system, the voice communications data took precedence over the LOP positional data. This meant that during complex operations, the LOP positional data was delayed due to high voice communications traffic. Due to the same data transmission problems, the digital system was also observed to be slow in updating the enemy positions on the LOP. Third, as a consequence of the problems discussed, it was found that users placed a low level of trust in the SA-related information presented by the digital system. In response to these flaws, Salmon et al. (2009a) reported that the teams observed resorted to using the old 'paper map' processes for mission planning and battle management activities. It was therefore concluded that the mission support system did not adequately support the acquisition and maintenance of DSA during the activities observed.

Flowchart

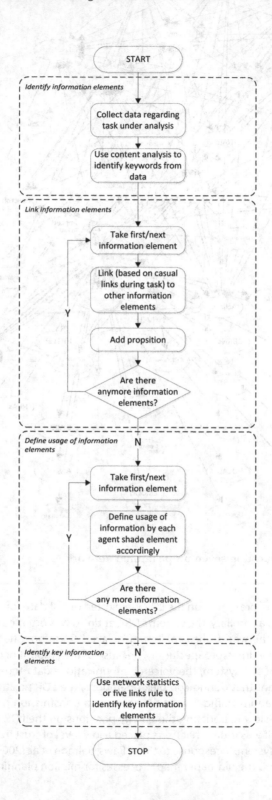

Chapter 8
Mental Workload Assessment Methods

The assessment of mental workload (MWL) is of crucial importance during the design and evaluation of complex systems. The increased role of technology and use of complex procedures has led to a greater level of demand being imposed on operators within complex systems. Individual operators possess a finite attentional capacity, and these attentional resources are allocated to the relevant tasks. MWL represents the proportion of resources demanded by a task or set of tasks. An excessive demand on resources imposed by the task attended to typically results in performance degradation. There has

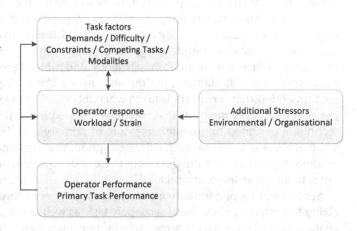

Figure 8.1   Framework of interacting stressors affecting MWL (adapted from Megaw, 2005)

been much debate as to the nature of MWL, with countless attempts at providing a definition. Rather than reviewing these (often competing) definitions, we opt for the approach proposed by Megaw (2005), which is to consider MWL in terms of a framework of interacting stressors on an individual (see Figure 8.1 above). The arrows indicate the direction of effects within this framework and imply that when we measure MWL, we are examining the impact of a whole host of factors on both performance and response. Clearly, this means that we are facing a multi-dimensional problem that is not likely to be amenable to single measures.

The construct of MWL has been investigated in a wide variety of domains, including aviation, ATC, military operations, driving and control room operation to name but a few. The assessment or measurement of MWL is used throughout the design life-cycle to inform system and task design, and to provide an evaluation of MWL imposed by existing operational systems and procedures. MWL assessment is also used to evaluate the workload imposed during the operation of existing systems. There are a number of different MWL assessment procedures available to the HF practitioner. Traditionally, using a single approach to measure operator MWL has proved inadequate, and as a result a combination of the methods available is typically used. The assessment of operator MWL typically requires the use of a battery of MWL assessment techniques, including primary task performance measures, secondary task performance measures (reaction times, embedded tasks), physiological measures (heart rate (HR), heart rate variability (HRV)), and subjective rating techniques (SWAT, NASA-TLX). A review of mental workload methods identified the following categories of MWL assessment techniques:

- primary and secondary task performance measures;
- physiological measures;
- subjective rating techniques; and
- quantitative evaluation of task demands.

A brief description of each category and also of each MWL assessment technique considered is given below.

Primary task performance measures of operator MWL involve the measurement of the operator's ability to perform the primary task under analysis, which is expected to diminish as MWL increases. Specific aspects of the primary task are assessed in order to measure performance. For example, in a study of driving with automation, Young and Stanton (2004) measured speed, lateral position and headway as indicators of performance on a driving task. According to Wierwille and Eggemeier (1993), primary tasks measures should be included in any assessment of operator MWL. The main advantages associated with the use of primary task measures for the assessment of operator MWL are their reported sensitivity to variations in workload (Wierwille and Eggemeier, 1993; Pretorius and Cilliers, 2007) and their ease of use, since performance of the primary task is normally measured anyway. There are a number of disadvantages associated with this method of MWL assessment, including the ability of operators to perform efficiently under high levels of workload, due to factors such as experience and skill. Similarly, performance may suffer during low workload parts of the task. It is recommended that great care is taken when interpreting the results obtained through primary task performance assessment of MWL. Mental Workload Index (MWLI: Pretorius and Cilliers, 2007) is a technique designed to provide a measure of primary task demands in the rail domain rather than a subjective analysis of task difficulty perception. The method measures three task elements and 11 moderating factors, and the data from these task aspects is analysed using a weighted formulae in order to derive an index of workload.

Secondary task performance measures of operator MWL involve the measurement of the operator's ability to perform an additional secondary task, as well as the primary task involved in the task under analysis. Typical secondary task measures include memory recall tasks, mental arithmetic tasks, reaction time measurement and tracking tasks. The use of secondary task performance measures is based upon the assumption that as operator workload increases, the ability to perform the secondary task will diminish due to a reduction in spare mental capacity, and so secondary task performance will suffer. The main disadvantages associated with secondary task performance assessment techniques are a reported lack of sensitivity to minor workload variations (Young and Stanton, 2004) and their intrusion on primary task performance. One way around this is the use of embedded secondary task measures, whereby the operator is required to perform a secondary task with the system under analysis. Since the secondary task is no longer external to that of operating the system, the level of intrusion is reduced. According to Young and Stanton (2004), researchers adopting a secondary task measurement approach to the assessment of MWL are advised to adopt discrete stimuli, which occupy the same attentional resource pools as the primary task. For example, if the primary task is a driving one, then the secondary task should be a visio-spatial one involving manual response. This ensures that the technique really is measuring spare capacity and not an alternative resource pool.

Physiological measures of MWL involve the measurement of those physiological aspects that may be affected by increased or decreased levels of workload. Heart rate, heart rate variability, eye movement and brain activity have all been used to provide a measure of operator workload. The main advantage associated with the use of physiological measures of MWL is that they do not intrude upon primary task performance and also that they can be applied in the field, as opposed to simulated settings. There are a number of disadvantages associated with the use of physiological techniques, including the high cost, physical obtrusiveness and reliability of the technology used, and the doubts regarding the construct validity and sensitivity of the techniques.

Subjective rating MWL assessment techniques are administered either during or post-task performance and involve participants providing ratings regarding their perceived MWL during task performance. Subjective-rating techniques can be categorised as either uni-dimensional or multi-dimensional, depending on the workload dimensions that they assess. Young and Stanton (2004) suggest that the data obtained when using uni-dimensional techniques is far simpler to analyse than the data obtained when using multi-dimensional techniques; however, multi-dimensional techniques possess a greater level of diagnosticity than uni-dimensional techniques. Subjective rating assessment techniques are attractive due to their ease and speed of application, and also the low cost involved. They are also unintrusive to primary task performance and can be used in the field in 'real-world' settings, rather than in simulated environments. That said, subjective MWL assessment techniques are normally only used when there is an operational system available and therefore it is difficult to employ them during the design process, as the system under analysis may not actually exist, and simulation can be extremely costly. There are also a host of problems associated with collecting subjective data post-trial. Often, MWL ratings correlate with performance on the task under analysis. Participants are also prone to forgetting certain parts of the task where variations in their workload may have occurred. A brief description of the subjective MWL assessment techniques reviewed is given below.

The NASA Task Load Index (NASA-TLX: Hart and Staveland, 1988) is a multi-dimensional subjective rating tool that is used to derive a MWL rating based upon a weighted average of six workload sub-scale ratings. The six sub-scales are mental demand, physical demand, temporal demand, effort, performance and frustration level. The TLX is the most commonly used subjective MWL assessment technique and there have been a number of validation studies associated with this technique. The subjective workload assessment technique (SWAT: Reid and Nygren, 1988) is a multi-dimensional tool that measures three dimensions of operator workload: time load, mental effort load and stress load. After an initial weighting procedure, participants are asked to rate each dimension and an overall workload rating is calculated. Along with the NASA-TLX technique of subjective workload, SWAT is probably the most commonly used of the subjective workload assessment techniques.

The DRA workload scale (DRAWS) uses four different workload dimensions to elicit a rating of operator workload. The dimensions used are input demand, central demand, output demand and time pressure. The technique is typically administered online and involves verbally querying the participant for a subjective rating between 0 and 100 for each dimension during task performance. The workload profile technique (Tsang and Velazquez, 1996) is based upon multiple resource theory (Wickens, Gordon and Lui, 1998) and involves participants rating the demand imposed by the task under analysis for each dimension proposed by multiple resource theory. The workload dimensions used are perceptual/central processing, response selection and execution, spatial processing, verbal processing, visual processing, auditory processing manual output and speech output. Participant ratings for each dimension are summed in order to determine an overall workload rating for the task under analysis.

The Modified Cooper Harper scale (MCH: Cooper and Harper, 1969) is a uni-dimensional measure that uses a decision tree to elicit a rating of operator mental workload. It is a modified version of the Cooper Harper scale that was originally developed as an aircraft handling measurement tool. The scales were used to attain subjective pilot ratings of the controllability of aircrafts. The output of the scale is based upon the controllability of the aircraft and also the level of input required by the pilot to maintain suitable control. The subjective workload dominance technique (SWORD: Vidulich and Hughes, 1991) uses paired comparison of tasks in order to provide a rating of workload for each individual task. Administered post-trial, participants are required to rate one task's dominance over another in terms of workload imposed. The Malvern capacity estimate (MACE) technique uses a rating scale to determine the remaining capacity of air traffic controllers. It is a very simple technique, involving querying air traffic controllers for subjective estimations of their remaining mental capacity during a simulated task. The Bedford scale (Roscoe and Ellis, 1990) uses a hierarchical decision tree to assess spare capacity whilst performing a task. Participants simply follow the decision tree to gain a workload rating for the task under analysis. The instantaneous self-assessment (ISA) of workload technique involves participants self-rating their workload during a task (normally every two minutes) on a scale of 1 (low) to 5 (high).

A more recent theme in the area of MWL assessment is the use of assessment techniques to predict operator MWL. Analytical techniques are those MWL techniques that are used to predict the level of MWL that an operator may experience during the performance of a particular task. Analytical techniques are typically used during system design when an operational version of the system under analysis is not yet available. Although literature regarding the use of predictive MWL is limited, a number of these techniques do exist. In the past, models have been used to predict operator workload, such as the timeline model or Wicken's multiple resource model. Subjective MWL assessment techniques such as Pro-SWORD have also been tested for their use in predicting operator MWL (Vidulich, Ward and Schueren, 1991). Although the use of MWL assessment techniques in a predictive fashion is limited, Salvendy (1997) reports that SME projective ratings tend to correlate well with operator subjective ratings. It is apparent that analytical mental or predictive workload techniques are particularly important in the early stages of system design and development. A brief description of the analytical techniques reviewed is given below.

Cognitive task load analysis (CTLA: Neerincx, 2003) is used to assess or predict the cognitive load of a task or set of tasks imposed upon an operator. CTLA is based upon a model of cognitive task load that describes the effects of task characteristics upon operator MWL. According to the model, cognitive (or mental) task load is comprised of percentage time occupied, level of information processing and the number of task-set switches exhibited during the task in question. Pro-SWAT is a variation of SWAT (Reid and Nygren, 1988) that has been used to predict operator MWL. SWAT is a multi-dimensional tool that uses three dimensions of operator workload; time load, mental effort load and stress load. SWORD is a SWAT that has been used both retrospectively and predictively. It uses a paired comparison of tasks in order to provide a workload rating for each individual task. Participants are required to rate one task's dominance over another in terms of workload imposed. When used predictively, tasks are rated for their dominance before the trial begins and are then rated post-test to check for the sensitivity of the predictions. Vidulich, Ward and Schueren (1991) report the use of the SWORD technique for predicting the workload imposed upon F-16 pilots by a new head-up display (HUD) attitude display system.

MWLI is a method designed to be used to both assess and predict workload levels focusing on task demands rather than on subjective evaluations of difficulty (Pretorious and Cilliers, 2007). It uses quantitative evaluation of three core task factors and 11 moderating factors to assess task demands. Each of these aspects are measured and the data collected is inputted into a mathematical formulae to derive a mental workload index.

Typical MWL assessments use a selection of techniques from each of the three categories described above. The multi-method approach to the assessment of MWL is designed to ensure comprehensiveness. The suitability of MWL assessment techniques can be evaluated on a number of dimensions. Wierwille and Eggemeier (1993) suggest that for a MWL assessment technique to be recommended for use in a test and evaluation procedure, it should possess the following properties:

- Sensitivity: the degree to which the technique can discriminate between differences in the levels of MWL imposed on a participant.
- Limited intrusiveness: the degree to which the assessment technique intrudes upon primary task performance.
- Diagnosticity: the degree to which the technique can determine the type or cause of the workload imposed on a participant.
- Global sensitivity: the ability to discriminate between variations in the different types of resource expenditure or factors affecting workload.
- Transferability: the degree to which the technique can be applied in different environments other than what it was designed for.
- Ease of implementation: the level of resources required to use the technique, such as technology and training requirements.

Wierwille and Eggemeier suggest that non-intrusive workload techniques that possess a sufficient level of global sensitivity are of the most importance in terms of test and evaluation applications. According to them, the most frequently used and therefore most appropriate test and evaluation scenarios are MCH, SWAT and the NASA-TLX technique. A summary of the MWL assessment techniques reviewed is presented in Table 8.1.

Table 8.1   Summary of mental workload assessment techniques

Method	Type of method	Domain	Training time	App time	Related methods	Tools needed	Validation studies	Advantages	Disadvantages
Primary task performance measures.	Performance measure.	Generic.	Low.	Low.	Physiological measures, subjective assessment techniques.	Simulator, computer.	Yes.	1) Primary task performance measures offer a direct index of performance. 2) Primary task performance measures are particularly effective when measuring workload in tasks that are lengthy in duration. 3) Can be easily used in conjunction with secondary task performance, physiological and subjective measures.	1) Primary task performance measures may not always distinguish between levels of workload. 2) Not a reliable measure when used in isolation.
Secondary task performance measures.	Performance measure.	Generic.	Low.	Low.	Physiological measures, subjective assessment techniques.	Simulator, computer.	Yes.	1) Sensitive to workload variations when performance measures are not. 2) Easy to use. 3) Little extra work is required to set up a secondary task measure.	1) Secondary task measures have been found to be sensitive only to gross changes in workload. 2) Intrusive to primary task performance. 3) Great care is required when designing the secondary task in order to ensure that it uses the same resource pool as the primary task.
Physiological measures.	Physiological measure.	Generic.	High.	Low.	Primary and secondary task performance measures, subjective assessment techniques.	Heart rate monitor, eye tracker, EEG.	Yes.	1) Various physiological measures have demonstrated sensitivity to variations in task demand. 2) Data is recorded continuously throughout the trial. 3) Can be used in 'real-world' settings.	1) Data is often confounded by extraneous interference. 2) Measurement equipment is temperamental and difficult to use. 3) Measurement equipment is physically obtrusive.
NASA-TLX.	Multi-dimensional subjective rating tool.	Generic.	Low.	Low.	Primary and secondary task performance measures, physiological measures.	Pen and paper.	Yes.	1) Quick and easy to use, requiring little training or cost. 2) Consistently performs better than SWAT. 3) TLX scales are generic, allowing the technique to be applied in any domain.	1) More complex to analyse than uni-dimensional tools. 2) The TLX weighting procedure is laborious. 3) Caters for individual workload only.

Table 8.1   Continued

Method	Type of method	Domain	Training time	App time	Related methods	Tools needed	Validation studies	Advantages	Disadvantages
MCH scales.	Uni-dimensional subjective rating tool.	Generic.	Low.	Low.	Primary and secondary task measures, physiological measures.	Pen and paper.	Yes.	1) Quick and easy to use, requiring little training or cost. 2) Widely used in a number of domains. 3) Data obtained is easier to analyse than multi-dimensional data.	1) Unsophisticated measure of workload. 2) Limited to manual control tasks. 3) Not as sensitive as TLX or SWAT.
SWAT.	Multi-dimensional subjective rating tool.	Generic (aviation).	Low.	Low.	Primary and secondary task performance measures, physiological measures.	Pen and paper.	Yes.	1) Quick and easy to use, requiring little training or cost. 2) Multi-dimensional. 3) SWAT sub-scales are generic, allowing the technique to be applied in any domain.	1) More complex to analyse than uni-dimensional tools. 2) A number of studies suggest that the NASA-TLX is more sensitive to workload variations. 3) MWL ratings may correlate with task performance.
PRO-SWAT.	Multi-dimensional predictive method.	Generic.	High.	High.	SWAT, Pro-SWORD.	Pen and paper, software.	No.	1) Simple and efficient method to predict workload. 2) Can be applied in any domain.	1) Has yet to be validated and is still in its infancy. 2) Low sensitivity to mental workload.
SWORD.	Subjective paired comparison technique.	Generic (aviation).	Low.	Low.	Primary and secondary task performance measures, physiological measures.	Pen and paper.	Yes.	1) Quick and easy to use, requiring little training or cost. 2) Very effective when comparing the MWL imposed by two or more interfaces.	1) More complex to analyse than uni-dimensional tools. 2) Data is collected post-trial. There are a number of problems with this, such as a correlation with performance.
PRO-SWORD.	Predictive method.	Generic (aviation).	Low.	Low.	Pro-SWAT, MCH, HTA	Pen and paper.	Yes.	1) Easy and quick to learn and apply. 2) High levels of face validity.	1) Based on a dated approach to workload. 2) Not widely used.
DRAWS.	Multi-dimensional subjective rating tool.	Generic (aviation).	Low.	Low.	Primary and secondary task performance measures, physiological measures.	Pen and paper.	No.	1) Quick and easy to use, requiring little training or cost.	1) More complex to analyse than uni-dimensional tools. 2) Data is collected post-trial. There are a number of problems with this, such as a correlation with performance. 3) Limited use and validation.

Table 8.1   Continued

Method	Type of method	Domain	Training time	App time	Related methods	Tools needed	Validation studies	Advantages	Disadvantages
MACE.	Uni-dimensional subjective rating tool.	ATC.	Low.	Low.	Primary and secondary task performance measures, physiological measures.	Pen and paper.	No.	1) Quick and easy to use, requiring little training or cost.	1) Data is collected post-trial. There are a number of problems with this, such as a correlation with performance. 2) Limited evidence of use or reliability and validity.
Workload Profile Technique.	Multi-dimensional subjective rating tool.	Generic.	Med.	Low.	Primary and secondary task performance measures, physiological measures.	Pen and paper.	Yes.	1) Quick and easy to use, requiring little training or cost. 2) Based upon sound theoretical underpinning (multiple resource theory).	1) More complex to analyse than uni-dimensional tools. 2) Data is collected post-trial. There are a number of problems with this, such as a correlation with performance. 3) More complex than other MWL techniques.
Bedford scale.	Multi-dimensional subjective rating tool.	Generic.	Low.	Low.	Primary and secondary task performance measures, physiological measures.	Pen and paper.	Yes.	1) Quick and easy to use, requiring little training or cost.	1) More complex to analyse than uni-dimensional tools. 2) Data is collected post-trial. There are a number of problems with this, such as a correlation with performance.
Mental Workload Index.	Primary task measure.	Rail.	High.	High.	N/A	Pen and paper.	No.	1) Insights into resource allocation. 2) Unobtrusive. 3) Sensitive to task changes.	1) Specific to rail domain at the present time. 2) Requires a high degree of participant access.

Primary and Secondary Task Performance Measures

Background and Applications

MWL assessment typically involves the use of a combination or battery of MWL assessment techniques. Primary task performance measures, secondary task performance measures and physiological measures are typically used in conjunction with post-trial subjective rating techniques. Primary task performance measures of MWL involve assessing suitable aspects of participant performance during the task under analysis, assuming that an increase in MWL will facilitate a performance decrement of some sort. Secondary task performance measures typically involve participants performing an additional task in addition to that of primary task performance. Participants are required to maintain primary task performance and also perform the secondary task as and when the primary task allows them to. The secondary task is designed to compete for the same resources as the primary task. Any differences in workload between primary tasks are then reflected in the performance of the secondary task. Examples of secondary tasks used in the past include tracking tasks, memory tasks, rotated figures tasks and mental arithmetic tasks.

Domain of Application

Generic.

Procedure and Advice

Step 1: Define the Primary Task under Analysis
The first step in an assessment of operator workload is to clearly define the task (or tasks) under analysis. It is recommended that for this purpose, a HTA is conducted for the task under analysis. When assessing the MWL associated with the use of a novel or existing system or interface, it is recommended that the task assessed are as representative of the system or interface under analysis as possible, i.e. the task is made up of tasks using as much of the system or interface under analysis as possible.

Step 2: Define Primary Task Performance Measures
Once the task (or tasks) under analysis is clearly defined and described, the analyst should next define those aspects of the task that can be used to measure participant performance. For example, in a driving task, Young and Stanton (2004) used speed, lateral position and headway as measures of primary task performance. The measures used may be dependent upon the equipment that is used during the analysis. The provision of a simulator that is able to record various aspects of participant performance is especially useful. The primary task performance measures used are dependent upon the task and system under analysis.

Step 3: Design Secondary Task and Associated Performance Measures
Once the primary task performance measures are clearly defined, an appropriate secondary task measure should be selected. Stanton and Young (2004) recommend that great care is taken to ensure that the secondary task competes for the same attentional resources as the primary task. For example, they used a visual-spatial task that required a manual response as their secondary task when analysing driver workload; this was designed to use the same attentional resource pool as the primary task of driving the car. As with the primary task, the secondary task used is dependent upon the system and task under analysis.

Step 4: Test Primary and Secondary Tasks
Once the primary and secondary task performance measures are defined, they should be thoroughly tested in order to ensure that they are sensitive to variations in task demand. The analyst should define a set of tests that are designed to ensure the validity of the primary and secondary task measures chosen.

Step 5: Brief Participants

Once the measurement procedure has been subjected to sufficient testing, the appropriate participants should be selected and then briefed regarding the purpose of the analysis and the data collection procedure employed. It may be useful to select the participants that are to be involved in the analysis prior to the data collection date. This may not always be necessary and it may suffice to simply select participants randomly on the day of analysis. However, if workload is being compared across rank or experience levels, then clearly effort is required to select the appropriate participants. Before the task (or tasks) under analysis is performed, all of the participants involved should be briefed regarding the purpose of the study, MWL, MWL assessment and the techniques that are being employed. It may be useful at this stage to take the participants through an example workload assessment analysis so that they understand how primary and secondary task performance measurement works and what is required of them as participants. If SWAT is also being used, participants should be briefed regarding the chosen technique.

Step 6: Conduct Pilot Run

Once the participants understand the data collection procedure, a small pilot run should be conducted to ensure that the process runs smoothly and efficiently. Participants should be instructed to perform a small task (separate from the task under analysis), and an associated secondary task. Upon completion of the task, the participants should be instructed to complete the appropriate SWAT. This acts as a pilot run of the data collection procedure and serves to highlight any potential problems. The participants should be instructed to ask any questions regarding their role in the data collection procedure.

Step 7: Begin Primary Task Performance

Once a pilot run of the data collection procedure has been successfully completed and the participants are comfortable with their role during the trial, the 'real' data collection procedure can begin. The participants should be instructed to begin the task under analysis and to attend to the secondary task when they feel that they can. The task should run for a set amount of time and the secondary task should run concurrently.

Step 8: Administer SWAT

Typically, SWATs, such as the NASA-TLX (Hart and Staveland, 1988) are used in conjunction with primary and secondary task performance measures to assess participant workload. The chosen technique should be administered immediately once the task under analysis is completed and participants should be instructed to rate the appropriate workload dimensions based upon the primary task that they have just completed.

Step 9: Analyse Data

Once the data collection procedure is completed, the data should be analysed appropriately. Young and Stanton (2004) used the frequency of correct responses on a secondary task to indicate the amount of spare capacity the participant had, i.e. the greater the correct responses on the primary task, the greater the participant's spare capacity was assumed to be.

Advantages

- When using a battery of MWL assessment techniques to assessment MWL, the data obtained can be cross-checked for reliability purposes.
- Primary task performance measures offer a direct index of performance.
- They are also particularly effective when measuring workload in tasks that are lengthy in duration (Young and Stanton, 2004).
- They are also useful when measuring operator overload.
- They require no further effort on behalf of the analyst to set up and record, as primary task performance is normally measured anyway.

- Secondary task performance measures are effective at discriminating between tasks when no difference was observed assessing performance alone.
- Primary and secondary task performance measures are easy to use, as a computer typically records the required data.

Disadvantages

- Primary task performance measures alone may not distinguish between different levels of workload, particularly minimal ones. Different operators may still achieve the same performance levels under completely different workload conditions.
- Young and Stanton (2004) suggest that primary task performance is not a reliable measure when used in isolation.
- Secondary task performance measures have been found to be only sensitive to gross changes in MWL.
- Secondary task performance measures are intrusive to primary task performance.
- Great care is required during the design and selection of the secondary task to be used. The analyst must ensure that the secondary task competes for the same resources as the primary task. According to Young and Stanton (2004), the secondary task must be carefully designed in order to be a true measure of spare attentional capacity.
- Extra work and resources are required in developing the secondary task performance measure.
- The techniques need to be used together in order to be effective.
- Using primary and secondary task performance measures may prove expensive, as simulators and computers are required.

Example

Young and Stanton (2004) describe the measurement of MWL in a driving simulator environment (Figure 8.2). Primary task performance measurement included recording data regarding speed, lateral position and headway (distance from the vehicle in front). A secondary task was used to assess spare attentional capacity. This was designed to compete for the same attentional resources as the primary task of driving the car. The secondary task was comprised of a rotated figures task (Baber, 1991) whereby participants were randomly presented with a pair of stick figures (one upright, the other rotated through 0°, 90°, 180° or 270°) holding one or two flags. The flags were made up of either squares or diamonds. Participants were required to make a judgment, via a button, as to whether the figures were the same or different, based upon the flags that they were holding. The participants were instructed to attend to the secondary task only when they felt that they had time to do so. Participants' correct responses were measured and it was assumed that the higher the frequency of correct responses, the greater participant spare capacity was assumed to be.

Related Methods

Primary and secondary task performance measures are typically used in conjunction with physiological measures and subjective workload techniques in order to measure operator MWL. A number of secondary task performance measurement techniques exist, including task reaction times, tracking tasks, memory recall tasks and mental arithmetic tasks. Physiological measures of workload include measuring participant heart rate, heart rate variability, blink rate and brain activity. SWATs are completed post-trial by participants and involve participants rating specific dimensions of workload. There are a number of subjective workload assessment techniques, including NASA-TLX (Hart and Staveland, 1988), SWAT (SWAT; Reid and Nygren, 1988) and the workload profile technique (Tsang and Velazquez, 1996).

Approximate Training and Application Times

The training and application times associated with both primary and secondary task performance measures of MWL are typically estimated to be low. However, substantial time is usually required for the development of an appropriate secondary task measure.

Reliability and Validity

According to Young and Stanton (2004), it is not possible to comment on the reliability and validity of primary and secondary performance

Figure 8.2   Screenshot of a driving simulator
Source: Young and Stanton, 2004.

measures of MWL, as they are developed specifically for the task and application under analysis. The reliability and validity of the techniques used can be checked to an extent by using a battery of techniques (primary task performance measures, secondary task performance measures, physiological measures and subjective assessment techniques). The validity of the secondary task measure can be ensured by making sure that the secondary task competes for the same attentional resources as the primary task.

Multiple studies have validated the use of secondary task measures through comparison with physiological methods. For example, Wester et al. (2008) used primary and secondary task performance to measure workload in driving. An auditory oddball task, in which random arrangements of standard, deviant and novel sounds were played, was used alongside a primary task of keeping in lane. Wester et al. compared the results of the secondary measure with EEG-measured brain activity and found that ERP amplitude (a measure of brain activity) was significantly lower in the secondary task when performed alongside the primary task than when it was performed alone.

Baldauf, Burgard and Wittman (2009) explored the utility of using time estimates (as a secondary task) in the measurement of mental workload within a simulated driving task. In order to ascertain the suitability of time estimates as a measure of workload, the technique was compared to a physiological measure of workload (electrodermal activity) and a subjective measure of workload (SWAT). The time interval measure involved tasking participants with pushing a button when they thought that 17 seconds has elapsed. The results revealed that all three measures increased as the task difficulty increased and that the time interval task did not interfere with the driving (primary) task. Baldauf et al. (2009) argued that the time measure was sensitive to variations in workload, with time estimates increasing with workload levels.

Tools Needed

The tools needed are dependent upon the nature of the analysis. For example, in the example described above, a driving simulator and a computer were used. The secondary task is normally presented separately from the primary task via a desktop or laptop computer. The simulator or a computer is normally used to record participant performance on the primary and secondary tasks.

Flowchart

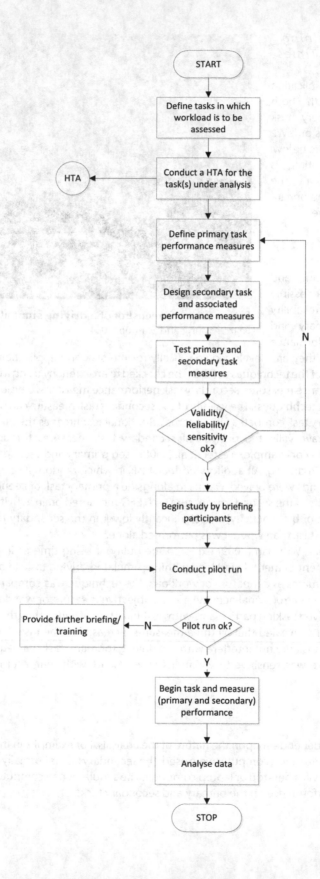

Physiological Measures

Background and Applications

Physiological or psychophysiological measures have also been used in the assessment of participant SA. Physiological measurement techniques are used to measure variations in participant physiological responses to the task under analysis. The use of physiological measures as indicators of MWL is based on the assumption that as task demand increases, marked changes in various participant physiological systems are apparent. There are a number of different physiological measurement techniques available to the HF practitioner. In the past, heart rate, heart rate variability, endogenous blink rate, brain activity, electrodermal response, eye movements, papillary responses and event-related potentials have all been used to assess operator MWL. Measuring heart rate is one of the most common physiological measures of workload. It is assumed that an increase in workload causes an increase in operator heart rate. Heart rate variability (HRV) has also been used as an indicator of operator MWL. According to Salvendy (1997), laboratory studies have reported a decrease in HRV (heart rhythm) under increased workload conditions. Endogenous eye blink rate has also been used in the assessment of operator workload. Increased visual demands have been shown to cause a decreased endogenous eye blink rate (Salvendy, 1997). According to Wierwille and Eggemeier (1993), a relationship between blink rate and visual workload has been demonstrated in the flight environment. It is assumed that a higher visual demand causes the operator to reduce his or her blink rate in order to achieve greater visual input. Measures of brain activity involve using EEG recordings to assess operator MWL. According to Wierwille and Eggemeier, measures of evoked potentials have demonstrated a capability of discriminating between levels of task demand.

Domain of Application

Generic.

Procedure and Advice

The following procedure offers advice on the measurement of heart rate as a physiological indicator of workload. When using other physiological techniques, it is assumed that the procedure is the same, only with different equipment being used.

Step 1: Define the Primary Task under Analysis

The first step in an assessment of operator workload is to clearly define the task under analysis. It is recommended that an HTA is conducted for the task under investigation. When assessing the MWL associated with the use of a novel or existing system or interface, it is recommended that the task assessed are as representative of the system or interface under analysis as possible, i.e. the task is made up of tasks using as much of the system or interface under analysis as possible.

Step 2: Select the Appropriate Measuring Equipment

Once the task (or tasks) under analysis is clearly defined and described, the analyst should select the appropriate measurement equipment. For example, when measuring MWL in a driving task, Young and Stanton (2004) measured heart rate using a Polar Vantage NV heart rate monitor. These are relatively cheap to purchase and comprise a chest belt and a watch. The type of measures used may be dependent upon the environment in which the analysis is taking place. For example, in infantry operations, it may be difficult to measure blink rate or brain activity.

Step 3: Conduct Initial Testing of the Data Collection Procedure

It is recommended that a pilot run of the data collection procedure is conduced in-house in order to test the measuring equipment used and the appropriateness of the data collected. Physiological measurement equipment is typically temperamental and difficult to use. Consequently, it may take some time for the analyst to become proficient in its use. It is recommended that the analyst involved practice using the equipment until he or she becomes proficient in its use.

Step 4: Brief Participants

Once the measurement procedure has been subjected to sufficient testing, the appropriate participants should be selected and briefed regarding the purpose of the study and the data collection procedure employed. It may be useful to select the participants that are to be involved in the analysis prior to the data collection date. This may not always be necessary and it may suffice to simply select participants randomly on the day of analysis. However, if workload is being compared across rank or experience levels, then clearly effort is required to select the appropriate participants. Before the task under analysis is performed, all of the participants involved should be briefed regarding the purpose of the study, MWL, MWL assessment and the physiological techniques employed. It may be useful at this stage to take the participants through an example workload assessment analysis so that they understand how the physiological measures in question work and what is required of them as participants. If a subjective workload assessment technique is also being used, participants should also be briefed regarding the chosen technique.

Step 5: Fit Measuring Equipment

Next, the participants should be fitted with the appropriate physiological measuring equipment. The heart rate monitor consists of a chest belt, which is placed around the participant's chest, and a watch, which the participant can wear on his or her wrist, or the analyst can hold. The watch collects the data and is then connected to a computer post-trial in order to download the data collected.

Step 6: Conduct Pilot Run

Once the participants understand the data collection procedure, a small pilot run should be conducted to ensure that the process runs smoothly and efficiently. Participants should be instructed to perform a small task (separate from the task under analysis) and an associated secondary task whilst wearing the physiological measurement equipment. Upon completion of the task, the participants should be instructed to complete the appropriate SWAT. This acts as a pilot run of the data collection procedure and serves to highlight any potential problems. The participants should be instructed to ask any questions regarding their role in the data collection procedure.

Step 7: Begin Primary Task Performance

Once a pilot run of the data collection procedure has been successfully completed and the participants fully understand their role in the trial, the data collection procedure can begin. The participants should be instructed to begin the task under analysis and to attend to the secondary task when they feel that they can. The task should run for a set amount of time and the secondary task should run concurrently. The heart rate monitor continuously collects participant heart rate data throughout the task. Upon completion of the task, the heart rate monitor should be turned off and removed from the participant's chest.

Step 8: Administer SWAT

Typically, subjective workload assessment techniques, such as the NASA-TLX (Hart and Staveland, 1988) are used in conjunction with primary and secondary task performance measures, and physiological measures to assess participant workload. The chosen technique should be administered immediately once the task under analysis is completed and participants should be instructed to rate the appropriate workload dimensions based upon the primary task that they have just completed.

Step 9: Download Collected Data
The heart rate monitor data collection tool (typically a watch) can now be connected to a laptop computer in order to download the data collected.

Step 10: Analyse Data
Once the data collection procedure is completed, the data should be analysed appropriately. It is typically assumed that an increase in workload causes an increase in operator heart rate. HRV has also been used as an indicator of operator MWL. According to Salvendy (1997), laboratory studies have reported a decrease in HRV (heart rhythm) under increased workload conditions.

Advantages

- Various physiological techniques have demonstrated a sensitivity to task demand variations.
- When using physiological techniques, data is recorded continuously throughout task performance.
- Physiological measurements can often be taken in a real-world setting, removing the need for a simulation of the task.
- Advances in technology have resulted in an increased level of accuracy and sensitivity in the various physiological measurement tools.
- Physiological measurement does not interfere with primary task performance.

Disadvantages

- The data is easily confounded by extraneous interference (Young and Stanton, 2004), such as temperature and movement (Baldauf, Burgard and Wittman, 2009).
- The equipment used to measure physiological responses is typically physically obtrusive.
- It is also typically expensive to acquire, temperamental and difficult to operate.
- Physiological data is very difficult to obtain and analyse.
- In order to use physiological techniques effectively, the analyst requires a thorough understanding of physiological responses to workload.
- It may be difficult to use certain equipment in the field, e.g. brain and eye measurement equipment.
- The measure is argued to be one conceptual step away from actual workload in that it is dependent variable of workload rather than a direct measure (Pretorious and Cilliers, 2007).
- Brookhuis and de Waard (2010) argue that taking physiological measures of workload requires good skills (both research and technical) and is time-consuming.

Example

Hilburn (1997) describes a study that was conducted in order to validate a battery of objective physiological measurement techniques when used to assess operator workload. The techniques were to be used to assess the demands imposed upon ATC controllers under free flight conditions. Participants completed an ATC task based upon the Maastricht-Brussels sector, during which heart rate variability, pupil diameter and eye scan patterns were measured. Participant HRV was measured using the Vitaport® system. Respiration was measured using inductive strain gauge transducers and an Observer® eye-tracking system was used to measure participant eye scan patterns. It was concluded that all three measures (pupil diameter in particular) were sensitive to varied levels of traffic load (Hilburn, 1997).

Related Methods

A number of different physiological measures have been used to assess operator workload, including heart rate, heart rate variability, and brain and eye activity. Physiological measures are typically used in conjunction with other MWL assessment techniques, such as primary and secondary task measures and SWATs. Primary task performance measures involve measuring certain aspects of participant performance on the task under analysis. Secondary task performance measures involve measuring participant performance on an additional task separate from the primary task under analysis. SWATs are completed post-trial by participants and involve participants rating specific dimensions of workload. There are a number of subjective workload assessment techniques, including the NASA-TLX (Hart and Staveland, 1988), SWAT (Reid and Nygren, 1988) and the workload profile technique (Tsang and Velazquez, 1996).

Approximate Training and Application Times

The training time associated with physiological measurement techniques is estimated to be high. The equipment is often difficult to operate and the data may also be difficult to analyse and interpret. The application time for physiological measurement techniques is dependent upon the duration of the task under analysis. For lengthy, complex tasks, the application time for a physiological assessment of workload may be high. However, it is estimated that the typical application time for a physiological measurement of workload would be low.

Reliability and Validity

According to Young and Stanton (2004), physiological measures of MWL are supported by a considerable amount of research, which suggests that HRV is probably the most promising approach. Whilst a number of studies have reported the sensitivity of a number of physiological techniques to variations in task demand, a number of studies have also demonstrated a lack of sensitivity to demand variations using the techniques.

Tools Needed

When using physiological measurements techniques, expensive equipment is often required. Monitoring equipment such as heart rate monitors, eye trackers, EEG measurement equipment and electro-oculographic measurement tools is needed, depending on the chosen measurement approach. A laptop computer is also typically used to transfer data from the measuring equipment.

Flowchart

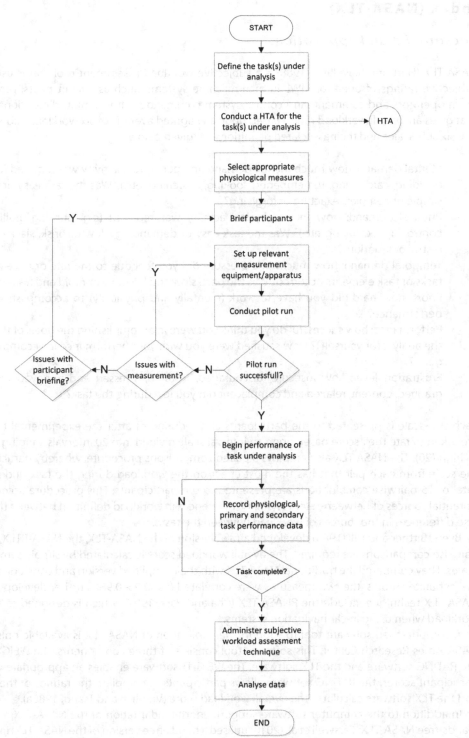

The National Aeronautics and Space Administration Task Load Index (NASA-TLX)

Background and Applications

NASA-TLX (Hart and Staveland, 1988) is a subjective workload assessment tool that is used to gather subjective ratings of operator MWL in man-machine systems, such as aircraft pilots, process control room operators and command and control system commanders. It is a multi-dimensional rating tool that gives an overall workload rating based upon a weighted average of six workload sub-scale ratings. The six sub-scales and their associated definitions are given below:

- Mental demand: how much mental demand and perceptual activity was required (e.g. thinking, deciding, calculating, remembering, looking, searching, etc.)? Was the task easy or demanding, simple or complex, exacting or forgiving?
- Physical demand: how much physical activity was required (e.g. pushing, pulling, turning, controlling, activating etc.)? Was the task easy or demanding, slow or brisk, slack or strenuous, restful or laborious?
- Temporal demand: how much time pressure did you feel due to the rate or pace at which the tasks or task elements occurred? Was the pace slow and leisurely or rapid and frantic?
- Effort: how hard did you have to work (mentally and physically) to accomplish your level of performance?
- Performance: how successful do you think you were in accomplishing the goals of the task set by the analyst (or yourself)? How satisfied were you with your performance in accomplishing these goals?
- Frustration level: how insecure, discouraged, irritated, stressed and annoyed versus secure, gratified, content, relaxed and complacent did you feel during the task?

Each sub-scale is presented to the participants either during or after the experimental trial and they are asked to rate their score based upon an interval scale divided into 20 intervals, ranging from low (1) to high (20). The NASA-TLX also employs a paired comparisons procedure, whereby participants select the scale from each pair that has the most effect on the workload during the task under analysis. A total of 15 pairwise combinations are presented to the participants. This procedure accounts for two potential sources of between-rater variability: differences in workload definition between the raters and also differences in the sources of workload between the tasks.

Byers, Bittner and Hill (1989) developed a 'raw' version of the NASA-TLX, the NASA-RTLX, in which no pairwise comparisons are required. The overall workload score is calculated by simply summing the six scales. They compared the traditional NASA-TLX with their simplified version and concluded that, across five separate studies, the two measures were correlated (Rs=0.96–0.98). Further developments of the NASA-TLX technique include the RNASA-TLX (Cha and Park, 1997), which is designed to assess driver workload when using in-car navigation systems.

A computerised software tool to support the application of NASA-TLX is available online from the NASA Ames Research Center. This software tool consists of three components: the WEIGHT software, the RATING software and the TLX software. The WEIGHT software enables an appropriate weighting of participant scores, the RATING software allows participants to complete the ratings of the dimensions and the TLX software calculates the overall workload score (Vidulich and Tsang, 1986a).

In addition to the computer software support, another adaptation of the NASA-TLX method is the 'hands-free' NASA-TLX. Carswell et al. (2010) utilised a hands-free version of the NASA-TLX, in which vocal responses to spoken probes were used instead of the paper or computer-based versions. Carswell et al. argue that the paper and computer-based versions may not always be appropriate, citing examples such as participants with limited motor, visual or literacy skills. The authors specifically focused on the development of a vocal measure for use in laparoscopic surgery, which involves complex procedures to be conducted using hand-based tools, which are not put down after each trial. The vocal NASA-TLX application was compared to the unweighted version NASA-RTLX and another vocal mental workload measure adaptation, the Multiple Resources Questionnaire. Carswell et al. concluded that acceptable levels of equivalence were found for the overall workload levels between the vocal and written versions of the standard NASA-TLX (r=0.85) and the NASA-RTLX (r=0.81). Variations were found in a number of the NASA-TLX sub-scales, particularly the 'effort' sub-scale.

The NASA-TLX is the most commonly used subjective MWL assessment technique and has been applied in numerous settings, including robotic assistance development (Stefanidis et al., 2010), civil and military aviation, driving (Horrey, Lesh and Garabet, 2008), nuclear power plant control room operation and ATC. The tool has its own website (http://human-factors.arc.nasa.gov/groups/TLX) where both versions (paper and computer) are available along with a list of publications. Hart (2006) presents a review of the 550 discussions and applications of NASA-TLX over the past 20 years. The reader is referred to this text for a comprehensive account of the contexts of use, adaptations and relationships with other variables that have been previously explored, ranging from space applications to mobile phone use.

Domain of Application

Generic.

Procedure and Advice (Computerised Version)

Step 1: Define the Task under Analysis
The first step in a NASA-TLX analysis (aside from the process of gaining access to the required systems and personnel) is to define the task (or tasks) that is to be subjected to analysis. The types of task analysed are dependent upon the focus of the analysis. For example, when assessing the effects on operator workload caused by a novel design or a new process, it is useful to analyse as representative a set of tasks as possible. To analyse a full set of tasks will often be too time-consuming and labour-intensive, and so it is pertinent to use a set of tasks that utilise all aspects of the system under analysis.

Step 2: Conduct an HTA for the Task under Analysis
Once the task (or tasks) under analysis is defined clearly, an HTA should be conducted for each task. This allows the analyst and participants to understand the task fully.

Step 3: Select Participants
Once the task (or tasks) under analysis is clearly defined and described, it may be useful to select the participants that are to be involved in the analysis. This may not always be necessary and it may suffice to simply select participants randomly on the day. However, if workload is being compared across rank or experience levels, then clearly effort is required to select the appropriate participants.

Step 4: Brief Participants
Before the task (or tasks) under analysis is performed, all of the participants involved should be briefed regarding the purpose of the study and the NASA-TLX technique. It is recommended that participants are given a workshop on workload and workload assessment. It may also be useful at this stage to take the participants through an example NASA-TLX application so that they understand how the technique works and what is required of them as participants. It may even be pertinent to get the participants to perform a small task and then get them to complete a workload profile questionnaire. This would act as a 'pilot run' of the procedure and would highlight any potential problems.

Step 5: Performance of Task under Analysis

Next, the participant should perform the task under analysis. The NASA-TLX can be administered during the trial or after the trial. It is recommended that the TLX is administered after the trial as online administration is intrusive to the primary task. If online administration is required, the TLX should be administered and completed verbally.

Step 6: Weighting Procedure

When the task under analysis is complete, the weighting procedure can begin. The WEIGHT software presents 15 pairwise comparisons of the six sub-scales (mental demand, physical demand, temporal demand, effort, performance and frustration level) to the participants. The participants should be instructed to select, from each of the 15 pairs, the sub-scale that contributed the most to the workload of the task. The WEIGHT software then calculates the total number of times each sub-scale was selected by the participant. Each scale is then rated by the software based upon the number of times it is selected by the participant. This is done using a scale of 0 (not relevant) to 5 (more important than any other factor).

Step7: NASA-TLX Rating Procedure

Participants should be presented with the interval scale for each of the TLX sub-scales (this is done via the RATING software). Participants are asked to give a rating for each sub-scale, between 1 (low) and 20 (high), in response to the associated sub-scale questions. The ratings provided are based entirely on the participant's subjective judgment.

Step 8: TLX Score Calculation

The TLX software is then used to compute an overall workload score. This is calculated by multiplying each rating by the weight given to that sub-scale by the participant. The sum of the weighted ratings for each task is then divided by 15 (the sum of weights). A workload score of between 0 and 100 is then provided for the task under analysis.

Advantages

- The NASA-TLX provides a quick and simple technique for estimating operator workload.
- Its sub-scales are generic, so the technique can be applied to any domain. In the past, it has been used in a number of different domains, such as aviation, ATC, command and control, nuclear reprocessing, petrochemical and automotive domains.
- It has been tested thoroughly in the past and has also been the subject of a number of validation studies, e.g. Hart and Staveland (1988).
- The provision of the TLX software package removes most of the work for the analyst, resulting in a very quick and simple procedure.
- For those without computers, it is also available in a pen-and-paper format (Vidulich and Tsang, 1986a).
- It is probably the most widely used technique for estimating operator workload.
- It is a multi-dimensional approach to workload assessment.
- A number of studies have shown its superiority over SWAT (Hart and Staveland, 1988; Nygren, 1991).
- When administered post-trial, the approach is non-intrusive to primary task performance.
- According to Wierwille and Eggemeier (1993), it has demonstrated sensitivity to demand manipulations in numerous flight experiments.

Disadvantages

- When administered online, the TLX can be intrusive to primary task performance.

- When administered after the fact, participants may have forgotten high workload aspects of the task.
- Workload ratings may be correlated with task performance, e.g. subjects who performed poorly on the primary task may rate their workload as very high and vice versa.
- The sub-scale weighting procedure is laborious and adds more time to the procedure.

Flowchart

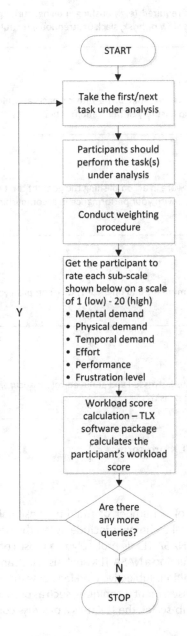

Example

An example of the NASA-TLX pro-forma is presented in Figure 8.3.

Mental Demand

How much mental and perceptual activity was required (e.g., thinking, deciding, calculating, remembering, looking, searching etc.)? Was the task easy or demanding, simple or complex, exacting or forgiving?

Low |————————————————————————————————————| High

Physical Demand

How much physical activity was required (e.g., pushing, pulling, turning, controlling, activating etc.)? Was the task easy or demanding, slow or brisk, slack or strenuous, restful or laborious?

Low |————————————————————————————————————| High

Temporal Demand

How much time pressure did you feel due to the rate or pace at which the tasks or task elements occurred? Was the pace slow and leisurely or rapid and frantic?

Low |————————————————————————————————————| High

Performance

How successful do you think you were in accomplishing the goals of the task set by the experimenter (or yourself)? How satisfied were you with your performance in accomplishing these goals?

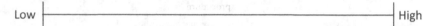

Low |————————————————————————————————————| High

Effort

How hard did you have to work (mentally and physically) to accomplish your level of performance?

Low |————————————————————————————————————| High

Frustration Level

How insecure, discouraged, irritated, stressed and annoyed versus secure, gratified, content, relaxed and complacent did you feel during the task?

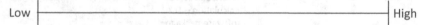

Low |————————————————————————————————————| High

Figure 8.3   Example NASA-TLX pro-forma

Related Methods

The NASA-TLX technique is one of a number of multi-dimensional subjective workload assessment techniques. Other multi-dimensional techniques include SWAT, the Bedford scale, DRAWS and the MACE technique. Along with SWAT, the NASA-TLX is probably the most commonly used subjective workload assessment technique. When conducting a NASA-TLX analysis, a task analysis (such as an HTA) of the task or scenario is often conducted. In addition, subjective workload assessment techniques are normally used in conjunction with other workload assessment techniques, such as primary and secondary task performance measures. In order to weight the sub-scales, the TLX uses a pairwise comparison weighting procedure.

Approximate Training Times and Application Times

The NASA-TLX technique is simple to use and quick to apply. The training times and application times are typically low. Rubio et al. (2004) report that in a study comparing the NASA-TLX, SWAT and workload profile techniques, the NASA-TLX took 60 minutes to apply.

Reliability and Validity

A number of validation studies concerning the NASA-TLX have been conducted (Hart and Staveland, 1988; Vidulich and Tsang, 1985, 1986b). Vidulich and Tsang (1985, 1986b) reported that the NASA-TLX produced more consistent workload estimates for participants performing the same task than the SWAT technique (Reid and Nygren, 1988) did. Hart and Staveland (1988) also reported that the NASA-TLX workload scores suffer from substantially less between-rater variability than one-dimensional workload ratings did. Luximon and Goonetilleke (2001) also reported that a number of studies have shown that the NASA-TLX is superior to SWAT in terms of sensitivity, particularly for low mental workloads (Hart and Staveland, 1988; Nygren, 1991). In a comparative study between the NASA-TLX, the RNASA-TLX, SWAT and the MCH scale, Cha (2001) reported that the RNASA-TLX is the most sensitive and acceptable when used to assess driver mental workload during in-car navigation based tasks.

Pretorious and Cilliers (2007) raise a number of objections to the reliability of the NASA-TLX, arguing that the method provides a measure of participants' subjective opinions of task difficulty and as such is impacted by factors such as experience, personality and even motivation, rather than being an objective measure of task demands.

Tools Needed

A NASA-TLX analysis can be conducted using either pen and paper or the software method. Both the pen-and-paper method and the software method can be purchased from NASA Ames Research Center.

The Modified Cooper Harper Scale (MCH)

Background and Applications

The MCH scale is a uni-dimensional measure that uses a decision tree flowchart to elicit subjective ratings of MWL. The Cooper Harper scales (Cooper and Harper, 1969) are a decision tree rating scale that were originally developed to measure aircraft handling capability. In their original form, the scales were used to elicit subjective pilot ratings of the controllability of aircrafts. The output of the scale was based upon the controllability of the aircraft and also the level of input required by the pilot to maintain suitable control. The MCH scale (Casali and Wierwille, 1983) works on the assumption that there is a direct relationship between the level of difficulty of aircraft controllability and pilot workload, and is presented in Figure 8.4 on the next page.

The MCH scale is administered post-trial and participants simply follow the decision tree, answering questions regarding the task and system under analysis in order to provide an appropriate MWL rating.

Domain of Application

Aviation.

Procedure and Advice

Step 1: Define the Task under Analysis
The first step in an MCH analysis (aside from the process of gaining access to the required systems and personnel) is to define the task (or tasks) that is to be subjected to investigation. The types of task

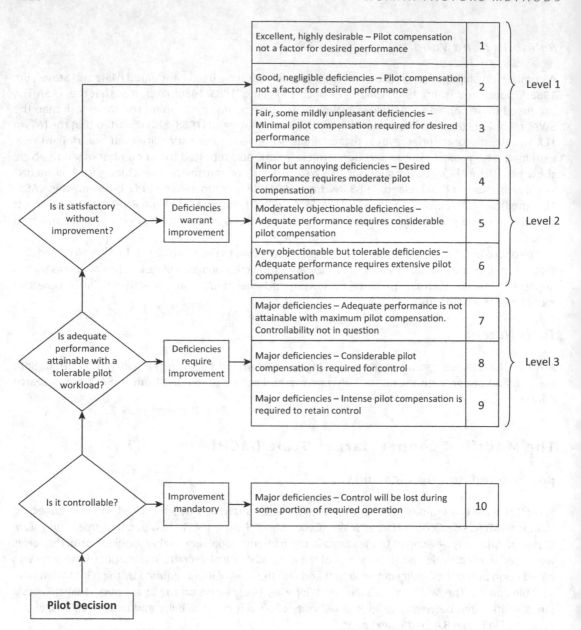

Figure 8.4   The MCH scale

analysed are dependent upon the focus of the analysis. For example, when assessing the effects on operator workload caused by a novel design or a new process, it is useful to analyse a set of tasks that are as representative of the full functionality of the interface, device or procedure as possible. To analyse a full set of tasks will often be too time-consuming and labour-intensive, and so it is pertinent to use a set of tasks that utilise all aspects of the system under analysis.

Step 2: Conduct an HTA for the Task under Analysis

Once the task (or tasks) under analysis is defined clearly, an HTA should be conducted for each task. This allows the analyst and participants to understand the task fully.

Step 3: Selection of Participants
Once the task (or tasks) under analysis is clearly defined and described, it may be useful to select the participants that are to be involved in the analysis. This may not always be necessary and it may suffice to simply select participants randomly on the day. However, if workload is being compared across rank or experience levels, then clearly effort is required to select the appropriate participants.

Step 4: Brief Participants
Before the task (or tasks) under analysis is performed, all of the participants involved should be briefed regarding the purpose of the study and the MCH technique. It is recommended that participants are also given a workshop on MWL and MWL assessment. It may also be useful at this stage to take the participants through an example MCH application. It may even be pertinent to get the participants to perform a small task and then to complete a workload profile questionnaire. This would act as a 'pilot run' of the procedure and would highlight any potential problems.

Step 5: Performance of the Task under Analysis
Next, the participant should perform the task under analysis. The MCH is normally administered post-trial.

Step 6: Completion of the MCH
Once the participants have completed the task in question, they should complete the MCH scale. To do this, they simply work through the decision tree to arrive at a MWL rating for the task under analysis. If there are further tasks, then the participants should repeat steps 5 and 6 until all tasks have been assigned a workload rating.

Advantages

- The MCH scale is very easy and quick to use, requiring only minimal training.
- It offers a non-intrusive measure of workload.
- A number of validation studies have been conducted using the MCH scales. Wierwinke (1974) reported a high coefficient between subjective difficulty rating and objective workload level.
- It has been widely used to measure workload in a variety of domains.
- According to Casali and Wierwille (1983), the MCH scales are inexpensive, unobtrusive, easily administered and easily transferable.
- It has a high face validity.
- According to Wierwille and Eggemeier (1993), the MCH technique has been successfully applied to workload assessment in numerous flight simulation experiments incorporating demand manipulations.
- The data obtained when using uni-dimensional tools is easier to analyse than when using multi-dimensional tools.

Disadvantages

- It is dated.
- It was originally developed to rate the controllability of aircrafts.
- It is limited to manual control tasks.
- Data is collected post-trial. This is subject to a number of problems, such as a correlation with performance. Participants are also poor at reporting past mental events.
- It is uni-dimensional.

Flowchart

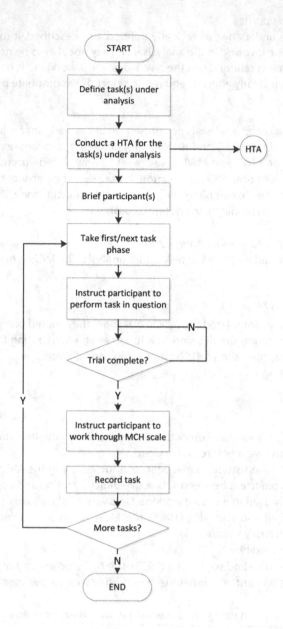

Related Methods

There are a number of other subjective MWL assessment techniques, including the NASA-TLX, SWAT, workload profile, DRAWS, MACE and Bedford scales. MCH is a uni-dimensional, decision tree-based workload assessment technique, which is similar to the Bedford scale workload assessment technique. It is also recommended that a task analysis (such as an HTA) of the task or scenario under analysis is conducted before the MCH data collection procedure begins.

Approximate Training and Application Times

The MCH scale is a very quick and easy procedure, so training and application times are both estimated to be very low. The application time is also dependent upon the length of the task (or tasks) under analysis.

Reliability and Validity

Wierwinke (1974) reported an extremely high coefficient between subjective task difficulty rating and objective workload level.

The Subjective Workload Assessment Technique (SWAT)

Background and Applications

SWAT (Reid and Nygren, 1988) is a workload assessment technique that was developed by the US Air Force's Armstrong Aerospace Medical Research Laboratory at the Wright Patterson Air Force Base. It was originally developed to assess pilot workload in cockpit environments, but more recently has been used predictively (Vidulich and Hughes, 1991). Along with the NASA-TLX technique of subjective workload, it is probably one the most commonly used of the subjective techniques to measure operator workload. Like the NASA-TLX, it is a multi-dimensional tool that uses three dimensions of operator workload: time load, mental effort load and stress load.

Time load is the extent to which a task is performed within a time limit and the extent to which a multiple tasks must be performed concurrently. Mental effort load is the associated attentional demands of a task, such as attending to multiple sources of information and performing calculations. Finally, stress load includes operator variables such as fatigue, level of training and emotional state. After an initial weighting procedure, participants are asked to rate each of the three dimensions on a scale of 1–3. A workload score is then calculated for each dimension and an overall workload score of between 1 and 100 is also calculated. SWAT uses a three-point rating scale for each dimension. This scale is shown in Table 8.2.

Table 8.2   SWAT three-point rating scale

Time load	Mental effort load	Stress load
1) Often have spare time: interruptions or overlap among other activities occur infrequently or not at all.	1) Very little conscious mental effort or concentration required: activity is almost automatic, requiring little or no attention.	1) Little confusion, risk, frustration or anxiety exists and can be easily accommodated.
2) Occasionally have spare time: interruptions or overlap among activities occur frequently.	2) Moderate conscious mental effort or concentration required: complexity of activity is moderately high due to uncertainty, unpredictability or unfamiliarity; considerable attention is required.	2) Moderate stress due to confusion, frustration or anxiety noticeably adds to workload: significant compensation is required to maintain adequate performance.
3) Almost never have spare time: interruptions or overlap among activities are very frequent or occur all of the time.	3) Extensive mental effort and concentration are necessary: very complex activity requiring total attention.	3) High to very intense levels of stress due to confusion, frustration or anxiety: high to extreme determination and self-control required.

As with the NASA-TLX, both verbal and simplified versions of SWAT have been developed. Morgan and Hancock (2011) used a simplified version of SWAT to explore the association between delayed adaptation in driver workload and transitions in task demands. The study analysed the performance of 32 drivers completing a set route in a driving simulator using a navigational aid which was programmed to fail at some point during the driving task. They used the Simplified Subjective Workload Assessment Technique (S-SWAT) (Luximon and Goonetilleke, 2001). Participants were verbally presented with the probes and verbally responded with their scores. Further variations of the SWAT technique have also been developed, including a predictive variation (Pro-SWAT) and a computerised version.

Domain of Application

Aviation.

Procedure and Advice

Step 1: Scale Development
First, participants are required to place in rank order all possible 27 combinations of the three workload dimensions (time load, mental effort load and stress load) according to their effect on workload. This 'conjoint' measurement is used to develop an interval scale of workload rating from 1 to 100.

Step 2: Performance of the Task under Analysis
Once the initial SWAT ranking has been completed, the participants should perform the task under analysis. SWAT can be administered during the trial or after the trial; however, it is recommended that it is administered after the trial, as online administration is intrusive to the primary task. If online administration is required, then the SWAT should be administered and completed verbally.

Step 3: SWAT Scoring
The participants are required to provide a subjective rating of workload by assigning a value of 1–3 to each of the three SWAT workload dimensions.

Step 4: SWAT Score Calculation
For the workload score, the analyst should take the scale value associated with the combination given by the participants. The scores are then translated into individual workload scores for each SWAT dimension. Finally, an overall workload score should be calculated.

Advantages

*   The SWAT technique provides a quick and simple technique for estimating operator workload.
*   Its workload dimensions are generic, so it can be applied to any domain. In the past, it has been used in a number of different domains, such as the aviation, ATC, command and control, nuclear reprocessing, petrochemical and automotive domains.
*   It is one of the most widely used and well-known subjective workload assessment techniques available and has been subjected to a number of validation studies (Hart and Staveland, 1988; Vidulich and Tsang, 1985, 1986b).
*   The Pro-SWAT variation allows the technique to be used predictively.
*   It is a multi-dimensional approach to workload assessment.
*   It is unobtrusive.

Disadvantages

*   SWAT can be intrusive if administered online.
*   In a number of validation studies, it has been reported that the NASA-TLX is superior to SWAT in terms of sensitivity, particularly for low mental workloads (Hart and Staveland, 1988; Nygren, 1991).
*   It has been constantly criticised for having a low sensitivity to mental workloads (Luximon and Goonetilleke, 2001).
*   The initial SWAT combination ranking procedure is time-consuming (Luximon and Goonetilleke, 2001).
*   Workload ratings may be correlated with task performance, e.g. subjects who performed poorly on the primary task may rate their workload as very high and vice versa. This is not always the case.
*   When administered after the fact, participants may have forgotten high or low workload aspects of the task.
*   The method appears to be an unsophisticated measure of workload. The NASA-TLX appears to be more sensitive.

Flowchart

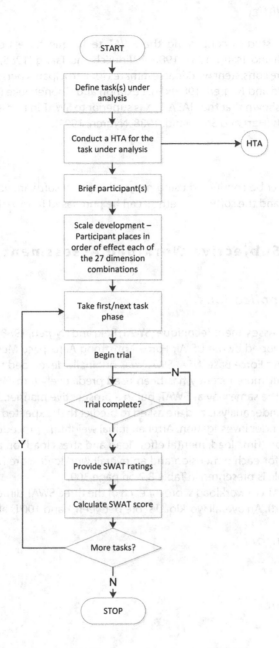

START

Define task(s) under analysis

Conduct a HTA for the task under analysis → HTA

Brief participant(s)

Scale development – Participant places in order of effect each of the 27 dimension combinations

Take first/next task phase

Begin trial

Trial complete? — N

Y

Provide SWAT ratings

Calculate SWAT score

More tasks?

Y

N

STOP

Related Methods

The SWAT technique is similar to a number of subjective workload assessment techniques, such as the NASA-TLX, the MCH scales and the Bedford scale. For predictive use, the Pro-SWORD technique is similar.

Approximate Training Times and Application Times

Whilst the scoring phase of the SWAT technique is very simple to use and quick to apply, the initial ranking phase is time-consuming and laborious. Thus, the training times and application times are estimated to be quite high.

Reliability and Validity

A number of validation studies concerning the SWAT technique have been conducted (Hart and Staveland, 1988; Vidulich and Tsang, 1985, 1986b). Vidulich and Tsang (1985, 1986b) reported that the NASA-TLX produced more consistent workload estimates for participants performing the same task than the SWAT technique (Reid and Nygren, 1988) did. Luximon and Goonetilleke (2001) also reported that a number of studies have shown that the NASA-TLX is superior to SWAT in terms of sensitivity, particularly for low mental workloads (Hart and Staveland, 1988; Nygren, 1991).

Tools Needed

A SWAT analysis can either be conducted using pen and paper. A software version also exists. Both the pen-and-paper method and the software method can be purchased from various sources.

The Projective Subjective Workload Assessment Technique (Pro-SWAT)

Background and Applications

The Subjective Workload Assessment Technique (SWAT; Reid and Nygren, 1988) is a workload assessment technique that was developed by the US Air Force Armstrong Aerospace Medical Research Laboratory at the Wright Patterson Air Force Base, USA. SWAT was originally developed to assess pilot workload in cockpit environments but more recently has been used predictively (Pro-SWAT; Salvendy, 1997). Pro-SWAT is administered in the same way as SWAT, but in a prospective manner. The participants are given a briefing of the system under analysis and are asked to predict their expected workload for undertaking a task using the system under investigation. After an initial weighting procedure, participants are asked to predict each dimension (time load, mental effort load and stress load) on a scale of 1–3. A workload score is then calculated for each dimension and an overall workload score between 1 and 100 is also calculated. The SWAT scale is presented in Table 8.2 on page 309.

The output of Pro-SWAT is a workload score for each of the three SWAT dimensions (time load, mental effort load and stress load). An overall workload score between 1 and 100 is also calculated.

Domain of Application

Aviation.

Procedure and Advice

Step 1: Scale Development
First, subject matter expert (SME) participants are required to place in rank order all possible 27 combinations of the three workload dimensions (time load, mental effort load and stress load) according to their effect on workload. This 'conjoint' measurement is used to develop an interval scale of workload rating from 1 to 100.

Step 2: Task Demo/Walkthrough
The SMEs should be given a walkthrough or demonstration of the task that they are to predict the workload for. Normally a verbal walkthrough will suffice.

Step 3: Workload Prediction
The SMEs should now be instructed to predict the workload imposed by the task under analysis. They should assign a value of 1 to 3 to each of the three SWAT workload dimensions.

Step 4: Performance of Task under Analysis
Once the initial SWAT ranking has been completed, the subject should perform the task under analysis. SWAT can be administered during the trial or after the trial. It is recommended that the SWAT is administered after the trial, as online administration is intrusive to the primary task. If online administration is required, then the SWAT should be administered and completed verbally.

Step 5: SWAT Scoring
The participants are required to provide a subjective rating of workload by assigning a value of 1 to 3 to each of the three SWAT workload dimensions.

Step 6: SWAT Score Calculation
Next, the analyst should calculate the workload scores from the SME predictions and also the participant workload ratings. For the workload scores, the analyst should take the scale value associated with the combination given by the participant. The scores are then translated into individual workload scores for each SWAT dimension. Finally, an overall workload score should be calculated.

Step 7: Compare Workload Scores
The final step is to compare the predicted workload scores to the workload scores provided by the participants who undertook the task under analysis.

Advantages

- The SWAT technique provides a quick and simple technique for estimating operator workload.
- The SWAT workload dimensions are generic, so the technique can be applied to any domain. In the past, the SWAT technique has been used in a number of different domains, such as the aviation, ATC, command and control, nuclear reprocessing, petrochemical and automotive domains.
- It is one of the most widely used and well-known subjective workload assessment techniques available, and has been subjected to a number of validation studies (Hart and Staveland, 1988; Vidulich and Tsang, 1985, 1986a).
- The Pro-SWAT variation allows the technique to be used predictively.
- SWAT is a multi-dimensional approach to workload assessment.
- It is unobtrusive.

Disadvantages

- SWAT can be intrusive if administered online.
- Pro-SWAT has yet to be validated thoroughly.
- In a number of validation studies, it has been reported that the NASA-TLX is superior to SWAT in terms of sensitivity, particularly for low mental workloads (Hart and Staveland, 1988; Nygren, 1991).
- It has been constantly criticised for having a low sensitivity to mental workloads (Luximon and Goonetilleke, 2001).
- The initial SWAT combination ranking procedure is very time-consuming (Luximon and Goonetilleke, 2001).
- Workload ratings may be correlated with task performance, e.g. subjects who performed poorly on the primary task may rate their workload as very high and vice versa. This is not always the case.

- When administered after the fact, participants may have forgotten high or low workload aspects of the task.
- It has been criticised as an unsophisticated measure of workload; the NASA-TLX appears to be more sensitive.
- The Pro-SWAT technique is still in its infancy.

Flowchart

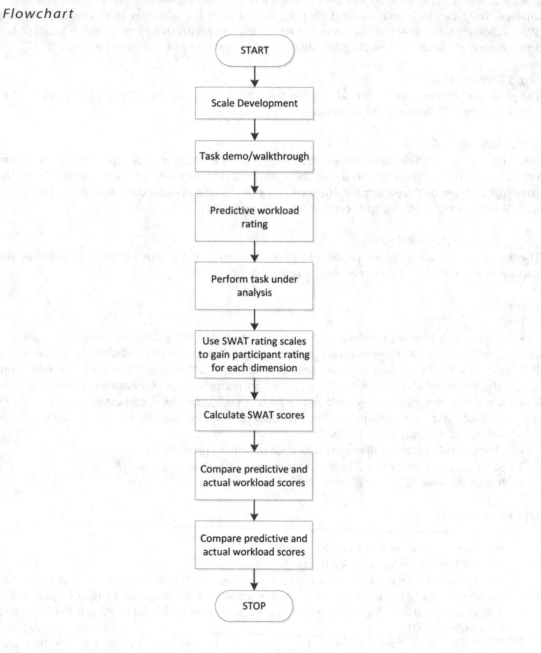

Related Methods

The Pro-SWAT technique is similar to a number of predictive subjective workload assessment techniques, such as the Pro-SWORD technique.

Approximate Training Times and Application Times

Whilst the scoring phase of the SWAT technique is very simple to use and quick to apply, the initial ranking phase is time-consuming and laborious. Thus, the training times and application times are estimated to be quite high.

Reliability and Validity

A number of validation studies concerning the SWAT technique have been conducted (Hart and Staveland, 1988; Vidulich and Tsang, 1985, 1986b). Vidulich and Tsang (1985, 1986b) reported that the NASA-TLX produced more consistent workload estimates for participants performing the same task than the SWAT technique (Reid and Nygren, 1988) did. Luximon and Goonetilleke (2001) also reported that a number of studies have shown that the NASA-TLX is superior to SWAT in terms of sensitivity, particularly for low mental workloads (Hart and Staveland, 1988; Nygren, 1991).

Tools Needed

A Pro- SWAT analysis can either be conducted using pen and paper. A software version also exists. Both the pen-and-paper method and the software method can be purchased from various sources.

The Subjective Workload Dominance Technique (SWORD)

Background and Applications

SWORD (Vidulich, 1989) is a subjective workload assessment technique that has been used both retrospectively and predictively (Pro-SWORD: Vidulich, Ward and Schueren, 1991). Originally designed as a retrospective workload assessment technique, it uses paired comparison of tasks in order to provide a rating of workload for each individual task. Administered post-trial, participants are required to rate one task's dominance over another in terms of workload imposed. When used predictively, tasks are rated for their dominance before the trial begins and are then rated post-test to check for the sensitivity of the predictions.

Domain of Application

Generic.

Procedure and Advice

Step 1: Define the Task under Analysis
The first step in a SWORD analysis (aside from the process of gaining access to the required systems and personnel) is to define the task (or tasks) that is to be subjected to analysis. The types of task analysed are dependent upon the focus of the analysis. For example, when assessing the effects on

operator workload caused by a novel design or a new process, it is useful to analyse as representative a set of tasks as possible. To analyse a full set of tasks will often be too time-consuming and labour-intensive, and so it is pertinent to use a set of tasks that utilise all aspects of the system under analysis.

Step 2: Conduct an HTA for the Task under Analysis
Once the task (or tasks) under analysis is defined clearly, an HTA should be conducted for each task. This allows the analyst and participants to understand the task fully.

Step 3: Create SWORD Rating Sheet
Once a task description (e.g. HTA) is developed, the SWORD rating sheet can be created. The analyst should list all of the possible combinations of tasks (e.g. A v B, A v C, B v C) and the dominance rating scale. An example of a SWORD rating sheet is presented in Figure 8.5.

Task	Absolute	Very Strong	Strong	Weak	EQUAL	Weak	Strong	Very Strong	Absolute	Task
A										B
A										C
A										D
A										E
B										C
B										D
B										E
C										D
C										E
D										E

Figure 8.5   Example SWORD rating sheet

Step 4: Selection of Participants
Once the task (or tasks) under analysis is defined, it may be useful to select the participants that are to be involved in the analysis. This may not always be necessary and it may suffice to simply select participants randomly on the day. However, if workload is being compared across rank or experience levels, then clearly effort is required to select the appropriate participants.

Step 5: Brief Participants
Before the task (or tasks) under analysis is performed, all of the participants involved should be briefed regarding the purpose of the study and the SWORD technique. It is recommended that participants are also given a workshop on workload and workload assessment. It may also be useful at this stage to take the participants through an example SWORD application, so that they understand how the technique works and what is required of them as participants. It may even be pertinent to get them to perform a small task and then to complete a workload profile questionnaire. This would act as a 'pilot run' of the procedure and would highlight any potential problems.

Step 6: Performance of the Task under Analysis
SWORD is normally applied post-trial. Therefore, the task under analysis should be performed first. As SWORD is applied after the task performance, intrusiveness is reduced and the task under analysis can be performed in its real-world setting.

Step 7: Administration of SWORD Questionnaire
Once the task under analysis is complete, the SWORD data collection process begins. This involves the administration of the SWORD rating sheet. The participant should be presented with the SWORD rating sheet immediately after task performance has ended. The SWORD rating sheet lists all possible paired comparisons of the tasks conducted in the scenario under analysis. A 17-point rating scale is used.

The 17 slots represent the possible ratings. These slots are made up of two nine-point scales which can represent preference for either task in addition to the equal point which is the same for both tasks (Vidulich, Ward and Schueren, 1991).The analyst has to rate the two tasks against each other (e.g. task A v B) in terms of their level of workload imposed. For example, if a participant feels that the two tasks imposed a similar level of workload, then he or she should mark the 'Equal' point on the rating sheet. However, if a participant feels that task A imposed a slightly higher level of workload than task B did, he or she would move towards task A on the sheet and would mark the 'Weak' point on the rating sheet. If a participant felt that task A imposed a much greater level of workload than task B, then he or she would move towards task A on the sheet and mark the 'Absolute' point on the rating sheet. This allows the participant to provide a subjective rating of one task's workload dominance over the over. This procedure should continue until all of the possible combinations of tasks in the scenario under analysis are exhausted and given a rating.

Step 8: Constructing the Judgment Matrix
Once all ratings have been elicited, the SWORD judgment matrix should be conducted. Each cell in the matrix should represent the comparison of the task in the row with the task in the associated column. The analyst should fill each cell with the participant's dominance rating. For example, if a participant rated tasks A and B as equal, a '1' is entered into the

	A	B	C	D	E
A	1	2	6	1	1
B	-	1	3	2	2
C	-	-	1	6	6
D	-	-	-	1	1
E	-	-	-	-	1

Figure 8.6   SWORD judgment matrix

appropriate cell. If task A is rated as dominant, then the analyst simply counts from the 'Equal' point to the marked point on the sheet and enters the number in the appropriate cell. An example SWORD judgment matrix is shown on the right.

The rating for each task is calculated by determining the mean for each row of the matrix and then normalising the means (Vidulich, Ward and Schueren, 1991).

Step 9: Matrix Consistency Evaluation
Once the SWORD matrix is complete, the consistency of the matrix can be evaluated by ensuring that there are transitive trends amongst the related judgments in the matrix. For example, if task A is rated twice as hard as task B, and task B is rated three times as hard as task C, then task A should be rated as six times as hard as task C (Vidulich, Ward and Schueren, 1991). Therefore, the analyst should use the completed SWORD matrix to check the consistency of the participant's ratings.

Advantages

- SWORD is easy to learn and use.
- It is non-intrusive.
- It appears to have a high face validity.
- It has been demonstrated to have a sensitivity to workload variations (Reid and Nygren, 1988).
- It is very quick in its application.

Disadvantages

- Data is collected post-trial.
- SWORD requires further validation.
- It has not been as widely used as other workload assessment techniques, such as SWAT and the NASA-TLX.

Related Methods

SWORD is one of a number of mental workload assessment techniques, including the NASA-TLX, SWAT, the MCH scale and DRAWS. A number of the techniques have also been used predictively, such as Pro-SWAT and MCH. Any SWORD analysis requires a task description of some sort, such as an HTA or a tabular task analysis.

Approximate Training and Application Times

Although no data is offered regarding the training and application times for the SWORD technique, it is apparent that the training time for such a simple technique would minimal. The application time associated with the technique would be based on the scenario under analysis. For large, complex scenarios involving a large number of tasks, the application time would be high as an initial HTA would have to be performed, then the scenario would have to performed and finally the SWORD technique. The actual application time associated purely with the administration of the SWORD technique is very low.

Reliability and Validity

Vidulich and Tsang (1986) reported that the SWORD technique was more reliable and sensitive than the NASA-TLX technique.

Tools Needed

The SWORD technique can be applied using pen and paper. The system or device under analysis is also required.

Flowchart

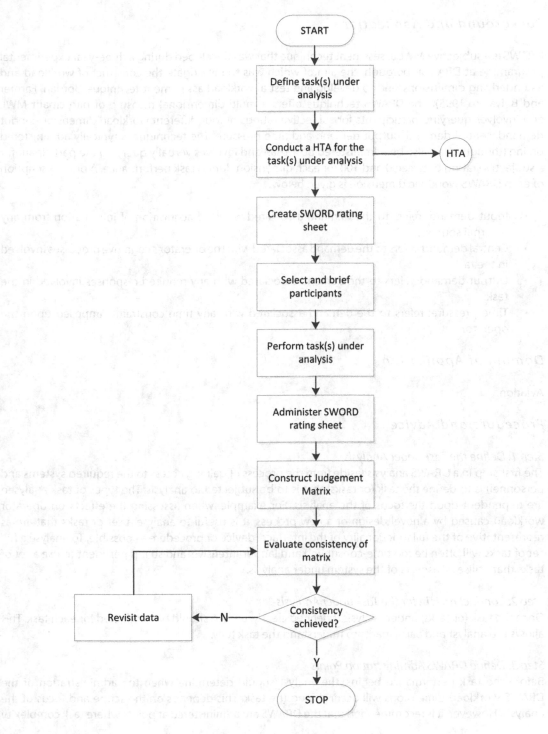

The DRA Workload Scales (DRAWS)

Background and Applications

DRAWS is a subjective MWL assessment technique that was developed during a three-year experimental programme at DRA Farnborough, the aim of which was to investigate the construct of workload and its underlying dimensions, and to develop and test a workload assessment technique (Jordan, Farmer and Belyavin, 1995). The DRAWS technique offers a multi-dimensional measure of participant MWL and involves querying participants for subjective ratings of four different workload dimensions: input demand, central demand, output demand and time pressure. The technique is typically administered online (though it can also be administered post-trial) and involves verbally querying the participant for a subjective rating between 0 and 100 for each dimension during task performance. A brief description of each DRAWS workload dimension is given below:

- Input demand: refers to the demand associated with the acquisition of information from any external sources.
- Central demand: refers to the demand associated with the operator's cognitive processes involved in the task.
- Output demand: refers to the demand associated with any required responses involved in the task.
- Time pressure: refers to the demand associated with any time constraints imposed upon the operator.

Domain of Application

Aviation.

Procedure and Advice

Step 1: Define the Task under Analysis
The first step in a DRAWS analysis (aside from the process of gaining access to the required systems and personnel) is to define the task (or tasks) that is to be subjected to analysis. The types of task analysed are dependent upon the focus of the analysis. For example, when assessing the effects on operator workload caused by a novel design or a new process, it is useful to analyse a set of tasks that are as representative of the full functionality of the interface, device or procedure as possible. To analyse a full set of tasks will often be too time-consuming and labour-intensive, and so it is pertinent to use a set of tasks that utilise all aspects of the system under analysis.

Step 2: Conduct an HTA for the Task under Analysis
Once the task (or tasks) under analysis is defined clearly, an HTA should be conducted for each task. This allows the analyst and participants to understand the task fully.

Step 3: Define DRAWS Administration Points
Before the task performance begins, the analyst should determine when the administration of the DRAWS workload dimensions will occur during the task. This depends on the scope and focus of the analysis. However, it is recommended that the DRAWS are administered at points where task complexity

is low, medium and high, allowing the sensitivity of the technique to be tested. Alternatively, it may be useful to gather the ratings at regular intervals, e.g. 10-minute intervals.

Step 4: Selection of Participants

Once the task (or tasks) under analysis is defined, it may be useful to select the participants that are to be involved in the analysis. This may not always be necessary and it may suffice to simply select participants randomly on the day. However, if workload is being compared across rank or experience levels, then clearly effort is required to select the appropriate participants.

Step 5: Brief Participants

Next, the participants should be briefed regarding the purpose of the analysis and the functionality of the DRAWS technique. In a workload assessment study (Jordan, Farmer and Belyavin, 1995), participants were given a half-hour introductory session. It is recommended that they be briefed regarding the DRAWS technique, including what it measures and how it works. It may be useful to demonstrate a DRAWS data collection exercise for a task similar to the one under analysis. This allows the participants to understand how the technique works and also what is required of them. It is also crucial at this stage that the participants have a clear understanding of the DRAWS workload scale being used. In order for the results to be valid, the participants should have the same understanding of each component of the DRAWS workload scale. It is recommended that they are taken through the scale and that examples of workload scenarios are provided for each level on the scale. Once the participants fully understand the DRAWS workload scale being used, the analysis can proceed to the next step.

Step 6: Pilot Run

Once the participants have a clear understanding of how the DRAWS technique works and what is being measured, it is useful to perform a pilot run of the experimental procedure. Whilst performing a small task, participants should be subjected to a DRAWS MWL data collection exercise. This allows them to experience the technique in a task performance setting. Participants should be encouraged to ask questions during the pilot run in order to fully understand the technique and the experimental procedure adopted.

Step 7: Performance of Task under Analysis

Once the participants clearly understand how the DRAWS technique works and what is required of them as participants, performance of the task under analysis should begin. The DRAWS are typically administered during task performance, but can also be administered after the post-trial upon completion of the task.

Step 8: Administer Workload Dimensions

Once the task performance has begun, the analyst should ask the participants to subjectively rate each workload dimension on a scale of 1–100 (1 = low, 100 = high). The point at which the participants are required to rate their workload is normally defined before the trial. The analyst should verbally ask the participants to subjectively rate each dimension at that point in the task. Participants should then call out a subjective rating for each DRAWS dimension for that point of the task under analysis. The frequency with which participants are asked to rate the four DRAWS dimensions is determined by the analyst. Step 8 should continue until sufficient data regarding the participant MWL is collected.

Step 9: Calculate Participant Workload Score
Once the task performance is completed and sufficient data is collected, the participants' MWL scores should be calculated. Typically, a mean value for each of the DRAWS workload dimensions is calculated for the task under analysis. Since the four dimensions are separate facets of workload, a total workload score is not normally calculated.

Advantages

- It offers a simple, quick and low-cost approach for assessing participant MWL.
- Data is obtained online during task performance and so the problems of collecting post-trial MWL data are removed.
- It has a high face validity.
- It has demonstrated a sensitivity to workload variation (Jordan, Farmer and Belyavin, 1995).
- The workload dimensions used in the technique were validated in a number of studies during its development.
- Although developed for application in the aviation domain, the workload dimensions are generic, allowing it to be applied in any domain.

Disadvantages

- It is intrusive to primary task performance.
- There are limited applications reported in the literature.
- The workload ratings may correlate highly with task performance at the point of administration.
- Limited validation evidence is available in the literature. The technique requires further validation.

Example

There is no evidence relating to the use of the DRAWS MWL assessment technique available in the literature.

Related Methods

The DRAWS technique is one of a number of subjective workload assessment techniques, such as NASA-TLX, SWAT and the MCH technique. Such techniques are normally used in conjunction with primary task measures, secondary task measures and physiological measures in order to assess operator workload. The DRAWS technique was developed through an analysis of the validity of existing workload dimensions employed by other workload assessment techniques, such as the NASA-TLX and the prediction of operator performance technique (POP: Farmer et al., 1995).

Approximate Training and Application Times

The DRAWS technique requires very little training (approximately half an hour) and is quick in its application, using only four workload dimensions. The total application time is ultimately dependent upon the amount of workload ratings that are required by the analysis and the length of time associated with performing the task under analysis.

Flowchart

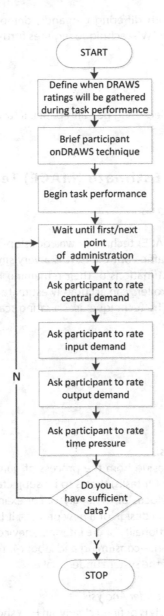

```
                    ( START )
                        │
                        ▼
            ┌───────────────────────┐
            │  Define when DRAWS    │
            │ ratings will be gathered│
            │ during task performance│
            └───────────────────────┘
                        │
                        ▼
            ┌───────────────────────┐
            │    Brief participant   │
            │   onDRAWS technique    │
            └───────────────────────┘
                        │
                        ▼
            ┌───────────────────────┐
            │  Begin task performance │
            └───────────────────────┘
                        │
                        ▼
            ┌───────────────────────┐
            │  Wait until first/next │
            │         point          │
            │   of  administration   │
            └───────────────────────┘
                        │
                        ▼
            ┌───────────────────────┐
            │ Ask participant to rate │
            │     central demand     │
            └───────────────────────┘
                        │
                        ▼
            ┌───────────────────────┐
            │ Ask participant to rate │
            │      input demand      │
            └───────────────────────┘
         N                │
                        ▼
            ┌───────────────────────┐
            │ Ask participant to rate │
            │     output demand      │
            └───────────────────────┘
                        │
                        ▼
            ┌───────────────────────┐
            │ Ask participant to rate │
            │     time pressure      │
            └───────────────────────┘
                        │
                        ▼
                  ◇ Do you      ◇
                 ◇ have sufficient ◇
                  ◇   data?     ◇
                        │ Y
                        ▼
                    ( STOP )
```

Reliability and Validity

During the development of the technique, nine workload dimensions were evaluated for their suitability for use in assessing operator workload. It was found that the four dimensions (input demand, central demand, output demand and time pressure) were capable of discriminating between the demands imposed by different tasks (Jordan, Farmer and Belyavin, 1995). Furthermore, Jordan, Farmer and Belyavin reported that scores for the DRAWS dimensions were found to be consistent with performance across tasks with differing demands, demonstrating a sensitivity to workload variation. It is apparent that the DRAWS technique requires further testing in relation to its reliability and validity.

Tools Needed

The DRAWS technique can be applied using pen and paper. If task performance is simulated, then the appropriate simulator is also required.

The Malvern Capacity Estimate (MACE) Technique

Background and Applications

The Malvern capacity estimate (MACE) technique was developed by DERA in order to measure the mental workload capacity of air traffic controllers. It is a very simple technique, involving asking air traffic controllers for subjective estimations of their remaining mental capacity during a simulated task. As such, it assumes that controllers can accurately estimate how much remaining capacity they possess during a task or scenario. The technique uses a rating scale designed to elicit ratings of spare capacity.

Domain of Application

ATC.

Procedure and Advice

Step 1: Define the Task under Analysis
The first step in a MACE analysis (aside from the process of gaining access to the required systems and personnel) is to define the task (or tasks) that is to be subjected to investigation. The types of task analysed are dependent upon the focus of the analysis. For example, when assessing the effects on operator workload caused by a novel design or a new process, it is useful to analyse a set of tasks that are as representative of the full functionality of the interface, device or procedure as possible. To analyse a full set of tasks will often be too time-consuming and labour-intensive, and so it is pertinent to use a set of tasks that utilise all aspects of the system under analysis.

Step 2: Conduct an HTA for the Task under Analysis
Once the task (or tasks) under analysis is defined clearly, an HTA should be conducted for each task. This allows the analyst and participants to understand the task fully.

Step 3: Selection of Participants
Once the task (or tasks) under analysis is defined, it may be useful to select the participants that are to be involved in the analysis. This may not always be necessary and it may suffice to simply select participants randomly on the day. However, if workload is being compared across rank or experience levels, then clearly effort is required to select the appropriate participants.

Step 4: Brief Participants
The participants should be briefed regarding the MACE technique, including what it measures and how it works. It may be useful to demonstrate a MACE data collection exercise for a task similar to the one under analysis. This allows the participants to understand how the technique works and also what is required of them. It is also crucial at this stage that the participants have a clear understanding of the MACE rating scale. In order for the results to be valid, the participants should have the same understanding of each level of the workload scale, i.e. what level of perceived workload constitutes a rating of 50 per cent on the MACE workload scale and what level constitutes a rating of 100 per cent. It is recommended that the participants are taken through the scale and that examples of workload scenarios are provided for each level on the scale. Once the participants fully understand the MACE rating scale, the analysis can proceed to the next step.

Step 5: Conduct Pilot Run
Once the participants have a clear understanding of how the MACE technique works and what is being measured, it is useful to perform a pilot run. Whilst performing a small task, participants should be subjected to the MACE data collection procedure, which allows them to experience the technique in a task performance setting. Participants should be encouraged to ask questions during the pilot run in order to understand the technique and the experimental procedure fully.

Step 6: Begin Task Performance
The participants can now begin performance of the task or scenario under analysis. The MACE technique is typically applied online during task performance in a simulated system.

Step 7: Administer MACE Rating Scale
The analyst should administer the MACE rating scale and ask the participants for an estimation of their remaining capacity. The timing of the administration of the MACE rating scale is dependent upon the analysis requirements. It is recommended that this is defined prior to the onset of the trial. Participants can be queried for their spare capacity any number of times during task performance. It is recommended that capacity ratings are elicited during low and high complexity portions of the task, and also during routine portions of the task.

Step 8: Calculate Capacity
Once the trial is complete and sufficient data is collected, participant spare capacity should be calculated for each MACE administration.

Example

According to Goillau and Kelly (1996), the MACE technique has been used to assess air traffic controller workload and the workload estimates provided showed a high degree of consistency. Goillau and Kelly also stated that the MACE approach has been tested and validated in a number of unpublished ATC studies. However, there are no outputs of the MACE analyses available in the literature.

Flowchart

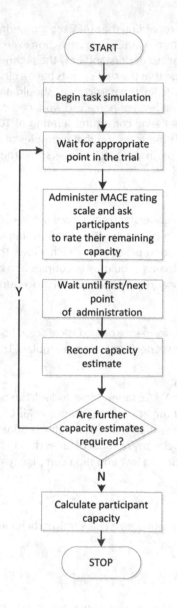

START

↓

Begin task simulation

↓

Wait for appropriate
point in the trial

↓

Administer MACE rating
scale and ask
participants
to rate their remaining
capacity

↓

Wait until first/next
point
of administration

↓

Record capacity
estimate

↓

Are further
capacity estimates
required?

Y ↑ (loop back to "Wait for appropriate point in the trial")

N ↓

Calculate participant
capacity

↓

STOP

Advantages

- The MACE technique offers a quick, simple and low-cost approach for assessing participant spare capacity.
- The output is potentially very useful, indicating when operators are experiencing mental overload and mental underload.
- It provides a direct measure of operator capacity.
- Online administration removes the problems associated with the post-trial collection of MWL (correlation with performance, forgetting certain portions of the task, etc.).

Disadvantages

- The technique is totally dependent upon the ability of participants to estimate their remaining capacity.
- It remains largely unvalidated.
- Its reliability and accuracy are questionable.
- It has only been used in simulators. It would be a very intrusive technique if applied online during task performance in the 'real world'.

Related Methods

The MACE technique is one of a number of subjective workload assessment techniques, including the NASA-TLX, SWAT and Bedford scales. However, it is unique in that it is used to elicit ratings of remaining operator capacity rather than a direct measure of perceived workload.

Approximate Training and Application Times

The MACE technique is a very simple and quick technique to apply. As a result, it is estimated that the training and application times associated with it would be very low. Application time is dependent upon the duration of the task under analysis.

Reliability and Validity

There is limited reliability and validity data associated with the MACE technique, Goillau and Kelly (1996) stressed that the technique requires further validation and testing. During initial testing of the technique, they reported that estimates of controllers' absolute capacity appeared to show a high degree of consistency and that peak MACE estimates were consistently higher than sustained MACE capacity estimates. However, they also reported that individual differences in MACE scores were found between controllers for the same task, indicating a potential problem with the reliability of the technique. The technique's reliance upon operators to subjectively rate their own spare capacity is certainly questionable.

The Workload Profile Technique

Background and Applications

The workload profile technique (Tsang and Velazquez, 1996) is a recently developed multi-dimensional subjective mental workload assessment technique that is based upon the multiple resources model of attentional resources proposed by Wickens (1987a). The workload profile technique is used to elicit ratings of demand imposed by the task under analysis for the following eight MWL dimensions:

- perceptual/central processing;
- response selection and execution;
- spatial processing;
- verbal processing;

- visual processing;
- auditory processing;
- manual output; and
- speech output.

Once the task (or tasks) under analysis is completed, participants provide a rating between 0 (no demand) and 1 (maximum demand) for each of the MWL dimensions. The ratings for each task are then summed in order to determine an overall MWL rating for the task under analysis. An example of the workload profile pro-forma is shown in Table 8.3.

Table 8. 3   Workload profile pro-forma

Workload dimensions								
	Stage of processing		Code of processing		Input		Output	
Task	Perceptual/central	Response	Spatial	Verbal	Visual	Auditory	Manual	Speech
1.1								
1.2								
1.3								
1.4								
1.5								
1.6								
1.7								

Domain of Application

Generic.

Procedure and Advice

Step 1: Define the Task under Analysis
The first step in a workload profile analysis (aside from the process of gaining access to the required systems and personnel) is to define the task (or tasks) that is to be subjected to investigation. The types of task analysed are dependent upon the focus of the analysis. For example, when assessing the effects on operator workload caused by a novel design or a new process, it is useful to analyse a set of tasks that are as representative of the full functionality of the interface, device or procedure as possible. To analyse a full set of tasks will often be too time-consuming and labour-intensive, and so it is pertinent to use a set of tasks that utilise all aspects of the system under analysis.

Step 2: Conduct an HTA for the Task under Analysis
Once the task (or tasks) under analysis is defined clearly, a HTA should be conducted for each task. This allows the analyst and participants to understand the task fully.

Step 3: Create Workload Profile Pro-forma
Once it is clear which tasks are to be analysed and which of those tasks are separate from one another, the workload profile pro-forma should be created. An example of a workload profile pro-forma is shown in Table 8.3. The left-hand column contains those tasks that are to be assessed. The workload dimensions, as defined by Wickens' multiple resource theory, are listed across the page.

Step 4: Selection of Participants
Once the task (or tasks) under analysis is defined, it may be useful to select the participants that are to be involved in the analysis. This may not always be necessary and it may suffice to simply select participants randomly on the day. However, if workload is being compared across rank or experience levels, then clearly effort is required to select the appropriate participants.

Step 5: Brief Participants
Before the task (or tasks) under analysis are performed, all of the participants involved should be briefed regarding the purpose of the study, MWL, multiple resource theory and the workload profile technique. It is recommended that participants are given a workshop on MWL, MWL assessment and also multiple resource theory. The participants used should have a clear understanding of multiple resource theory and of each dimension used in the workload profile technique. It may also be useful at this stage to take the participants through an example workload profile analysis so that they understand how the technique works and what is required of them as participants.

Step 6: Conduct Pilot Run
Once the participants have a clear understanding of how the workload profile technique works and what is being measured, it is useful to perform a pilot run. The participants should perform a small task and should then be instructed to complete a workload profile pro-forma. This allows them to experience the technique in a task performance setting. Participants should be encouraged to ask questions during the pilot run in order to understand the technique and the experimental procedure fully.

Step 7: Task Performance
Once the participants fully understand the workload profile techniques and the data collection procedure, they are free to undertake the task under analysis as normal.

Step 8: Completion of Workload Profile Pro-forma
Once the participants have completed the relevant task, they should provide ratings for the level of demand imposed by the task for each dimension. They should assign a rating between 0 (no demand) and 1 (maximum demand) for each MWL dimension. If there are any tasks requiring analysis left, the participants should then move on to the next task.

Step 9: Calculate Workload Ratings for Each Task
Once the participants have completed and rated all of the relevant tasks, the analyst should calculate MWL ratings for each of the tasks under investigation. In order to do this, the individual workload dimension ratings for each task are summed in order to gain an overall workload rating for each task (Rubio et al., 2004).

Advantages

- The technique is based upon sound underpinning theory (multiple resource theory: Wickens, 1987a).
- It is quick and easy to use, requiring minimal analyst training.
- As well as offering an overall task workload rating, the output also provides a workload rating for each of the eight workload dimensions.
- It is a multi-dimensional MWL assessment technique.
- As it is applied post-trial, it can be applied in real-world settings.

Disadvantages

- It may be difficult for participants to rate workload on a scale of 0 to 1. A more sophisticated scale may be required in order to gain a more appropriate measure of workload.
- The post-trial collection of MWL data has a number of associated disadvantages, including a potential correlation between MWL ratings and task performance, and participants 'forgetting' different portions of the task when workload was especially low.
- There is little evidence of the actual usage of the technique.
- There is limited validation evidence associated with the technique.
- Participants require an understanding of MWL and multiple resource theory.
- The dimensions used by the technique may not be fully understood by participants with limited experience of psychology and human factors. In a study comparing the NASA-TLX, SWAT and workload profile techniques, Rubio et al. (2004) reported that there were problems with some of the participants understanding the different dimensions used in the workload profile technique.

Example

A comparative study was conducted in order to test the workload profile, the Bedford scale (Roscoe and Ellis, 1990) and psychophysical techniques for the following criteria (Tsang and Velazquez, 1996):

- sensitivity to manipulation in task demand;
- concurrent validity with task performance; and
- test-retest reliability.

Sixteen subjects completed a continuous tracking task and a Sternberg memory task. The tasks were performed either independently of one another or concurrently. Subjective workload ratings were collected from participants post-trial. Tsang and Velazquez (1996) reported that the workload profile technique achieved a similar level of concurrent validity and test-retest reliability to the other workload assessment techniques tested. Furthermore, the workload profile technique also demonstrated a level of sensitivity to different task demands.

Related Methods

The workload profile techniques is one of a number of multi-dimensional subjective MWL assessment techniques. Other multi-dimensional MWL assessment techniques include the NASA-TLX (Hart and Staveland, 1988), SWAT (Reid and Nygren, 1988), and DRAWS. When conducting a workload profile analysis, a task analysis (such as HTA) of the task or scenario is normally required. In addition, subjective MWL assessment techniques are normally used in conjunction with other MWL measures, such as primary and secondary task measures.

Approximate Training and Application Times

The training time for the workload profile technique is estimated to be low, as it is a very simple technique to understand and apply. The application time associated with the technique is based upon the number and duration of the task (or tasks) under analysis. The application time is also lengthened somewhat by the requirement of a multiple resource theory workshop to be provided for the participants. In a study using the workload profile technique (Rubio et al., 2004), it was reported that the administration time was 60 minutes.

Flowchart

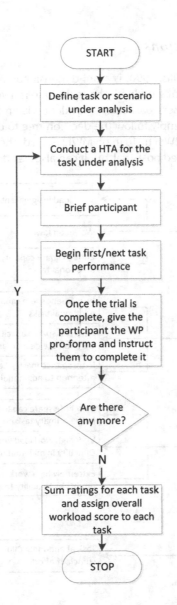

Reliability and Validity

Rubio et al. (2004) conducted a study in order to compare the NASA-TLX, SWAT and workload profile techniques in terms of intrusiveness, diagnosticity, sensitivity, validity (convergent and concurrent) and acceptability. It was found that the workload profile technique possessed a higher sensitivity than the NASA-TLX and SWAT techniques. In addition, it possessed a high level of convergent validity and diagnosticity. In terms of concurrent validity, it was found to have a lower correlation with performance than the NASA-TLX technique.

Tools Needed

The workload profile technique is applied using pen and paper.

The Bedford Scale

Background and Applications

The Bedford scale (Roscoe and Ellis, 1990) is a uni-dimensional MWL assessment technique that was developed by DERA to assess pilot workload. The technique is a very simple one, involving the use of a hierarchical decision tree to assess participant workload via an assessment of spare capacity whilst performing a task. Participants simply follow the decision tree to derive a workload rating for the task under analysis. A scale of 1 (low MWL) to 10 (high MWL) is used. The Bedford scale is presented in Figure 8.7. The scale is normally completed post-trial, but it can also be administered during task performance.

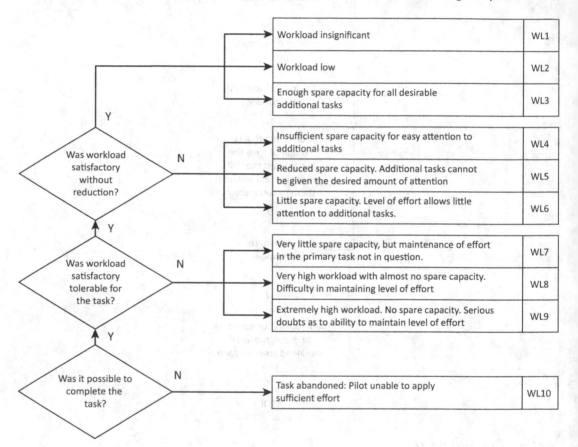

Figure 8.7   Bedford scale (Roscoe and Ellis, 1990)

Domain of Application

Aviation.

Procedure and Advice

Step 1: Define the Task under Analysis
The first step in a Bedford scale analysis (aside from the process of gaining access to the required systems and personnel) is to define the task (or tasks) that is to be subjected to analysis. The types of task analysed

are dependent upon the focus of the analysis. For example, when assessing the effects on operator MWL caused by a novel design or a new process, it is useful to analyse a set of tasks that are as representative of the full functionality of the interface, device or procedure as possible. To analyse a full set of tasks will often be too time-consuming and labour-intensive, and so it is pertinent to use a set of tasks that utilise all aspects of the system under analysis.

Step 2: Conduct an HTA for the Task under Analysis
Once the task (or tasks) under analysis is defined clearly, an HTA should be conducted for each task. This allows the analyst and participants to understand the task fully.

Step 3: Selection of Participants
Once the task (or tasks) under analysis is defined, it may be useful to select the participants that are to be involved in the analysis. This may not always be necessary and it may suffice to simply select participants randomly on the day. However, if workload is being compared across rank or experience levels, then clearly effort is required to select the appropriate participants.

Step 4: Brief Participants
Before the task (or tasks) under analysis is performed, all of the participants involved should be briefed regarding the purpose of the study and the Bedford scale technique. It is recommended that participants are given a workshop on MWL and MWL assessment. It may also be useful at this stage to take the participants through an example Bedford scale analysis so that they understand how the technique works and what is required of them as participants. It may even be pertinent to get them to perform a small task and then to complete a Bedford scale questionnaire. This acts as a 'pilot run' of the procedure highlighting any potential problems.

Step 5: Task Performance
Once the participants fully understand the Bedford scale technique and the data collection procedure, they are free to undertake the task under analysis as normal.

Step 6: Completion of the Bedford Scale
Once the participanst have completed the relevant task, they should be given the Bedford scale and instructed to work through it, based on the task that they have just completed. Once they have finished working through the scale, a rating of participant MWL is derived. If there are any tasks requiring analysis left, the participants should then move on to the next task and repeat the procedure.

Advantages

- The Bedford scale is very quick and easy to use, requiring minimal analyst training.
- It is generic and so can easily be applied in different domains.
- It may be useful when used in conjunction with other techniques of MWL assessment.
- It offers a low level of intrusiveness.

Disadvantages

- There is little evidence of actual use of the technique.
- There is limited validation evidence associated with the technique.
- Its output is limited.
- Participants are not efficient at reporting mental events 'after the fact'.

Flowchart

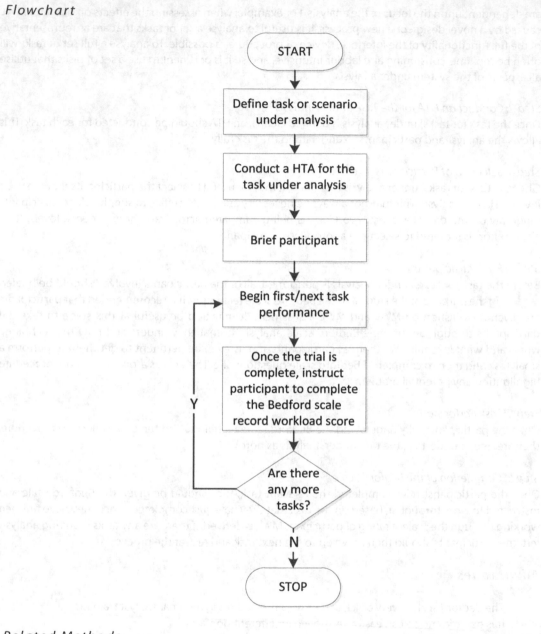

Related Methods

The Bedford scale technique is one of a number of subjective MWL assessment techniques. Other subjective MWL techniques include the NASA-TLX, MCH, SWAT, DRAWS and MACE. It is especially similar to the MCH technique, as it uses a hierarchical decision tree in order to derive a measure of participant MWL. When conducting a Bedford scale analysis, a task analysis (such as an HTA) of the task or scenario is normally required. Also, subjective MWL assessment techniques are normally used in conjunction with other MWL assessment techniques, such as primary and secondary task measures.

Approximate Training and Application Times

The training and application times for the Bedford scale are estimated to be very low.

Reliability and Validity

There is no data regarding the reliability and validity of the technique available in the literature.

Tools Needed

The Bedford scale technique is applied using pen and paper.

Instantaneous Self-Assessment (ISA)

Background and Applications

The ISA workload technique is another very simple subjective MWL assessment technique that was developed by NATS (formerly National Air Traffic Services) for use in the assessment of air traffic controller MWL during the design of future ATM systems (Kirwan et al., 1997). It involves participants self-rating their workload during a task (normally every two minutes) on a scale of 1 (low) to 5 (high). Kirwan et al. (1997) used the following ISA scale in Table 8.4 to assess air traffic controllers' workload.

Table 8.4 Example ISA workload scale

Level	Workload heading	Spare capacity	Description
5	Excessive	None	Behind on tasks; losing track of the full picture.
4	High	Very little	Non-essential tasks suffering. Could not work at this level very long.
3	Comfortable busy pace	Some	All tasks well in hand. Busy but stimulating pace. Could keep going continuously at this level.
2	Relaxed	Ample	More than enough time for all tasks. Active on ATC task less than 50 per cent of the time.
1	Under-utilised	Very much	Nothing to do. Rather boring.

Source: Kirwan et al., 1997.

Typically, the ISA scale is presented to the participants in the form of a colour-coded keypad. The keypad flashes when a workload rating is required, and the participant simply pushes the button that corresponds to his or her perceived workload rating. Alternatively, the workload ratings can be requested and acquired verbally. The ISA technique allows a profile of operator workload throughout the task to be constructed and allows the analyst to ascertain excessively high or low workload parts of the task under analysis. The appeal of the ISA technique lies in its low resource usage and its low level of intrusiveness.

Domain of Application

Generic. It has mainly been used in ATC.

Procedure and Advice

Step 1: Construct a Task Description
The first step in any workload analysis is to develop a task description for the task or scenario under investigation. It is recommended that an HTA is used for this purpose.

Step 2: Brief Participants
The participants should be briefed regarding the ISA technique, including what it measures and how it works. It may be useful to demonstrate an ISA data collection exercise for a task similar to the one under analysis. This allows the participants to understand how the technique works and also what is required of them. It is also crucial at this stage that the participants have a clear understanding of the ISA workload scale being used. In order for the results to be valid, the participants should have the same understanding of each level of the workload scale, i.e. what level of perceived workload constitutes a rating of 5 on the ISA workload scale and what level constitutes a rating of 1. It is recommended that the participants are taken through the scale and examples of workload scenarios are provided for each level on the scale. Once the participants fully understand the ISA workload scale being used, the analysis can proceed to the next step.

Step 3: Pilot Run
Once the participants have a clear understanding of how the ISA technique works and what is being measured, it is useful to perform a pilot run. Whilst performing a small task, participants should be subjected to the ISA technique. This allows them to experience the technique in a task performance setting. Participants should be encouraged to ask questions during the pilot run in order to understand the technique and the experimental procedure fully.

Step 4: Begin Task Performance
Next, the participants should begin the task under analysis. Normally, a simulation of the system under analysis is used; however, this is dependent upon the domain of application. ISA can also be used during task performance in a 'real-world' setting, although it has mainly been applied in simulator settings. Simulators are also useful as they can be programmed to record the workload ratings throughout the trial.

Step 5: Request and Record Workload Rating
The analyst should request a workload rating either verbally or through the use of flashing lights on the workload scale display. The frequency and timing of the workload ratings should be determined beforehand by the analyst. Typically, a workload rating is requested every two minutes. It is crucial that the provision of a workload rating is as unintrusive to the participant's primary task performance as possible. Step 4 should continue at regular intervals until the task is completed. The analyst should make a record of each workload rating given.

Step 6: Construct Task Workload Profile
Once the task is complete and the workload ratings are collected, the analyst should construct a workload profile for the task under analysis. Typically, a graph is constructed, highlighting the high and low workload points of the task under analysis. An average workload rating for the task under analysis can also be calculated.

Advantages

- ISA is a very simple technique to learn and use.
- The output allows a workload profile for the task under analysis to be constructed.
- It is very quick in its application as data collection occurs during the trial.
- It has been used extensively in numerous domains.
- It requires very little in the way of resources.
- Whilst it is obtrusive to the primary task, it is probably the least intrusive of the online workload assessment techniques.
- It is a low-cost technique.

Disadvantages

- ISA is intrusive to primary task performance.
- There is limited validation evidence associated with the technique.
- It is a very simplistic technique, offering only a limited assessment of operator workload.
- Participants are not very efficient at reporting mental events.

Related Methods

ISA is a subjective workload assessment technique of which there are many, such as the NASA-TLX, MACE, MCH, DRAWS and Bedford scales. To ensure comprehensiveness, it is often used in conjunction with other subjective techniques, such as the NASA-TLX.

Approximate Training and Application Times

It is estimated that the training and application times associated with the ISA technique are very low. Application time is dependent upon the duration of the task under analysis.

Reliability and Validity

Djokic, Lorenz and Fricke (2010) used ISA to explore the ATC operators' subjective workload in a simulated experiment involving the Central European Air Traffic Services (CEATS) airspace. The subjective measure of workload was compared to ATC complexity components (identified using principal component analysis) and controller activity metrics in order to assess the relative contribution of each to workload. The results found that complexity components alone could not predict workload as additional factors contributed to workload level such as communication load.

Engelmann et al. (2011) used ISA to explore the impact of providing surgeons with additional levels of intermittent breaks during operations. Measures of stress hormones and amylase were drawn from the surgeons' saliva both during and after the operation, as were measures of ISA, EEG and musculoskeletal strain. They found that cortisol and amylase levels dropped and that error rates and musculoskeletal strain were reduced, as were workload levels.

Tools Needed

ISA can be applied using pen and paper.

Flowchart

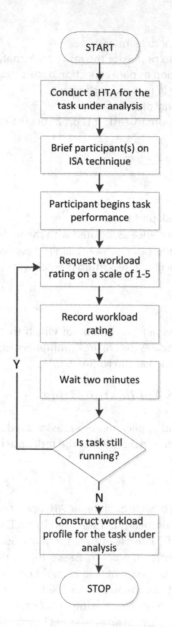

Cognitive Task Load Analysis (CTLA)

Background and Applications

CTLA is a technique used to assess or predict the cognitive load of a task or set of tasks imposed upon an operator. It is typically used early in the design process to aid the provision of an optimal cognitive load for the system design in question and has been used in its present format in the naval domain (Neerincx, 2003). It is based on a model of cognitive task load (Neerincx, 2003) that describes the effects of task characteristics upon operator mental workload. According to the model, cognitive (or mental) task load is comprised of the percentage of time occupied, level of information processing and the number of task set switches exhibited during the task. According to Neerincx (2003), the operator should not be occupied by one task for more than 70–80 per cent of the total time. The level of information processing is defined using the SRK framework (Rasmussen, 1986). Finally, task set switches are defined by changes of applicable task knowledge on the operating and environmental level exhibited by the operators under analysis (Neerincx, 2003). The three variables of time occupied, level of information processing and task set switches are combined to determine the level of cognitive load imposed by the task. High ratings for the three variables equate to a high cognitive load imposed on the operator by the task.

Domain of Application

Maritime.

Procedure and Advice

The following procedure is adapted from Neerincx (2003).

Step 1: Define the Task or Scenario under Analysis
The first step in analysing operator cognitive load is to define the task or scenario under analysis.

Step 2: Data Collection
Once the task or scenario under analysis is clearly defined, specific data should be collected regarding the task. Observation, interviews, questionnaires and surveys are typically used.

Step 3: Task Decomposition
The next step involves defining the overall operator goals and objectives associated with each task under analysis. Task structure should also be described fully.

Step 4: Create Event List
Next, a hierarchical event list for the task under analysis should be created. According to Neerincx (2003), this should describe the event classes that trigger task classes, providing an overview of any situation-driven elements.

Step 5: Describe Scenario
Once the event classes are described fully, the analyst should begin to describe the scenarios involved in the task under analysis. This description should include sequences of events and their consequences. Neerincx (2003) recommends that this information is displayed on a timeline.

Step 6: Describe Basic Action Sequences (BAS)
BAS describe the relationship between event and task classes. These action sequences should be depicted in action sequence diagrams.

Step 7: Describe Compound Action Sequences (CAS)
CAS describe the relationship between event and task instances for situations and the associated interface support. The percentage of time occupied, the level of information processing and the number of task set switches are elicited from the CAS diagram.

Step 8: Determine Percentage Time Pccupied, Level of Information Processing and Number of Task Set Switches
Once the CAS are described, the analyst should determine the operator's percentage of time occupied, level of information processing and number of task set switches exhibited during the task or scenario under analysis.

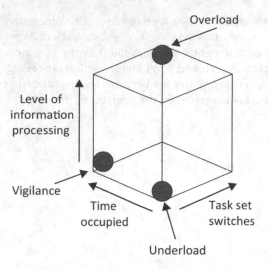

Figure 8.8   Three-dimensional model of cognitive task load (Neerincx, 2003)

Step 9: Determine Cognitive Task Load
Once the percentage of time occupied, the level of information processing and the number of task set switches are defined, the analyst should determine the operator's cognitive task load. The three variables should be mapped onto the model of cognitive task load shown in Figure 8.8 on the left.

Advantages

- The technique is based on sound theoretical underpinning.
- It can be used during the design of systems and processes to highlight tasks or scenarios that impose especially high cognitive task demands.
- It seems to be suited to analysing control room-type tasks or scenarios.

Disadvantages

- The technique appears to be quite complex
- Such a technique would be very time-consuming in its application.
- A high level of training would be required.
- There is no guidance on the rating of cognitive task load. It would be difficult to give task load a numerical rating based upon the underlying model.
- The initial data collection stage would be very time-consuming.
- The technique requires validation.
- Evidence of its use is limited.

Related Methods

The CTLA technique uses action sequence diagrams, which are very similar to operator sequence diagrams. In the data collection phase, techniques such as observation, interviews and questionnaires are used.

Approximate Training and Application Times

It is estimated that the training and application times associated with the CTLA technique would both be very high.

Reliability and Validity

No data regarding the reliability and validity of the technique is offered in the literature.

Tools Needed

Once the initial data collection phase is complete, CTLA can be conducted using pen and paper. The data collection phase would require video and audio recording equipment and a computer.

Flowchart

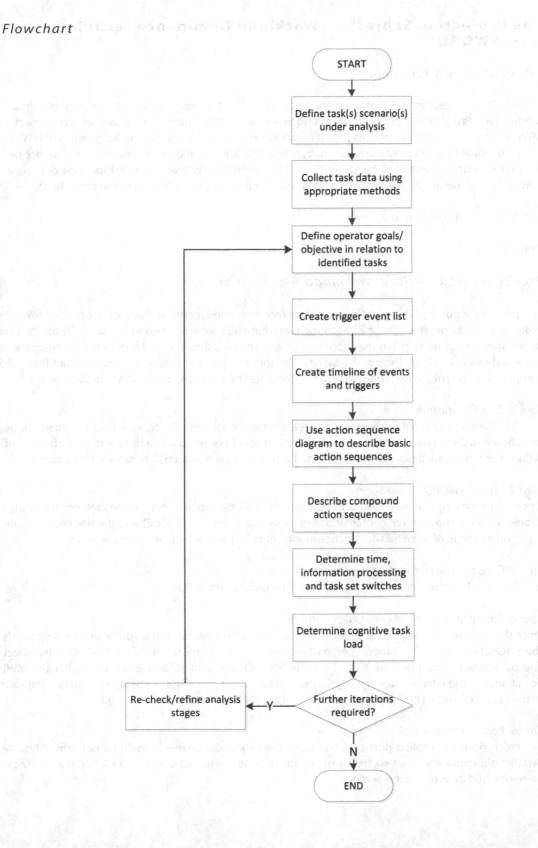

The Projective Subjective Workload Dominance Technique (Pro-SWORD)

Background and Applications

SWORD is a subjective MWL assessment technique that has been used both retrospectively and predictively (Pro-SWORD) (Vidulich, Ward and Schueren, 1991). Originally designed as a retrospective MWL assessment technique, it uses paired comparison of tasks in order to provide a rating of MWL for each individual task. Administered post-trial, participants are required to rate one task's dominance over another in terms of workload imposed. When used predictively, tasks are rated for their dominance before the trial begins and are then rated post-test to check for the sensitivity of the predictions.

Domain of Application

Generic.

Procedure and Advice – Workload Assessment

The procedure outlined below is the procedure recommended for an assessment of operator MWL. In order to predict operator MWL, it is recommended that SMEs are employed to predict MWL for the task under analysis before step 3 in the procedure below. The task should then be performed and operator workload ratings obtained using the SWORD technique. The predicted MWL ratings should then be compared to the subjective ratings in order to calculate the sensitivity of the MWL predictions made.

Step 1: Task Description
The first step in any SWORD analysis is to create a task or scenario description of the scenario under investigation. Each task should be described individually in order to allow the creation of the SWORD rating sheet. Any task description can be used for this step, such as an HTA or tabular task analysis.

Step 2: Create SWORD Rating Sheet
Once a task description (e.g. HTA) is developed, the SWORD rating sheet can be created. The analyst should list all of the possible combinations of tasks (e.g. A v B, A v C, B v C) and the dominance rating scale. An example of a SWORD dominance rating sheet is shown in Figure 8.5 on page 316.

Step 3: Conduct Walkthrough of the Task
A walkthrough of the task under analysis should be given to the SMEs.

Step 4: Administration of SWORD Questionnaire
Once the SMEs have been given an appropriate walkthrough or demonstration of the task under analysis, the SWORD data collection process begins. This involves the administration of the SWORD rating sheet. The participant should be presented with the SWORD rating sheet and asked to predict the MWL dominance of the interface under analysis. The SWORD rating sheet lists all possible paired comparisons of the tasks conducted in the scenario under analysis. A 17-point rating scale is used.

Step 5: Performance of Task
SWORD is normally applied post-trial. Therefore, the task under analysis should be performed first. As SWORD is applied after the task performance, intrusiveness is reduced and the task under analysis can be performed in its real-world setting.

Step 6: Administration of SWORD Questionnaire
Once the task under analysis is complete, the SWORD data collection process begins. This involves the administration of the SWORD rating sheet, and example of which can be seen on page 316 (Figure 8.5). The participant should be presented with the SWORD rating sheet immediately after task performance has ended. As in step four, the SWORD rating sheet lists all possible paired comparisons of the tasks conducted in the scenario under analysis. A 17-point rating scale is used.

The 17 slots represent the possible ratings. The analyst has to rate the two tasks against each other (e.g. task A v B) in terms of their level of MWL imposed. For example, if a participant feels that the two tasks imposed a similar level of MWL, then he or she should mark the 'EQUAL' point on the rating sheet. However, if a participant feels that task A imposed a slightly higher level of MWL than task B did, he or she would move towards task A on the sheet and mark the 'weak' point on the rating sheet. If a participant felt that task A imposed a much greater level of workload than task B, then he or she would move towards task A on the sheet and mark the 'Absolute' point on the rating sheet. This allows the participant to provide a subjective rating of one task's MWL dominance over the over. This procedure should continue until all of the possible combinations of tasks in the scenario under analysis are exhausted and given a rating.

Step 7: Constructing the Judgment Matrix
Once all ratings have been elicited, the SWORD judgment matrix should be conducted. Each cell in the matrix should represent the comparison of the task in the row with the task in the associated column. The analyst should fill in each cell with the participant's dominance rating. For example, if a participant rated tasks A and B as equal, a '1' is entered into the appropriate cell. If task A is rated as dominant, then the analyst simply counts from the 'Equal' point to the marked point on the sheet and enters the number in the appropriate cell. An example SWORD judgment matrix is shown in Figure 8.6 on page 317. The rating for each task is calculated by determining the mean for each row of the matrix and then normalising the means (Vidulich, Ward and Schueren, 1991).

Step 8: Matrix Consistency Evaluation
Once the SWORD matrix is complete, the consistency of the matrix can be evaluated by ensuring that there are transitive trends amongst the related judgments in the matrix. For example, if task A is rated twice as hard as task B, and task B is rated three times as hard as task C, then task A should be rated as six times as hard as task C (Vidulich, Ward and Schueren, 1991). Therefore the analyst should use the completed SWORD matrix to check the consistency of the participant's ratings.

Step 9: Compare Predicted Ratings to Retrospective Ratings
The analyst should now compare the predicted MWL ratings against the ratings offered by the participants post-trial.

Advantages

- SWORD is easy to learn and use.
- It is non-intrusive
- It has a high face validity
- It has been demonstrated to have a sensitivity to workload variations (Ried and Nygren, 1988).
- It is very quick in its application.

Disadvantages

- Data is collected post-task.
- It is a dated approach to workload assessment.

- Workload projections are more accurate when domain experts are used.
- Further validation is required.
- It has not been as widely used as other workload assessment techniques, such as SWAT, MCH and the NASA-TLX.

Example

Vidulich, Ward and Schueren (1991) tested the SWORD technique for its accuracy in predicting the MWL imposed upon F-16 pilots by a new HUD attitude display system. Participants included F-16 pilots and college students, who were divided into two groups. The first group (F-16 pilots experienced with the new HUD display) retrospectively rated the tasks using the traditional SWORD technique, whilst the second group (F-16 pilots who had no experience of the new HUD display) used the Pro-SWORD variation to predict the MWL associated with the HUD tasks. A third group (college students with no experience of the HUD) also used the Pro-SWORD technique to predict the associated MWL. In conclusion, it was reported that the pilot Pro-SWORD ratings correlated highly with the pilot SWORD (retrospective) ratings (Vidulich, Ward and Schueren, 1991). Furthermore, the Pro-SWORD ratings correctly anticipated the recommendations made in an evaluation of the HUD system. Vidulich, Ward and Schueren (1991) also report that the SWORD technique was more reliable and sensitive than the NASA-TLX technique.

Related Methods

SWORD is one of a number of MWL assessment techniques, including the NASA-TLX, SWAT, MCH and DRAWS. A number of these techniques have also been used predictively, such as Pro-SWAT and MCH. A SWORD analysis requires a task description of some sort, such as an HTA or tabular task analysis.

Approximate Training and Application Times

Although no data is offered regarding the training and application times for the SWORD technique, it is apparent that the training time for such a simple technique would be minimal. The application time associated with the technique would be based upon the scenario under analysis. For large, complex scenarios involving a great number of tasks, the application time would be high as an initial HTA would have to be performed, then the scenario would have to performed and finally the SWORD technique. The actual application time associated purely with the administration of the SWORD technique is very low.

Reliability and Validity

Vidulich, Ward and Schueren (1991) tested the SWORD technique for its accuracy in predicting the MWL imposed upon F-16 pilots by a new HUD attitude display system. In conclusion, it was reported that the pilot Pro-SWORD ratings correlated highly with the pilot SWORD (retrospective) ratings (Vidulich, Ward and Schueren, 1991). Furthermore, the Pro-SWORD ratings correctly anticipated the recommendations made in an evaluation of the HUD system. Vidulich and Tsang (1987) also reported that the SWORD technique was more reliable and sensitive than the NASA-TLX technique.

Tools Needed

The SWORD technique can be applied using pen and paper. Of course, the system or device under analysis is also required.

Flowchart

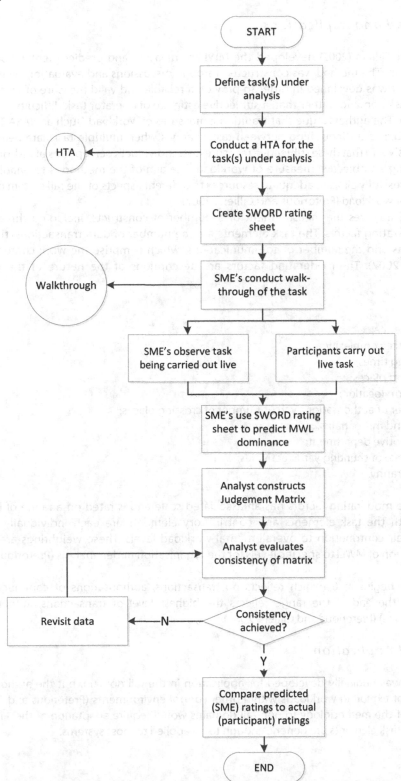

The Mental Workload Index (MWLI)

Background and Applications

Pretorius and Cilliers (2007) developed the MWLI to measure and predict mental workload within the rail domain. The method was constructed through discussions and evaluations with SMEs in the rail industry. It was developed in order to provide a reliable and valid measure of what the authors label 'pure' task demands rather than a subjective estimate of operator task difficulty (Pretorious and Cilliers, 2007). The authors argue that traditional measures of workload (such as NASA-TLX and MCH) are not capable of distinguishing between task changes when multiple tasks are being conducted simultaneously and that these subjective measures rate individual's perceptions of task difficulty rather than providing an objective measure of workload. The aim of the method is to enable managerial levels to successfully allocate adequate resources to different aspects of the rail system depending on predictions of workload (Pretorious and Cilliers, 2007).

The MWLI involves the measurement of a number of constructs, including three task factors and 11 moderating factors. The task elements are the number of data transactions, the number of authorisations and the number of communications, which comprise the work content (Pretorious and Cilliers, 2007). The moderating factors are descriptions of the nature of the work and its complexity:

1. shift;
2. experience;
3. interface complexity;
4. running times;
5. crossing places;
6. platform location;
7. number of authorisations versus number of crossing places;
8. type and mix of trains;
9. locomotive department;
10. presence of shunting yards/activities;
11. topography.

Each of these moderating factors has an associated scale and is rated on a range of between 1–2 and 1–4. Both the task elements and contributory elements are each individually weighted to represent their contribution to overall mental workload levels. These weightings are adapted for each application of MWLI to specific levels for the organisation under analysis (Pretorious and Cilliers, 2007).

The MWLI begins at 0 (which reflects no transactions, authorisations or communications) and is unlimited (the end of the range reflects the highest level of transactions, authorisations and communications) (Pretorious and Cilliers, 2007).

Domain of Application

The method was originally developed for application in the rail domain, but the authors argue that it is capable of exploring workloads within other control environments (Preterious and Cilliers, 2007). Application of the methodology to non-rail domains would require adaptation of the 11 moderators; however, the task elements are generic enough to be applied to most systems.

Procedure and Advice

Step 1: Define the Task and System under Analysis
The first stage of the analysis must involve the analyst clearly defining the task under investigation. This definition will guide the analysis and ensure that appropriate and relevant data are collected.

Step 2: Recruit Participants
A representative sample of participants should be identified and recruited.

Step 3: Brief Participants
The participants should be fully briefed on the aims of the analysis and the tasks they will be asked to conduct.

Step 4: Participants Undertake Work Scenarios
The participants recruited should undertake their normal work activities. This stage lasts as long as their normal day at work would last.

Step 5: Measurement of Task Elements
During the work scenario, the analyst should ensure that each of the task elements is measured. Comprehensive data must be collected on the number of data transactions, the number of authorisations and the number of communications that occurred during the work scenario. A frequency value for each of these transactions should be calculated.

Step 6: Measurement of Moderating Factors
The analyst must also ensure that throughout the work scenario the 11 moderating factors are systematically measured.

Step 7: Data Analysis
Once the task is complete and all data has been collected, data analysis can begin. This involves applying the appropriate weighting to each factor and then calculating the MWLI using the following formula:

MWLI = (sum of weighted task elements) x (product of weighted moderating factors) or:

$$I = \left(\sum_{i=1}^{3} Ti \; x \; Wi \right) x \left(\prod_{j=1}^{11} Mj \; x \; Vj \right)$$

where I is the MWLI, Wi, are the weights of the three task elements, Ti, i.e. i=1, 2, 3 and Vj, are the weights of the 11 moderating factors Mj, with j=1, 2, …, 11.

Advantages

- The method provides insights into resource allocation requirements (Pretorius and Cilliers, 2007).
- It is unobtrusive as analysis is performed offsite and data collection does not involve the participant directly (Pretorius and Cilliers, 2007).
- Pretorius and Cillers (2007) argue that the method provides sensitive evaluation of changes in task demands and difficulties.

Disadvantages

- A high level of access to participants within their natural work environment is required.
- Use of the method outside of the rail domain would require adaptation of the moderators.

Tools Needed

In order to develop a MWLI, a high degree of access to participants and their work domain is required. The method can be conducted using pen and paper.

Reliability and Validity

Research by the authors of the method has shown it to have high levels of sensitivity, selectivity, diagnostic ability and operator acceptance (Pretorius and Cilliers, 2007), which bodes well for its reliability and validity.

Approximate Application and Training Times

The method requires a high level of data collection, which may be time-consuming both to collect and analyse.

Example

Pretorius and Cilliers (2007) applied the MWLI to the analysis of 36 rail control centres. Data was collected on the task performance occurring during a 'shift' at each of the control centres under analysis. Various forms of data collection were engaged to capture task performance, including the recording of voice lockers to evaluate the number and content of communications engaged in, and examination of authorisation sheets (a log of all authorisations made). The initial data collection phase can be summarised by completing a data collection sheet in which simple raw figures are presented, as given in Table 8.5.

Table 8.5   Example MWLI data template

	Control centre 1	Control centre 2	Control centre 3
Task element			
Data transaction count			
Data authorisation count			
Communications count			
Moderatory element			
Shift			
Experience			
Interface complexity			
Running times			
Crossing places			
Platform location			
Number of authorisations versus number of crossing places			
Type and mix of trains			
Locomotive department			
Presence of shunting yard trains			
Topography			

Once the raw data was collected, values were then assigned for each of the three task factors and 11 contributory factors for all rail control centres. Examination of the data enabled identification of the specific task demands and levels of demands at each control centre. The data collected enabled Pretorius and Cilliers (2007) to derive mental workload indexes for each rail centre, which ranged from 89 to 5,789 across those analysed. The analysis enabled them to identify those control centres with the highest and lowest task demands both through deriving a mental workload index and by assessing the data in more detail.

Flowchart

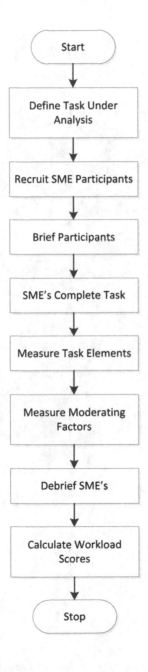

Chapter 9

Team Assessment Methods

Introduction

An increased use of teams of actors within complex systems has led to the emergence of various approaches for the assessment of different features associated with team performance. According to Savoie (1998; cited in Salas, 2004), the use of teams has risen dramatically, with reports of 'team presence' from workers rising from five per cent in 1980 to 50 per cent in the mid-1990s. Over the last two decades, the performance of teams in complex systems has received considerable attention from the HF community and a number of techniques have been developed in order to assess and evaluate team performance. Research into team performance is currently being undertaken in a number of areas, including the aviation, the military ATC and the emergency services domains.

A team can be defined in simple terms as a group of actors working collaboratively within a system. According to Salas (2004), a team consists of two or more people dealing with multiple information sources who are working to accomplish a shared goal of some sort. With regard to the roles that teams take within complex systems, Cooke (2004) suggests that teams are required to detect and interpret cues, remember, reason, plan, solve problems, acquire knowledge and make decisions as an integrated and coordinated unit. Team-based activity in complex systems comprises two components: teamwork and taskwork. Teamwork refers to those instances where actors within a team or network coordinate their behaviour in order to achieve tasks related to the team's goals, while taskwork refers to those tasks that are conducted by team members individually or in isolation of one another.

The complex nature of team-based activity ensures that sophisticated assessment techniques are required for team performance assessment. Team-based activity involves multiple actors with multiple goals performing both teamwork and taskwork activity. The activity is typically complex (hence the requirement for a team) and may be dispersed across a number of different geographical locations. Consequently, there are a number of different team performance techniques available to the HF practitioner, each designed to assess certain aspects of team performance in complex systems. The team performance techniques considered in this review can be broadly classified into the following categories:

- team task analysis (TTA) techniques;
- team cognitive task analysis techniques;
- team communication assessment techniques;
- team behavioural assessment techniques; and
- team MWL assessment techniques.

A brief description of each team methods category is given below, along with a brief outline of the techniques considered in the review.

TTA techniques are used to describe team performance in terms of requirements (knowledge, skills and attitudes) and the tasks that require either teamwork or individual (taskwork) performance (Burke, 2004). According to Baker, Salas and Cannon-Bowers (1998), TTA refers to the analysis of team tasks and also the assessment of a team's teamwork requirements (knowledge, skills and abilities). TTA outputs are typically used in the development of team-training interventions, such as crew resource management training programmes, for the evaluation of team performance and also to identify operational and teamwork skills required within teams (Burke, 2004). According to Salas (2004), optimising team performance and effectiveness involves understanding a number of components surrounding the use of teams such as communication and task requirements, team environments and team objectives. The TTA techniques reviewed in this document attempt to analyse such components. Groupware Task Analysis (Van Welie and Van Der Veer, 2003) is a TTA technique that is used to study and evaluate group or team activities in order to inform the design and analysis of similar team systems. HTA(T) (Annett, 2004) is an adaptation of HTA that caters for team performance in complex systems.

Team cognitive task analysis (CTA) techniques are used to elicit and describe the cognitive processes associated with team decision-making and performance. According to Klein (2000), a team CTA provides a description of the cognitive skills required for a team to perform a task. Team CTA techniques are used to assess team performance and then to inform the development of strategies designed to improve team performance. The output of team CTA techniques is typically used to aid the design of team-based technology, the development of team-training procedures, task allocation within teams and also the organisation of teams. It is a technique that is used to describe the cognitive skills that a team or group of individuals are required to undertake in order to perform a particular task or set of tasks. The decision requirements exercise is a technique that is very similar to team CTA and is used to specify the requirements or components (difficulties, cues and strategies used, errors made) associated with decision-making in team scenarios. The Cockpit Management Attitudes Questionnaire (CMAQ: Helmreich, 1984; Helmreich, Wilhelm and Gregorich, 1988) is a questionnaire developed to provide insights into the attitudes of flight crews regarding team processes. The questionnaire focuses on communication, coordination and leadership.

Communication between team members is crucial to the overall success of team performance. Team communication assessment techniques are used to assess the content, frequency, efficiency, technology used and nature of communication between the actors within a particular team. The output of team communication assessment techniques can be used to determine procedures for effective communication, to specify appropriate technology to use in communications, to aid the design of team training procedures and team processes, and to assess existing communication procedures. The comms usage diagram (CUD: Watts and Monk, 2000) approach is used to analyse and represent communications between actors dispersed across different geographical locations. The output of a CUD analysis describes how, why and when communications between team members occur, which technology is involved in the communication, and the advantages and disadvantages associated with the technology used. Social Network Analysis (SNA: Driskell and Mullen, 2005; Wasserman and Faust, 1994) is used to analyse and represent the relationships between actors within a social network which can be considered analogous to the concept of a team. It uses mathematical methods from graph theory to analyse these relationships and can be employed to identify key agents and other aspects of a particular social network.

Team behavioural assessment techniques are used to assess team performance or behaviours exhibited by teams during a particular task or scenario. Behavioural assessment techniques have typically been used in the past to evaluate the effectiveness of team training interventions such as crew resource management programmes. Behavioural observation scales (BOS: Baker, 2004) are a general class of observer-rating approaches that are used to assess different aspects of team performance. Coordination demands analysis (CDA: Burke, 2004) is used to rate the level of coordination between team members during task performance. The task and training requirements analysis methodology (TTRAM: Swezey et al., 2000) uses a number of techniques to identify team-based task training requirements and also to evaluate any associated training technologies that could potentially be used in the delivery of team training. Questionnaires for distributed assessment of team mutual awareness (MacMillan et al., 2004) comprise a series of self-rating questionnaires designed to assess team member mutual awareness (individual awareness and team awareness). Targeted acceptable responses to generated events or tasks (TARGETs: Fowlkes et al., 1994) is a behavioural assessment method in which a team scenario is derived that incorporates a range of trigger events. Pre-defined responses to these events are given to observers to rate their occurrence during task performance.

The assessment of team MWL has previously received only minimal attention. Team MWL assessment techniques are used to assess the MWL imposed on both the actors within a team and also on the team as a whole during task performance. The team workload technique is an approach to the assessment of team workload described by Bowers and Jentsch (2005) that involves the use of a modified NASA-TLX (Hart and Staveland, 1988). As we saw in Chapter 7, there is also some interest in studying SA in teams. A summary of the team performance analysis techniques considered in this review is presented in Table 9.1.

Behavioural Observation Scales (BOS)

Background and Applications

BOS are a general class of observer-rating techniques used to assess different aspects of team performance in complex systems. Observer-rating approaches work on the notion that appropriate SMEs can accurately rate participants on externally exhibited behaviours based on an observation of the task or scenario under analysis. Observer-rating techniques have been used to measure a number of different constructs, including situation awareness (e.g. SABARS: Endsley, 2000, Matthews and Beal, 2002), shared cognition (Jarmasz et al., 2009) and CRM skills (e.g. NOTECHS: Flin et al., 1998). BOS techniques involve appropriate SMEs observing team-based activity and then providing ratings of various aspects of team performance using an appropriate rating scale. According to Baker (2004), BOS techniques are typically used to provide performance feedback during team training exercises. However, it is apparent that they can be used for a number of different purposes, including analysing team performance, situation awareness, error, CRM-related skills and command and control activity.

Domain of Application

Generic. Provided that an appropriate rating scale is used, BOS techniques can be applied in any domain.

Procedure and Advice

The following procedure describes the process of conducting an analysis using a pre-defined BOS. For an indepth description of the procedure involved in the development of a BOS, the reader is referred to Baker (2004).

Step 1: Define the Task under Analysis
First, the task (or tasks) and team (or teams) under analysis should be defined clearly. Once the task under analysis is clearly defined, it is recommended that an HTA be conducted. This allows the analyst to gain a complete understanding of the task and also of the types of behaviour that are likely to be

Table 9.1   Summary of team performance analysis techniques

Method	Type of method	Domain	Training time	App time	Related methods	Tools needed	Validation studies	Advantages	Disadvantages
BOS.	Team performance analysis.	Generic (military).	Med–High.	High	Behavioural rating scale, observation.	Pen and paper.	No.	1) Can be used to assess multiple aspects of team performance. 2) Seems suited to use in analysis of C4i analysis. 3) Easy to use.	1) There is a limit to what can be accurately assessed through observing participant performance. 2) A new BOS scale may need to be developed 3) Reliability is questionable.
CUD.	Comms analysis.	Generic (medical).	Low.	Med.	OSD, HTA, observation.	Pen and paper, video and audio recording equipment.	No.	1) Output provides a comprehensive description of task activity. 2) The technology used is analysed and recommendations are offered. 3) Seems suited to use in the analysis of C4i activity.	1) Limited reliability and validity evidence. 2) Neither time nor error occurrence are catered for. 3) Could be time-consuming and difficult to construct for large, complex tasks.
CDA.	Coordination analysis.	Generic.	Low.	Med.	HTA, observation.	Pen and paper.	No.	1) Very useful output, providing an assessment of team coordination. 2) Seems suited to use in the analysis of C4i activity.	1) Requires SMEs. 2) Rating procedure is time-consuming and laborious.
DRX.	Decision-making assessment.	Generic (military).	Med.	Med–High.	CDM, observation.	Pen and paper, video and audio recording equipment.	No.	1) Output is very useful, offering an analysis of team decision-making in a task or scenario. 2) Based upon actual incidents, removing the need for simulation. 3) Seems suited to use in the analysis of C4i activity.	1) Data is based upon past events, which may be subject to memory degradation. 2) Reliability is questionable. 3) May be time-consuming.
GTA.	Design.	Generic.	Med.	High.	N/A	Pen and paper.	No.	1) The output specifies information requirements and the potential technology to support task performance.	1) Limited use. 2) Resource-intensive. 3) A number of analysts are required.
HTA (T).	Team performance analysis.	Generic.	Med.	Med.	HEI, task analysis.	Pen and paper.	Yes.	1) Team HTA based upon extensively used HTA technique. 2) Caters for team-based tasks.	1) Of limited use.

Table 9.1   Continued

Method	Type of method	Domain	Training time	App time	Related methods	Tools needed	Validation studies	Advantages	Disadvantages
Questionnaires for Distributed Assessment of Team Mutual Awareness.	Team awareness, workload assessment.	Generic.	Low.	Med.	Questionnaires, NASA-TLX.	Pen and paper.	No.	1) Provides an assessment of team awareness and team workload. 2) Low cost, easy to use, requiring little training.	1) Data is collected post-trial. 2) Of limited use.
SNA.	Team analysis.	Generic.	High.	High.	Observation.	Pen and paper.	No.	1) Highlights the most important relationships and roles within a team. 2) Seems suited to use in the analysis of C4i activity.	1) Difficult to use for complex tasks involving multiple actors. 2) Data collection could be time-consuming.
TCTA.	Team cognitive task analysis.	Generic (military).	High.	High.	Observation, interviews, CDM.	Pen and paper, video and audio recording equipment.	Yes.	1) Can be used to elicit specific information regarding team decision-making in complex environments. 2) Seems suited to use in the analysis of C4i activity. 3) Output can be used to develop effective team decision-making strategies.	1) Reliability is questionable. 2) Resource-intensive. 3) A high level of training and expertise is required in order to use the technique properly.
Team Communications Analysis.	Comms analysis.	Generic.	Med.	Med.	Observation, checklists, frequency counts.	Pen and paper, computer.	No.	1) Provides an assessment of communications taking place within a team. 2) Suited to use in the analysis of C4i activity. 3) Can be used effectively during training.	1) Coding of data is time-consuming and laborious. 2) Initial data collection may be time-consuming.
TTA.	Team task analysis.	Generic.	Med.	Med.	Coordination demand analysis, observation.	Pen and paper.	No.	1) Output specifies the knowledge, skills and abilities required during task performance. 2) Useful for team training procedures. 3) Specifies which of the tasks are team-based and which are individual-based.	1) Time-consuming in its application. 2) SMEs are required throughout the procedure. 3) Considerable skill is required on behalf of the analysts.

Table 9.1   Continued

Method	Type of method	Domain	Training time	App time	Related methods	Tools needed	Validation studies	Advantages	Disadvantages
Team Workload Assessment	Workload assessment	Generic	Low	Low	NASA-TLX	Pen and paper	No	1) Output provides an assessment of both individual and team workload. 2) Quick and easy to use, requiring little training or cost. 3) Based upon the widely used and validated NASA-TLX measure	1) Extent to which team members can provide an accurate assessment of overall team workload and other team member workload is questionable. 2) Requires much further testing. 3) Data is collected post-trial
TTRAM	Training analysis	Generic	High	High	Observation, interview, questionnaire	Pen and paper	No	1) Useful output, highlighting those tasks that are prone to skill decay. 2) Offers training solutions	1) Time-consuming in its application. 2) SMEs required throughout. 3) Requires a high level of training
CMAQ	Team cognitive task analysis	Aviation	Low	Low	SAQ, ICUMAQ, ORMAQ	Pen and paper	Yes	1) Enables large amounts of information to be collected from large populations. 2) Provides a baseline and enables continual assessment	1) Data collected is limited. 2) Participant responses are forced into one of five categories
TARGETs	Team performance analysis	Generic	Low	High	Observation	Pen and paper, work domain SMEs	No	1) Enables a naturalistic evaluation of teams. 2) Minimises subjective judgment and enables controlled assessment	1) Requires a high level of SME input. 2) Scenario development can be complex and time-consuming

exhibited during the task. A number of different data collection procedures may be adopted during the development of the HTA, including observational study, interviews and questionnaires.

Step 2: Select or Develop Appropriate BOS

Once the task and team under analysis are clearly defined and described, an appropriate BOS scale should be selected. If an appropriate scale does not already exist, then one should be developed. It may be that an appropriate BOS already exists, and if this is the case, the scale can be used without modification. Typically, an appropriate BOS is developed from scratch to suit the analysis requirements. According to Baker (2004), the development of a BOS scale involves the following key steps:

1. conduct critical incident analysis;
2. develop behavioural statements;
3. identify teamwork dimensions;
4. classify behavioural statements into teamwork categories;
5. select appropriate metric, e.g. five-point rating scale (1 = almost never, 5 = almost always), checklist, etc.; and
6. pilot test BOS.

Step 3: Select Appropriate SME Raters

Once the BOS has been developed and tested appropriately, the SME raters who will use the technique to assess team performance during the task under analysis should be selected. The appropriate SMEs should possess an indepth knowledge of the task under analysis and also of the various different types of behaviours exhibited during performance of the task under investigation. The number of raters used is dependent upon the type and complexity of the task and also the scope of the analysis effort.

Step 4: Train Raters

Once an appropriate set of SME raters is selected, the raters should be given adequate training in the BOS technique. Baker (2004) recommends that a combination of behavioural observation training (BOT: Thornton and Zorich, 1980) and frame of reference training (FOR: Bernardin and Buckley, 1981) be used for this purpose. BOT involves teaching raters how to accurately detect, perceive, recall and recognise specific behavioural events during the task performance (Baker, 2004). FOR training involves teaching raters a set of standards for evaluating team performance. The raters should be encouraged to ask any questions during the training process. It may also be useful for the analyst to take the raters through an example BOS rating exercise.

Step 5: Assign Participants to Raters

Once the SME raters fully understand the BOS technique, they should be informed which of the participants they are to observe and rate. It may be that the raters are observing the team as a whole or that they are rating individual participants.

Step 6: Begin Task Performance

Once the raters fully understand how the BOS works and what is required of them, the data collection phase can begin. Prior to task performance, the participants should be briefed regarding the nature and purpose of the analysis. Performance of the task under analysis should then begin and the raters should observe their assigned team members. It is recommended that the raters make additional notes regarding the task performance in order to assist the rating process. It may also be useful to record the task using a video recorder. This allows the raters to consult footage of the task if they are unsure of a particular behaviour or rating.

Step 7: Rate Observable Behaviours
Ratings can be made either during task performance or post-trial. If a checklist approach is being used, raters simply check those behaviours observed during the task performance.

Step 8: Calculate BOS Scores
Once task performance is complete and all ratings and checklists are compiled, appropriate BOS scores should be calculated. The scores calculated depend upon the focus of the analysis. Typically, scores for each behaviour dimension (e.g. communication, information exchange) and an overall score are calculated. Baker (2004) recommends that BOS scores are calculated by summing all behavioural statements within a BOS. Each team's overall BOS score can then be calculated by summing each of the individual team member scores.

Advantages

- BOS techniques offer a simple approach to the assessment of team performance.
- They are low cost and easy to use.
- BOS can be used to provide an assessment of observable team behaviours exhibited during task performance, including communication, information exchange, leadership, teamwork and taskwork performance.
- It seems to be suited for use in the assessment of team performance in C4i environments.
- The output can be used to inform the development of team training exercises and procedures.
- It can be used to assess both teamwork and taskwork.
- It is a generic procedure and can be used to assess multiple features of performance in a number of different domains.

Disadvantages

- Existing scales may require modification for use in different environments. Scale development requires considerable effort on behalf of the analyst involved.
- Observer-rating techniques are limited in terms of what they can accurately assess. For example, the BOS can only be used to provide an assessment of observable behaviour exhibited during task performance. Other pertinent facets of team performance, such as SA, MWL and decision-making cannot be accurately assessed using a BOS.
- A typical BOS analysis is time-consuming to conduct, requiring the development of the scale, training of the raters, observation of the task under analysis and rating of the required behaviours. Even for a small-scale analysis, considerable time may be required.
- The reliability and validity of such techniques remains a concern.

Approximate Training and Application Times

It is estimated that the total application time for a BOS analysis would be high. A typical BOS analysis involves training the raters in the use of the technique, observing the task performance and then completing the BOS sheet. According to Baker (2004), rater training could take up to four hours and the application time may require up to three hours per team.

Flowchart

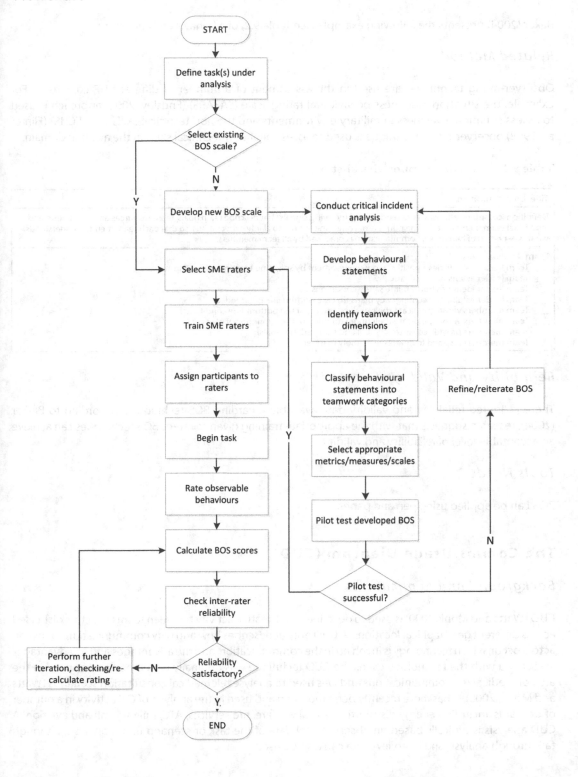

START

Define task(s) under analysis

Select existing BOS scale?

N

Y

Develop new BOS scale

Conduct critical incident analysis

Select SME raters

Develop behavioural statements

Train SME raters

Identify teamwork dimensions

Assign participants to raters

Classify behavioural statements into teamwork categories

Refine/reiterate BOS

Begin task

Y

Select appropriate metrics/measures/scales

Rate observable behaviours

Pilot test developed BOS

Calculate BOS scores

N

Check inter-rater reliability

Pilot test successful?

Perform further iteration, checking/re-calculate rating

N

Reliability satisfactory?

Y

END

Example

Baker (2004) presents the following example (see Table 9.2) of a behavioural checklist.

Related Methods

Observer-rating techniques are used in the assessment of a number of different HF constructs. For example, the situation awareness behavioural rating scale (SABARS) (Endsley, 2000) approach is used to assess situation awareness in military environments and the non-technical skills (NOTECHS) (Flin et al., 1998) observer-rating technique is used to assess pilot non-technical skills in the aviation domain.

Table 9.2   Communication checklist

Title: Communication
Definition: communication involves sending and receiving signals that describe team goals, team resources and constraints, and individual team member tasks. The purpose of communication is to clarify expectations, so that each team member understands what is expected of him or her. Communication is practised by all team members.
Example behaviours: ___ Team leader establishes a positive work environment by soliciting team members' input. ___ Team leader listens non-evaluatively. ___ Team leader identifies bottom-line safety conditions. ___ Team leader establishes contingency plans (in case bottom line is exceeded). ___ Team members verbally indicate their understanding of the bottom-line conditions. ___ Team members verbally indicate their understanding of the contingency plans. ___ Team members provide consistent verbal and non-verbal signals. ___ Team members respond to queries in a timely manner.

Reliability and Validity

There is limited reliability and validity data available regarding BOS techniques. According to Barker (2004), research suggests that with the appropriate training given to raters, BOS techniques can achieve an acceptable level of reliability and validity.

Tools Needed

BOS can be applied using pen and paper.

The Comms Usage Diagram (CUD)

Background and Applications

CUD (Watts and Monk, 2000) is used to describe collaborative activity between teams of actors dispersed across different geographical locations. A CUD output describes how and why communications between actors occur, which technology is involved in the communication, and the advantages and disadvantages associated with the technology used. The CUD technique was originally developed and applied in the area of medical telecommunications and was used to analyse telemedical consultation scenarios (Watts and Monk, 2000). It has more recently been modified and used in the analysis of C4i activity in a number of domains, including energy distribution, naval warfare, fire services, ATC, military, rail and aviation. A CUD analysis is typically based on observational data of the task or scenario under analysis, although talkthrough analysis and interview data can also be used.

Domain of Application

Generic. Although the technique was originally developed for use in the medical domain, the fact that it is generic means that it can be applied in any domain that involves distributed activity.

Procedure and Advice

Step 1: Define the Task or Scenario under Analysis
The first step in a CUD analysis is to clearly define the task or scenario under analysis. It may be useful to conduct an HTA of the task under analysis for this purpose. A clear definition of the task under analysis allows the analyst to prepare for the data collection phase.

Step 2: Data Collection
Next, the analyst should collect specific data regarding the task or scenario under analysis. A number of data collection procedures may be used for this purpose, including observational study, interviews and questionnaires. It is recommended that specific data regarding the activity conducted, the actors and individual task steps involved, the communication between actors, the technology used and the different geographical locations should be collected.

Step 3: Create Task or Scenario Transcript
Once sufficient data regarding the task under analysis has been collected, a transcript of the task or scenario should be created using the data collected as its input. The transcript should contain all of the date required for the construction of the CUD, i.e. the communications between different actors and the technology used.

Step 4: Construct CUD
The scenario transcript created during step 3 is then used as the input into the construction of the CUD. The CUD contains a description of the activity conducted at each geographical location, the communication between the actors involved, the technology used for the communications and the advantages and disadvantages associated with that technology medium, as well as a recommended technology if there is one. Arrows are used to represent the communication and direction of communication between personnel at each of the different locations. For example, if person A at site A communicates with person B at site B, the two should be linked with a two-way arrow. Column three of the CUD output table specifies the technology used in the communication, while column four lists any advantages and disadvantages associated with the particular technology used during the communication. In column five, recommended technology mediums for similar communications are provided. The advantages, disadvantages and technology recommendations are based on the subjective judgment of the analyst.

Advantages

- The CUD technique is simple to use and requires only minimal training.
- Its output is particularly useful, offering a description of the task under analysis and also a description of the communications between actors during the task, including the order of activity, the personnel involved, the technology used and the associated advantages and disadvantages.
- Its output is particularly useful for highlighting communication flaws in a particular network.
- The technique is particularly useful for the analysis of teamwork, distributed collaboration and C4i activity.
- It is also flexible and could potentially be modified to make it comprehensive. Factors such as time, error and workload could potentially be incorporated, ensuring that a much more exhaustive analysis is produced.

- Although it was developed and originally used in the medical domain, it is a generic technique and could potentially be applied in any domain involving distributed collaboration or activity.

Disadvantages

- For large, complex tasks involving multiple actors, conducting a CUD analysis may become time-consuming and laborious.
- The initial data collection phase of the CUD technique is also time-consuming and labour-intensive, potentially including interviews, observational analysis and talk-through analysis. As the activity is dispersed across different geographical locations, a team of analysts is also required for the data collection phase.
- No validity or reliability data is available for the technique.
- Application of the technique appears to be limited.
- Limited guidance is offered to analysts using the technique. For example, the advantages and disadvantages of the technology used and the recommended technology sections are based entirely upon the subjective judgment of the analyst.

Example

The CUD technique has recently been used as part of the Event Analysis of Systemic Teamwork (EAST: Baber and Stanton, 2004) method in the analysis of C4i activity in the fire service, naval warfare, aviation, energy distribution (Stanton, Baber and Harris, 2008), ATC and rail (Walker, Gibson, Stanton, Baber, Salmon and Green, 2006) domains. The following example is a CUD analysis of an energy distribution task. The task involved the return from isolation of a high-voltage circuit. The data collection phase involved an observational study of the scenario using two observers. The first observer was situated at the (NGT) National Operations Centre (NOC) and observed the activity of the NOC control room operator (CRO). The second observer was situated at the substation and observed the activity of the senior authorised person (SAP) and authorised person (AP) who completed work required to return the circuit from isolation. From the observational data obtained, an HTA of the scenario was developed. This acted as the main input for the CUD. The CUD analysis for the energy distribution task is presented in Figure 9.1.

Related Methods

The CUD data collection phase may involve the use of a number of different procedures, including observational study, interviews, questionnaires and walk-through analysis. It is also useful to conduct an HTA of the task under analysis prior to performing the CUD analysis. The CUD technique has also recently been integrated with a number of other techniques (HTA, observation, coordination demands analysis, social network analysis, operator sequence diagrams and propositional networks) to form the EAST (Baber and Stanton, 2004) methodology, which has been used for the analysis of C4i activity.

Approximate Training and Application Times

The training time for the CUD technique is minimal, normally no longer than one to two hours, assuming that the practitioner involved is already proficient in data collection techniques such as interviews and observational study. The application time for the technique is also minimal, providing the analyst has access to an appropriate drawing package such as Microsoft Visio. For the C4i scenario presented in the example section, the associated CUD application time was approximately two hours.

Flowchart

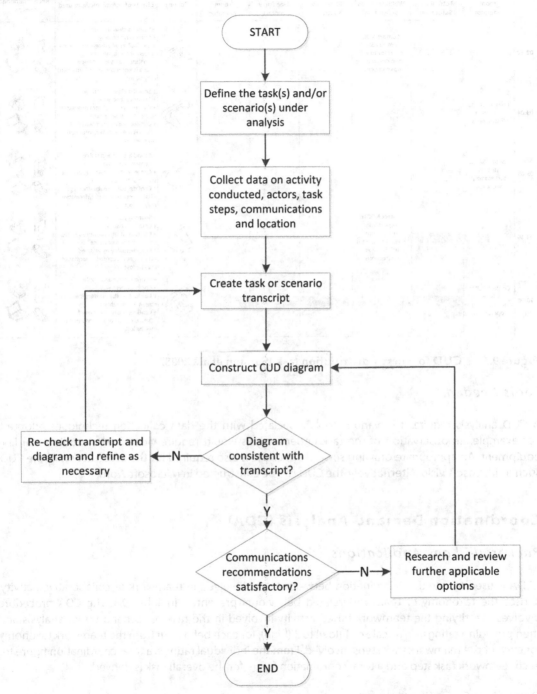

Reliability and Validity

No data regarding the reliability and validity of the technique is available in the literature.

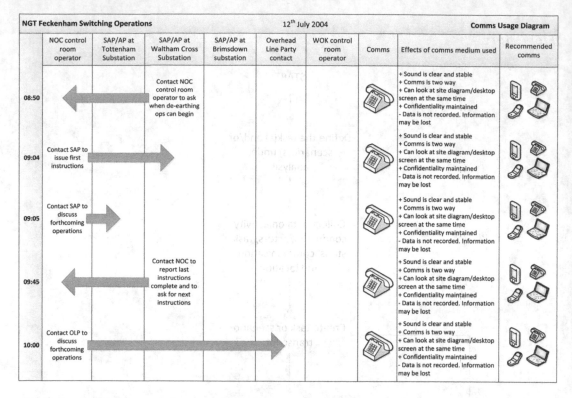

Figure 9.1   CUD for energy distribution task (Salmon et al., 2008)

Tools Needed

A CUD analysis requires the various tools associated with the data collection techniques adopted. For example, an observation of the task under analysis would require video and/or audio recording equipment. An appropriate drawing software package is also required for the construction of the CUD, such as Microsoft Visio. Alternatively, the CUD can be constructed in Microsoft Word.

Coordination Demand Analysis (CDA)

Background and Applications

CCDA is used to rate the coordination between actors involved in teamwork or collaborative activity. It uses the taxonomy of teamwork-related behaviours presented in Table 9.3. The CDA procedure involves identifying the teamwork-based activity involved in the task or scenario under analysis and then providing ratings, on a scale of 1 (low) to 3 (high), for each behaviour from the teamwork taxonomy for each of the teamwork task steps involved. From the individual ratings, a total coordination figure for each teamwork task step and a total coordination figure for the overall task is derived.

Domain of Application

The CDA technique is generic and can be applied to any task that involves teamwork or collaboration.

Table 9.3   A teamwork taxonomy

Coordination dimension	Definition
Communication	Includes sending, receiving and acknowledging information among crew members
Situation awareness (SA)	Refers to identifying the source and nature of problems, maintaining an accurate perception of the aircraft's location relative to the external environment, and detecting situations that require action
Decision-making (DM)	Includes identifying possible solutions to problems, evaluating the consequences of each alternative, selecting the best alternative and gathering information needed prior to arriving at a decision
Mission analysis (MA)	Includes monitoring, allocating and coordinating the resources of the crew and aircraft; prioritising tasks; setting goals and developing plans to accomplish the goals; creating contingency plans
Leadership	Refers to directing activities of others, monitoring and assessing the performance of crew members, motivating members and communicating mission requirements
Adaptability	Refers to the ability to alter one's course of action as necessary, maintaining constructive behaviour under pressure and adapting to internal or external changes
Assertiveness	Refers to the willingness to make decisions, demonstrating initiative and maintaining one's position until convinced otherwise by facts
Total coordination	Refers to the overall need for interaction and coordination among crew members

Source: Burke, 2005.

Procedure and Advice

Step 1: Define the Task or Scenario under Analysis
The first step in a CDA is to define the task or scenario that will be analysed. This is dependent upon the focus of the analysis. It is recommended that if team coordination in a particular type of system (e.g. command and control) is under investigation, then a set of scenarios that are representative of all aspects of team performance in the system under analysis should be used. If time and financial constraints do not allow this, then a task that is as representative as possible of team performance in the system under analysis should be used.

Step 2: Select Appropriate Teamwork Taxonomy
Once the task (or tasks) under analysis is defined, an appropriate teamwork taxonomy should be selected. Again, this may depend upon the purpose of the analysis. However, it is recommended that the taxonomy used covers all aspects of teamwork in the task under investigation. A generic CDA teamwork taxonomy is presented in Table 9.3.

Step 3: Data Collection Phase
The next step involves collecting the data that will be used to inform the CDA. Typically, observational study of the task or scenario under analysis is used as the primary data source for a CDA. It is recommended that specific data regarding the task under analysis should be collected during this process, including information regarding each task step, each team member's role and all communications made. It is also recommended that particular attention is given to the teamwork activity involved in the task under investigation. Further, it is recommended that video and audio recording equipment are used to record any observations or interviews conducted during this process.

Step 4: Conduct an HTA for the Task under Analysis
Once sufficient data regarding the task under analysis has been collected, an HTA should be conducted.

Step 5: Construct CDA Rating Sheet
Once an HTA for the task under analysis is completed, a CDA rating sheet should be created. This should include a column containing each bottom-level task step as identified by the HTA. The teamwork

behaviours from the taxonomy should run across the top of the table. An extract of a CDA rating sheet is presented in Table 9.4.

Step 6: Taskwork/Teamwork Classification

Only those task steps that involve teamwork are rated for the level of coordination between the actors involved. The next step of the CDA procedure involves the identification of teamwork and taskwork task steps involved in the scenario under analysis. Those task steps that are conducted by individual actors involving no collaboration are classified as taskwork, whilst those task steps that are conducted collaboratively and involving more than one actor are classified as teamwork.

Step 7: SME Rating Phase

Appropriate SMEs should then rate the extent to which each teamwork behaviour is required during the completion of each teamwork task step. This involves presenting the task step in question and discussing the role of each of the teamwork behaviours from the taxonomy in the completion of the task step. An appropriate rating scale should be used, e.g. low (1), medium (2) and high (3).

Step 8: Calculate Summary Statistics

Once all of the teamwork task steps have been rated according to the teamwork taxonomy, the final step is to calculate appropriate summary statistics. In its present usage, a total coordination value and mean coordination value for each teamwork task step are calculated. The mean coordination is simply an average of the ratings for the teamwork behaviours for the task step in question. A mean overall coordination value for the entire scenario is also calculated.

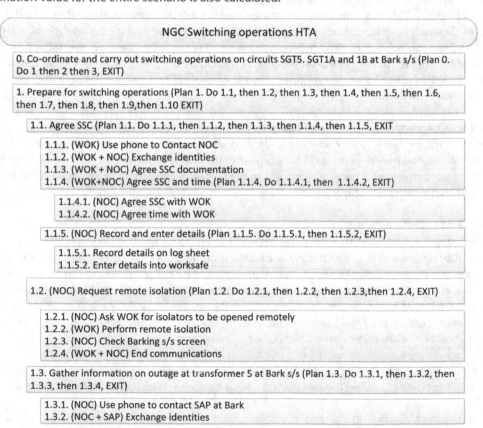

NGC Switching operations HTA

0. Co-ordinate and carry out switching operations on circuits SGT5. SGT1A and 1B at Bark s/s (Plan 0. Do 1 then 2 then 3, EXIT)

1. Prepare for switching operations (Plan 1. Do 1.1, then 1.2, then 1.3, then 1.4, then 1.5, then 1.6, then 1.7, then 1.8, then 1.9,then 1.10 EXIT)

1.1. Agree SSC (Plan 1.1. Do 1.1.1, then 1.1.2, then 1.1.3, then 1.1.4, then 1.1.5, EXIT

1.1.1. (WOK) Use phone to Contact NOC
1.1.2. (WOK + NOC) Exchange identities
1.1.3. (WOK + NOC) Agree SSC documentation
1.1.4. (WOK+NOC) Agree SSC and time (Plan 1.1.4. Do 1.1.4.1, then 1.1.4.2, EXIT)

1.1.4.1. (NOC) Agree SSC with WOK
1.1.4.2. (NOC) Agree time with WOK

1.1.5. (NOC) Record and enter details (Plan 1.1.5. Do 1.1.5.1, then 1.1.5.2, EXIT)

1.1.5.1. Record details on log sheet
1.1.5.2. Enter details into worksafe

1.2. (NOC) Request remote isolation (Plan 1.2. Do 1.2.1, then 1.2.2, then 1.2.3,then 1.2.4, EXIT)

1.2.1. (NOC) Ask WOK for isolators to be opened remotely
1.2.2. (WOK) Perform remote isolation
1.2.3. (NOC) Check Barking s/s screen
1.2.4. (WOK + NOC) End communications

1.3. Gather information on outage at transformer 5 at Bark s/s (Plan 1.3. Do 1.3.1, then 1.3.2, then 1.3.3, then 1.3.4, EXIT)

1.3.1. (NOC) Use phone to contact SAP at Bark
1.3.2. (NOC + SAP) Exchange identities

Figure 9.2   Extract of an HTA for the NGT switching scenario

Table 9.4   Extract of a CDA rating sheet

Task step	Agent	Step no.	Taskwork	Teamwork	Comm	SA	DM	MA	Lead	Ad	Ass	TOT CO-ORD mode	TOT CO-ORD mean
1.1.1	WOK control room operator	Use phone to contact NOC	1										
1.1.2	WOK control room operator	Exchange identities		1	3	3	1	1	1	1	1	1.00	1.57
	NOC control room operator												
1.1.3	WOK control room operator	Agree SSC documentation		1	3	3	3	1	1	1	1	1.00	1.86
	NOC control room operator												
1.1.4.1	NOC control room operator	Agree SSC with WOK		1	3	3	3	1	1	1	1	1.00	1.86
1.1.4.2	NOC control room operator	Agree time with WOK		1	3	3	3	1	1	1	1	1.00	1.86
1.1.5.1	NOC control room operator	Record details on log sheet	1										
1.1.5.2	NOC control room operator	Enter details into WorkSafe	1										
1.2.1	NOC control room operator	Ask for isolators to be opened remotely		1	3	3	1	2	2	1	1	1.00	1.86
1.2.2	WOK control room operator	Perform remote isolation	1										
1.2.3	NOC control room operator	Check barking s/s screen	1										
1.2.4	WOK control room operator	End communications		1	3	1	1	1	1	1	1	1.00	1.29
	NOC control room operator												
1.3.1	NOC control room operator	Use phone to contact SAP at Barking	1										
1.3.2	NOC control room operator	Exchange identities		1	3	3	1	1	1	1	1	1.00	1.57
	SAP at Barking												

Source: Salmon et al., 2008.

Table 9.5   CDA results

Category	Result
Total task steps	314
Total taskwork	114 (36%)
Total teamwork	200 (64%)
Mean total coordination	1.57
Modal total coordination	1.00
Minimum coordination	1.00
Maximum coordination	2.14

Source: Salmon et al., 2008.

Example

The CDA technique has recently been used as part of the EAST (Baber and Stanton, 2004) framework for the analysis of C4i activity in the fire service, naval warfare, aviation, energy distribution (Salmon et al., 2005), ATC and rail (Walker et al., 2006) domains. The following example is an extract of a CDA analysis of an energy distribution task. The task involved the switching out of three circuits at a high-voltage electricity substation. Observational data from the substation and the remote control centre was used to derive an HTA of the switching scenario. Each bottom-level task step in the HTA was then defined by the analyst as either taskwork or teamwork. Each teamwork task was then rated using the CDA taxonomy on a scale of 1 (low) to 3 (high). Over the previous few pages, an extract of the HTA for the task is presented in Figure 9.2, an extract of the CDA rating sheet is presented in Table 9.4 and on the right, the overall CDA results are presented in Table 9.5.

The CDA indicated that of the 314 individual task steps involved in the switching scenario, 64 per cent were classified as teamwork-related and 36 per cent were conducted individually. A mean total coordination figure of 1.57 (out of 3) was calculated for the teamwork task steps involved in the switching scenario. This represents a medium level of coordination between the actors involved.

Advantages

- The output of a CDA is very useful, offering an insight into the use of teamwork behaviours and also a rating of coordination between actors in a particular network or team.
- Coordination can be compared across scenarios, different teams and different domains.
- CDA is particularly useful for the analysis of C4i activity.
- The teamwork taxonomy presented by Burke (2004) covers all aspects of team performance and coordination. The taxonomy is also generic, allowing the technique to be used in any domain without modification.
- Providing the appropriate SMEs are available, it is simple to apply and requires only minimal training.
- The taskwork/teamwork classification of the task steps involved is also useful.
- It provides a breakdown of team performance in terms of task steps and the level of coordination required.
- It is generic and can be applied to teamwork scenarios in any domain.

Disadvantages

- The CDA rating procedure is time-consuming and laborious. The initial data collection phase and the creation of an HTA for the task under analysis also add further time to the analysis.
- For the technique to be used properly, the appropriate SMEs are required. It may be difficult to gain sufficient access to SMEs for the required period of time.
- Intra-analyst and inter-analyst reliability is questionable. Different SMEs may offer different teamwork ratings for the same task (intra-analyst reliability), whilst SMEs may provide different ratings on different occasions.

Flowchart

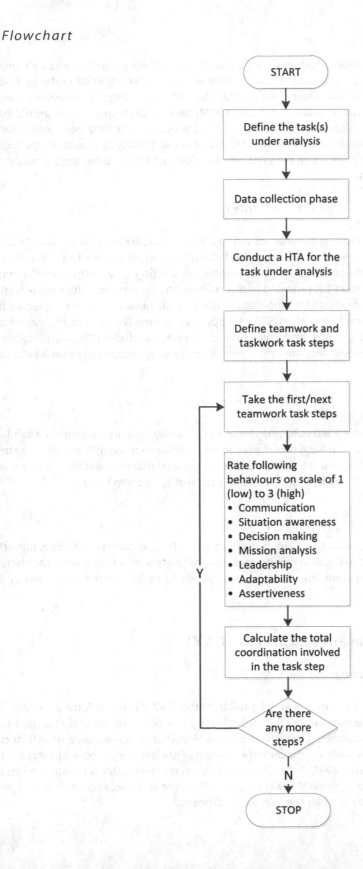

Related Methods

In conducting a CDA analysis, a number of other HF techniques are used. Data regarding the task under analysis is typically collected using observational study and interviews. An HTA for the task under analysis is normally conducted, the output of which feeds into the CDA. A Likert-style rating scale is also normally used during the team behaviour rating procedure. Burke (2004) also suggests that a CDA should be conducted as part of an overall TTA procedure. The CDA technique has also recently been integrated with a number of other techniques (HTA, observation, CUD, social network analysis, operator sequence diagrams and propositional networks) to form the EAST methodology, which has been used to analyse C4i activity in a number of domains.

Approximate Training and Application Times

The training time for the CDA technique is minimal, requiring only that the SMEs used understand each of the behaviours specified in the teamwork taxonomy and also the rating procedure. The application time is high, involving observation of the task under analysis, conducting an appropriate HTA and the lengthy ratings procedure. In the CDA provided in the analysis, the ratings procedure alone took approximately four hours. This represents a low application time in itself; however, when coupled with the data collection phase and completion of an HTA, the application time is high. For the example presented, the overall analysis, including data collection, development of the HTA, identification of teamwork and taskwork task steps, and the rating procedure took approximately two weeks to complete.

Reliability and Validity

There is no data regarding the reliability and validity of the technique available in the literature. Certainly, both the intra-analyst and inter-analyst reliability of the technique may be questionable, and this may be dependent upon the type of rating scale used; for example, it is estimated that the reliability may be low when using a scale of 1–10, whilst it may be improved using a scale of 1–3 (low to high).

Tools Needed

During the data collection phase, video (e.g. camcorder) and audio (e.g. recordable mini-disc player) recording equipment are required in order to make a recording of the task or scenario under analysis. Once the data collection phase is complete, the CDA technique can be conducted using pen and paper.

The Decision Requirements Exercise (DRX)

Background and Applications

The DRX (Klinger and Hahn, 2004) technique is an adaptation of the CDM (Klein and Armstrong, 2004) that is used to highlight critical decisions made by a team during task performance and also used to analyse the factors surrounding decisions, e.g. why the decision was made, how it was made, what factors affected the decision, etc. It was originally used during the training of nuclear power control room crews as a debriefing tool (Klinger and Hahn, 2004). Typically, a decision requirements table is constructed and a number of critical decisions are analysed within a group interview-type scenario. According to Klinger and Hahn (2004), the DRX should be used for the following purposes:

- to calibrate a team's understanding of its own objectives;
- to calibrate understanding of roles, functions and the requirements of each team member;
- to highlight any potential barriers to information flow; and
- to facilitate the sharing of knowledge and expertise between team members.

Domain of Application

The DRX technique was originally developed for use in nuclear power control room training procedures. However, it is generic and can be used in any domain.

Procedure and Advice

Step 1: Define the Task under Analysis
The first step in a DRX analysis involves clearly defining the type of task (or tasks) under analysis. This allows the analyst to develop a clear understanding of the task under analysis and also the types of decisions that are likely to be made. It is recommended that an HTA is conducted for the task under analysis. A number of data collection procedures may be used for this purpose, including observational study, interviews and questionnaires.

Step 2: Select Appropriate Decision Probes
It may be useful to select the types of factors surrounding the decisions that are to be analysed before the analysis begins. This is often dependent upon the scope and nature of the analysis. For example, Klinger and Hahn (2004) suggest that difficulty, errors, cues used, factors used in making the decision, information sources used and strategies are all common aspects of decisions that are typically analysed. The chosen factors should be given a column in the decision requirements table and a set of appropriate probes should be created. These probes are used during the DRX analysis in order to elicit the appropriate information regarding the decision under analysis. An example set of probes are presented in step 7 of this procedure.

Step 3: Describe Task and Brief Participants
Once the task (or tasks) is clearly defined and understood, the analyst should gather appropriate information regarding the performance of the task. If a 'real-world' task is being used, then typically observational data is collected (it is recommended that video/audio recording equipment is used to record any observations made). If a training scenario is being used, then a task description of the scenario will suffice. Once the task under analysis has been performed and/or adequately described, the team members involved should be briefed regarding the DRX technique and what is required of them as participants in the study. It may be useful to take them through an example DRX analysis or even perform a pilot run for a small task. Participants should be encouraged to ask questions regarding the use of the technique and their role in the data collection process. Only when all participants fully understand the technique can the analysis proceed to the next step.

Step 4: Construct Decision Requirements Table
The analyst should next gather all of the team members at one location. Using a whiteboard, the analyst should then construct the decision requirements table (Klinger and Hahn, 2004).

Step 5: Determine Critical Decisions
Next, the analyst should 'walk' the team members through the task, asking for any critical decisions that they made. Each critical decision elicited should be recorded. No further discussion regarding the

decisions identified should take place at this stage, and this step should only be used to identify the critical decisions made during the task.

Step 6: Select Appropriate Decisions
Typically, numerous decisions are made during the performance of a team-based task. The analyst should use this step to determine which of the decisions gathered during step 5 are the most critical. According to Klinger and Hahn (2004), four or five decisions are normally selected for further analysis, although the number selected is dependent upon the time constraints imposed on the analysis. Each decision selected should be entered into the decision requirements table.

Step 7: Analyse Selected Decisions
The analyst should take the first decision and begin to analyse the features of the decision using the probes selected during step 2 of the procedure. Participant responses should be recorded in the decision requirements table. A selection of typical DRX probes is presented below

- Why was the decision difficult?
- What is difficult about making this decision?
- What can get in the way when you make this decision?
- What might a less experienced person have trouble with when making this decision?

Common errors
- What errors have you seen people make when addressing this decision?
- What mistakes do less experienced people tend to make in this situation?
- What could have gone wrong (or did go wrong) when making this decision?

Cues and factors
- What cues did you consider when you made this decision?
- What were you thinking about when you made the decision?
- What information did you use to make the decision?
- What made you realise that this decision had to be made?

Strategies
- Is there a strategy you used when you made this decision?
- What are the different strategies that can be used for this kind of decision?
- How did you use various pieces of information when you made this decision?

Information sources
- Where did you get the information that helped you make this decision?
- Where did you look to get the information to help you here?
- What about sources, such as other team members, individuals outside the team, technologies and mechanical indicators, and even tools like maps or diagrams?

Suggested changes
- How could you do this better next time?
- What would need to be changed with the process or the roles of team members to make this decision easier next time?
- What will you pay attention to next time to help you with this decision?

Example

The following example was developed as part of the analysis of C4i activity in the fire service. Observational study of fire service training scenarios was used to collect required data. A hazardous chemical incident was described as part of a fire service-training seminar. Students on a hazardous materials course at the Fire Service Training College participated in the exercise, which consisted of a combination of focus group discussion and paired activity to define appropriate courses of action to deal with a specific incident. The incident involved the report of possible hazardous materials on a remote farm. Additional information was added to the incident as the session progressed, e.g. reports of casualties, problems with labelling on hazardous materials, etc. The exercise was designed to encourage experienced fire-fighters to consider risks arising from hazardous materials and the appropriate courses of action they would need to take, e.g. in terms of protective equipment, incident management, information seeking activity, etc. In order to investigate the potential application of the DRX technique in the analysis of C4i activity, a team DRX was conducted for the hazardous chemical incident, based on the observational data obtained. An extract of the DRX is presented in Table 9.6.

Advantages

- Specific decisions are analysed and recommendations made regarding the achievement of effective decision-making in future similar scenarios.
- The output seems to be very useful for team training purposes.
- The analyst can control the analysis, selecting the decisions that are analysed and also the factors surrounding the decisions that are focused upon.
- The DRX can be used to elicit specific information regarding team decision-making in complex systems.
- The incidents which the technique considers have already occurred, removing the need for costly, time-consuming to construct observations or event simulations.
- Real-life incidents are analysed using the DRX, ensuring a more comprehensive, realistic analysis than simulation techniques.

Disadvantages

- The reliability of such a technique is questionable. Klein and Armstrong (2004) suggest that methods that analyse retrospective incidents are associated with concerns of data reliability, due to evidence of memory degradation.
- The DRX may struggle to create an exact description of an incident.
- It is a resource-intensive technique, typically incurring a high application time.
- A high level of expertise and training is required in order to use it to its maximum effect (Klein and Armstrong, 2004).
- It relies upon interviewee verbal reports in order to reconstruct incidents. How far a verbal report accurately represents the cognitive processes of the decision-maker is questionable. Facts could be easily misrepresented by the participants and glorification of events can potentially occur.
- It may be difficult to gain sole access to team members for the required period of time.
- After-the-fact data collection has a number of concerns associated with it, including memory degradation, and a correlation with task performance.

Table 9.6   Extract of the DRX for a hazardous chemicals incident

Decision	What did you find difficult when making this decision?	What cues did you consider when making this decision?	Which information sources did you use when making this decision?	Were any errors made whilst making this decision?	How could you make a decision more efficiently next time?
Level of protection required when conducting search activity	- The level of protection required is dependent upon the nature of the chemical hazard within the farmhouse. This was unknown at the time - There was also significant pressure from the hospital for a positive ID of the substance	- Urgency of diagnosis required by the hospital - Symptoms exhibited by the child in hospital - Time required to get into full protection suits	- Correspondence with hospital personnel. - Police officer - Fire control	- Initial insistence upon full suit protection before identification of chemical type	- Diagnose chemical type prior to arrival, through communication with farmhouse owner - Consider urgency of chemical diagnosis as critical
Determine type of chemical substance found and relay information to hospital	- The chemical label identified substance as a liquid, but the substance was in powder form	- Chemical drum labels - Chemical form, e.g. powder, liquid - Chemdata information - Chemsafe data	- Chemical drum - Chemdata database - Fire control (Chemsafe database)	- Initial chemical diagnosis made prior to confirmation with Chemdata and Chemsafe databases	- Use Chemdata and Chemsafe resources prior to diagnosis - Contact farmhouse owner en route to farmhouse

Flowchart

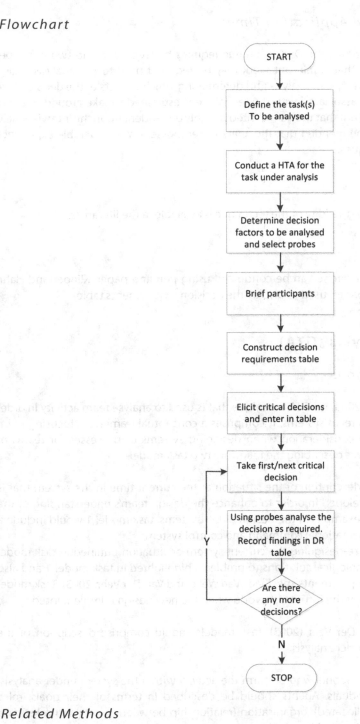

Related Methods

The DRX is an adaptation of the CDM technique (Klein and Armstrong, 2004) for use in the analysis of team performance. It uses a group interview or focus group-type approach to analyse critical decisions made during task performance. Task analysis techniques (such as HTA) may also be used in the initial process of task definition.

Approximate Training and Application Times

According to Klinger and Hahn (2004), the DRX technique requires between one and two hours per scenario. However, it is apparent that significant work may be required prior to the analysis phase, including observation, task definition, task analysis and determining which aspects of the decisions are to be analysed. The training time associated with the technique is estimated to take around one day. However, it is worthwhile pointing out that the data elicited is highly dependent upon the interview skills of the analyst. Therefore, it is recommended that the analyst used possesses considerable experience and skill in interview-type techniques.

Reliability and Validity

No data regarding the reliability and validity of the technique is available in the literature.

Tools Needed

The team decision requirements exercise can be conducted using pen and paper. Klinger and Hahn (2004) recommend that a whiteboard is used to display the decision requirements table.

Groupware Task Analysis (GTA)

Background and Applications

GTA (Van Welie and Van Der Veer, 2003) is a TTA technique that is used to analyse team activity in order to inform the design and analysis team systems. It comprises a conceptual framework focusing upon the relevant aspects that require consideration when designing systems or processes for teams or organisation. The technique involves describing the following two task models:

*   Task model 1: this offers a description of the situation at the current time in the system that is being designed. This is developed in order to enhance the design team's understanding of the current work situation. For example, in the design of C4i systems, task model 1 would include a description of the current operational command and control system.
*   Task model 2: this involves re-designing the current system or situation outlined in task model 1. This should include technological solutions to problems highlighted in task model 1 and also technological answers to requirements specified (Van Welie and Van Der Veer, 2003). Task model 2 should represent a model of the future task world when the new design is implemented.

According to Van Welie and Van Der Veer (2003), task models should comprise description of the following features of the system under analysis:

*   Agents: this refers to the personnel who perform the activity within the system under analysis, including teams and individuals. Agents should be described in terms of their goals, roles (which tasks the agent is allocated), organisation (relationship between agents and roles) and characteristics (agent experience, skills, etc.).
*   Work: the task or tasks under analysis should also be described, including unit and basic task specification (Card, Moran and Newell, 1983). It is recommended that an HTA is used for this aspect of task model 1. Events (triggering conditions for tasks) should also be described.
*   Situation: the situation description should include a description of the environment and any objects in the environment.

The techniques used when conducting a GTA are determined by the available resources. For guidelines on which techniques to employ, see Van Welie and Van Der Veer (2003). Once the two task models are completed, the design of the new system can begin, including specification of functionality and also the way in which the system is presented to the user. According to Van Welie and Van Der Veer, the task model can be used to answer the following design questions:

- What are the critical tasks?
- Are they always performed by the same user?
- Which roles do they have?
- Which tasks should be possible to undo?
- Which errors can be expected?
- How can prevention be effective?
- How frequently are those tasks performed?
- Which types of user are there?
- Which tasks belong to which roles?
- Which tasks have effects that cannot be undone?
- What are the error consequences for users?

Domain of Application

Generic.

Procedure and Advice

Step 1: Define the System under Analysis
The first step in a GTA is to define the system (or systems) under analysis. For example, in the design of C4i systems, existing command and control systems would be analysed, including railway, ATC, security and gas network command and control systems.

Step 2: Data Collection Phase
Before task model 1 can be constructed, specific data regarding the existing systems under analysis should be collected. Traditional techniques should be used during this process, including observational analysis, interviews and questionnaires. The data collected should be as comprehensive as possible, including information regarding the task (specific task steps, procedures, interfaces used, etc.), the personnel (roles, experience, skills, etc.) and the environment.

Step 3: Construct Task Model 1
Once sufficient data regarding the system or type of system under analysis has been collected, task model 1 should be constructed. This should completely describe the situation as it currently stands, including the agents, work and situation categories outlined above.

Step 4: Construct Task Model 2
The next stage of the GTA is to construct task model 2. This model involves re-designing the current system or situation outlined in task model 1. The procedure used for constructing task model 2 is determined by the design teams, but may include focus groups, scenarios and brainstorming sessions.

Step 5: Re-design the System
Once task model 2 has been constructed, the system re-design should begin. Obviously, this procedure is dependent upon the system under analysis and the design team involved. The reader is referred to Van Welie and Van Der Veer (2003) for guidelines.

Advantages

- GTA output provides a detailed description of the system requirements and highlights specific issues that need to be addressed in the new design.

- Task model 2 can potentially highlight the technologies required and their availability.
- GTA provides the design team with a detailed understanding of the current situation and problems.
- It seems to be suited to the analysis of existing command and control systems.
- It is able to cope with multiple users in complex settings (Goschnick, Balbo and Sonenberg, 2008).

Disadvantages

- GTA appears to be extremely resource-intensive and time-consuming in its application.
- There is limited evidence of its use in the literature.
- It provides limited guidance for its application.
- A large team of analysts would be required in order to conduct a GTA analysis.

Flowchart

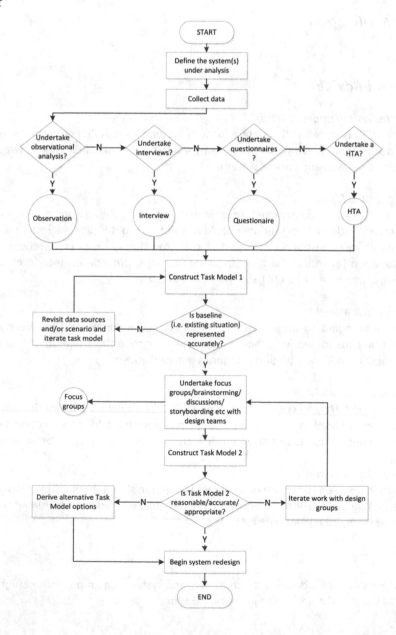

Related Methods

GTA analysis is a team task analysis technique and so is related to CUD, SNA and TTA. The data collection phase may involve the use of a number of approaches, including observational study interviews, surveys, questionnaires and an HTA.

Approximate Training and Application Times

It estimated that the training and application times for the GTA technique would be very high.

Reliability and Validity

There is no data regarding the reliability and validity of the GTA technique available in the literature.

Tools Needed

Once the initial data collection phase is complete, GTA can be conducted using pen and paper. The data collection phase would require video and audio recording devices and a computer. Caffiau et al. (2010) state that there are two software tools developed to aid in GTA: the GTA tool (which they stress is no longer supported) and Euterpe.

HTA for Teams (HTA(T))

Background and Applications

Traditionally, task analysts have used HTA to describe the goals of individual workers, but Annett and others have argued that HTA can provide sub-goal hierarchies at many levels within a system. The analyst can choose to focus on the human agents, machine agents or the entire system. Annett (2004) shows how an adaptation of HTA can produce an analysis of team-based activity (HTA(T)). The method is based on the notion that a goal is the same when achieved by an individual or a team, but that the processes through which this goal is achieved differ within a team context (Annett, 1997).

According to Annett, Cunningham and Mathias-Jones (2000), building upon the traditional HTA analysis, HTA(T) uses a model of teamwork to explore the teamwork processes required to meet goals within a team context. This model focuses on three core aspects of teamwork: communication, coordination and cooperation (Annett, 1997; Annett, Cunningham and Mathias-Jones, 2000). It involves identifying the goals at a team level and then distinguishing which individual team members must work together to achieve these goals. Each task step is allocated a teamwork dimension in order to represent the team processes required to achieve it.

In addition to providing a representation of goals, plans and tasks, HTA(T) provides an exploration of the team processes required to achieve these within a team context. Smith et al. (2008) employed HTA(T) to explore the communication and coordination strategies employed by ad hoc crime scene examination teams in the UK. The authors concluded that the method enabled insights into the different investigative practices employed by these teams and a comparison of these practices to formal descriptions of work. The insights gained from the study enabled a series of recommendations to be derived regarding the development of technologies to assist decision-making in ad hoc crime scene examination teams.

Domain of Application

Generic.

Procedure and Advice

The basic heuristics for conducting an HTA are as follows (Stanton, 2005a).

Step 1: Define the Purpose of the Analysis
Although the case has been made that HTA can be all things to all people, the level or re-description and the associated information collected might vary depending upon the purpose. Examples of different purposes for HTA would include system design, analysis of workload and manning levels, and training design. The name, contact details and a brief biography of the analyst should also be recorded. This will enable future analysts to check with the HTA originator if they plan to re-use or adapt the HTA.

Step 2: Define the Boundaries of the System Description
Depending upon the purpose, the system boundaries may vary. If the purpose of the analysis is to analyse coordination and communication in teamwork, then the entire set of tasks of a team of people would be analysed. If the purpose of the analysis is to determine the allocation of system function to humans and computers, then the whole system will need to be analysed.

Step 3: Try to Access a Variety of Sources of Information about the System to be Analysed
All task analysis guides stress the importance of multiple sources of information to guide, check and validate the accuracy of the HTA. Sources such as observation, SMEs, interviews, operating manuals, walkthroughs and simulations can all be used as a means of checking the reliability and validity of the analysis. Careful documentation and recording of the sources of data needs to be archived so that the analyst or others may refer back and check if they need to.

Step 4: Describe the System Goals and Sub-goals
As proposed in the original principles for HTA, the overall aim of the analysis is to derive a sub-goal hierarchy for the tasks under scrutiny. As goals are broken down and new operations emerge, sub-goals for each of the operations need to be identified. As originally specified, it is not the operations that are being described, but their sub-goals. All of the lower level sub-goals are a logical expansion of the higher ones. A formal specification for the statement of each of the sub-goals can be derived, although most analyses do not go such lengths.

Step 5: Try to Keep the Number of Immediate Sub-goals under any Super-ordinate Goal to a Small Number (i.e. between 3 and 10)
There is an art to HTA, which requires that the analysis does not turn into a procedural list of operations. The goal hierarchy is determined by looking for clusters of operations that belong together under the same goal. This normally involves several iterations of the analysis. Whilst it is accepted that there are bound to be exceptions, for most HTAs, any super-ordinate goal with have 3–10 immediate subordinates. It is generally good practice to continually review the sub-goal groupings in order to check if they are logical. HTA does not permit single subordinate goals.

Step 6: Link Goals to Sub-goals and Describe the Conditions under which Sub-goals are Triggered
Plans are the control structures that enable the analyst to capture the conditions which trigger the sub-goals under any super-ordinate goal. They are read from the top of the hierarchy down to the sub-goals that are triggered and back up the hierarchy again as the exit conditions are met. As each of the sub-goals (and the plans that trigger them) are contained within higher goals (and higher plans), the considerable complexity of tasks within systems can be analysed and described. The plans contain the

context under which particular sub-goals are triggered. This context might include time, environmental conditions, completion of other sub-goals, system state, receipt of information and so on. For each goal, the analyst has to question how each of its immediate subordinates is triggered. As well as identifying the sub-goal trigger conditions, it is also important to identify the exit condition for the plan that will enable the analyst to trace his or her way back up the sub-goal hierarchy to avoid being stuck in a control loop with no obvious means of exiting.

Step 7: Stop Re-describing the Sub-goals When You Judge the Analysis is Fit for Purpose

When to stop the analysis has been identified as one of the more conceptually troublesome aspects of HTA. The proposed P x C (probability versus cost) stopping rule is a rough heuristic, but analysts may have trouble quantifying the estimates of P and C. The level of description is likely to be highly dependent upon the purpose of the analysis, so it is conceivable that a stopping rule could be generated at this point in the analysis. For example, in analysing teamwork, the analysis could stop at the point where sub-goals dealt with the exchange of information (e.g. receiving, analysing and sending information from one agent to another). For practical purposes, the stopping point of the analysis is indicated by underlining the lowest-level sub-goal in the hierarchical diagram or ending the sub-goal description with a double forward slash (i.e. '//') in the hierarchical list and tabular format. This communicates to the reader that the sub-goal is not re-described further elsewhere in the document.

Step 8: Attribute Agents to Goals

When the HTA is complete, using a tabular format as shown in Table 9.7 below, the goal hierarchy in the left-hand column should be listed out, then the goals should be broken down into a goal statement, associated plan and criterion for successful task completion in the right-hand column. The analyst must decide at this point which goals are related to teamwork and what goals rely only on 'taskwork'. This format should be used to systematically attribute the agent to the teamwork-related goals expressed in the left-hand column.

Step 9: Try to Verify the Analysis with SMEs

It is important to check the HTA with SMEs. This can both help to verify the completeness of the analysis and help the experts develop a sense of ownership of the analysis.

Step 10: Be Prepared to Revise the Analysis

HTA requires a flexible approach to achieve the final sub-goal hierarchy with plans and notes. The first pass analysis is never going to be sufficiently well developed to be acceptable, no matter what the purpose. The number of revisions will depend upon the time available and the extent of the analysis, but simple analyses (such as the analysis of the goals of extracting cash from an automatic cash machine) may require at least three interactions, whereas more complex analyses (such as the analysis of the emergency services responding to a hazardous chemical incident) might require at least 10 iterations. It is useful to think of the analysis as a working document that only exists in the latest state of revision. Careful documentation of the analysis will mean that it can be modified and re-used by other analysts as required.

Advantages

- Smith et al. (2008) argue that the method allows a comparison of team performance with formal rules.
- HTA(T) elicits useful data on the individuals and wider system (Smith et al., 2008).

Disadvantages

• The method requires a high degree of direct access to participants (Smith et al., 2008).

Related Methods

HTA for teams is one of a number of adaptations of HTA, which include methods for the analysis of investigative design decisions, the analysis of human machine interactions, the prediction of error, allocation of function and the assessment of interface design.

Approximate Training and Application Times

Stanton and Young (1999a) report that the training and application time for HTA is substantial. The application time associated with it is dependent upon the size and complexity of the task under analysis. For large, complex tasks, the application time for HTA would be high. As HTA for teams involves a considerable amount of additional analysis on top of the standard HTA procedure, it can be assumed that both the training and application time of the method are considerable.

The application of HTA(T) to ad hoc crime scene teams was estimated to involve three days for initial development of the HTA and a further day for comparison to formal procedures (Stanton, Baber and Harris, 2008).

Reliability and Validity

There is no data regarding the reliability and validity of HTA used for TTA purposes available in the literature. That said, however, SMEs have commented favourably on the ecological validity of the method and representation.

Tools Needed

HTA can be carried out using only pencil and paper, although there are software tools, such as those developed by the HFI-DTC and others, which can make the processes of developing, editing and re-using the goal and plan structure less laborious.

Example

The HTA(T) was based on the analysis of the emergency services responses to a hazardous chemical incident. In the scenario analysed, some youths had broken into a farm and disturbed some chemicals in sacking. One of the youths had been taken to the hospital with respiratory problems, whilst the others were still at the scene. The police were sent to investigate the break-in at the farm. They called in the fire service to identify the chemical and clean up the spillage.

The overall analysis shows four main sub-goals: receive notification of an incident, gather information about the incident, deal with the chemical incident and resolve the incident. Only part of the analysis is presented in order to illustrate HTA(T). As multiple agencies and people are involved in the team task, they have been identified under each of the sub-goals. Police control, fire control, the hospital and the police officer have all been assigned to different sub-goals.

The overview of the HTA for teams is presented in Figure 9.3. Only some of these goals are further re-described in Table 9.7, as they are the ones involving teamwork. Any goals that do not involve teamwork do not have to be entered into the table.

Table 9.7   Tabular form of selected team work operations

1. Deal with chemical incident. Plan: Wait until 1, then do 2 then do 3 – If [hazard], then 4, then exit – otherwise exit.	Goal: deal with safely with the chemical incident. Teamwork: this is a multi-agency task involving the police and the fire service as well as the hospital with a possible casualty. Plan: determine nature of incident and then call in appropriate agencies, avoid any further casualties. Criterion measure: chemical incident cleared up with no further injuries.
2. [Police control] gather information about incident. Plan 2: Do 2.1 at any time if appropriate Do 2.2, then 2.3. Then exit.	Goal: gather information about the nature of the incident. Teamwork: to decide who to send to the site and gather information and liaise with other agencies as necessary. Plan: send requests to other agencies for information and send a patrol out to the site to search the scene for physical evidence and suspects. Criterion measure: appropriate response with minimal delay. A hospital may call in about a casualty at any time, but it has to be linked with this incident. The police officer has to find his/her way to the scene of the incident.
2.2. [Police control] get a police officer to search scene of incident. Plan 2.2: do 2.1.1, then 2.2.2, then 2.2.3. Until [suspects] or [hazards] then exit.	Goal: to get the office to search the scene of the incident for evidence of the hazard or suspects. Teamwork: police control has to direct the office to the hazard and provide details about the incident. If police control receives information about the incident from other agencies, this information needs to be passed on to the office at the scene. Plan: once at the scene of the incident, the office needs to search for hazards and suspects. Criterion measure: the police officer may have to find a remote location based on sketchy information. The police officer has to search for signs of a break-in and hazards.
2.3. [Police control] get police officer to report nature of incident. Plan 2.3: If [suspects], then 2.3.1. If [suspects], then 2.3.2, then 2.3.3. Then 2.3.4, then exit. Otherwise exit.	Goals: detailed information on the nature of the incident and the degree of potential hazard present and report this information to police control. Teamwork: incident details need to be passed on so that the clean-up operation can begin. Plan: if the officer at the scene identifies a hazard, he/she has to report it to police control; if he/she identifies a suspect, he/she has to interview the suspect and report the results to police control. Criterion measure: any potential hazard needs to be identified, including the chemical ID number. Any suspects on the scene need to be identified. Suspects need to be questioned about the incident.

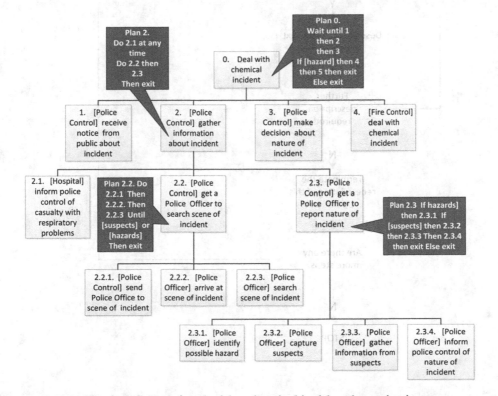

Figure 9.3   HTA(T) of goals associated with a chemical incident investigation

Flowchart

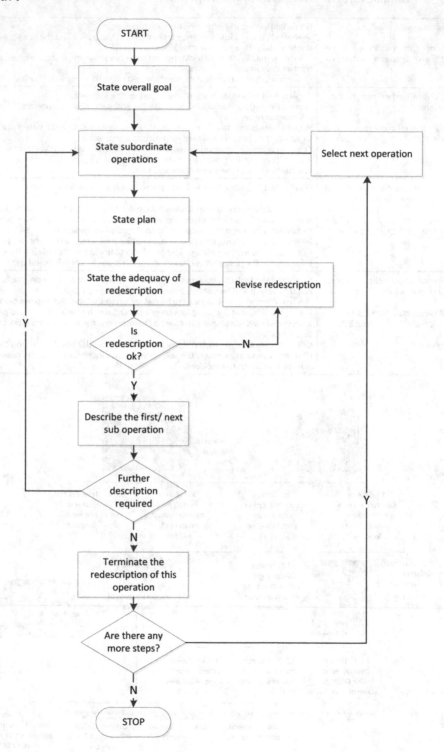

Team Cognitive Task Analysis (TCTA)

Background and Applications

TCTA (Klein, 2000) is used to describe the cognitive skills and processes that a team or group of actors employ during task performance. It uses semi-structured interviews and pre-defined probes to elicit data regarding the cognitive aspects of team performance and decision-making. The approach is based on the CDM technique that is used to analyse the cognitive aspects of individual task performance. According to Klein (2000), the TCTA technique addresses the following team cognitive processes:

- control of attention;
- shared situation awareness;
- shared mental models;
- application of strategies/heuristics to make decisions, solve problems and plan; and
- metacognition.

According to Klein (2000), a TCTA allows the analyst to capture each of the processes outlined above and also to represent the findings to others. TCTA outputs are used to enhance team performance through informing the development and application of team training procedures, the design of teams and also the design and development of team procedures and processes.

Domain of Application

Generic.

Procedure and Advice

Step 1: Specify the Desired Outcome
According to Klein (2000), it is important to specify the desired outcome of the analysis before any data collection is undertaken. The desired outcome is dependent upon the purpose and scope of the analysis effort in question. Typical desired outcomes of TCTA include reducing errors, cutting costs, speeding up reaction times, increasing readiness and reducing team personnel. Other desired outcomes may be functional allocation, task allocation, improved overall performance or testing the effects of a novel design or procedure.

Step 2: Define the Task under Analysis
Once the desired outcome is specified, the task (or tasks) under analysis should be clearly defined and described. This is normally dependent upon the focus of the analysis. For example, it may be that an analysis of team performance in specific emergency scenarios is required. Once the nature of the task is defined, it is recommended that an HTA be conducted. This allows the analyst to gain a deeper understanding of the task under investigation.

Step 3: Observational Study of the Task under Analysis
Observational study and semi-structured interviews are typically used as the primary data collection tools in a TCTA. The task under analysis should be observed and recorded. It is recommended that video and audio recording equipment are used to record the task, and that the analyst involved takes relevant

notes during the observation. Klein (2000) suggests that observers should record any incident related to the five team cognitive processes presented above (control of attention, shared situation awareness, shared mental models, application of strategies/heuristics to make decisions, solve problems and plan, and metacognition). The time of each incident and personnel involved should also be recorded. An observational transcript of the task under analysis should then be created, including a timeline and a description of the activity involved, and any additional notes that may be pertinent.

Step 4: Perform CDM Interviews

The TCTA technique involves the use of CDM-style interviews with the different team members involved. It is recommended that interviews with each team member be conducted. Interviews are used to gather more information regarding the decision-making incidents collected during the observation phase. Using a CDM (Klein and Armstrong, 2004) approach, the interviewee should be probed regarding the critical decisions recorded during the observation. The analyst should ask the participant to describe the incident in detail, referring to the five cognitive processes outline above. CDM probes should also be used to analyse the appropriate incidents. A set of generic CDM probes is presented in Table 9.8. It may be useful to create a set of specific team CTA probes prior to the analysis, although this is not always necessary.

Table 9.8   CDM probes

Goal specification	What were your specific goals at the various decision points?
Cue identification	What features were you looking for when you formulated your decision? How did you know that you needed to make the decision? How did you know when to make the decision?
Expectancy	Were you expecting to make this sort of decision during the course of the event? Describe how this affected your decision-making process.
Conceptual	Are there any situations in which your decision would have turned out differently? Describe the nature of these situations and the characteristics that would have changed the outcome of your decision.
Influence of uncertainty	At any stage, were you uncertain about either the reliability of the relevance of the information that you had available? At any stage, were you uncertain about the appropriateness of the decision?
Information integration	What was the most important piece of information that you used to formulate the decision?
Situation Awareness	What information did you have available to you at the time of the decision?
Situation assessment	Did you use all of the information available to you when formulating the decision? Was there any additional information that you might have used to assist in the formulation of the decision?
Options	Were there any other alternatives available to you other than the decision you made?
Decision blocking – stress	Was there any stage during the decision-making process in which you found it difficult to process and integrate the information available? Describe precisely the nature of the situation.
Basis of choice	Do you think that you could develop a rule, based on your experience, which could assist another person to make the same decision successfully? Why/why not?
Analogy/generalisation	Were you at any time reminded of previous experiences in which a similar decision was made? Were you at any time reminded of previous experiences in which a different decision was made?

Source: O'Hare et al., 2000.

Step 5: Record Decision Requirements
The key decision requirements involved in each incident should be determined and recorded. In a study focusing on Marine Corps command posts, Klein et al. (1996) reported 40 decision requirements that included critical decisions, reasons for difficulty, common errors and cues/strategies for effective decision-making. Klinger and Hahn (2004) describe an approach to the analysis of team decision requirements. The categories proposed include why the decision was difficult, common errors made when making the decision, environmental cues used when making the decision, factors known prior to the decision, strategies and information sources used when addressing the decision and recommendations for better decision-making.

Step 6: Identify Decision-Making Barriers
The next step involves identifying any barriers to effective decision-making that were evident during the incident under analysis. Barriers to decision-making may include the use of inappropriate technology, poor communication, mismanagement of information, etc. Each barrier identified should be recorded.

Step 7: Create Decision Requirements Table
A decision requirements table should be created, detailing each critical decision, its associated decision requirements and strategies for effective decision-making in similar scenarios. An extract of a decision requirements table is presented in the example section below.

Advantages

* TCTA can be used to elicit specific information regarding team decision-making in complex systems.
* Its output can be used to inform teams of effective decision-making strategies.
* Decision-making barriers identified can be removed from the system of process under analysis, facilitating improved team performance.
* The incidents that the technique analyses have already occurred, removing the need for costly, time-consuming to construct event simulations.
* Once the analyst is familiar with the technique, it is easy to apply.
* CDM has been used extensively in a number of domains and has the potential to be used anywhere.
* Real-life incidents are analysed using TCTA, ensuring a more comprehensive, realistic analysis than simulation techniques.
* The cognitive probes used in the CDM have been used for a number of years and are efficient at capturing the decision-making process (Klein and Armstrong, 2004).

Disadvantages

* The reliability of such a technique is questionable. Klein and Armstrong (2004) suggest that methods that analyse retrospective incidents are associated with concerns of data reliability due to evidence of memory degradation.
* The quality of the data collected using such techniques is entirely dependent upon the skill of the interviewer and also the participant involved.
* TCTA is a resource-intensive technique, including observation and interviews, both of which require significant effort.

- A high level of expertise and training is required in order to use TCTA to its maximum effect (Klein and Armstrong, 2004).
- It relies upon interviewee verbal reports in order to reconstruct incidents. The accuracy of verbal reports is questionable and there are various problems associated with such data, including misrepresentation and glorification of facts.
- Collecting subjective data post-task performance also has a number of associated problems, such as memory degradation and a correlation with performance.

Example

A study of Marine Corps command posts was conducted by Klein et al. (1996) as part of an exercise to improve the decision-making process in command posts. Three data collection phases were used during the exercise. First, four regimental exercises were observed and any decision-making-related incidents were recorded. As a result, over 200 critical decision-making incidents were recorded. Second, interviews with command post personnel were conducted in order to gather more specific information regarding the incidents recorded during the observation. Third, a simulated decision-making scenario was used to test participant responses. Klein et al. (1996) present 40 decision requirements, including details regarding the decision, reasons for difficulty in making the decision, errors and cues, and strategies used for effective decision-making. The decision requirements were categorised into the following groups: building and maintaining situational awareness, managing information and deciding on a plan. Furthermore, a list of 30 'barriers' to effective decision-making were also presented. A summary of the barriers identified is presented in Table 9.9.

Table 9.9   Summary of decision-making barriers (adapted from Klein, 2000)

Decision requirements category	Barriers
Building and maintaining SA	Information presented on separate map-boards Map-boards separated by location, furniture and personnel System of overlays archaic and cumbersome Over-reliance upon memory whilst switching between maps Erroneous communication
Managing information	Sending irrelevant messages Inexperienced personnel used to route information Commanders critical information requirements (CCIR) concept misapplied
Deciding on a plan	Communication systems unreliable Too many personnel to coordinate information with

From the simulated decision-making exercise, it was found that the experienced personnel (colonels and lieutenant colonels) required only 5–10 minutes to understand a situation. However, majors took over 45 minutes to study and understand the same situation (Klein et al., 1996). In conclusion, Klein et al. reported that there were too many personnel in the command post, which made it more difficult to complete the job in hand. They suggested that reduced staffing at the command posts would contribute to speed and quality improvements in the decisions made.

Related Methods

TCTA is based on observational study of the task under analysis and also semi-structured interview data derived from interviews with the actors involved. The semi-structured interview approach adopted in a TCTA is based on the CDM technique (Klein and Armstrong, 2004) that is used for individual CTA purposes. It is also recommended that an initial HTA of the task under analysis be conducted.

Flowchart

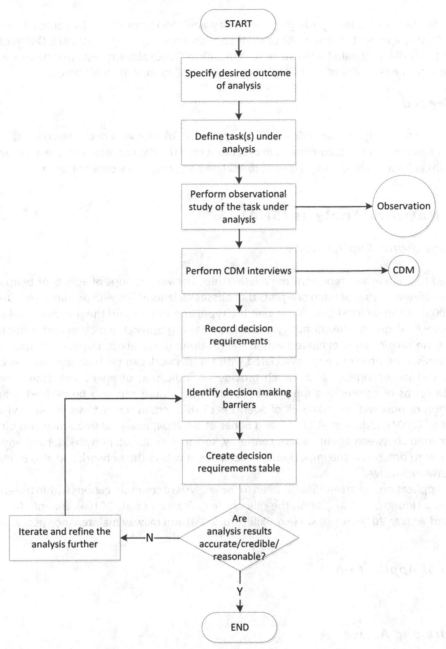

Approximate Training and Application Times

For analysts without experience in the conduct of interviews, the training time associated with TCTA would be high. For those analysts already skilled in interview techniques, the associated training time would be minimal, requiring only that they become familiar with the CDM probes being used. The typical application time for a CDM-type interview is between one and two hours. Since TCTA requires that CDM interviews are conducted with all of the team members involved, it is estimated that the total application time for TCTA would be high.

Reliability and Validity

There are no data available regarding the reliability and validity of the TCTA approach outlined by Klein (2000). It is apparent that the reliability of such an approach is questionable. Different analysts might elicit very different data for the same scenario. Klein (2000) also suggests that there are concerns associated with the reliability of the CDM due to evidence of memory degradation.

Tools Needed

The observational study and interview components of TCTA require video (camcorders) and audio (mini-disc recorder) recording equipment in order to record the data collected. It is recommended that Microsoft Excel (or a similar package) is used to analyse and present the data obtained.

Social Network Analysis (SNA)

Background and Applications

SNA is used to analyse and represent the relationships between groups of agents or teams. A social network is defined as a 'set or team of agents that possess relationships with one another' (Driskell and Mullen, 2005). It can be used to demonstrate the type, importance and the number of relationships within a specified group. The output typically provides a graphical depiction and a mathematical analysis of the relationships exhibited within the group under analysis. Depending upon the focus of the analysis, a number of facets associated with the network can be analysed, such as centrality, closeness and betweenness, all of which provide an indication of agent importance within the network in terms of communications. A network density figure can also be derived, which gives an indication of how well the network of agents is distributed. In the analysis of C4i environments, Salmon et al. (2004), Walker et al. (2004) and Baber et al. (2004) analysed frequency and direction of communications between agents, agent centrality, sociometric status, network density and network type in order to determine the importance of each agent within the network and also to classify the type of network involved.

Recent applications have seen SNA applied to the analysis of communications within the emergency services (e.g. Houghton et al., 2006), the military (e.g. Stanton et al., 2010a), friendly fire (Rafferty, Stanton and Walker, 2012), terrorism (e.g. Skillicorn, 2004) and railway maintenance (e.g. Walker et al., 2006).

Domain of Application

Generic.

Procedure and Advice

Step 1: Define the Network or Group
The first step in a SNA involves defining the network of agents or group of networks that are to be analysed. For example, in analysing C4i networks, the authors specified a number of different C4i agent networks, including the emergency services (fire and police), the military, civil energy distribution, ATC, railway signalling and naval warfare networks.

Step 2: Define Scenarios
Typically, networks are analysed over a number of different scenarios. Once the type of network under analysis has been defined, the scenarios within which they will be analysed should be defined. For a thorough analysis of the networks involved, it is recommended that a number of different scenarios be analysed. For example, in the analysis of naval warfare C4i activity (Stanton et al., 2006b), the following scenarios were defined and analysed:

- air threat scenario;
- surface threat scenario;
- subsurface threat scenario.

Step 3: Data Collection
Once the network and scenarios under analysis are defined clearly, the data collection phase can begin. This phase typically involves conducting an observational study of the scenarios under analysis. It is recommended that specific data regarding the relationship (e.g. communications) between the agents involved in the scenario is collected. Typically, the frequency, direction and content of any communications between agents in the network are recorded. Additional data collection techniques may also be employed in order to gather supplementary data, such as interviews and questionnaires.

Step 4: Construct Agent Association Matrix
Once sufficient data regarding the scenario under analysis is collected, the data analysis component of the SNA can begin. The first step in this process involves the construction of an agent association matrix. The matrix represents the frequency of associations between each agent within the network. Example matrices of association are presented in Table 9.10 on the right.

Step 5: Construct Social Network Diagram
Once the matrix of association is completed, the social network diagram should be constructed. This depicts each agent in the network and the communications that occurred between them during the scenario under analysis. Within the diagram, communications between agents are represented by directional arrows linking the agents involved, and the frequency of communications is presented in numeric form.

Step 6: Calculate Agent Centrality
Agent centrality is calculated in order to determine the central or key agent (or agents) within the network. There are a number of different centrality calculations that can be made. For example, agent centrality can be calculated using Bavelas-Leavitt's index. The mean centrality + standard deviation can then be used to define key agents within the network. Those agents who possess a centrality figure that exceeds the mean + standard deviation figure are defined as key agents for the scenario under analysis.

Table 9.10   SNA matrices

Chain	A	B	C	D	E
A	-	1	-	-	-
B	1	-	1	-	-
C	-	1	-	1	-
D	-	-	1	-	1
E	-	-	-	1	-

Y	A	B	C	D	E
A	-	-	1	-	-
B	-	1	1	-	-
C	1	-	-	1	-
D	-	-	1	-	1
E	-	-	-	1	-

Start	A	B	C	D	E
A	-	1	1	1	1
B	1	-	-	-	-
C	1	-	-	-	-
D	1	-	-	-	-
E	1	-	-	-	-

Circle	A	B	C	D	E
A	-	1	-	-	-
B	1	-	1	-	-
C	-	1	-	1	-
D	-	-	1	-	1
E	1	-	-	1	-

All-connected	A	B	C	D	E
A	-	1	1	1	1
B	1	-	1	1	1
C	1	1	-	1	1
D	1	1	1	-	1
E	1	1	1	1	-

Chain

Position	Degree	Betweenness	Closeness
A	1	0	0.4
B	2	6	0.6
C	2	8	0.7
D	2	6	0.6
E	1	0	0.4

Nodes	Edges	Density	Cohesion	Diameter
5	4	0.4	0.4	4

Y

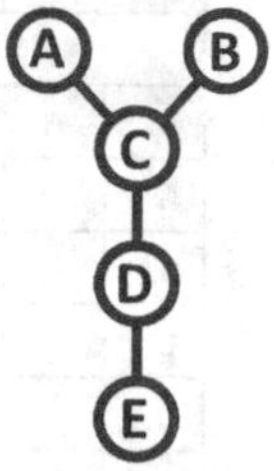

Position	Degree	Betweenness	Closeness
A	1	0	0.5
B	1	0	0.5
C	3	10	0.8
D	2	6	0.7
E	1	0	0.4

Nodes	Edges	Density	Cohesion	Diameter
5	4	0.4	0.4	3

Star

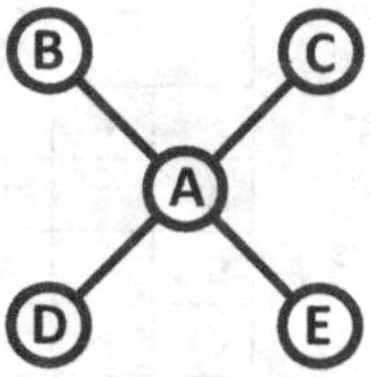

Position	Degree	Betweenness	Closeness
A	4	12	1
B	1	0	0.6
C	1	0	0.6
D	1	0	0.6
E	1	0	0.6

Nodes	Edges	Density	Cohesion	Diameter
5	4	0.4	0.4	2

Cricle

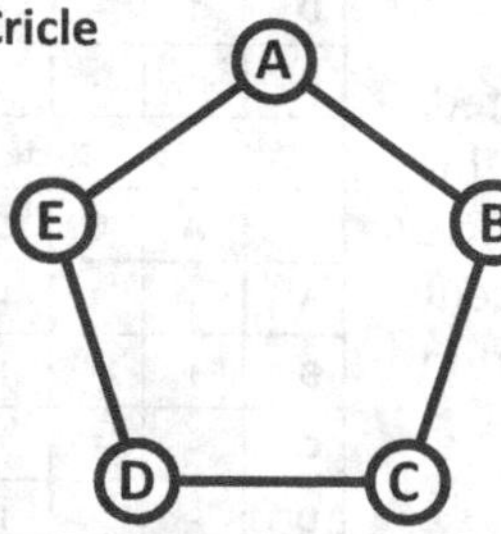

Position	Degree	Betweenness	Closeness
A	2	2	0.7
B	2	2	0.7
C	2	2	0.7
D	2	2	0.7
E	2	2	0.7

Nodes	Edges	Density	Cohesion	Diameter
5	5	0.5	0.5	2

All-connected

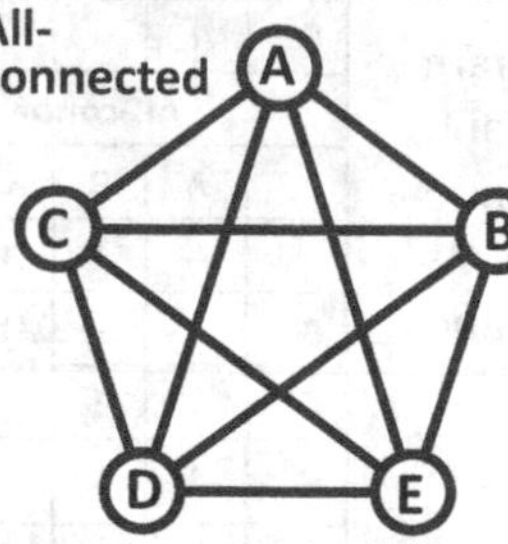

Position	Degree	Betweenness	Closeness
A	4	0	1
B	4	0	1
C	4	0	1
D	4	0	1
E	4	0	1

Nodes	Edges	Density	Cohesion	Diameter
5	10	1	1	1

Figure 9.4   Five-person networks, with indices of centrality

Step 7: Calculate Sociometric Status
The sociometric status of each agent refers to the number of communications received and emitted, relative to the number of nodes in the network. The mean sociometric status + standard deviation can also be used to define key agents within the network. Those agents who possess a sociometric status figure that exceeds the mean + standard deviation figure can be defined as key agents for the scenario under analysis.

Step 8: Calculate Network Density
Network density is equal to the total number of links between the agents in the network divided by the total number of possible links. Low network density figures are indicative of a well-distributed network of agents. High-density figures are indicative of a network that is not well distributed.

Advantages

- SNA can be used to determine the importance of different agents within a team or group of agents.
- It offers a comprehensive analysis of the network in question. The key agents within the network are identified, as are the frequency and direction of communications within the network. Further classifications include network type and network density. There are also additional analyses that can be calculated, such as betweenness, closeness and distance calculations.
- Networks can be classified according to their structure. This is particularly useful when analysing networks across different domains.
- It is well suited to the analysis of C4i scenarios.
- It has been used extensively in the past for the analysis of various social networks.
- It is simple to learn and easy to use.
- The Agna SNA software package reduces application time considerably.
- It is a generic technique that could potentially be applied in any domain involving team-based or collaborative activity.

Disadvantages

- For large, complex networks, it may be difficult to conduct SNA. Application time is a function of network size, and large networks may incur lengthy application times.
- The data collection phase involved in a typical SNA is resource-intensive.
- Some knowledge of mathematical techniques is required.
- It is difficult to collect comprehensive data for SNA. For example, a dispersed network of 10 agents would require at least 10 observers in order to accurately and comprehensively capture the communications made between all agents.
- Without the provision of the Agna SNA software package, the technique may be time-consuming to apply.

Flowchart

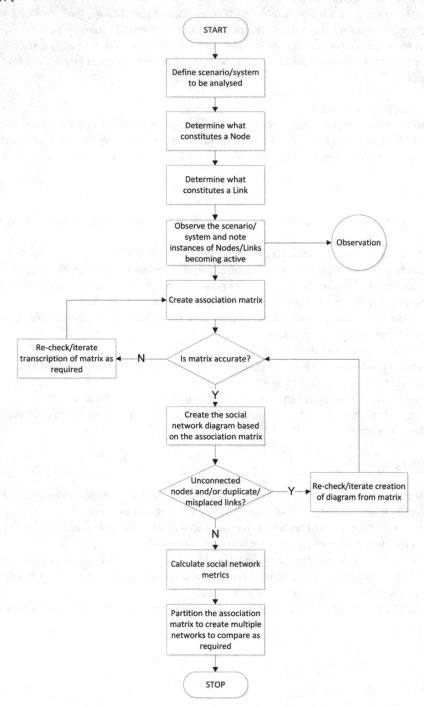

Related Methods

SNA is based on observational data of the scenario under analysis. Additional HF data collection techniques, such as interviews and questionnaires, might also be used to gather supplementary data.

Approximate Training and Application Times

The training time associated with the SNA technique is typically low. Although some knowledge of mathematical analysis is required, the basic SNA procedure is a simple one. The associated application time is also minimal and can be reduced considerably with the provision of the AGNA SNA software support package, which can be used for the data analysis part of an SNA. The application time is of course dependent upon the size and complexity of the network under analysis, and small simple networks may only incur limited application times. However, larger, more complex networks would incur a considerable application time. In a recent study of C4i activity in the energy distribution domain (Salmon et al., 2004), the SNA conducted typically took around one to three hours.

Reliability and Validity

No data regarding the reliability and validity of the SNA technique is available.

Tools Needed

Once the initial data collection phase is completed, an SNA can be conducted using pen and paper. The tools required during the data collection phase for an SNA would be dependent upon the type of data collection techniques used. Observational analysis, interviews and questionnaires would normally require visual and audio recording equipment (video cameras, audio recorder, computer, etc.). For the data analysis component, various forms of software exist on the Internet.

Questionnaires for Distributed Assessment of Team Mutual Awareness

Background and Applications

MacMillan et al. (2004) describe a set of self-rating questionnaires designed to assess team member mutual awareness. Based on a model of team mutual awareness, the methodology proposed by MacMillan et al. (2004) comprises three questionnaires: the task mutual awareness questionnaire, the team workload awareness questionnaire and the teamwork awareness questionnaire. The task mutual awareness questionnaire involves participants recalling salient events that occurred during the task under analysis and then describing the tasks that they were performing during these events and also the tasks that they think the other team members were performing during these events. The team workload awareness questionnaire is a subjective MWL assessment technique based on the NASA-TLX (Hart and Staveland, 1988) and is used to elicit subjective ratings of team member MWL on the following dimensions: mental demand, temporal demand, performance effort and frustration. Team members also provide an overall rating of other team members' MWL and also a rating of each TLX dimension for the team as a whole. The teamwork awareness questionnaire is used to rate the team on four components of teamwork processes. Team members provide subjective ratings of the team's performance on the following team behaviours: communication, back-up, coordination and information management, and leadership/team orientation. Each of the questionnaires is administered post-trial in order to gain a measure of 'team mutual awareness'.

Procedure and Advice

Step 1: Define the Task to be Analysed
The first step is to clearly define the task (or set of tasks) that is to be analysed. This allows the analyst to gain a clear understanding of the task content and also allows for the modification of the behavioural

rating scale, whereby any behaviours missing from the scale that may be evident during the task are added. It is recommended that an HTA is conducted for the task under analysis. This allows the analysts involved to gain a thorough understanding of the task under investigation.

Step 2: Select the Team to be Observed
Once the task (or tasks) under analysis are clearly defined and described, and the analyst has gained a full understanding of the task under analysis, the participants that are to be observed can be selected. This may be dependent upon the purpose of the analysis. Typically, the team (or teams) under analysis are defined by the nature of the task under analysis.

Step 3: Brief Participants
In most cases, it is appropriate to brief the participants involved regarding the purpose of the study and also the techniques used during the procedure. The participants involved should be instructed in the completion of each of the three questionnaires. It may be useful to conduct a walkthrough of an example analysis using the three questionnaires. This procedure should continue until all of the team members fully understand how the techniques work and also what is expected of them as participants in the trial.

Step 4: Begin Task Performance
The questionnaires are typically administered post-trial. The team should be instructed to perform the task under analysis as normal.

Step 5: Completion of Task Mutual Awareness Questionnaire
Once task performance is completed, the data collection phase can begin. The task mutual awareness questionnaire involves the participants recalling salient events that occurred during the task performance. Once an appropriate event is recalled, participants are required to describe the tasks that they were performing during the recalled event and those tasks that they thought the other team members were performing. An appropriate SME is then used to classify the responses into task categories.

Step 6: Completion of Team Workload Awareness Questionnaire
The team workload awareness questionnaire involves participants rating their own workload across the five NASA-TLX workload dimensions: mental demand, temporal demand, performance, effort and frustration. The participant should then rate the other team members' workload and also the overall team's workload across the five dimensions described above.

Step 7: Completion of Teamwork Awareness Questionnaire
In completing the teamwork awareness questionnaire, team members subjectively rate team performance on four teamwork behaviours: communication, coordination, information management and leadership/team orientation.

Step 8: Calculate Questionnaire Scores
Once all of the questionnaires have been completed by all of the team members, the data analysis phase can begin. Each questionnaire has its own unique scoring procedure. In scoring the mutual awareness questionnaires, the task category reported by each team member is compared to the task category that he or she was performing as reported by the other team members. The number of category matches for each individual are then summed and a percentage agreement (congruence score) is calculated for each item. In scoring the mutual awareness workload questionnaire, a convergence measure that reflects the difference between each team member's self-reported workload and the estimate of his or her workload provided by the other team members is calculated. Scoring of the teamwork awareness questionnaire involves calculating a mean score of each rating across the team. According to MacMillan et al. (2004),

this score reflects how well the team are performing. Agreement scores within the team should also be calculated.

Flowchart

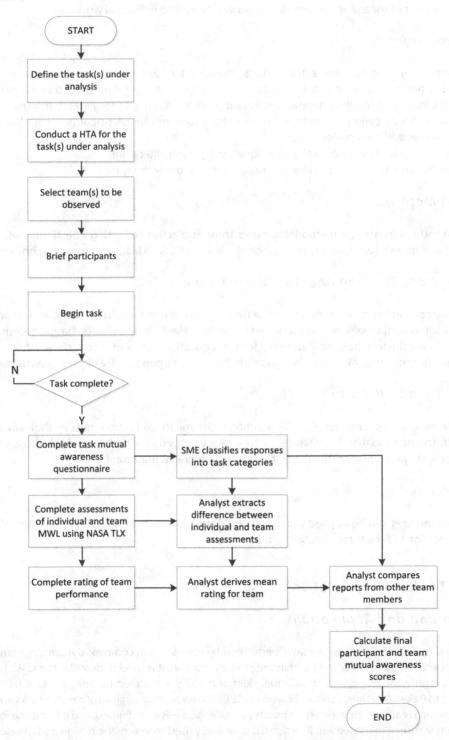

START

Define the task(s) under analysis

Conduct a HTA for the task(s) under analysis

Select team(s) to be observed

Brief participants

Begin task

Task complete?   N

Y

Complete task mutual awareness questionnaire

SME classifies responses into task categories

Complete assessments of individual and team MWL using NASA TLX

Analyst extracts difference between individual and team assessments

Complete rating of team performance

Analyst derives mean rating for team

Analyst compares reports from other team members

Calculate final participant and team mutual awareness scores

END

Advantages

- The questionnaire techniques used are quick, low-cost and easy to apply.
- Minimal training is required in order to use the technique effectively.
- A number of measures are provided, including team and individual workload.

Disadvantages

- Each technique uses subjective ratings provided by participants once the task is complete. There are a number of problems associated with this form of data collection. Participants are not efficient at recalling mental events and have a tendency to forget certain aspects (such as low-workload periods of the task). There is also a tendency for participants to correlate workload measures with task performance.
- There is limited evidence of the technique's usage in the literature.
- There is limited validation evidence associated with the technique.

Related Methods

The team mutual awareness methodology uses three subjective self-rating questionnaires. The team workload awareness questionnaire is based on the NASA-TLX workload assessment technique.

Approximate Training and Application Times

It is estimated that the training times for the three questionnaires would be minimal. The application time for each questionnaire is also estimated to be low. MacMillan et al. (2004) reported that several minutes of introductory training is required for each questionnaire and that each questionnaire takes around five minutes to complete, although this is dependent upon the size of the teams under analysis.

Reliability and Validity

MacMillan et al. (2004) reported that the validity of the measures is supported by their correlation to team performance and that the measures possess face validity due to their focus upon those observable aspects of team performance that the team members define as important.

Tools Needed

The questionnaires can be applied using pen and paper. MacMillan et al. (2004) have also developed software versions of the three questionnaires.

Team Task Analysis (TTA)

Background and Applications

TTA is used to describe and analyse tasks performed by teams within complex, dynamic systems, such as infantry operations and C4i activity within the military domain. It is used to describe the tasks performed and also identify the associated knowledge, skills and abilities required for effective task performance. According to Baker, Salas and Cannon-Bowers (1998), TTA refers to the analysis of a team's tasks and also the assessment of a team's teamwork requirements (knowledge, skills and abilities), and it forms the foundation for all team resource management functions. It is typically used to inform the design and development of team training interventions, such as CRM training, the design of teams and their associated processes, and also for team performance evaluation. It defines the following aspects of team-based activity:

- teamwork: those tasks related to the teams goals that involve interaction or collaboration between the actors within the team or network; and
- taskwork: those tasks that are performed individually by the actors within the team.

According to Burke (2004), the TTA procedure has not yet been widely adopted by organisations, with the exception of the US military and aviation communities. Although a set procedure for TTA does not exist, Burke (2004) attempted to integrate the existing TTA literature into a set of guidelines for conducting a TTA.

Domain of Application

Generic.

Procedure and Advice

Step 1: Conduct Requirements Analysis
First, a requirements analysis should be conducted. This involves clearly defining the task scenario to be analysed, including describing all duties involved and also conditions under which the task is to be performed. Burke (2004) also suggests that when conducting the requirements analysis, the methods of data collection to be used during the TTA should be determined. Typical TTA data collection methods include observational study, interviews, questionnaires and surveys. The requirements analysis also involves identifying the participants that will be involved in the data collection process, including occupation and number.

Step 2: Task Identification
Next, the tasks involved in the scenario under analysis should be defined and described clearly. Burke (2004) recommends that interviews with SMEs, observation and source documents should be used to identify the full set of tasks. Once each individual task step is identified, a task statement should be written (for the component task), including the following information:

- task name;
- task goals;
- what the individual has to do to perform the task;
- how the individual performs the task;
- which devices, controls and interfaces are involved in the task; and
- why the task is required.

Step 3: Identify Teamwork Taxonomy
Once all of the tasks involved in the scenario under analysis have been identified and described fully, a teamwork taxonomy should be selected for use in the analysis (Burke, 2004). According to Burke, several teamwork taxonomies exist in the literature. A generic teamwork taxonomy is presented in Table 9.11.

Step 4: Conduct a Coordination Analysis
Once an appropriate teamwork taxonomy is selected, a coordination demands analysis should be conducted. The TTA involves classifying the tasks under analysis into teamwork and taskwork activity, and then rating each teamwork task step for the level of coordination between team members for each behaviour identified in the teamwork taxonomy.

Table 9.11   Teamwork taxonomy

Coordination dimension	Definition
Communication	Includes sending, receiving and acknowledging information among crew members
Situation awareness (SA)	Refers to identifying the source and nature of problems, maintaining an accurate perception of the aircraft's location relative to the external environment, and detecting situations that require action
Decision-making (DM)	Includes identifying possible solutions to problems, evaluating the consequences of each alternative, selecting the best alternative and gathering information needed prior to arriving at a decision
Mission analysis (MA)	Includes monitoring, allocating and coordinating the resources of the crew and aircraft; prioritising tasks; setting goals and developing plans to accomplish the goals; creating contingency plans
Leadership	Refers to directing activities of others, monitoring and assessing the performance of crew members, motivating members and communicating mission requirements
Adaptability	Refers to the ability to alter one's course of action as necessary, maintaining constructive behaviour under pressure and adapting to internal or external changes
Assertiveness	Refers to the willingness to make decisions, demonstrating initiative and maintaining one's position until convinced otherwise by facts
Total coordination	Refers to the overall need for interaction and coordination among crew members

Source: Burke, 2005.

Step 5: Determine Relevant Taskwork and Teamwork Tasks

The next step in the TTA procedure involves determining the relevance of each of the component tasks involved in the scenario under analysis, including both teamwork and taskwork tasks. Burke (2004) recommends that a Likert scale questionnaire is used for this step and that the following task factors should be rated:

- importance to train;
- task difficulty;
- importance to job.
- task frequency;
- difficulty of learning; and

It is recommended that the task indices used should be developed based on the overall aims and objectives of the TTA.

Step 6: Translation of Tasks into KSAs

Next, the knowledge, skills, abilities and attitudes (KSAs) for each of the relevant task steps should be determined. Normally, interviews or questionnaires are used to elicit the required information from an appropriate set of SMEs.

Step 7: Link KSAs to Team Tasks

The final step of a TTA is to link the KSAs identified in step 6 to the individual tasks. According to Burke (2004), this is most often achieved through the use of surveys completed by SMEs.

Advantages

- TTA goes further than individual task analysis techniques by specifying the KSAs required to complete each task step.
- The output from TTA can be used in the development of team training procedures such as CRM training programmes, and also in the design of teams and their associated procedures.
- The TTA output states which of the component tasks involved are team-based and which are performed individually.

Disadvantages

- TTA is a time-consuming technique to apply.
- Appropriate SMEs and domain experts are required throughout the procedure. Access to sufficient SMEs is often difficult to obtain.
- There is no rigid procedure for the TTA technique. As a result, the reliability of the technique may be questionable.
- Great skill is required on behalf of the analyst in order to elicit the required information throughout the TTA procedure.

Flowchart

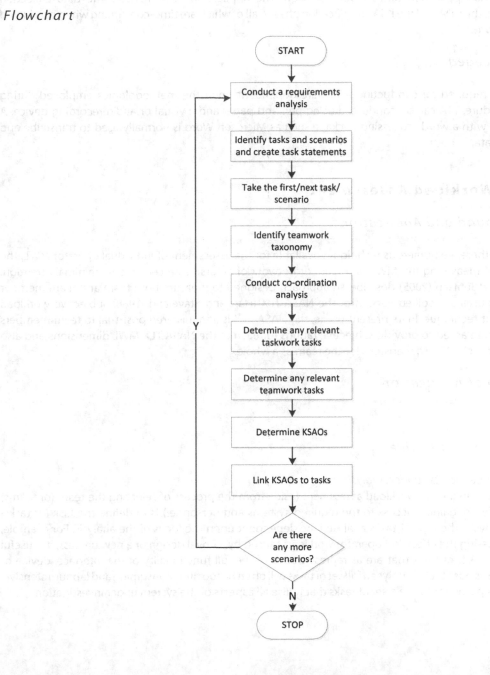

Related Methods

There are a number of different approaches to TTA, such as TTRAM, CUD and SNA. TTA also utilises a number of traditional HF data collection techniques, such as observational study, interviews, questionnaires and surveys.

Approximate Training and Application Times

Due to the exhaustive nature of the TTA procedure outlined above, it is estimated that the associated training and application times would be high. The application time includes the data collection procedure, the CDA and the KSA identification phases, all of which are time-consuming when conducted on their own.

Tools Needed

The tools required for conducting a TTA are dependent upon the methodologies employed during the procedure. TTA can be conducted using pen and paper and a visual or audio recording device. A computer with a word processing package such as Microsoft Word is normally used to transcribe and sort the data.

Team Workload Assessment

Background and Applications

Although there are numerous techniques available for the assessment of individual operator MWL, the concept of measuring the MWL of a team or network of actors has received only minimal attention. Bowers and Jentsch (2005) describe an approach designed to measure both team and team member MWL that uses a modified version of the NASA-TLX (Hart and Staveland, 1988) subjective workload assessment technique. In its present usage, the NASA-TLX is administered post-trial to team members who are then asked to provide subjective ratings for each of the NASA-TLX MWL dimensions and also ratings for each of the dimensions for the team as a whole.

Domain of Application

Generic.

Procedure and Advice

Step 1: Define the Task under Analysis
The first step in a team workload assessment (aside from the process of selecting the team (or teams) to be analysed, gaining access to the required systems and personnel) is to define the task (or tasks) under analysis. The type of tasks analysed are dependent upon the focus of the analysis. For example, when assessing the effects on operator workload caused by a novel design or a new process, it is useful to analyse a set of tasks that are as representative of the full functionality of the interface, device or procedure as possible. To analyse a full set of tasks will often be too time-consuming and labour-intensive, and so it is pertinent to use a set of tasks that utilise all aspects of the system under investigation.

Step 2: Conduct an HTA for the Task under Analysis
Once the task (or tasks) under analysis is defined clearly, an HTA should be conducted for each task. This allows the analyst and participants to understand the task fully.

Step 3: Brief Participants
Before the task (or tasks) under analysis is performed, all of the participants involved should be briefed regarding the purpose of the study and the NASA-TLX technique. It is recommended that participants are given a workshop on MWL and MWL assessment. It may also be useful at this stage to take the participants through an example team MWL assessment so that they understand how the technique works and what is required of them as participants.

Step 4: Conduct Pilot Run
Before the 'real' data collection procedure begins, it is useful to conduct a pilot run. The team should perform a small task and then complete a NASA-TLX for themselves and for the team as a whole. This highlights any potential problems in the data collection procedure.

Step 5: Performance of Task Under Analysis
The NASA-TLX is typically administered post-trial. The team should be instructed to perform the task or scenario in question as normal.

Step 6: Weighting Procedure
When the task under analysis is complete, the weighting procedure can begin. A computerised software tool to support the application of the NASA-TLX is available online from the NASA Ames Research Center. This software tool consists of three components: the WEIGHT software, the RATING software and the TLX software. The WEIGHT software enables an appropriate weighting of participant scores, the RATING software allows participants to complete the ratings of the dimensions and the TLX software calculates the overall workload score (Vidulich and Tsang, 1986a). The WEIGHT software presents 15 pairwise comparisons of the six sub-scales (mental demand, physical demand, temporal demand, effort, performance and frustration level) to the participant. Participants should be instructed to select, from each of the 15 pairs, the sub-scale that contributed the most to the workload of the task. The WEIGHT software then calculates the total number of times each sub-scale was selected by the participant. Each scale is then rated by the software based on the number of times it is selected by the participant. This is done using a scale of 0 (not relevant) to 5 (more important than any other factor).

Step 7: NASA-TLX Rating Procedure
Participants should be presented with the interval scale for each of the NASA-TLX sub-scales. Participants are asked to provide a subjective rating for each sub-scale, between 1 (low) and 20 (high), in response to the associated sub-scale questions. This is based entirely on the subjective judgment of the participants. Participants should be instructed to complete a NASA-TLX for themselves and also for the team as a whole.

Step 8: NASA-TLX Score Calculation
A workload score is then calculated for each team member and also for the team as a whole. This is calculated by multiplying each rating by the weight given to that sub-scale by the participant. The sum of the weighted ratings for each task is then divided by 15 (the sum of weights). A workload score of between 0 and 100 is then provided for the task under analysis.

Advantages

- The technique offers a quick and easy approach for measuring team and team member MWL.
- It is low cost and easy to apply, requiring only minimal training.
- The NASA-TLX technique is the most commonly used MWL assessment technique available and has been subjected to numerous validation studies.
- The NASA TLX sub-scales are generic, so the technique can be applied in any domain.
- It offers a multi-dimensional assessment of workload.
- Team MWL ratings can be compared across team members to ensure reliability.

Disadvantages

- The extent to which team members can provide an accurate assessment of overall team workload is questionable and requires further testing.
- A host of problems are associated with collecting data post-trial. Participants may have forgotten high or low workload aspects of the task and workload ratings may also be correlated with task performance, e.g. subjects who performed poorly on the primary task may rate their workload as very high and vice versa.
- Bowers and Jentsch (2005) report that the approach is cumbersome and also highlight the fact that the technique that does not provide separate estimates for teamwork versus taskwork.

Approximate Training and Application Times

Due to the simplistic nature of the NASA-TLX technique, the training time associated with the technique is estimated to be very low. Similarly, the application time associated with the technique is also estimated to be very low. Bowers and Jentsch (2005) report that the individual and team measures take about 10 minutes each to complete.

Reliability and Validity

There is limited reliability and validity data available regarding this approach to the assessment of MWL. The reliability of such an approach certainly is questionable. The extent to which individuals can accurately provide a measure of team workload is also questionable. Bowers and Jentsch (2005) describe a study designed to test the validity of the approach whereby team performance was compared to MWL ratings. It was found that the lowest individual MWL rating was the best predictor of performance, in that the higher the lowest reported individual MWL rating was, the poorer the team's performance. It is apparent that such approaches to the assessment of team MWL require further testing in terms of reliability and validity. How to test the validity of such techniques is also a challenge, as there are problems associated with associating workload and performance. In other words, it may be that team performance was poor and team members rated the overall team workload as high, due to a correlation with performance. However, this may not be the case and it may be that teams with low workload perform poorly, due to factors other than workload.

Tools Needed

The NASA-TLX can be applied using pen and paper.

Flowchart

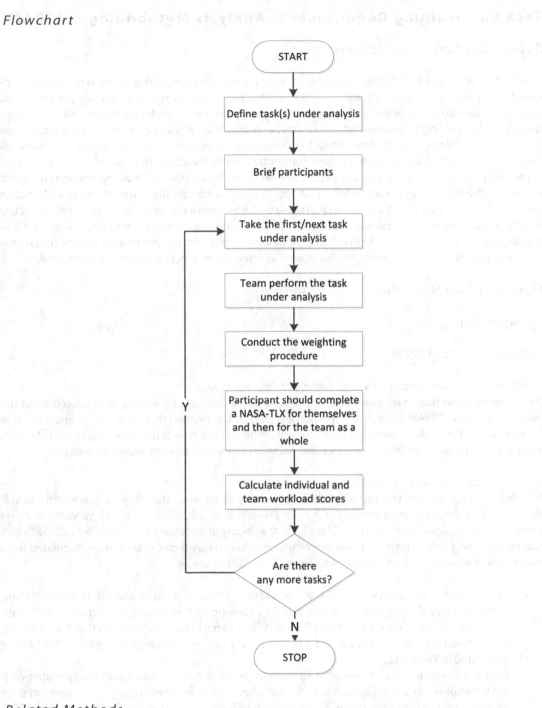

START

↓

Define task(s) under analysis

↓

Brief participants

↓

Take the first/next task under analysis

↓

Team perform the task under analysis

↓

Conduct the weighting procedure

↓

Participant should complete a NASA-TLX for themselves and then for the team as a whole

↓

Calculate individual and team workload scores

↓

Are there any more tasks?

Y

N

↓

STOP

Related Methods

The team MWL assessment technique uses the NASA-TLX subjective MWL assessment technique. A number of other multi-dimensional subjective MWL assessment techniques exist, such as SWAT (Reid and Nygren, 1988) and the workload profile technique (Tsang and Velazquez, 1996).

Task and Training Requirements Analysis Methodology (TTRAM)

Background and Applications

The TTRAM (Swezey et al., 2000) technique comprises a number of methods that are used to identify team-based task training requirements and also to evaluate any associated training technologies that could potentially be used in the delivery of team training. The method was developed for the military aviation domain and, according to Swezey et al. (2000), has been shown to be effective at discriminating tasks that are prone to skill decay, tasks that are critical to mission success, tasks that require high levels of teamwork (internal and external) and tasks that require further training intervention. The TTRAM technique is used to identify current training and practice gaps, and then to determine potential training solutions designed to address the training and practice gap identified. In order to identify the current training and practice gaps, a skill decay analysis and a practice analysis is conducted, and a skill decay index score and a practice effectiveness index score are derived. The two scores are then compared in order to identify practice and training gaps. For example, a task high skill decay index score compared to a low practice effectiveness index score would demonstrate a requirement for additional training and practice for the task under analysis.

Domain of Application

Military aviation.

Procedure and Advice

Step 1: Perform a Task Analysis for the Scenario or Task under Analysis
It is recommended that a task analysis for the task (or set of tasks) under investigation should act as the initial input to the TTRAM analysis. For this purpose, it is recommended that an HTA is the most suitable for this. A number of data collection procedures may be used to collect the data required for the HTA, including interviews with SMEs and observational study of the task or scenario under analysis.

Step 2: Conduct Skill Decay Analysis
The skill decay analysis is conducted in order to identify those tasks that may be susceptible to skill degradation without sufficient training or practice (Swezey et al., 2000). This analysis involves identifying the difficulty associated with each task, identifying the degree of prior learning associated with each task and determining the frequency of task performance. A skill decay index score is then calculated from these three components. Each component is described further below:

- Task difficulty: the analyst should rate each task in terms of its associated difficulty, including difficulty in performing the task and also in acquiring and retaining the required skills. Task difficulty is rated as low (1), medium (2) or high (3). Swezey et al. (2000) suggest that task difficulty be assessed via SME interviews and a behaviourally anchored rating scale (BARS). The BARS is presented in Table 9.12.
- Degree of prior learning: the analyst should assess the degree of prior learning associated with each task under analysis. SMEs and BARS are also used to gather these ratings. The degree of prior learning for a task is rated as low (3), medium (2) or high (1). The degree of prior learning BARS is presented in Table 9.13.
- Frequency of task performance: the analyst should rate the frequency of performance of each task. This is rated as infrequent, frequent or very frequent. The frequency of task performance assessment scale is shown in Table 9.14, which maps onto the scoring algorithm of low (3),

medium (2) or high (1), with low representing infrequent, medium representing frequent and high representing very frequent.

Table 9.12   Task difficulty BARS

Question: how difficult is this task to perform?	
Difficulty levels	**Associated task characteristics**
Low	Virtually no practice is required. Most trained individuals (i.e. 90%) will be able to perform this task with minimal exposure or practice on the operational equipment. Consists of very few procedural steps and each step is dependent upon proceeding steps.
Medium	Individuals can accomplish most of the activity following baseline instruction. The majority of trained individuals (i.e. 60%) will be able to perform this task with minimal exposure or practice on the operational equipment. This activity does require moderate practice to sustain competent performance at the desired level of proficiency. Consists of numerous complex steps.
High	Requires extensive instruction and practice to accomplish the activity. Very few trained individuals (i.e. 10%) will be able to perform this task with minimal exposure or practice on the operational equipment. Consists of a large number of complex steps and there is little if any dependency among the task steps.

Source: Swezey et al., 2000.

Table 9.13   Degree of prior learning BARS

Question: what level of training is required to maintain an adequate level of proficiency on this task?	
Proficiency levels	**Associated task characteristics**
Low	A high level of training is required to maintain proficiency on this task. Individual cannot be expected to perform the task without frequent recurrency training. Individual fails to meet task performance standards without frequent recurrency training.
Medium	A moderate level of training is required to maintain proficiency. Individual can perform the task in the trainer under a restricted set of task conditions; however, needs more practice in the actual job setting under varying task conditions and under supervision. Individual meets minimum performance standards without frequent recurrency training.
High	Minimal training is required to maintain proficiency. Individual can perform the task completely and accurately without supervision across varying task conditions; has achieved mastery level proficiency. Individual exceeds performance standards.

Source: Swezey et al., 2000.

Table 9.14   Frequency of task performance BARS

Question: how often is this task performed in the context of your job (across different missions)? Do not factor in time spent training: limit responses to the frequency with which the task is inherently performed as part of the operational setting.	
Frequency levels	**Associated task characteristics**
Infrequent	Very little time is spent performing the task. Task is infrequently performed.
Frequent	A moderate amount of time is spent performing this task. Task is performed frequently.
Very frequent	This task comprises a large amount of time. Task is performed very frequently.

Source: Swezey et al., 2000.

Step 3: Calculate Skill Decay Index Score
Once ratings for task difficulty, degree of prior learning and frequency of task performance are obtained, the skill decay index score should be calculated. This is calculated by summing the individual scores

for each of the three components (task difficulty, degree of prior learning and frequency of task performance). A skill decay index score of 3–9 is derived.

Step 4: Conduct Practice Analysis
The practice analysis is conducted in order to determine the current levels of task and skill practice associated with the task under analysis. It is comprised of the following:

- Amount of practice: the amount of practice associated with each task is determined using SME interviews and rated on a scale of 1 (low) to 3 (high).
- Frequency of practice: the frequency with which the tasks are practised is rated as high (3), medium (2) or low (1).
- Quality of practice: the quality of the practice undertaken for each task is also rated using a scale of high (3), medium (2) and low (1). A team skill training questionnaire and a simulator capability and training checklist are also used. The team skill training questionnaire is presented in Table 9.15.
- Simulator capability: the analyst is also required to assess the capability of any simulators used in the provision of practice.

Table 9.15   Team skill training questionnaire

Extent to which training allows team members to practise coordinated activities required by the task (both internal and external to the simulator).	
Extent to which training provides practice for improving the effectiveness of communication among crew members.	
Extent to which training incorporates objective measures for evaluating crew performance.	
Level of feedback provided by training on how well the aircrew performed as a team.	

Source: Swezey et al., 2000.

Step 5: Calculate Practice Effectiveness Index
Once the ratings for the four components outlined above are determined, the practice effectiveness index score is calculated. This is derived by summing the four values derived for practice analysis components (amount of practice, frequency of practice, quality of practice and simulator capability) during step 4. For each task that is using the TTRAM technique, a skill decay index score and a practice effectiveness index score should be derived.

Step 6: Compare Skill Decay Index and Practice Effectiveness Index Scores
The next step involves comparing the associated skill decay index and practice effectiveness scores for each of the tasks under analysis. Those tasks with higher skill decay index scores possess a greater potential for skill degradation, whilst those tasks with higher practice effectiveness scores indicate too great a level of task support.

Step 7: Identify Training Gaps
Once the analyst has determined those tasks that are not adequately supported by training or practice and that have the potential for skill decay, the nature of the training gaps should be determined. According to Swezey et al. (2000), gaps represent areas of task practice or training in which task skills are not addressed or are inadequately addressed by current training schemes.

Step 8: Identify Potential Training Intervention
For each training gap specified, the analyst should attempt to determine potential training solutions, such as simulations and computer-based training interventions.

Step 9: Perform Training Technology Analysis
The training technology analysis is conducted in order to identify alternative and appropriate training or practice interventions for those training gaps identified. This analysis involves the following components:

- Identification of task skill requirements: a behavioural classification system (Swezey et al., 2000) is used to categorise tasks in terms of their underlying process.
- Identification of task criticality level: SMEs should be used to rate the criticality of each task under analysis. A task criticality table is used for this purpose (see Table 9.16).
- Identification of task teamwork level: the extent to which the task requires coordinated activity and interaction amongst individuals is also assessed using SMEs. A teamwork assessment scale is used for this purpose (see Table 9.17).

Specification of Training Media and Support Recommendations

Table 9.16   Task criticality table

Question: how critical is this task to successful mission performance?	
Criticality levels	**Associated task characteristics**
Low	Errors are unlikely to have any negative consequences to overall mission success. Task is not a critical/important component of the overall duty/mission. Task can be ignored for long periods of time.
Medium	Errors or poor performance would have moderate consequences and may jeopardise mission success. Task is somewhat critical/important to overall duty/mission. Task requires attention, but does not demand immediate action.
High	Errors would most likely have serious consequences; failing to execute the task correctly would lead to mission failure. Task is a critical/important component of the overall duty/mission. Task requires immediate attention and action.

Source: Swezey et al., 2000.

Table 9.17   Teamwork assessment scale

Question: what level of teamwork is required in order to perform this task? Assign two ratings: one for internal crew member teamwork and a second for external teamwork.	
Criticality levels	**Associated task characteristics**
Low	Task can be accomplished on the basis of individual performance alone; the task can be performed in isolation of other tasks. Virtually no interaction or coordination among team members is required. Task can be performed in parallel with other team member tasks.
Medium	Requires a moderate degree of information exchange about internal/external resources, and some task interdependencies among individuals exist. Some coordination among team members is required if the task is to be successfully completed. Some sequential dependencies among sub-tasks are required.
High	Involves a dynamic exchange of information and resources among team members. Response coordination and sequencing of activities of activities among team members is vital to successful task performance (activities must be synchronised and precisely timed). Actions are highly dependent upon the performance of other team members.

Source: Swezey et al., 2000.

Flowchart

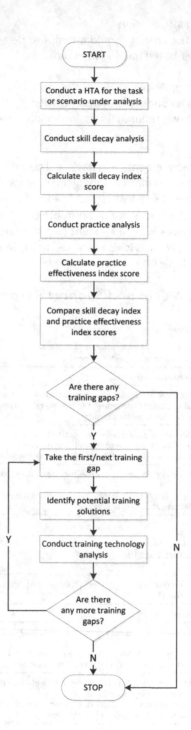

START

Conduct a HTA for the task or scenario under analysis

Conduct skill decay analysis

Calculate skill decay index score

Conduct practice analysis

Calculate practice effectiveness index score

Compare skill decay index and practice effectiveness index scores

Are there any training gaps?

Y

Take the first/next training gap

Identify potential training solutions

Conduct training technology analysis

Are there any more training gaps?

N

Y

N

STOP

Advantages

- The output of a TTRAM analysis is extremely useful for a number of different purposes. Tasks prone to skill decay are identified and training solutions are offered. Training gaps are also identified, as are the underlying skills associated with each task. TTRAM also rates the level of teamwork required for task steps.

- The TTRAM procedure is very exhaustive.

Disadvantages

- TTRAM is time-consuming in its application.
- SMEs are required for a TTRAM analysis; access to these may prove difficult.
- It is resource-intensive.

Related Methods

The TTRAM technique uses a number of different approaches, including interviews, BARS, classification schemes and checklists.

Approximate Training and Application Times

The training time associated with the TTRAM technique is estimated to be high. It is likely that a practitioner with no prior experience of the techniques used would require in excess of one to two days of training for the technique. The application time for the technique would also be high, considering that it uses a number of interviews, as well as a number of rating scales and checklists.

Reliability and Validity

No data regarding the reliability and validity of the TTRAM technique is offered by the authors.

Tools Needed

The tools required for a TTRAM analysis would include those required for any interview-type analysis, such as a computer with Microsoft Excel and video and audio recording equipment. Each of the TTRAM behavioural rating scales would also be required, along with the task process classification scheme. The analyst would also require some access to the simulators, simulations and software that are used for training purposes in the establishment under analysis.

The Cockpit Management Attitudes Questionnaire (CMAQ)

Background and Applications

CMAQ (Helmreich, 1984; Helmreich, Wilhelm and Gregorich, 1988) is a questionnaire designed to measure flight crews' attitudes regarding communication, coordination and leadership. The survey was developed based on CRM research conducted by NASA in addition to the analysis of accident reports from the National Transportation Safety Board (NTSB) and the Aviation Safety Reporting System (ASRS) (Helmreich, 1984).

The CMAQ scale consists of 25 items and a five-point Likert scale ranging from 1 (strongly disagree) to 5 (strongly agree), which participants use to rate each item. The items are based around three factors: communications and team coordination; leadership and authority; and recognition of stressors (Gregorich, Helmreich and Wilhelm, 1990). The results of the CMAQ provide a baseline to evaluate the effectiveness of CRM or other training interventions and offer an insight into the areas in which teams need to improve (Gregorich, Helmreich and Willheim, 1990; Crichton, 2005).

Domain of Application

The questionnaire items are specific to the aviation domain. According to Sexton, Thomas and Grillo (2003), the items can be amended slightly to enable the utilisation of the method in other domains such as intensive care units within the medical domain.

Procedure and Advice

Step 1: Define the Task under Analysis
The first stage of analysis involves clearly defining the task, system and team under analysis. This definition is what guides the analysis and ensures that appropriate and relevant information is collected.

Step 2: Recruit Participants
Once the task is clearly defined, relevant participants should be identified and then recruited. The participants must be a representative sample of the population under analysis.

Step 3: Experimental Scenario Design
At this point, the analyst must design an appropriate experimental scenario for the participants to undertake. This scenario should match the task under analysis and represent the participants' normal work activities.

Step 4: Brief Participants
The analyst must now brief the participants on why the study is being undertaken. The participants should be informed of exactly what will be asked of them. Before moving on to the next stage, they should be comfortable and happy with the analysis.

Step 5: Complete Task Scenario
The participants should now undertake the task scenario. This stage is complete once all participants have completed the task scenario.

Step 6: Application of the CMAQ
Once the task scenario is complete, CMAQs should be distributed and completed fully by all participants. This stage of the analysis is not complete until all participants have completed all questionnaire items.

Step 7: Analysis of Data
The results of the CMAQs should now be collated by the analyst. The results can be analysed qualitatively by examining the data for core themes or quantitatively by deriving statistical values and graphs.

Advantages

- The CMAQ provides a baseline of teamwork and enables a continued assessment of training interventions (Gregorich, Helmreich and Willheim, 1990: Crichton, 2005).
- It is an effective technique for quickly gaining large amounts of information from large populations.

Disadvantages

- The data collected can be limited.
- The survey only captures subjective information from team members.
- Participant responses are forced into one of five response options.

Related Methods

According to Sexton, Thomas and Grillo (2003), there are numerous variations of the CMAQ, including: intensive care (Intensive Care Unit Management Attitudes Questionnaire: ICUMAQ); operating rooms (Operating Rooms Management Attitudes Questionnaire); and health care (Safety Attitudes Questionnaire: SAQ). Versions are also under development for pharmacies, emergency departments, and labour and delivery units (Sexton, Thomas and Grillo, 2003).

Approximate Training and Application Times

The method is simple and quick to apply; application times to complete a CMAQ are around 10–15 minutes (Sexton, Thomas and Grillo, 2003).

Reliability and Validity

The method has been shown to have high levels of reliability and validity, and is capable of predicting behaviour (Helmreich et al., 1986).

Tools Needed

In order to complete the CMAQ, a pen and paper are required.

Example

Helmreich (1984) applied the CMAQ to 245 Boeing 737 pilots, consisting of captains and first officers. Table 9.18 represents an example of the items included in the CMAQ.

Table 9.18   Example CMAQ items

Item 1	The pilot flying the aircraft should verbalise his plans for manoeuvres and should be sure that the information is understood and acknowledged by the other pilot
Item 2	It is important to avoid negative comments about the procedures and techniques of other crew members
Item 3	Overall successful flight deck management is a primary function of the flying proficiency of the captain
Item 4	The captain should take control and fly the aircraft in emergency and non-standard situations
Item 5	First officers should not question the decisions or actions of the captain except when they threaten the safety of the flight.
Item 6	Captains should encourage their first officers to question procedures during normal flight operations and in emergencies
Item 7	There are no circumstances (except total incapacity) where the first officer should assume command of the aircraft

Source: Helmreich, 1984.

Helmreich (1984) was interested in exploring the level of agreement between the ratings provided by captains and first officers. Using the participant responses, Helmreich calculated the level of agreement between participants, both within the sub-groups of pilots and first officers and then between the groups. The results of the CMAQ would be in a format similar to that presented in Table 9.19. Once the pilot in-group level of agreement is derived, this can be compared to the first officer in-group level to identify similarities and differences. In this way, Helmreich (1984) was able to provide insights into what the pilots and the first officers felt was appropriate on the flight deck, as well as identifying areas of discrepancy and convergence.

Table 9.19   Example of participant response sheet to CMAQ items

Response Item 1					
	Strongly disagree	Slightly disagree	Neutral	Slightly agree	Strongly agree
Pilot A	1				
Pilot B					1
Pilot C	1				
Pilot D		1			
Agreement Level	50%	0%	0%	0%	0%

Source: Helmreich, 1984.

Flowchart

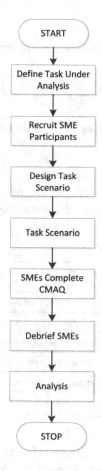

Targeted Acceptable Responses to Generated Events or Tasks (TARGETs)

Background and Applications

TARGETs (Fowlkes et al., 1994) is a form of structured observation to assess team skills. Task events are introduced to teams in order to trigger specific team responses. Acceptable responses are determined before the teams perform the tasks and during task performance, the teams' actual responses are scored in relation to these acceptable responses. The method is based on observers counting the occurrence of pre-defined behavioural responses to trigger events.

Domain of Application

The authors of the method believe that it is applicable to any team scenarios within the aviation domain as well as to wider team tasks.

Procedure and Advice

Step 1: Define the Task under Analysis
The initial stage of analysis involves the definition of the task under analysis. This definition guides the analysis and ensures that all data collected is relevant and appropriate.

Step 2: Develop Scenario Scripts
The analyst should now develop an appropriate and realistic task scenario. During this development, the analyst should consider when each trigger event will occur. Fowlkes et al. (1994) posit that a detailed script should be developed outlining when each trigger event should occur in order to ensure that future testing of the scenario maintains reliability.

Step 3: Recruit SMEs
Next, specific SMEs are identified and recruited. A set of specific requirements should be developed to guide SME selection in order to ensure that the appropriate SMEs are identified.

Step 4: Use SMEs to Define Acceptable Responses
Interviews with SMEs are required in order to elicit the acceptable responses for individual task steps that make up the task scenarios.

Step 5: Recruit Participants
Appropriate participants are identified, recruited and briefed.

Step 6: Introduce Participants to Task Events
Participants are then asked to complete the task scenarios.

Step 7: Rate Appropriate Responses to Task Events as Present or Absent
Raters observe the task performance and complete a rating chart. The rating chart contains each trigger event with the corresponding possible subsequent actions. Next to the subsequent actions is a space for the observer to annotate and rate the actions that take place. Each action that occurs is scored against the criteria on a rating scale of 0, 1 or 2: 2 if correctly done, 1 if partially correct and 0 if the action was not carried out.

Step 8: Analysis
Once the rating scores have been completed and compiled, overall performance scores can be obtained (proportion of TARGETs hit), as well as scores for specific task segments of specific groupings of TARGETs.

Advantages

- The technique allows for the evaluation of team performance within an environment similar to that in the real-world context (Fowlkes et al., 1994).
- The use of predefined behaviours and responses minimises subjective rater judgments.

- It allows for controlled observation across multiple teams.
- Fowlkes et al. (1994) argue that the method has enabled reliable and sensitive performance scores to be obtained from six military aircrews.
- The analyst can control which behaviours are elicited and when.
- An observer without expertise can observe and determine if specific behaviours are present or absent. Observers are not asked to rate the quality of responses, merely if they occur or not.

Disadvantages

- The method requires a high level of SME input.
- The scenario development involved in TARGETs is complex and time-consuming for complex tasks.
- The checklists developed to score teams are specific to the scenario under analysis.
- It is difficult to develop scenarios with comparable levels of complexity.

Approximate Application and Training Times

The training time for the method is low, as little subjective judgment is required. The application time depends upon the complexity of the task under analysis, but should be relatively short. The time to derive the rating scales and develop the scenarios could be considerable and complex.

Reliability and Validity

Previous applications of the method have shown reliable scores and good internal consistency (Fowlkes et al., 1994). Since the method minimises observer judgment, there should be a high level of inter-observer reliability (Annett, Cunningham and Mathias-Jones, 2000).

Tools Needed

In order to conduct the analysis access to the work domain and SMEs are required. The rating stage of the analysis can be conducted with pen and paper, although tables and graphs are best supported by software packages such as Microsoft Excel.

Example

Fowlkes et al. (1994) employed the TARGETs method to explore teamwork within dual-piloted military cargo helicopters. Helicopter crews of varying levels of skill and experience took part in the task scenario and were observed by a number of TARGETs raters. Fowlkes et al. developed a realistic task scenario for the crews to undertake within a simulated environment. Relevant behaviours to trigger were determined, as were appropriate behavioural responses. An example of a target behaviour utilised by Fowlkes et al. is 'Pilot question unsafe navigation procedure'. The corresponding appropriate behavioural response or 'Critical aircrew coordination behaviour' was defined by the authors as 'State opinions on decisions/procedures (under assertiveness)'.

Observer raters were then provided with observation rating sheets following the template shown below in Table 9.20. Every time the raters observed a target behaviour occurring, they ticked the 'hit' box.

Table 9.20   Example of target behaviours observation rating sheet template

Category	Event A	Target behaviour	Critical aircrew coordination behaviour	Hit
Description	A brief description of event A	A brief description of target A	A brief description of behaviour A	

Source: Fowlkes et al., 1994.

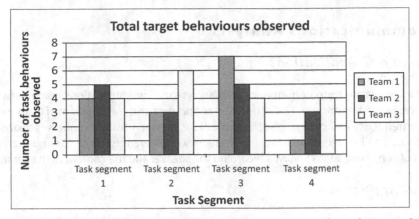

Figure 9.5 Example of analytical output illustrating the proportion of targeted behaviours observed for three teams across four task segments

Once the analysis was complete, Fowlkes et al. held data regarding the proportion of target behaviours observed during the flight scenario as a whole, and in individual task segments, for each team analysed. An example illustration of the analytical output is displayed below in Figure 9.5.

Flowchart

Team Communications Analysis

Background and Applications

One way of investigating team performance is to analyse the communications made between different team members. Jentsch and Bowers (2004) describe an approach to the assessment of team communication whereby the frequency and pattern of communications between team members is analysed. A simple frequency count can be used to measure the frequency of different types of communication. Analysing the patterns of communications involves recording the speaker and the content of the communication.

Domain of Application

Generic.

Procedure and Advice

Step 1: Define the Task under Analysis
The first step in a communications analysis is to define clearly the task (or tasks) under analysis. This enables the analyst to gain a clear understanding of the task content and also allows for the modification of the behavioural rating scale, whereby any behaviours missing from the scale that may be evident during the task are added. It is recommended that an HTA is conducted for the task under analysis. It is also worthwhile at this stage to select the team (or teams) who are to be observed.

Step 2: Create Observation Sheet
Before the data collection process begins, the analyst should create an appropriate data collection sheet for use during observation of the task under analysis. This should include sections to record a timeline, the communication type, the individuals involved, the content of the communication and the communication medium used. The aspects of communication recorded are dependent upon the nature and focus of the analysis. It may be that further categories are required, depending on the purpose of the analysis. An example data collection sheet is presented in Table 9.21.

Table 9.21   Example data collection sheet

Time	Communication type	Individuals involved	Communication content	Communications resource used

Step 3: Brief Participants
Before task performance begins, the participants should be briefed regarding the purpose of the analysis.

Step 4: Begin Task Performance
The communications analysis data collection process begins when the task under analysis starts. The observers should use the rating sheet to record communication type, content and speaker. It is recommended that the task under analysis is recorded using a video recorder so that the observers can watch the task afterwards in order to check the data for comprehensiveness and errors.

Step 5: Record All Communications Made
All of the communications made between team members should be recorded. The amount of information regarding each communication recorded is dependent upon the focus of the analysis. It is recommended that for each communication observed, the following data is recorded:

- the communication content;
- personnel involved;
- type of communication (SA, location, planning, etc.);
- related task component;
- technology used;
- error.

Step 6: Code Communications
According to Jenstch and Bowers (2004), communications should be coded using a content categorisation approach. Different categories of communication content should be determined and each communication should be subsequently coded. Example categories could be location, instruction, SA, planning, etc.

Step 7: Analyse Data
Once the data is coded correctly, the analyst should proceed to analyse the data as required. Jentsch and Bowers (2004) suggest that a lag-sequential or Markov-chain analysis is used to identify the pattern sizes and a contingency table is created, containing each chain of communications for each pattern size.

Advantages

- Communications analysis provides an assessment of the communications occurring in the team under analysis. This can be useful for training purposes, understanding errors in communication, analysing the importance of individuals within a team and analysing the communication resources used.
- The output can help provide a better understanding the team's communications requirements in terms of content and technology.
- It can be used to highlight redundant roles in the team.
- The output can be effectively used during training procedures.

Disadvantages

- The coding of communications is time-consuming and laborious.
- Recording communications during task performance is also tedious.

Related Methods

The initial data collection involved during a communications analysis uses observation as the primary data collection technique. Frequency counts or checklists are typically used when recorded the communications.

Approximate Training and Application Times

According to Jentsch and Bowers (2004), considerable training (10 hours plus) should be given to the analysts in order to ensure the reliability of the coding procedure. The application time includes an observation of the task (or tasks) under analysis and so could be quite considerable.

Reliability and Validity

According to Jentsch and Bowers (2004), acceptable reliability is achieved through the provision of appropriate analyst training.

Tools Needed

In order to record the task under analysis, video (camcorder) and audio (mini-disc recorder) recording equipment is required. The coding procedure can be completed using pen and paper. The data analysis phase requires Microsoft Excel and a statistical software package, such as SPSS.

Flowchart

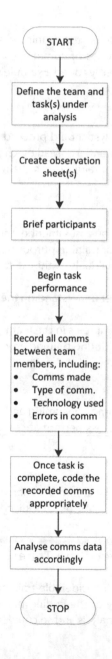

Chapter 10

Interface Analysis Methods

Introduction

Interface analysis methods are used to assess the man–machine interface of a particular system, product or device. Interface analysis techniques can be used to assess a number of different aspects associated with a particular interface, including usability, user satisfaction, error, layout, labelling and the controls and displays used. The output of interface analysis methods is then typically used to improve the interface in question through re-design. Such techniques are used to enhance design performance through improving the device or system's usability, user satisfaction, and reducing user errors and interaction time. ISO9241-11 requires that the usability of software is considered along three dimensions: effectiveness (how well does the product performance meet the tasks for which it was designed?); efficiency (how much time or effort is required to use the product in order to perform these tasks?) and attitude (e.g. how favourably do users respond to the product?). It is important to note that it is often necessary to conduct separate evaluations for each dimension rather than using one method and hoping that it can capture all aspects.

According to ISO13407, it is important to apply interface analysis techniques throughout a product's life-cycle, either in the design stage to evaluate design concepts or in the operational stage to evaluate effects on performance. In particular, this standard calls for the active involvement of users in the design process in order to gain an appropriate understanding of requirements and an appropriate allocation of function between users and the technology. It assumes that the design process is both multi-disciplinary and iterative. This suggests that there is a need to have a clear and consistent set of representations that can be shared across the design team and revised during the development of the design. In this chapter, we review methods that can fulfil these requirements. Most of the methods considered in this review require at least some form of interface, ranging from paper-based functional diagrams to the operational product itself, and most methods normally use end-users of the system, product or device under analysis.

A number of different types of interface analysis technique are available, such as usability assessment, error analysis, interface layout analysis and general interface assessment techniques. Indeed, it could be argued that the interface analysis category covers a number of methods described previously in this review, such as HEI. Usability assessment techniques are used to assess the usability (effectiveness, learnability, flexibility and attitude) of a particular interface. As Baber (2005a) notes, a significant element of evaluation lies in defining an appropriate referent model; it is not sufficient to simply conduct an evaluation of a product in isolation because this does not provide adequate grounds for making a judgment of the quality. Consequently, it is necessary to either make a comparison with another product or to define a target against which to make a judgment. Some questionnaire techniques such as the software usability measurement inventory (SUMI), the Post-Study System Usability Questionnaire (PSSUQ), the Purdue Usability Testing Questionnaire (PUTQ), the Usefulness, Satisfaction and Ease of use (USE) questionnaire, the questionnaire for user interface satisfaction (QUIS) and the system usability scale (SUS) provide scores that can be judged against some notion of a 'baseline'. Typically, these are completed by potential end-users based upon user trials with the device or system under analysis. Checklists such as Ravden and Johnson's (1989) HCI usability checklist are also employed to assess the usability of an interface.

The layout of an interface can also be assessed using techniques such as link analysis and layout analysis. As the names suggest, these methods are used to assess the layout of the interface and its effects upon task performance. More general interface analysis techniques such as heuristic evaluation (including those of Neilsen and Schneiderman) and user trials are used to assess the interface as a whole, and are flexible in that the focus of the analysis is determined by the analyst. The advantages associated with the use of interface analysis techniques lie in their simplistic nature and the usefulness of their outputs. Most of the techniques are simple to apply, requiring minimal time and costs, and also require only minimal training. The utility of the outputs is also ensured, as most approaches offer interface re-designs based upon end-user opinions. The only significant disadvantages associated with the use of interface analysis techniques are that the data analysis procedures may be time-consuming and laborious, and also that much of the data obtained is subjective. A brief description of the interface analysis methods considered in this review is given below.

Checklists offer a simplistic and low-cost approach to interface assessment. When using a checklist, the analyst checks the product or system interface against a pre-defined set of criteria in order to evaluate its usability. Conducting a checklist analysis is a matter of simply inspecting the device against each point on the chosen checklist. A number of checklists are available, including Ravden and Johnson's (1989) HCI checklist and Woodson, Tillman and Tillman's (1992) human engineering checklists. Heuristics analysis is one of the simplest interface analysis techniques available, involving simply obtaining subjective opinions of the analyst based on his or her interactions with a particular device or product. In conducting a heuristic analysis, an analyst or end-user should perform a user trial with the device or product under analysis and make observations regarding the usability, quality and error potential of the design. Schneiderman's eight golden rules and Nieslen's 10 heuristics are argued to be the most prominent examples of generic heuristic analysis guidance. Both authors present a series of criteria to guide the analyst in their assessment of the device or product. Interface surveys (Kirwan and Ainsworth, 1992) are a group of surveys that are used to assess the interface under analysis in terms of controls and displays used, their layout, labelling and ease of use. Each survey is completed after a user trial and conclusions regarding the usability and design of the interface are made.

Link analysis is used to evaluate and re-design an interface in terms of the nature, frequency and importance of links between elements of the interface in question. A link analysis defines links (hand or eye movements) between elements of the interface under analysis. The interface is then re-designed based upon these links, with the most-often-linked elements of the interface relocated to increase their proximity to one another. Layout analysis is also used to evaluate and re-design the layout of the interface in question. This involves arranging the interface components into functional groupings and then organising these groups by importance of use, sequence of use and frequency of use. The layout analysis output offers a re-design based upon the user's model of the task.

SUMI, QUIS, PSSUQ, PUTQ, USE and SUS are all examples of usability questionnaires. Typically, participants perform a user trial with the system or device under analysis and then complete the appropriate questionnaire. Overall usability scores and specific sub-scale scores for the system or device under analysis are then calculated.

Repertory grid analysis has also been used as an interface analysis technique (Stanton and Young, 1999a) and involves assessing user perceptions of the interface under analysis. A grid consisting of elements, constructs and opposites is formed and used to rate the interface elements. Walkthrough

Table 10.1   Summary of interface analysis techniques

Method	Type of method	Domain	Training time	App time	Related methods	Tools needed	Validation studies	Advantages	Disadvantages
Checklists	Subjective interface analysis	Generic	Low	Low	User trials	Pen and paper	Yes	1) Easy to use, low cost, requires little training. 2) Based upon established knowledge of human performance. 3) Offers a direct assessment of the system or device under analysis	1) Context is ignored when using checklists. 2) Data is subjective. 3) Inconsistent
Heuristic evaluation	Subjective interface analysis	Generic	Low	Low	User trials	Pen and paper	Yes	1) Easy to use, low cost, requires little training. 2) Output is immediately useful	1) Poor reliability and validity statistics. 2) Data is subjective. 3) Unstructured approach
Schneiderman's eight golden rules	Subjective interface analysis	Generic	Low	Low	User trials			1) Simple and quick. 2) Valuable output	1) Subjective. 2) Poor levels of reliability and validity
Nielson's 10 heuristics	Subjective interface analysis	Generic	Low	Low	User trials	Pen and paper	No	1) Simple and quick. 2) Produces immediate and valuable insights	1) Levels of reliability and validity are low. 2) The method is subjective
Interface surveys	Survey	Generic	Low	High	Surveys, user trials	Pen and paper	No	1) Easy to use, low cost, requires little training. 2) Potentially exhaustive. 3) Based upon traditional HF guidelines and standards	1) Time-consuming in application. 2) Surveys are dated. 3) Requires an operational system
Link analysis	Layout analysis	Generic	Low	Low	Observation, HTA	Pen and paper	Yes	1) Easy to use, low cost, requires little training. 2) Output is very useful, offering a logical redesign of the interface in question. 3) Can be used throughout the design process in order to evaluate design concepts (can be applied to functional diagrams)	1) Preliminary data collection involved, e.g. observation, HTA. 2) Does not consider cognitive processes. 3) Output is not easily quantifiable
Layout analysis	Layout analysis	Generic	Low	Low	Observation, HTA		Yes	1) Easy to use, low cost, requires little training. 2) Offers a redesign of the interface based upon importance, frequency and sequence of use. 3) Can be used throughout the design process in order to evaluate design concepts (can be applied to functional diagrams)	1) Poor reliability and validity statistics. 2) Preliminary data collection involved, e.g. observation, HTA. 3) May be difficult to use when considering complex interfaces
QUIS	Usability questionnaire	HCI	Low	Low	Questionnaires, user trials	Pen and paper	Yes	1) Quick and easy to use, involving little training and cost. 2) Output offers an assessment of the usability of the interface in question	1) May require substantial development to be used in the analysis of C4i. 2) Data is subjective

Table 10.1   Continued

Method	Type of Method	Domain	Training Time	App Time	Related Methods	Tools Needed	Validation Studies	Advantages	Disadvantages
Repertory grid analysis	Usability questionnaire	Product design	Med	High	N/A	Pen and paper	Yes	1) Structured, thorough procedure. 2) Easy to use. 3) Assesses end-user opinions	1) Procedure is a long and drawn-out one. 2) Does not always produce usable factors. 3) Reliability and validity questionable
SUMI	Usability questionnaire	HCI	Low	Low	Questionnaires, user trials	Pen and paper	Yes	1) Quick and easy to use, involving little training and cost. 2) Output offers an assessment of the usability of the interface in question. 3) Encouraging reliability and validity statistics	1) May require substantial development to be used in the analysis of C4i. 2) Data is subjective. 3) Only available commercially, costing over €1,000
SUS	Usability questionnaire	Generic	Low	Low	Questionnaires, user trials	Pen and paper	Yes	1) Quick and easy to use involving little training and cost. 2) Offers a usability score for the device under analysis	1) Output is limited. 2) Unsophisticated. 3) Data is subjective
User Trials	User trial	Generic	Low	High	Questionnaires, workload SA assessment techniques	Pen and paper	No	1) Can be used to assess anything from workload to usability. 2) Powerful insight into how the end product will potentially be used	1) Can be time-consuming. 2) Requires access to end-users
Walkthrough analysis	Task analysis	Generic	Low	Low	Talkthrough analysis	Pen and paper	No	1) Quick and easy to use, involving little training and cost. 2) Allows the analyst to understand the physical actions involved in the performance of a task. 3) Very flexible	1) SMEs required. 2) Access to the system under analysis is required. 3) Reliability is questionable
USE	Usability questionnaire	Generic	Low	Low	Questionnaires	Pen and paper	No	1) Provides usability metrics. 2) Enables tracking of usability	1) Exclusively positively phrased questions could bias results
PUTQ	Usability questionnaire	Generic	Low	Low	Questionnaires	Pen and paper	Yes	1) Levels of validity. 2) Strong theoretical basis	1) Does not explore enjoyment. 2) Restricted to traditional input devices
PSSUQ	Usability questionnaire	Generic	Low	Low	Questionnaires	Pen and paper		1) Valid and reliable measure. 2) Sensitive. 3) Can track usability progress	1) Exclusively positively phrased questions could bias results

analysis is a very simple procedure used by designers whereby experienced system operators or analysts perform a walkthrough or demonstration of a task or set of tasks using the system under analysis in order to provide an evaluation of the interface in question. User trials involve the potential system or device end-users performing trials with the interface under analysis and providing an assessment in terms of usability, user satisfaction, interaction times and error. A summary of the interface analysis techniques reviewed is presented in Table 10.1.

Checklists

Background and Applications

Checklists offer a quick, easy and low-cost approach to interface assessment. Typical checklist approaches involve analysts checking features associated with a product or interface against a checklist containing a pre-defined set of criteria. Checklist style evaluation can occur throughout the life-cycle of a product or system, from paper drawings to the finished product. Checklists can be used to evaluate the usability and design of a device or system in any domain. In the past, they have been used to evaluate product usability in the HCI (Ravden and Johnson, 1990), automotive (Stanton and Young, 1999a) and ATC domains. When using checklists, the analyst should have some level of skill or familiarity with the device under evaluation. Performing a checklist analysis is a matter of simply inspecting the device against each point on an appropriate checklist. Checklists are also very flexible in that they can be adapted or modified by the analyst according to the demands of the investigation. For example, Stanton and Young (1999a) used a section of Ravden and Johnson's HCI checklist in order to evaluate the design of in-car radios.

Domain of Application

Generic. Although checklist techniques originated in the HCI domain, they are typically generic and can be applied in any domain.

Procedure and Advice

Step 1: Select the Appropriate Checklist
First, the analyst must decide which form of checklist is appropriate for the product or system under investigation. The checklist used may be simply an existing one or the analyst may choose to adapt an existing checklist to make it more appropriate for the system under analysis. Alternatively, if a suitable checklist is not available, the analyst may choose to create a new checklist specifically for the system/product in question.

Step 2: Check Item on Checklist Against Product
The analyst should take the first point on the checklist and check it against the product or system under analysis. For example, the first item in Ravden and Johnson's checklist asks: 'Is each screen clearly identified with an informative title or description?' The analysts should then proceed to check each screen and its associated title and description. The options given are 'Always', 'Most of the time', 'Some of the time' and 'Never'. Using subjective judgment, the analyst should rate the device under analysis according to the checklist item. Step 2 should be repeated until each item on the checklist has been dealt with.

Flowchart

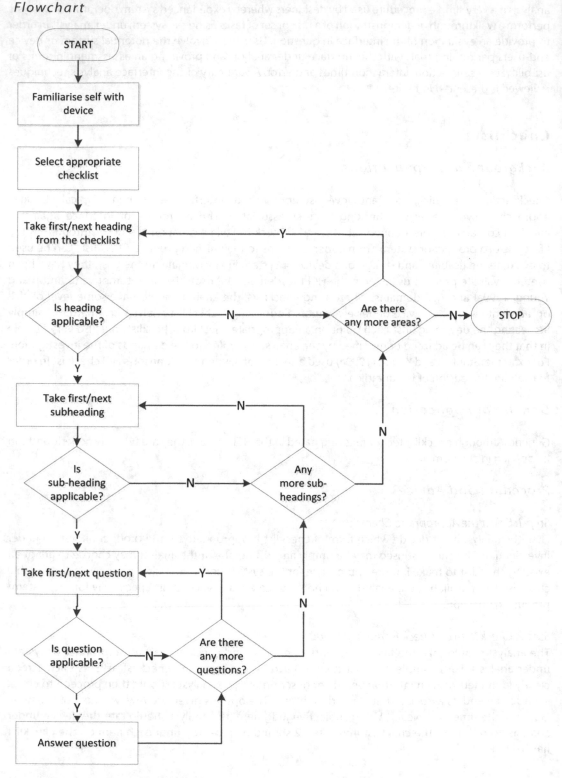

Advantages

- Checklists are quick and simple to apply, and incur only a minimal cost.
- They offer an immediately useful output.
- They are based upon established knowledge about human performance (Stanton and Young, 1999a).
- The technique requires very little training.
- Resource usage is very low.
- They are very adaptable and can easily be modified in order to use for other devices/systems. Stanton and Young (1999a) suggest that Ravden and Johnson's checklist), originally designed for HCI, can be easily adapted to cater for the usability of other devices, such as in-car stereos.
- A number of different checklists are available to the HF practitioner.

Disadvantages

- A checklist-type analysis does not account for errors or cognitive problems associated with the device.
- Context is ignored by checklists.
- Checklist data is totally subjective. What one analyst classes as bad design may be classed as suitable by another.
- Checklists offer a low level of consistency.
- They are not a very sophisticated approach to system evaluation.

Example

The following example (Table 10.2) is an extract of a checklist analysis of a Sony Ericcson t68i mobile phone using Ravden and Johnson's HCI checklist.

Table 10.2   Extract of checklist analysis

Section 1: Visual Clarity	A	M	S	N	Comments
1. Is each screen clearly identified with an informative title or description?				✓	Some screens lack titles.
2. Is important information highlighted on the screen (e.g. cursor position, instructions, errors)?			✓		
When the user enters information on the screen, is it clear: - where the information should be entered? - in what format it should be entered?			✓		
4. Where the user overtypes information on the screen, does the system clear the previous information, so that it does not get confused with the updated input?					N/A
5. Does information appear to be organised logically on the screen (e.g. menus organised by probable sequence of selection or alphabetically)?			✓		
6. Are different types of information clearly separated from each other on the screen (e.g. instructions, control options, data displays)?				✓	Different information is often grouped into lists.
7. Where a large amount of information is displayed on one screen, is it clearly separated into sections on the screen?			✓		
8. Are columns of information clearly aligned on the screen (e.g. columns of alphanumerics left-justified, columns of integers right-justified)?					
9. Are bright or light colours displayed on a dark background, and vice versa?			✓		
10. Does the use of colours make the displays clear?			✓		

Key: A = Always, M = Most of the time, S = Some of the time, N = Never.

Table 10.2   Continued

Section 1: Visual Clarity	A	M	S	N	Comments
11. Where colour is used, will aspects of the display be easy to see if used on a monochrome or low-resolution screen, or if the user is colour-blind?					
12. Is the information on the screen easy to see and read?					
13. Do screens appear uncluttered?			✓		
14. Are schematic and pictorial displays (e.g. figures and diagrams) clearly drawn and annotated?			✓	✓	
15. Is it easy to find the required information on a screen?			✓		Easy to get lost in menu system.

Key: A = Always, M = Most of the time, S = Some of the time, N = Never.

Related Methods

There are a number of checklists available to the HF practitioner, such as Woodson, Tillman and Tillman's (1992) human engineering checklists and Ravden and Johnson's (1989) HCI checklist.

Approximate Training and Application Times

Checklists require only minimal training time. Similarly, the application time associated with checklist techniques is minimal. In an analysis of 12 ergonomics methods, Stanton and Young (1999a) reported that checklists are one of the quickest techniques to train, practice and apply.

Reliability and Validity

Whilst Stanton and Young (1999a) reported that checklists performed quite poorly on intra-rater reliability, they also report that inter-rater reliability and predictive validity of checklists was good.

Tools Needed

Checklists can be applied using pen and paper only; however, for a checklist analysis, the analyst must have access to some form of the device under analysis. This could either be the finished article, paper drawings or a prototype version. An appropriate checklist is also required.

Heuristic Analysis

Background and Applications

Heuristic analysis techniques offer a quick and simple approach to interface evaluation. Heuristic analysis involves analysts providing subjective opinions based upon their interaction with a particular design, device or product. It is a flexible approach that can be used to assess a number of features associated with a particular product or interface, including usability, error potential, MWL and overall design quality. To conduct a heuristic analysis, an analyst (or team of analysts) performs a series of interactions with the product or interface under analysis, recording his or her observations as he or she proceeds. Heuristic-type analyses are typically conducted throughout the design process in order to evaluate design concepts and propose remedial measures for any problems encountered. The popularity of heuristic analysis lies in its simplicity and the fact that it can be conducted easily and with only minimal resource usage at any stage throughout the design process.

Domain of Application

Generic.

Procedure and Advice

Step 1: Define the Task under Analysis
The first step in a heuristic analysis is to define a representative set of tasks or scenarios for the system or device under analysis. It is recommended that heuristic analyses are based upon the analyst performing an exhaustive set of tasks with the device in question. The tasks defined should then be placed in a task list. It is normally useful to conduct an HTA for this purpose, based on the operation of the device in question. The HTA then acts as a task list for the heuristic analysis.

Step 2: Define Heuristic List
In some cases it may be fruitful to determine which aspects are to be evaluated before the analysis begins. Typically, usability (ease of use, effectiveness, efficiency and comfort) and error potential are evaluated.

Step 3: Familiarisation Phase
To ensure that the analysis is as comprehensive as possible, it is recommended that the analyst involved spends some time familiarising himself or herself with the device in question. This might involve consultation with the associated documentation (e.g. instruction manual), watching a demonstration of the device being operated or being taken through a walkthrough of the operation of the device.

Step 4: Perform Tasks
Once familiar with the device under analysis, the analyst should then perform each task from the task list developed during steps 1 and 2, and offer opinions regarding the design and the heuristic categories required. During this stage, any good points or bad points associated with the participant's interactions with the device should be recorded. If the analysis concerns a design concept, then a task walkthrough is sufficient. Each opinion offered should be recorded.

Step 5: Propose Remedies
Once the analyst has completed all of the tasks from the task list, remedial measures for any of the problems recorded should be proposed and recorded.

Advantages

- Heuristic analysis offers a quick, simple and low-cost approach to usability assessment.
- Due to its simplicity, only minimal training is required.
- It can be applied to any form of product, including paper-based diagrams, mock-ups, prototype designs and functional devices.
- The output derived is immediately useful, highlighting problems associated with the device in question.
- It involves the use of very few resources.
- It can be used repeatedly throughout the design life-cycle.

Disadvantages

- Heuristic analysis offers poor reliability, validity and comprehensiveness.

- It requires SMEs in order for the analysis to be worthwhile.
- It is subjective.
- It is totally unstructured.
- The consistency of such a technique is questionable.
- It is difficult to perform (Bennett and Stephens, 2009)

Example

The following example is taken from Stanton and Young (1999a). Heuristic analyses of Ford and Sharp in-car radio devices were conducted in order to assess the interface in terms of ease of skill acquisition, effectiveness on task, comfort/satisfaction and flexibility on task. The following heuristic analysis notes were recorded:

Ford radio
- Large on/off/volume button is very good.
- Pre-set buttons are large and clear; their positioning along the bottom of the unit is very good.
- Rocker seek button is satisfactory, good size and well located.
- Menu button a little small and awkward, also does not react enough when operated – could be more sensitive.
- News/TA buttons are well labelled and easy to operate.
- Pressing CD buttons for auto-reverse function is a little unconventional, but a good way of saving buttons.
- CD, AM/FM and Dolby buttons are well labelled and easy to use.
- Eject button is clear, easy to use and well positioned in relation to CD door.
- Very good consistency – all buttons have uniform size and labelling.
- PTY function is not very good; allocating generic titles to stations does not work very well.
- Display is well positioned and easy to read – informative and clear.
- RDS functions are a little obscure – necessary to read manual before initial operation.

Sharp radio
- On/off/volume control is a tad small and awkward, combined with difficult balance control.
- Push-button operation would be more satisfactory for on/off, as volume stays at preferred level.
- Fader control is particularly small and awkward.
- Both of the above points are related to the fact that a single button location has multiple functions – this is too complex.
- Treble and bass controls also difficult and stiff, although these functions are rarely adjusted once set.
- Station pre-set buttons are satisfactory; quite large and clear.
- Band selector button and FM mono-stereo button should not have two functions on each button – could result in confusion if the wrong function occurs. These buttons are the only buttons on the radio which are not self-explanatory – the user must consult the manual to discover their function
- Tuning seek and tuning scan buttons are easier to understand and use, although there are still two functions for one button.
- Auto-reverse function is not obvious, although it is an accepted standard.
- Illumination – is daytime/night-time illumination satisfactory? A dimmer control would probably aid matters.

Flowchart

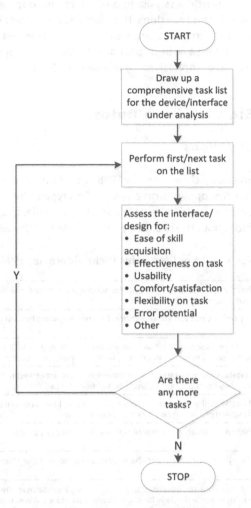

Approximate Training and Application Times

The technique requires very little, if any, training and the associated application time is also typically low.

Comparing a number of techniques in the analysis of in-car interfaces, Harvey and Stanton (in press) concluded that heuristic analysis required fewer time resources than methods such as HTA, SHERPA and CPA. They estimated that the method required one hour for data collection and one hour for analysis.

Reliability and Validity

In conclusion to their comparison of 12 HF methods, Stanton and Young (1999b) reported that the unstructured nature of the technique led to very poor results for reliability and predictive validity. Both intra- and inter-analyst reliability for the technique are questionable, due to its unstructured nature.

Tools Needed

Heuristic analysis is conducted using pen and paper only. The device under analysis is required in some form, e.g. functional diagrams, the actual device or paper drawings.

There are many forms of heuristic analysis, two of which are outlined in the following sections: Schneiderman's eight golden rules and Nieslen's 10 heuristics. These methods were chosen due to their prominence in the literature. Cronholm (2009) argues that both Nielsen and Schneiderman are frequently cited within the literature, illustrating a large impact on interface development. The importance of these two sets of heuristics is continually emphasised by researchers (Singh and Wesson, 2009).

Schneiderman's Eight Golden Rules

Background and Applications

Schneiderman's eight golden rules (Schneiderman, 1998) are a form of heuristic analysis. The rules guide the analyst in providing subjective opinions on a design. This type of heuristic analysis provides a quick and easy technique to evaluate the usability, quality and error potential of a design. Heuristic analysis can and should be conducted throughout the design life-cycle. Table 10.3 presents Schneiderman's heuristics.

Table 10.3   Schneiderman's eight golden rules (Schneiderman, 1998)

1. Strive for consistency	Ensure that throughout the system similar actions are required in similar situations; maintain consistency throughout processes and language
2. Enable frequent users to use shortcuts	Develop shortcuts such as abbreviations, function keys and hidden commands for expert users
3. Offer informative feedback	Ensure that users receive feedback for actions that they make. The feedback given should match the nature of the task, i.e. for small actions, feedback need only be modest
4. Design dialogue to yield closure	Ensure that the system organisation involves groups of actions with determinable ends. Feedback should be given once a set of actions is complete to build user satisfaction
5. Offer simple error handling	Ensure that the ability for a user to make an error is as low as possible. Ensure that the system is capable of detecting and providing support if errors are made
6. Permit easy reversal of actions	Ensure that users are able to reverse actions by following a simple procedure
7. Support internal locus of control	Design the system to make users the initiators of actions rather than the responders
8. Reduce short-term memory load	The limitation of human information processing in short-term memory requires that displays be kept simple, multiple-page displays be consolidated, window-motion frequency be reduced and sufficient training time be allotted for codes, mnemonics and sequences of actions

It is recommended that a team of analysts should be used when conducting heuristic analysis. Furthermore, SMEs or domain experts should also be used to enhance validity.

Domain of Application

Schneiderman's heuristics are generic and can be applied to any interface.

Procedure and Advice

Step 1: Define the Scenario or Task under Analysis
The first step in a heuristic analysis is to define a representative set of tasks or scenarios for the system or device under analysis.

Step 2: Define Task List
In some cases it may be fruitful to determine which aspects are to be evaluated before the analysis begins. Typically, usability (ease of use, effectiveness, efficiency and comfort) and error potential are evaluated.

Step 3: Perform Task
The analyst should then perform each task from the task list and offer opinions regarding the design and the heuristic categories required. If the analysis concerns a design concept, then a task walkthrough is sufficient. Each opinion offered should be recorded. It may be useful to record the session, ensuring that no data is missed.

Step 4: Propose Remedies
Once all tasks have been analysed, design remedies for each negative point highlighted should be proposed and recorded.

Advantages

- This is a very simple technique to apply, requiring very little training.
- Heuristic analysis can be very quick.
- It offers valuable output that is immediately useful.
- It involves the use of very few resources.
- It can be used repeatedly throughout the design life-cycle.

Disadvantages

- The technique offers poor reliability, validity and comprehensiveness.
- It requires a team of SMEs in order to be worthwhile.
- It is a subjective method based on the analyst's judgment.
- The consistency of such a technique is questionable.

Example

Schneiderman's heuristics have been applied to a number of interfaces in order to illustrate the utility of the method. Below are a number of screen shots representing interfaces that correlate well with Schneiderman's heuristics.

Strive for Consistency

Consistent sequences of actions should be required in similar situations; identical terminology should be used and consistent commands should be employed throughout, see below (Schneiderman, 1998).

Figure 10.1   Strive for consistency example

Adding new keyframes to the timeline within Adobe After Effects is the same for all parameters to which a keyframe can be added.

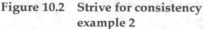

Figure 10.2   Strive for consistency   Figure 10.3   Enable frequent users to
                example 2   use shortcuts example

Within Adobe After Effects, when a new effect is added, its options and parameters area can be seen and adjusted within the panel on the left on the screen.

Enable Frequent Users to Use Shortcuts

Abbreviations, function keys, hidden commands and macro facilities are very helpful to an expert user (Schneiderman, 1998).

Adobe Photoshop CS4 allows users to create their own procedures using an Action function to reduce receptiveness of operations that are used frequently.

Offer Informative Feedback

For every operator action, there should be some system feedback. For frequent and minor actions, the response can be modest, while for infrequent and major actions, the response should be more substantial (Schneiderman, 1998).

In Adobe Photoshop CS4, when the selection lasso tool is selected, feedback is presented to the user by showing him or her where he or she chose to put each anchor point by drawing a faint line between them.

Once this action has been completed, the selection lines turn into marching ants (see Figure 10.4).

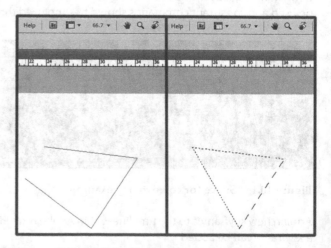

Figure 10.4   Offer informative feedback example

Design Dialogue to Yield Closure

Sequences of actions should be organised into groups with a beginning, middle and end. The informative feedback at the completion of a group of actions gives the operator the satisfaction of accomplishment, a sense of relief, the signal to drop contingency plans and options from his or her mind, and an indication that the way is clear to prepare for the next group of actions (Schneiderman, 1998). Autodesk 3ds Max is a 3D software package where three-dimensional objects can be constructed. Figure 10.5 shows the three-stage process of creating a cone, using the left mouse to select and moving it to adjust the shape.

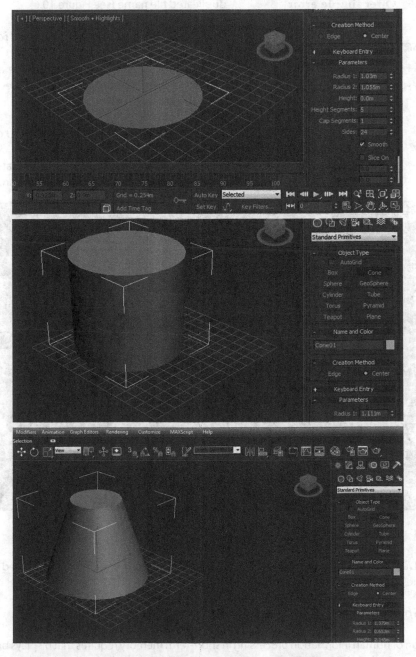

Figure 10.5   Design dialogue to yield closure examples

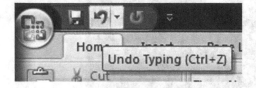

Figure 10.6   Offer simple error handling example

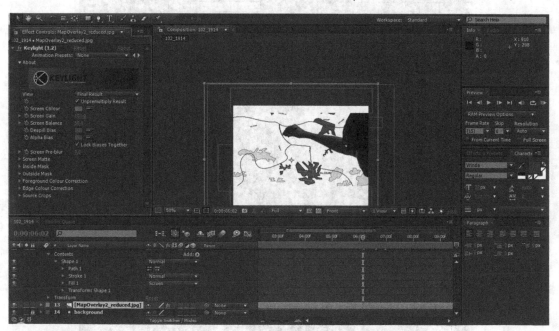

Figure 10.7   Permit easy reversal of actions example

Offer Simple Error Handling

As much as possible, the system should be designed so that the user cannot make a serious error. If an error is made, the system should be able to detect the error and offer simple, comprehensible mechanisms for handling it (Schneiderman, 1998). Windows Vista questions the user for confirmation when he or she has asked for a duplicated file name (see Figure 10.6).

Permit the Easy Reversal of Actions

The units of reversibility may be a single action, a data entry or a complete group of actions (see Figure 10.7) (Schneiderman, 1998).

Support Internal Locus of Control

Design the system to make users the initiators of actions rather than the responders (Schneiderman, 1998).

Figure 10.8   Support locus of control example

All Adobe Design software requires the user to do something; there are no automated functions to develop a design.

Reduce Short-Term Memory Load

The limitation of human information processing in short-term memory requires that displays be kept simple, multiple-page displays be consolidated, window-motion frequency be reduced and sufficient training time be allotted for codes, mnemonics and sequences of actions (Schneiderman, 1998). Adobe Bridge CS4 shows images of the documents with the document name instead of just the document name (see Figure 10.9).

Approximate Training and Application Times

The technique requires no training and the application time is very low.

Reliability and Validity

Heuristic analysis tends to have low levels of reliability and validity.

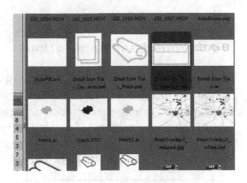

Figure 10.9 Reduce short-term memory load example

Tools Needed

The method requires pen and paper for application. It can be conducted on paper diagrams of the system under analysis (it is available online at http://faculty.washington.edu/jtenenbg/courses/360/f04/sessions/schneidermanGoldenRules.html).

Flowchart

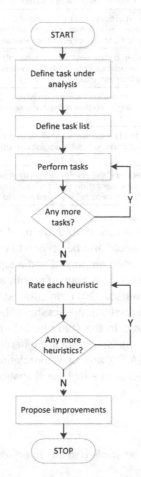

Nielson's 10 Heuristics

Background and Applications

Nielson's 10 heuristics (Nielsen, 1994a, 1994b) are a form of heuristic analysis. The set of 10 heuristics guide the analyst in deriving his or her subjective opinion of a design concept or product. This quick and easy method can be conducted at any stage of the design life-cycle in order to evaluate the usability, quality and error potential of the design. Nielsen's heuristics are given in Table 10.4.

Table 10.4   Nielsen's 10 Heuristics (Nielsen, 1994a, 1994b)

Visibility of system status	The system should always keep users informed about what is going on through appropriate feedback within a reasonable time
Match between system and the real world	The system should speak the user's language, with words, phrases and concepts familiar to the user, rather than system-oriented terms. It should follow real-world conventions, making information appear in a natural and logical order
User control and freedom	Users often choose system functions by mistake and will need a clearly marked 'emergency exit' to leave the unwanted state without having to go through an extended dialogue. Support undo and redo
Consistency and standards	Users should not have to wonder whether different words, situations or actions mean the same thing. Follow platform conventions
Error prevention	Even better than good error messages is a careful design which prevents a problem from occurring in the first place. Either eliminate error-prone conditions or check for them and present users with a confirmation option before they commit to the action
Recognition rather than recall	Minimise the user's memory load by making objects, actions and options visible. The user should not have to remember information from one part of the dialogue to another. Instructions for use of the system should be visible or easily retrievable whenever appropriate
Flexibility and efficiency of use	Accelerators – unseen by the novice user – may often speed up the interaction for the expert user such that the system can cater to both inexperienced and experienced users. Users should be permitted to tailor frequent actions
Aesthetic and minimalist design	Dialogues should not contain information which is irrelevant or rarely needed. Every extra unit of information in a dialogue competes with the relevant units of information and diminishes their relative visibility
Help users recognise, diagnose and recover from errors	Error messages should be expressed in plain language (no codes), should precisely indicate the problem and should constructively suggest a solution
Help and documentation	Even though it is better if the system can be used without documentation, it may be necessary to provide help and documentation. Any such information should be easy to search, focused on the user's task, list concrete steps to be carried out and not be too large

It is recommended that a team of analysts should be used when conducting heuristic analysis. Furthermore, SMEs or domain experts should also be utilised to enhance validity.

Nielsen's heuristics have previously been employed and adapted for multiple purposes, including website evaluation (McDonnell et al., 2009), evaluation of robotic arms (Tsui et al., 2009) and of driver-car interfaces (Brostrom, Bengtsson and Axelsson, 2011). In contrast to adaptations of Nielsen's heuristics, Singh and Wesson (2009) suggest that Nielsen's heuristics should be used alongside heuristics developed specifically for the system under analysis. In the study by Brostrom, Bengtsson and Axelsson (2011), Neilsens heuristics were unable to distinguish significant differences between two types of in-vehicle information systems, whereas the NASA-TLX and task completion times did distinguish between the systems. This suggests that the heuristics offer a high-level analysis that may benefit from the addition of more indepth analytical methods.

Domain of Application

Nielsen's heuristics are generic and can be applied to any interface or system.

Procedure and Advice

Step 1: Define the Scenario or Task under Analysis
The first step in a heuristic analysis is to define a representative set of tasks or scenarios for the system or device under analysis.

Step 2: Define Task List
In some cases it may be fruitful to determine which aspects are to be evaluated before the analysis begins. Typically, usability (ease of use, effectiveness, efficiency and comfort) and error potential are evaluated.

Step 3: Perform Task
The analyst should then perform each task from the task list and offer opinions regarding the design and the heuristic categories required. If the analysis concerns a design concept, then a task walkthrough is sufficient. Each opinion offered should be recorded. It may be useful to record the session, ensuring that no data is missed.

Step 4: Propose Remedies
Once all tasks have been analysed, design remedies for each negative point highlighted should be proposed and recorded.

Advantages

- The method is a very simple technique to apply requiring very little training.
- Heuristic analysis can be very quick.
- It produces valuable output that is immediately useful.
- It involves the use of very few resources.
- It can be used repeatedly throughout the design life-cycle.

Disadvantages

- The method offers poor reliability, validity and comprehensiveness.
- It requires a team of SMEs in order to be worthwhile.
- It is totally subjective.
- It is totally unstructured.
- The consistency of such a technique is questionable.
- The generic nature of the heuristics may mean that specific aspects of usability are not assessed (Singh and Wesson, 2009) and that adaptations are required (Tsui et al., 2009).

Example

The following examples represent interfaces which comply with Nielsen's (1994) heuristics.

Visibility of System Status
The system should always keep users informed about what is going on, through appropriate feedback within reasonable time (Nielsen, 1994a, 1994b). Here it is clear to see that the Bold function within the Windows Office software package has been enabled, as it is highlighted with a square box around the 'B' (see Figure 10.10).

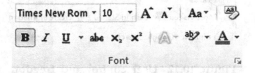

Figure 10.10   Visibility of system status example

Match between System and the Real World

The system should speak the user's language, with words, phrases and concepts familiar to the user, rather than system-oriented terms. The system should also follow real-world conventions, making information appear in a natural and logical order (Nielsen, 1994a, 1994b).

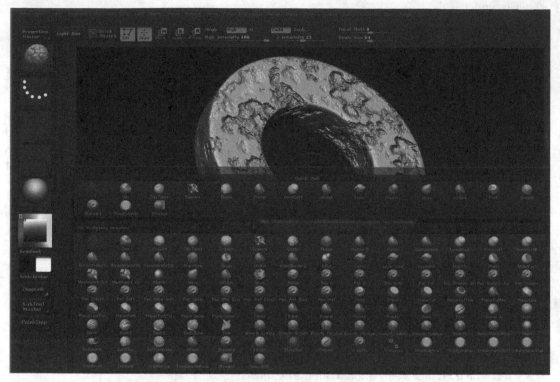

Figure 10.11   Match between system and real-world example

Within the Zbrush 3D software package, many of the panels are known as palettes, relating to areas where tools and brushes can be found, associating functionality with artists' and sculptors' terminology.

User Control and Freedom

Users often choose system functions by mistake and will need a clearly marked 'emergency exit' to leave the unwanted state without having to go through an extended dialogue. Support undo and redo (Nielsen, 1994a, 1994b).

The FireFox web browser provides the user with an undo function to enable quick navigation out of an unwanted state (see Figure 10.12 below).

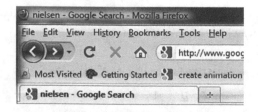

Figure 10.12   User control and freedom example

Consistency and Standards

Users should not have to wonder whether different words, situations, or actions mean the same thing. Follow platform conventions (Nielsen, 1994a, 1994b).

Windows Office software shows consistency between all of its products by presenting the majority of tool options within the Ribbon at the top of the document space (see Figure 10.13 on the next page).

Error Prevention

Even better than good error messages is a careful design which prevents a problem from occurring in the first place. Either eliminate error-prone conditions or check for them and present users with a confirmation option before they commit to the action (Nielsen, 1994a, 1994b).

Windows Vista uses 'pop-up' windows with a mouse hover function to enable the user to see what the document is before opening it (see Figure 10.14).

Recognition Rather than Recall

Minimise the user's memory load by making objects, actions, and options visible. The user should not have to remember information from one part of the dialogue to another. Instructions for use of the system should be visible or easily retrievable whenever appropriate (Nielsen, 1994a, 1994b).

Adobe Bridge CS4 shows images of the documents with the document name instead of just the document name (see Figure 10.15).

Flexibility and Efficiency of Use

Accelerators – unseen by the novice user – may often speed up the interaction for the expert user such that the system can cater to both inexperienced and experienced users (see Figure 10.16). Allow users to tailor frequent actions (Nielsen, 1994a, 1994b).

Aesthetic and Minimalist Design

Dialogues should not contain information which is irrelevant or rarely needed. Every extra unit of information in a dialogue competes with the relevant units of information and diminishes their relative visibility (Nielsen, 1994a, 1994b).

Within the directory browser within Windows Vista, when a document is hovered over, the document's details appear (see Figure 10.17 on the following page).

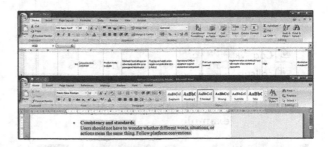

Figure 10.13   Consistency and standards example

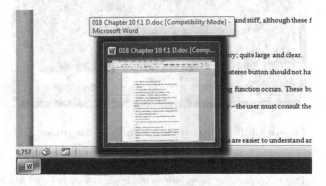

Figure 10.14   Error prevention example

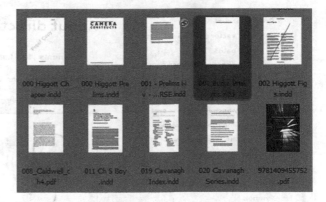

Figure 10.15   Recognition rather than recall example

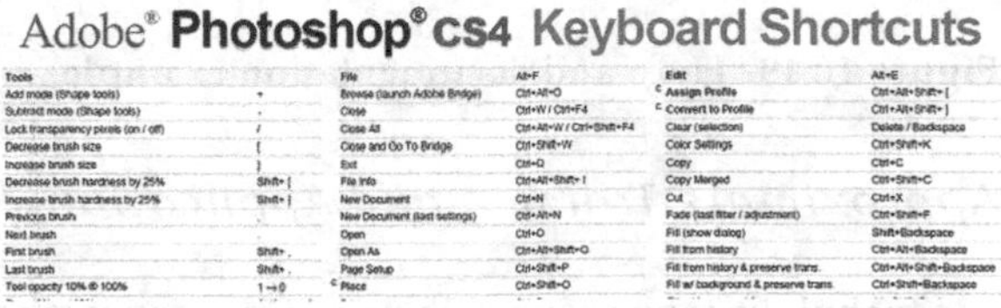

Adobe® **Photoshop® CS4** Keyboard Shortcuts

Figure 10.16   Flexibility and efficiency of use example

Figure 10.17 **Aesthetic and minimalist design example**

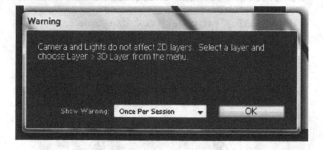

Figure 10.18 **Help users recognise, diagnose and recover from errors example**

Help Users Recognise, Diagnose and Recover from Errors

Error messages should be expressed in plain language (no codes), should precisely indicate the problem and should constructively suggest a solution (Nielsen, 1994a, 1994b).

The warning message in Figure 10.18 appears in Adobe After Effects CS4 if a 2D layer has not been set to be a 3D layer before a camera is added.

Help and Documentation

Even though it is better if the system can be used without documentation, it may be necessary to provide help and documentation. Any such information should be easy to search, should be focused on the user's task, should list concrete steps to be carried out and should not be too large (Nielsen, 1994a, 1994b).

The help section seen below in Figure 10.19 is from Abobe Illustrator CS4 and is accessed via the help tool.

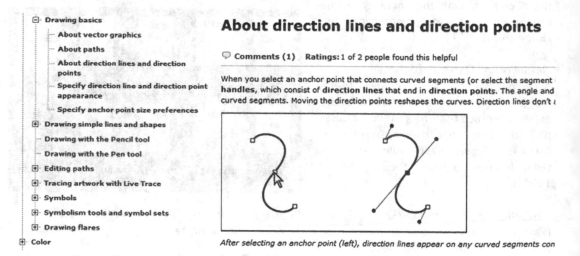

Figure 10.19 **Help and documentation example**

Approximate Training and Application Times

The technique requires no training and the application time is very low for a skilled HF analyst.

Reliability and Validity

Heuristic analysis tends to have poor levels of reliability and validity as the method is based upon the analyst's subjective judgment. Tsui et al. (2009) compared Nielsen's heuristics to a bespoke set they developed and found that the bespoke set was able to identify 67 per cent of problems not found by Nielsen. The specific nature of such adapted heuristics makes them more able to identify problems.

Tools Needed

The method requires pen and paper for application. It can be conducted on paper diagrams of the system under analysis. It is available online at http://www.useit.com/papers/heuristic/heuristic_list.html.

Flowchart

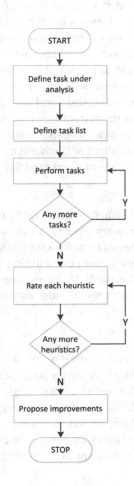

```
        ( START )
            │
            ▼
   ┌──────────────────┐
   │ Define task under│
   │     analysis     │
   └──────────────────┘
            │
            ▼
   ┌──────────────────┐
   │  Define task list│
   └──────────────────┘
            │
            ▼
   ┌──────────────────┐
   │   Perform tasks  │◄──────┐
   └──────────────────┘       │
            │                 │
            ▼            Y    │
        ◇ Any more ◇──────────┘
        ◇ tasks?   ◇
            │
            │ N
            ▼
   ┌──────────────────┐
   │ Rate each heuristic│◄─────┐
   └──────────────────┘       │
            │                 │
            ▼            Y    │
        ◇ Any more ◇──────────┘
        ◇ heuristics?◇
            │
            │ N
            ▼
   ┌──────────────────┐
   │Propose improvements│
   └──────────────────┘
            │
            ▼
        ( STOP )
```

Interface Surveys

Background and Applications

Kirwan and Ainsworth (1992) describe the interface survey technique, which is used to assess the physical aspects of man–machine interfaces. The technique involves the use of survey-based analysis to consider the following interface aspects:

- controls and displays;
- labelling;
- coding consistency;
- operator modification;

- sightline; and
- environmental aspects.

The interface surveys are used to pinpoint design inadequacies for an interface or design concept. A brief summary of each of the survey techniques is given below.

Control and Display Survey
A control and display survey is used to evaluate the controls and displays provided by a particular interface. According to Kirwan and Ainsworth (1992), the analyst should first record all of those parameters that can be controlled and displayed, and then create a list containing a description of each control used. Developing this list involves examining each control, recording exactly what the control is controlling, its location, type of control and any other relevant details, such as movement (e.g. up/down, rotary, left to right, etc.). Likewise, each display should be investigated in the same manner, e.g. display type, what is being displayed, location, etc. According to Kirwan and Ainsworth, the list should then be sorted into a hierarchical list containing the system, sub-system and parameter. The control and display list can then be used as a checklist to ensure that the system user is presented with adequate information and provided with the appropriate controls in order to perform the task. If required (depending upon the scope of the analysis), the appropriate guidelines or standards can also be applied in order to check that the system controls and displays adhere to the relevant guidelines/standards.

Labelling Surveys
Labelling surveys are used to examine the labelling provided by the interface under analysis. According to Kirwan and Ainsworth (1992), the following aspects of each label are recorded: reference, wording, size, position and colour. It may also be useful to make a subjective judgment on the clarity and ease of identification of each label identified. Any missing or confusing labels should also be recorded. Again, depending upon available resources, the labels identified can also be compared to the associated labelling standards and guidelines for the system under analysis. An extract of a labelling survey is presented in Table 10.4.

Coding Consistency Survey
Coding surveys are used to analyse any coding used on the interface under analysis. Typical types of coding used are colour coding (e.g. green for go, red for stop), positional coding, size coding and shape coding (Kirwan and Ainsworth, 1992). The coding analysis is used to highlight ambiguous coding and also where any additional coding may be required (Kirwan and Ainsworth, 1992). The analyst should systematically work through the interface, recording each use of coding, its location, the feature that is coded, description, relevance, instances where coding could be used but is not, instances of counteracting coding and any suggested revisions in terms of coding to the interface.

Operator Modification Survey
The end-users of systems often add temporary modifications to the interface in order to eradicate design inadequacies. Typically, operators use labels or markings to highlight where specific controls should be positioned or place objects such as paper cups over redundant controls. The modifications made by the end-users offer an intriguing insight into the usability of the interface, often highlighting bad design, poor labelling and simpler procedures (i.e. missing out one or two actions). Kirwan and Ainsworth (1992) suggest that such information can be gathered quickly through a survey of the operational system. The information collected can be used to inform the design of similar systems or interfaces. Conducting an operator modification survey simply involves observing a representative set of tasks being performed using the system under analysis and recording any instances of operator modification. The use of interviews is also useful in order to help understand why the modification occurred in the first place.

Sightline Surveys
A sightline survey involves an analysis of operator sightlines in terms of distance, angle and obstructions. Typically, a line is drawn from the operator's eye position to the display under analysis. If the line is interrupted, the obstruction should be recorded. Distance and angle of the sightline are also typically recorded. The output of a sightline survey can be presented in tabular or diagrammatic form.

Environmental Durvey
According to Kirwan and Ainsworth (1992), environmental surveys measure the state of the ambient environment, e.g. noise, illumination, temperature and humidity levels.

Domain of Application

Generic.

Procedure and Advice

Step 1: Select the Appropriate Survey and Prepare Data Collection Sheets
The first step in an interface survey-type analysis is to select the appropriate surveys that will be used during the analysis effort. This is dependent upon the focus of the analysis. Once the appropriate surveys are selected, data collection sheets should be created for each of the chosen surveys.

Step 2: Data Collection
The data collection phase involves completing each survey for the system under analysis. There are a number of ways to accomplish this. Access to the system under analysis is normally required, although Kirwan and Ainsworth (1992) suggest that the relevant data can sometimes be collected from drawings of the system under investigation. It is recommended that a walkthrough of the system under analysis is conducted, involving as representative a set of tasks of the full functionality of the system as possible. Observational study of task performance with the system under analysis is also very useful. For the operator modification surveys, interviews with system operators are required and for the environmental survey, online access to the operating system is required.

Step 3: Complete Appropriate Surveys
Once the data collection phase is complete, the appropriate surveys should be completed and analysed accordingly. The results are normally presented in tabular form.

Step 4: Propose Remedial Measures
Once the surveys are completed, it is often useful to propose any remedial measures designed to remove any problems highlighted by the surveys. Such recommendations might offer countermeasures for the system under analysis in terms of design inadequacies, error potential, poor coding, operator modifications, etc.

Advantages

- Each of the surveys described are easy to apply, requiring very little training.
- The surveys are generic and can be applied in any domain.
- Their output offers a useful analysis of the interface under analysis, highlighting instances of bad design and problems arising from the man–machine interaction.
- Standards and guidelines can be used in conjunction with the techniques in order to ensure comprehensiveness.
- If all of the surveys are applied, the interface in question is subjected to a very exhaustive analysis.

Disadvantages

- The application of the surveys is time-consuming.
- It is questionable whether such dated survey techniques will be useful in the analysis of contemporary complex, dynamic systems.
- An operational system is required for most of the techniques. The use of such techniques during the design process would be limited.
- Their reliability is questionable.
- Whilst they address the design inadequacies of the interface, no assessment of performance is given.

Example

Control and display and labelling surveys were conducted on the autopilot panel of a civil aircraft (Marshall et al., 2003) to examine their potential for use in the identification of design-induced pilot error. Extracts of each survey are presented in Tables 10.5 and 10.6.

Related Methods

Interface surveys are conducted on the basis of observational study and walkthrough analysis data of the task under investigation. Additionally, interviews are often used to inform interface survey analysis.

Approximate Training and Application Times

The training time for each of the surveys is estimated to be very low. However, the application time of survey techniques such as control and display survey and sightline survey is estimated to be very

Table 10.5   Extract of control and display survey for aircraft X autopilot panel

Control			Parameter	Display		
Name	Type	Comments		Name	Type	Comments
Speed/Mach selector	Rotary knob	Speed/Mach knob is very similar to the heading knob. These potentially could be confused with each other	Airspeed	Speed/Mach window	Digital, numerical	Window is very small and located in close proximity to the heading window. It is possible that the two may be confused
Heading/track selector knob	Rotary knob		Heading	Heading window	Digital, numerical	Window is very small and located in close proximity to the airspeed window. It is possible that the two may be confused

Table 10.6   Extract of labelling survey for aircraft X autopilot panel

Label	Description	Clarity	Error potential
Heading/track selector knob.	Blue triangle within white rotary knob.	Label is clear. No description of control included within the label.	Control is easily confused with other knobs in close proximity, e.g. speed/Mach selector knob.
Speed/Mach selector knob.	No label. White rotary knob. Same knob as the heading/track selector knob.	N/A No description of control.	Control is easily confused with other knobs in close proximity, e.g. heading/track selector knob.
Localiser.	White 'LOC' text located within black push-button control.	Very clear. LOC easily translated into localiser.	Control is similar in form to a number of others. However, the label is clear and identification of the control is immediate.

high. For example, a control and display survey involves recording each control and display used by the interface and then recording certain features regarding each control or display. This is a very time-consuming and laborious process.

Reliability and Validity

No data regarding the reliability and validity of the technique is presented in the literature. There may be problems associated with the intra- and inter-rater reliability of the technique. For example, different analysts may derive different results for the same interface, and also the same analyst may derive different results when using the technique for the same device or system on different occasions.

Tools Needed

Most of the surveys described can be applied using pen and paper. The environmental survey requires the provision of equipment capable of measuring the relevant environmental conditions, including noise, temperature, lighting and humidity levels. The sightline survey requires the appropriate measurement equipment, such as a tape measures and rulers.

Flowchart

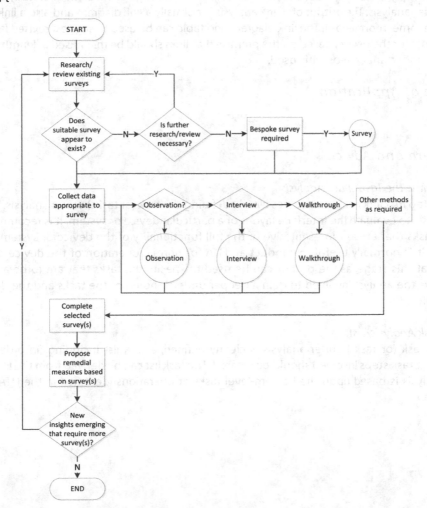

Link Analysis

Background and Applications

Link analysis is an interface evaluation method that is used to identify and represent 'links' in a system between interface components and operations, and to determine the nature, frequency and importance of these links. Links are defined as movements of attentional gaze or position between parts of the system, or communication with other system elements. For example, if an actor is required to press button A and then button B in sequence to accomplish a particular task, a link between buttons A and B is recorded. Link analysis uses spatial diagrams to represent the links within the system or device under analysis, with each link represented by a straight line between the 'linked' interface elements. Specifically aimed at aiding the design of interfaces and systems, the most obvious use of link analysis is in the area of workspace-layout optimisation (Stanton and Young, 1999a), i.e. the placement of controls and displays according first to their importance, then to their frequency of use, then to their function within the system and finally to their sequence of use. Link analysis was originally developed for use in the design and evaluation of process control rooms (Stanton and Young, 1999a), but it can be applied to any system where the user exhibits hand or eye movements, including driving, control room operation, aviation and ATC. When conducting a link analysis, establishing the links between system/interface components is normally achieved through a walkthrough or observational study of the task (or tasks) under analysis. The output of a link analysis is normally a link diagram and also a link table (both depict the same information). The link diagram and table can be used to suggest revised layouts of the components for the device, based on the premise that links should be minimised in length, particularly if they are important or frequently used.

Domain of Application

Generic.

Procedure and Advice

Step 1: Define the Task under Analysis
The first step in a link analysis involves clearly defining the task (or tasks) under analysis. When using link analysis to evaluate the interface layout of a particular device or system, it is recommended that a set of tasks that are as representative of the full functionality of the device or system as possible are used. It is normally useful to conduct an HTA for normal operation of the device or system in question at this stage, as the output can be used to specify the tasks that are to be analysed and also allows the analyst involved to gain a deeper understanding of the tasks and the device under investigation.

Step 2: Task Analysis/List
Once the task (or tasks) under analysis is clearly defined, a task list including (in order) all of the component task steps involved should be created. The task list can be derived from the HTA. Typically, a link analysis is based upon the bottom-level tasks or operations identified in the HTA developed during step 1.

Step 3: Data Collection

The analyst should then proceed to collect data regarding the tasks under analysis. This normally includes performing a walkthrough of the task steps contained in the task list and also observational study of the task in question. The analyst should record which components are linked by hand/eye movements and how many times these links occur during the tasks performed.

Step 4: Construct Link Diagram

Once the data collection phase is complete, construction of the link diagram can begin. This involves creating a schematic layout of the device/system/interface under analysis and adding the links between interface elements recorded during the data collection phase. Links are typically represented in the form of lines joining the linked interface elements or components. The frequency of the links is represented by the number of lines linking each interface element, e.g. seven lines linking interface elements A and B represents a total of seven links between the two interface elements during the task under analysis.

Step 5: Link Table

The link diagram is accompanied by a link table, which displays the same information as the link diagram, only in a tabular format. Components take positions at the heads of the rows and columns, and the numbers of links are entered in the appropriate cells.

Step 6: Re-design Proposals

Although not compulsory as part of a link analysis, a re-design for the interface under analysis is normally offered, based on the links defined between the interface elements during the analysis. The re-design is designed to reduce the distance between the linked interface components, particularly the most important and frequently used linked components.

Advantages

- Link analysis is a very simple technique that requires only minimal training.
- It is a quick technique that offers an immediately useful output.
- Its output helps to generate design improvements.
- It has been used extensively in the past in a number of domains.
- Its output prompts logical re-design of system interfaces.
- It can be used throughout the design process to evaluate and modify design concepts.

Disadvantages

- Link analysis requires preliminary data collection, including observational study and a walkthrough analysis of the task under analysis.
- The development of an HTA adds considerable time to the analysis.
- It only considers the basic physical relationship between the user and the system. Cognitive processes and error mechanisms are not accounted for.
- Its output is not easily quantifiable.

Flowchart

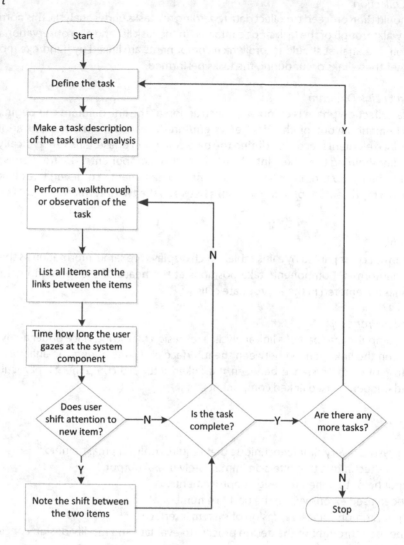

Example

The following example presents the results of a link analysis performed on the Sharp RG-F832E in-car radio, also see Tables 10.7 and 10.8 and Figures 10.20 and 10.21 on the opposite page (Stanton and Young, 1999a):

- task list;
- switch unit on;
- adjust volume;
- adjust bass;
- adjust treble;
- adjust balance;
- choose new pre-set;
- use seek, then store station;
- use manual search, then store station;

- insert CD;
- skip songs forwards and backwards;
- eject CD and switch off.

Related Methods

A link analysis normally requires an initial task description to be created for the task under analysis, such as an HTA. In addition, an observation or walkthrough analysis of the task under analysis should be performed in order to establish the links between components in the system.

Approximate Training and Application Times

In conclusion to their comparison of 12 ergonomics methods, Stanton and Young (1999a) reported that the link analysis technique is relatively fast to train and practice, and also that execution time is moderate compared to the other techniques (e.g. SHERPA, layout analysis, repertory grids, checklists and TAFEI).

Reliability and Validity

Stanton and Young (1999a) reported that link analysis performed particularly well on measures of intra-rater reliability and predictive validity. They also reported, however, that the technique was let down by poor inter-rater reliability.

Tools Needed

When conducting a link analysis, the analyst should have the device under analysis, pen and paper, and a stopwatch. For the observation part of the analysis, a video recording device is required. An eye-tracker device can also be used to record fixations during the task performance.

Table 10.7   Ford in-car radio components and functions (Stanton and Young, 1999a)

A = On/off/volume/balance/fader	H = CD eject button
B = Treble bass	I = CD compartment
C = Station preset buttons	J = Skip forward/programme buttons
D = FM mono stereo button	K = Tuning up/down buttons
E = DX-local button	L = Tuning scan/seek buttons
F = Band selector button	M = Tuning scan/seek buttons
G = ASPM/preset memory scan button	

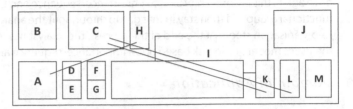

Figure 10.20   Link diagram for Ford in-car radio (Stanton and Young, 1999a)

Table 10.8   Link table for Ford in-car radio (Stanton and Young, 1999a)

	A	B	C	D	E	F	G	H	I	J	K	L	M
A	X												
B		X											
C			X										
D				X									
E					X								
F						X							
G							X						
H	1							X					
I								X					
J									X				
K		1									X		
L		1										X	
M													X

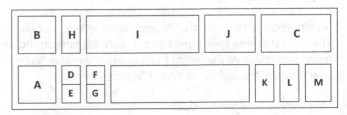

Figure 10.21   Revised design for Ford in-car radio (Stanton and Young, 1999a)

Layout Analysis

Background and Applications

Layout analysis is similar to link analysis in that it is based on spatial diagrams of the product and its output directly addresses interface design. It is used to analyse existing designs and suggests improvements to the interface arrangements based on functional grouping. The theory behind layout analysis is that the interface should mirror the user's structure of the task and the conception of the interface as a task map greatly facilitates design (Easterby, 1984). A layout analysis begins by simply arranging all of the components of the interface into functional groupings. These groups are then organised by their importance of use, sequence of use and frequency of use. The components within each functional group are then re-organised; once again, this is done according to importance, sequence and frequency of use. The components within a functional group will then stay in that group throughout the analysis and cannot move anywhere else in the re-organisation stage. At the end of the process, the analyst has re-designed the device in accordance with the user's model of the task based upon importance, sequence and frequency of use.

Domain of Application

Generic.

Procedure and Advice

Step 1: Schematic Diagram
First, the analyst should create a schematic diagram for the device under analysis. This diagram should contain each (clearly labelled) interface element.

Step 2: Arrange Interface Components into Functional Groupings
The analyst begins by arranging the interface components into functional groupings. Each interface element should be grouped according to its function in relation to the device under analysis. For example, the interface components of a Ford in-car radio were arranged into the functional groups radio and cassette (Stanton and Young, 1999a). This part of the analysis is based entirely upon the subjective judgment of the analyst involved.

Step 3: Arrange Functional Groupings into Importance of Use
Next, the analyst should arrange the functional groupings into importance of use. The analyst may want to make the most important functional group the most readily available on the interface. Again, this is based entirely on the analyst's subjective judgment.

Step 4: Arrange Functional Groupings into Sequence of Use
The analyst should then repeat step 3, only this time arranging the functional groupings based on their sequence of use.

Step 5: Arrange Functional Groupings into Frequency of Use
The analyst should then repeat step 3, only this time arranging the functional groupings based on their frequency of use. At the end of the process, the analyst has re-designed the device according to the end-user's map of the task (Stanton and Young, 1999a).

Step 6: Re-design the Interface
Once the functional groups have been organised based on their importance, sequence and frequency of use, the process is repeated for within functional group items. The analyst should base the interface re-

design on the three categories (importance, sequence and frequency of use). For example, the analyst may wish to make the most important and frequently used aspect of the interface the most readily available.

Flowchart

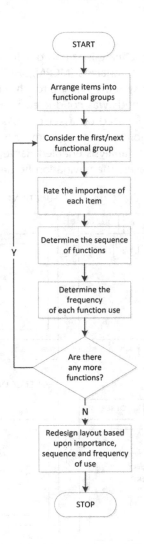

Advantages

- Layout analysis offers a quick, easy to use and low-cost approach to interface design and evaluation.
- It has a low level of resource usage.
- It requires only minimal training.
- Itan be applied to paper diagrams of the device/interface under analysis.
- Its output provided is immediately useful, offering a re-design of the interface under analysis based on importance, sequence and frequency of use of the interface elements.

Disadvantages

- Layout analysis offers poor reliability and validity (Stanton and Young 1999a).
- The output of the technique is very limited, i.e. it only caters for layout. Errors, MWL and task performance times are ignored.

- Literature regarding layout analysis is extremely sparse.
- If an initial HTA is required, application time can rise dramatically.
- Conducting a layout analysis for complex interfaces may be very difficult and time-consuming.

Example

Harvey and Stanton (2012) used layout analysis as part of a five-method evaluation. Heuristic analysis, layout analysis, multimodal critical path analysis (MCPA), SHERPA and HTA were each applied to the analysis of two in-vehicle information systems (IVISs): a touch-screen interface and a remote-controlled interface. The application of the methods was conducted by HF analysts after expert walkthroughs had taken place. The walkthroughs focused on a set of tasks which were representative of typical IVIS functions, including set navigation destination from system memory, reduce fan speed by two increments and play radio station. The analysis enabled Harvey and Stanton to identify a number of problems associated with IVIS interfaces. The navigation home menu screen in an IVIS which used a

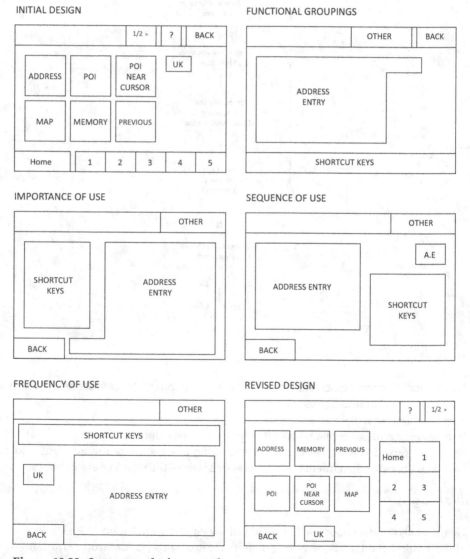

Figure 10.22 Layout analysis example

remote-control input device was identified as a suitable candidate for layout analysis: it had previously been identified by the heuristic analysis as being cluttered, with too much detail displayed on a single screen. Figure 10.22 on the previous page presents an illustration of the initial and revised interfaces along with the analysis conducted.

The frequency, sequence and importance of use rules were used to re-design the layout of this menu screen. The back button was considered to have relatively high importance and was therefore relocated to make it more visible; in the original design, it was partially obscured by the screen surround. The 'other' functions were relocated to the back button's original position as these were considered less important and would be used less frequently. The shortcut keys were considered to have relatively high frequency of use because they provide instant access to pre-programmed destinations: they therefore grouped closer together to speed up cursor-based navigation. The 'home' function was located with the shortcut keys because it is also a pre-programmed function and this placing therefore seemed most logical.

Related Methods

Layout analysis is very similar to link analysis in its approach to interface design.

Approximate Training and Application Times

In conclusion to their comparison study of 12 ergonomics methods, Stanton and Young (1999a) reported that little training is required for layout analysis and that it is amongst the quickest of the 12 techniques to apply. It is therefore estimated that the training and application times associated with the technique are low. However, if an initial HTA is required, the application time would rise considerably.

Comparing a number of techniques in the analysis of in-vehicle car interfaces, Harvey and Stanton (in press) concluded that layout analysis was quicker to use than methods such as HTA, SHERPA and CPA. They estimated that the method required one to two hours for data collection and an hour for analysis.

Reliability and Validity

Stanton and Young (1999a) reported poor statistics for intra-rater reliability and predictive validity for layout analysis.

Tools Needed

Layout analysis can be conducted using pen and paper, providing the device or pictures of the device under analysis are available.

The Questionnaire for User Interface Satisfaction (QUIS)

Background and Applications

The QUIS is a technique that is used to assess user acceptance and opinions of human–computer interfaces. It is designed to elicit subjective user opinions on all usability-related aspects of an interface, including ease of use, system capability, consistency and learning. There are a number of different versions of the QUIS technique available and it has been applied in various domains, including computer-supported collaborative learning (Vatrapu, Suthers and Medina, 2008) and the development of search engine interfaces (Chau and Wong, 2010). It uses questions relating to the use of human–computer interfaces. Each question has an associated rating scale, typically ascending from

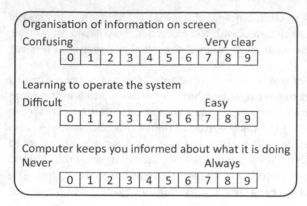

Figure 10.23   Example QUIS statements
Source: Chin, Diehl and Norman, 1988.

1 to 10. Examples of the QUIS statements are presented in Figure 10.23 on the right.

Procedure and Advice

Step 1: Identify User Sample

The first step in a QUIS analysis is to identify the user sample that will be used in the investigation. It is recommended that the user sample utilised represents a portion of the typical users of the software system or type of software system under analysis. It might also be useful to present the system to the users, so that they can experience examples of the tasks that could be performed using that system.

Step 2: Define Representative Task List for the System under Analysis

Once the participant sample has been defined, the analyst should develop a representative task list for the software system under investigation. This task list should be exhaustive, representing every possible task that can be performed using the system under analysis. This list represents the set of tasks that the participants will perform during the analysis. If the task list is too great (i.e. it requires more time to complete than is allowed by the scope of the analysis), the analyst should pick as representative a set of tasks as possible. It is recommended that an HTA for the software system under analysis be used to develop the task list.

Step 3: QUIS Briefing Session

Before the task performance step of the QUIS analysis, the participants should be briefed regarding the purpose of the analysis and how to complete the QUIS questionnaire. It may be useful for the analyst to run through the task list and the QUIS questionnaire, explaining any statements that may cause confusion. In some cases, a demonstration of the tasks required may be pertinent. The participants should be encouraged to ask any questions regarding the completion of the QUIS questionnaire and the task list at this point.

Step 4: Task Performance

Once the participant sample and task list have been defined, and the participants fully understand the tasks that they are required to perform and also how to complete the QUIS questionnaire, the task performance can begin. The participants should now be given the task list and be instructed to perform, as normal, the tasks in the order that they are specified using the system under analysis. It is important that no conferring between participants takes place during the task performance and also that no help is administered by the analyst. The task performance should go on as long as is required for each participant to complete the required task list.

Step 5: Administer QUIS

QUIS is normally administered post-trial. Once all of the participants have completed the task list for the software system under analysis, the QUIS should be administered. After a brief demonstration of how to complete the QUIS, the participants should be instructed to complete the questionnaire, basing their responses on the tasks that they have just carried out with the interface in question. Again, no conferring between participants is permitted during this step, although the analyst may assist the participants with statements that they do not understand.

Step 6: Calculate Global and Sub-scale QUIS Scores
Once all of the QUIS questionnaires are completed and handed in, the scoring process begins. The analyst may choose to calculate a global QUIS score and scores for each of the separate QUIS sub-scales (e.g. system capability, learning, screen, terminology and system information). These scores can then be averaged across participants in order to obtain mean scores for the system under analysis.

Advantages

- QUIS is a very quick and easy technique to use, requiring only minimal training.
- Its output is immediately useful, offering an insight into the system users' attitudes regarding the usability of the interface under analysis.
- If the correct sample is used, the data obtained is potentially very powerful, offering an end-user rating of system usability.
- Once an operational system is available, the speed, ease and usefulness of QUIS allow it to be used repeatedly throughout the design life-cycle to evaluate and modify design concepts.
- It provides encouraging reliability and validity statistics.
- It can be modified to suit analysis needs. For example, QUIS statements can be added and removed in order to make the analysis more suitable for the software system under analysis.
- It can be used effectively even with small sample sizes.

Disadvantages

- QUIS is limited to the analysis of HCI devices.

Related Methods

The QUIS technique is a questionnaire technique used for the analysis of HCI interfaces. There are a number of similar techniques which use attitude scales to assess the usability of a device or system, such as SUMI and the SUS technique.

Approximate Training and Application Times

The training time for the QUIS technique is very low, with little or no training required. In terms of application time, it is estimated that the QUIS questionnaire would take 5–20 minutes to complete. However, the total time for a QUIS analysis is dependent upon the length of the task performance stage, i.e. the number of tasks that the participants are required to perform before they complete the QUIS questionnaire. It is estimated that the total application time for QUIS is low and that even in scenarios where the task list is very large, the total application time would probably not exceed two hours.

Reliability and Validity

Chin, Diehl and Norman (1988) report a number of studies designed to assess the reliability and validity of the QUIS. In a study using QUIS versions 3.0 and 4.0 to assess the interactive batch-run IBM mainframe and an interactive syntax-directed editor programme environment, QUIS was found to have a high level of reliability (Version 3.0, 0.94, Version 4.0, 0.89, Cronbach's alpha). In another study, QUIS version 5.0 was used by participants to evaluate a product that they liked, a product that they disliked, MS-DOS and another comparable software product (e.g. WordStar™, WordPerfect™, Lotus™, Dbase™, etc.). Overall reliability was reported as 0.94 (Cronbach's alpha). Establishing the validity of the technique has proved to be more difficult and there is limited data regarding this available in the literature. According to Chin, Diehl and Norman (1988), there are two reasons for the difficulty in establishing the validity of the technique: first, there is a lack of theoretical constructs regarding HCI with which to test QUIS; and, secondly, there is a lack of established questionnaires that are available for cross-validating purposes.

Tools Needed

QUIS is normally applied using pen and paper. An operational version of the interface under analysis is also required for the task performance component of the QUIS analysis. A computerised version is also available.

Flowchart

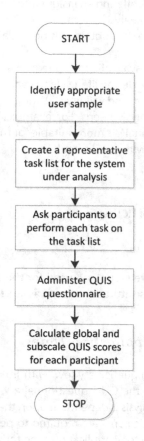

Repertory Grid Analysis

Background and Applications

The repertory grid technique is an interview-based technique that can be used to analyse participant perceptions or views regarding a set of similar things (these could be people, products, systems or devices). The technique was developed by Kelly (1955) to support his theory of personal 'constructs'. This theory assumes that people seek to develop a view of the world that allows them to combine their experiences and emotions into a set of 'constructs'. These can then be used to evaluate future experiences in terms of how positively or negatively they relate to a construct. Repertory grids have since been employed for a number of different purposes, including for product evaluations of in-car radio players (Stanton and Young, 1999a) and microwave ovens (Baber, 1996), the evaluation of different text types (Dillon and McKnight, 1990), the evaluation of consumer behaviour (Baber, 1996) and the evaluation of collaboration (Schuler et al., 1990). Repertory grid analysis can be used either in the early design life-cycle in order to provide an insight into how potential users think about the product in question and to specify product requirements and design preferences, or to evaluate existing product designs in terms of user attitudes. The technique involves presenting a single participant with a set of similar products or proposed product designs and eliciting constructs and contrasts for these items. Each construct and contrast is then analysed in relation to each product under analysis, and a set of factors for the product group is specified. The output of a repertory grid can either be qualitative or quantitative.

Domain of Application

Generic.

Procedure and Advice

The following procedure is adapted from Baber (2005b).

Step 1: Determine Products/Devices to be Compared
The first step in a repertory grid analysis is to determine which products or devices will be compared. If the analysis is based on an early design concept, a number of different design concepts may be compared. If it is based on the evaluation of existing products or devices, then each item should possess a common feature of some sort. It is useful at this stage to provide a description that caters for all of the items together in order to clarify the relation between the items. Baber (2005b) described a repertory grid analysis of 'wearable technology items'. The three items compared in this analysis were wristwatches, head-mounted displays and global positioning system (GPS) units. While the items might be very different, it is important to present them to the participant in terms of an overall category, such as 'wearable technology items', rather than as separate and unrelated items.

Step 2: Brief the Participant
Once the items under analysis are defined, the participant should be briefed regarding his or her role in the analysis, i.e. the participant should be informed that he or she is required to indicate how two of the items are similar and differ from the third, and then provide a short word or phrase to justify that item's selection.

Step 3: Determine Constructs
Using the items under analysis (or photographs of them), the analyst should present each item to the participant. The participant should then be encouraged to decide which of the two items are the most similar and then to describe how the third item is different. According to Baber (2005b), it is crucial that the participant provides a reason for his or her selection(s) and it is also useful for the analyst and participant to agree on a short word or phrase that describes the construct for the triad. Baber (1996) suggests that this step is repeated until the participant is unable to offer any new constructs.

Step 4: Construct Repertory Grid Table
Once all of the constructs are noted, the analyst should construct the repertory grid table. Each product or device should appear across the top row and the constructs should appear down the right-hand column of the table.

Step 5: Define Contrasts or Opposites
Next, the analyst has to gather a contrast or opposite for each construct identified from the participant. This involves probing the participant for an opposite for each construct in the repertory grid table.

Table 10.9   Constructs and contrasts for two in-car radio players

Constructs	Contrasts
Mode dependent	Separate functions
Push-button operation	Knob-turn operation
Bad labelling	Clear labelling
Easy controls	Fiddly controls
Poor functional grouping	Good functional grouping
Good illumination	Poor illumination

Source: Stanton and Young, 1999a.

Table 10.9 on the left displays the constructs and their contrasts identified by Stanton and Young (1999a) in a product evaluation of two in-car radio players.

Step 6: Relate Constructs to Items

Next, the participant should be asked to state whether or not each construct 'fits' each item or product, i.e. does radio A have bad labelling or good labelling? Only a 'yes' or 'no' answer should be given by the participant. Each 'yes' is represented by a '1' in the repertory grid table, while each 'no' is represented by a '0' in the repertory grid table.

Step 7: Review Repertory Grid Table

Step 7 acts as a check that each construct in the repertory grid table is scored correctly. The analyst simply takes the participant through each product and construct in the repertory grid table and ensures that the participant agrees with the scores given.

Step 8: Perform First Pass Analysis

The 'first pass' (Baber, 2005b) involves constructing a template in order to determine the membership of the group. For each product or device included in the analysis, the columns should be summed (0's and 1's). The value for each column is then entered into the appropriate cell at the foot of the column. To convert these totals into a template, the analyst should define a numerical cut-off point in the totals. The aim is to divide the items into approximately two groups. Each column total equal to or below the cut-off point is assigned a 0, whilst each column total above the cut-off total is assigned a 1. The template values should then be added to their appropriate columns, underneath the relevant column totals.

Step 9: Compare Template with Constructs

Once the template is entered into the repertory grid table, the analyst should compare the template with each construct. To do this, the template is overlaid upon each construct line in the repertory grid table. For all those cases where the template value matches the corresponding construct value, a 1 should be added to the score. If the number of matches is less than half of the total number of comparisons, then a negative score is assigned. If not, the total should be calculated and entered into the table. Once the table is complete, the analyst should proceed to step 11.

Step 10: Reflection Phase

If the construct value is not matched to the template value, then the analyst should reverse the construct/contrast. The reflection phase is used to remove any negative scores from the repertory grid.

Step 11: Define Groups

Next, the analyst should define the groups of constructs. The method assumes binomial distribution of responses, i.e. a statistical majority is required before a group can be defined.

Step 12: Name Factors

The participant is then asked to provide names for any factors that have been identified, i.e. in an analysis of microwave ovens (Baber, 1996), the group of constructs comprising touch pad, digital clock, >90 minute timer, memory and delay start were named 'technical sophistication'.

Example

The following example is a repertory grid analysis of the factors influencing consumer decisions involved in the selection of a microwave oven (Baber, 1996). Eight microwave ovens were used, along with one female participant.

The construct elicitation phase was conducted using photographs of the products combined with additional details provided by the manufacturers. The subject was presented with three photographs and was asked to provide a construct to define a pair. The resultant constructs and their contrasts are shown in Table 10.10.

The participant then rated each item and associated construct. A 1 value represents an agreement between the item and construct, whilst a 0 value represents an agreement between the item and its contrast. Next, each column is summed and a template is created. The cut-off point defined for this analysis is 7. This gives 4 scores of 7 or less (7, 6, 7, 5) and 4 scores of more than 7 (10, 8, 9, 10). All scores above 7 are scored as 1, whilst all equal to or less than 7 are scored as 0. This forms the template for the next stage of analysis (i.e. 0 0 0 1 1 1 0 1). The template is then overlaid on each row in the table and the number of matches between the values in the repertory grid table and the template are calculated. For example, taking the template and comparing with the first row (Construct = Dials), we can see that 1 1 1 0 0 0 0 0 compared with 0 0 0 1 1 1 0 1 has only one number in common (the '0' in the 6th position). Column Fla in Table 10.11 shows the total number of matches or 'hits'. If there are more misses than hits, a negative score is placed in the Fla column. So, for our example, there are 7 mismatches, so the Fla for the first row is -7. If there are more hits than misses, a positive score is assigned. If there are an equal number of hits and misses, a 0 is entered into the column. Table 10.11 shows the initial repertory grid table and first pass analysis.

Next, the 'reflection phase' is used to remove any negative scores from the repertory grid table (Baber, 1996). This is achieved by reversing the construct/contrast. Table 10.12 on the next page shows the repertory grid after the reflection phase.

Once the reflection phase is complete, common constructs are extracted from the repertory grid. According to Baber (1996), binomial theorem is used to determine the

Table 10.10 Constructs and contrasts for microwave ovens (adapted from Baber, 1996)

Construct	Contrast
Dials	Touch pad
<800W	>800W
Clock	No clock
White	Black
Timer (90 min)	<90 min
Memory	No memory
Grill	No grill
<5 settings	>5 settings
Defrost	No defrost
Button (door)	Lever (door)
<£130	>£130
Fitted plug	No plug
Delay start	No delay
<0.8ft^3 capacity	>0.8ft^3 capacity

Table 10.11 Initial repertory grid table and first pass analysis

Item number										
1	2	3	4	5	6	7	8	Construct	Contrast	Fla
1	1	1	0	0	0	0	0	Dials	Touch pad	-7
1	1	1	1	0	1	0	1	<800W	>800W	0
0	0	0	1	1	1	1	1	Clock	No clock	7
1	1	1	1	1	1	0	1	White	Black	5
0	0	0	1	1	1	1	1	Timer (90 min)	<90 min	7
0	0	0	1	1	1	1	1	Memory	No memory	8
0	0	0	0	0	0	0	1	Grill	No grill	5
1	1	0	0	1	0	0	1	<5 settings	>5 settings	0
0	0	0	0	1	1	1	1	Defrost	No defrost	6
1	0	1	1	1	1	1	0	Button (door)	Lever (door)	0
1	1	1	1	0	0	0	0	<£130	>£130	-6
0	0	1	1	0	0	0	0	Fitted plug	No plug	0
0	0	0	1	1	1	1	1	Delay start	No delay	7
1	1	1	1	0	1	0	1	<0.8ft^3 capacity	>0.8ft^3 capacity	0
7	6	7	10	8	9	5	10			
0	0	0	1	1	1	0	1			

Table 10.12 Modified repertory grid table

1	2	3	4	5	6	7	8	Construct	Contrast	Fla

1	2	3	4	5	6	7	8	Construct	Contrast	Fla
1	1	1	0	0	0	0	0	Touch pad	Dials	7
1	1	1	1	0	1	0	1	<800W	>800W	0
0	0	0	1	1	1	1	1	Clock	No clock	7
1	1	1	1	1	1	0	1	White	Black	5
0	0	0	1	1	1	1	1	Timer (90 min)	<90 min	7
0	0	0	1	1	1	0	1	Memory	No memory	8
0	0	0	0	0	0	0	1	Grill	No grill	5
1	1	0	0	1	0	0	1	<5 settings	>5 settings	0
0	0	0	0	1	1	1	1	Defrost	No defrost	6
1	0	1	1	1	1	1	0	Button (door)	Lever (door)	0
1	1	1	1	0	0	0	0	>£130	<£130	6
0	0	1	1	0	0	0	0	Fitted plug	No plug	0
0	0	0	1	1	1	1	1	Delay start	No delay	7
1	1	1	1	0	1	0	1	<0.8ft³ capacity	>0.8ft³ capacity	0
7	6	7	10	8	9	5	10			
0	0	0	1	1	1	0	1			

Table 10.13 Construct groups and their labels

Pass	No. of reflections	Constructs	Group label
1	1	Touch pad	
		Digital clock	
		>90 min timer	
		Memory	
		Delay start	'technical sophistication'
2	1	Power settings	
		Fitted plug	'electrics'
3	2	<800W	
		Defrost	
		<£130	'buying points'
4	0	<0.8ft³ capacity	'size'
5	0	White	
		Grill	'appearance'

probability of matches between reference and construct rows occurring by chance. The following sets of related constructs were extracted in this case.

> Group 1 Touch pad
> Digital clock
> >90 min timer
> Memory
> Delay start

A further four passes were conducted, and the following four factors were defined.

> Group 2 Power settings, plug
> Group 3 <800W, defrost, <£130
> Group 4 <0.8ft3 capacity
> Group 5 White, grill

Following this, a label for each construct group factor was provided by the subject. Table 10.13 shows the construct groups and their labels.

The resultant groupings reflect the participant's consideration of the products used in the analysis (Baber, 1996).

Advantages

- The method has a structured and thorough procedure.
- It is generic and can be applied in any domain for any product, system or device.
- It is a very easy technique to use.
- It can be used in the early design life-cycle in order determine user opinions on what the design should include, or with existing products for evaluation purposes.
- Its output is very useful, providing an insight into user perceptions and attitudes.
- Little training is required for its application.

Disadvantages

- The repertory grid procedure is a long and drawn-out one.
- It is tedious and time-consuming in its application.
- According to Baber (1996), it does not always produce usable factors.

- If quantitative analysis is required, additional training is also needed.
- The reliability and validity of the technique is questionable.
- Knowledge of statistics is required.

Flowchart

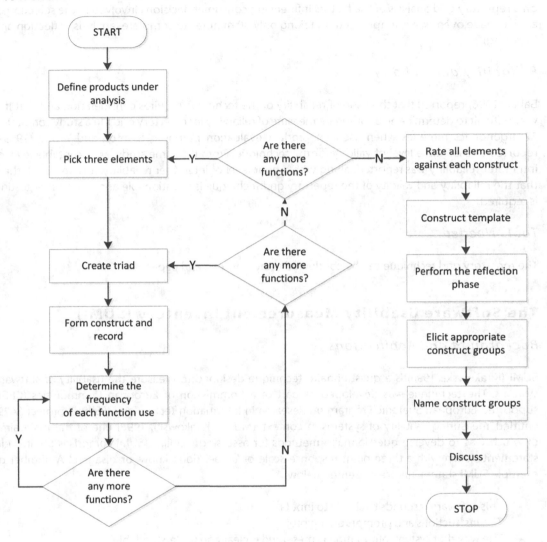

Related Methods

The repertory grid analysis technique is an interview-based knowledge elicitation technique. According to Baber (2005b), there are numerous techniques available to the HF practitioner for knowledge elicitation purposes, such as collegial verbalisation, the concurrent observer narrative technique (CONT) and the CDM.

Approximate Training and Application Times

Stanton and Young (1999a) reported a moderate training and application time for the repertory grid technique. According to Baber (2005b), analysts can become proficient users of the technique within two to three hours. However, both Stanton and Young (1999a) and Baber (2005b) suggested that further practice for the repertory grid technique is very useful. For its application time, Baber (1996) reported that the time taken for the construct elicitation phase depends upon a number of variables, such as the willingness of the subject, the similarity between the items and the number of items used. However, Baber (1996) suggested that the technique is a very quick one to use, with an example based on a repertory grid analysis of the factors influencing consumer decisions involved in the selection of a microwave oven (see example section) taking only 30 minutes to complete (analysis, reflection and factoring)

Reliability and Validity

Baber (1996) reported that the issue of reliability of the technique requires consideration and that it is very difficult to determine an appropriate measure of reliability for repertory grids. In a study comparing 12 ergonomics methods when used for product evaluation purposes, Stanton and Young (1999a) reported a reasonable level of validity for the technique. From the same study, an acceptable level of intra-rater reliability was reported, along with a poor level of inter-rater reliability. It is apparent, then, that the reliability and validity of the repertory grid technique is questionable and that further testing is required.

Tools Needed

The repertory grid technique can be conducted using only pen and paper.

The Software Usability Measurement Inventory (SUMI)

Background and Applications

SUMI (Kirakowski, 1996) is a questionnaire technique designed to measure the usability of software systems. The technique was developed as part of a Commission of European Communities (CEC)-supported European Strategic Program on Research in Information Technology (ESPRIT) project 5429, entitled 'measuring usability of systems in context' (MUSIC: Kirakowski, 1996) one of the main aims of which was to develop questionnaire methods for assessing usability. SUMI comprises 50 attitude statements, each with a three-point response scale of 'agree', 'don't know' or 'disagree'. A number of example SUMI statements are presented below:

- This software responds to slowly to inputs.
- The instructions and prompts are helpful.
- The way that system information is presented is clear and understandable.
- I would not like to use this software every day.

According to Kirakowski (1996), SUMI can be applied to any software system that has a display, a keyboard (or other data entry device) and a peripheral memory device such as a disk drive. When using SUMI, the sample group is given a representative set of tasks to perform with the system under analysis and is then asked to complete the SUMI questionnaire. In scoring the participant responses, Kirakowski (1996) reports that the SUMI technique provides the following results:

- Global usability score: This represents an overall rating of the system's usability.

- The following five usability subscale scores:
  - affect: this represents the user's general reaction to the software;
  - efficiency: this represents the degree to which the user feels that the software has assisted them in his or her task;
  - helpfulness: this represents the degree to which the software is self-explanatory;
  - control: this represents the extent to which the user feels in control;
  - learnability: this represents the speed with which the user has been able to master the system.
- Item consensual analysis: this represents a method of questionnaire analysis that was developed especially for the SUMI questionnaire. It involves using a database to generate expected response patterns for each SUMI item. The expected response patterns are then compared to the actual response patterns in order to determine those aspects that are unique to the system under analysis and also those aspects that require further development.

SUMI has been extensively used in the past for a number of different purposes. Kirakowski (1996) describes the following uses of SUMI:

- assessing new products during product evaluation;
- product comparisons;
- to set targets for future application development;
- to set verifiable goals for quality of use attainment;
- to track achievement of targets during product development; and
- to highlight the good and bad points of interfaces.

Procedure and Advice

Step 1: Identify User Sample
The first step in a SUMI analysis is to identify the user sample that will be used in the analysis. Kirakowski (1996) suggests that a minimal sample size of 10–12 participants should be used. It is recommended that the user sample used in the analysis represents a portion of the typical users of the software system or type of software system under analysis.

Step 2: Define Representative Task List for the System under Analysis
Once the participant sample has been defined, the analyst should develop a representative task list for the software system under investigation. This task list should be exhaustive, representing every possible task that can be performed using the system under analysis. This task list represents the set of tasks that the participants will perform during the analysis. If the task list is too great (i.e. it requires more time to complete than is allowed by the scope of the analysis), then the analyst should pick as representative a set of tasks as possible. It is recommended that an HTA for the software system under analysis be used to develop the task list.

Step 3: SUMI Briefing Session
Before the task performance step of the SUMI analysis, the participants should be briefed regarding the purpose of the analysis and how to complete the SUMI questionnaire. It may be useful for the analyst to quickly run through the task list and the SUMI questionnaire, explaining any statements that may cause confusion. In some cases, a demonstration of the tasks required may be pertinent. The participants should be encouraged to ask any questions regarding the completion of the SUMI questionnaire and the task list at this point.

Step 4: Task Performance

Once the participant sample and task list have been defined, and the participants fully understand the tasks that they are required to perform and also how to complete the SUMI questionnaire, the task performance can begin. The participants should now be given the task list and should be asked to begin performing the tasks in the order that they are specified. It is important that no conferring between participants takes place during the task performance and also that no help is administered by the analyst. The task performance should go on as long as is required for each participant to complete the required task list.

Step 5: Administer SUMI Questionnaire

Once all of the participants have completed the task list for the software system under analysis, the SUMI questionnaire is administered. After a brief demonstration of how to complete the SUMI questionnaire, the participants should be instructed to complete the questionnaire. Again, no conferring between participants is permitted during this step, although the analyst may assist them with statements that they do not understand.

Step 6: Calculate Global SUMI Score

Once all of the SUMI questionnaires are completed and handed in, the scoring process begins. The first score to be calculated for the software system under analysis is a global SUMI score for each participant. The global score represents and overall subjective rating of the system's usability.

Step 7: Calculate SUMI Sub-scale Scores

Next, the analyst should calculate the SUMI sub-scale scores for each participant. Scores for efficiency, affect, helpfulness, control and learnabilty should be calculated for each participant.

Step 8: Perform Item Consensual Analysis

Next, the analyst should use the SUMI database to generate expected response patterns for each SUMI item. These should then be compared to the actual responses gained during the analysis.

Advantages

- SUMI is a very quick and easy technique to use, requiring almost no training.
- Its output is very useful, offering an insight into the system users' attitudes regarding the systems usability.
- If the correct sample is used, the system's potential users are in effect rating the usability of the system.
- Once an operational system is available, the speed, ease and usefulness of SUMI mean that it can be used again and again to evaluate and modify the design concept.
- It offers encouraging reliability and validity statistics.
- SUMI statements can be added and removed in order to make the analysis more suitable for the software system in question.
- It can be used effectively even with small sample sizes.
- The scoring process is computerised.
- It is recognised by the ISO as a method for testing user satisfaction.

Disadvantages

- Developed specifically for software systems containing a display, a data input device and a peripheral memory device. If the system under analysis does not possess all of these facets, some modification of the SUMI statements would be required.
- The technique is only available commercially and costs over €1,000 to purchase.
- Wechsung and Naumann (2008) argue that it is not applicable to systems with multi-modal input functions.

Example

Kirakowski (1996) describes three examples of the successful application of the SUMI technique:

- Analysis A: SUMI was used by a company to evaluate the existing software systems that were currently in use in its offices. The results of the analysis highlighted the software that needed to be replaced.
- Analysis B: SUMI was used by a company that was about to purchase a new data entry system in order to evaluate the two possible systems. The company wished to involve the end-users in the selection process. A number of members of staff were used to evaluate both of the systems using the SUMI technique. Low learnability profile scores for the chosen system prompted the company to bid for an improved training and support system for the software.
- Analysis C: SUMI was used by a company to evaluate a new GUI version of a software package and the old version. In most of the SUMI sub-scales, the new interface was rated as worse than the old one. The company checked this and discovered that the interface for the new system became too complicated and also took too long to operate. As a result, the release of the new version was postponed and a further re-design based on the SUMI evaluation results was undertaken.

Related Methods

The SUMI technique is a questionnaire technique used for the assessment of the software systems. There are a number of similar techniques which use attitude scales to assess the usability of a device or system, such as CUSI, which SUMI is based on, the QUIS technique and the SUS technique.

Approximate Training and Application Times

The training time for the SUMI technique is very low, with little or no training required. In terms of application time, according to Kirakowski (1996), the SUMI questionnaire should take no longer than five minutes to complete. In the worst case scenario, it is estimated that the application time for the SUMI questionnaire is 10 minutes. Of course, the total time for a SUMI analysis is dependent upon the length of the task performance stage. The time associated with task performance depends on the length of the task list used. It is estimated that the total application time for SUMI is very low and that even in scenarios where the task list is very large, the total application time would not exceed two hours.

Flowchart

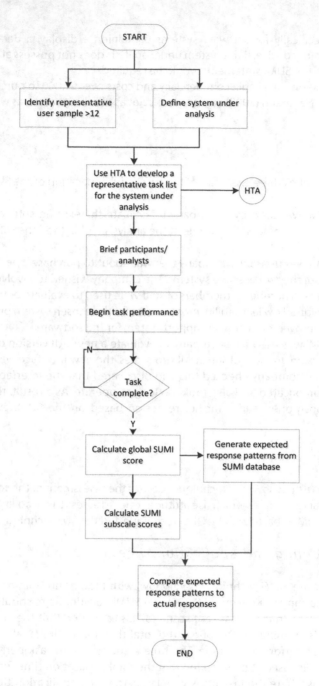

Reliability and Validity

Kirakowski (1996) describes a study reported by Kennedy (1992), where SUMI was used to compare two address book-type databases. The first version of the database was an older version that used 'old-fashioned' language and concepts, whilst the second version was a more modern database that used a more user-orientated language and concepts. The two versions were thus labelled the 'unfriendly' and 'friendly' versions. An expert user group (experienced with similar databases) and a casual user group (which had no experience) conducted a small set of commands using one of the database versions. Upon completion of the designated task, participants completed a SUMI questionnaire. The results demonstrated that SUMI was able to differentiate between different levels of usability. The friendly version was rated as more usable by the expert group than the casual group, and both groups rated the friendly version as more efficient. Both groups also disliked the unfriendly version in terms of effect. The casual group rated the friendly version as more helpful than the expert group did.

Wechsung and Naumann (2008) compared four usability questionnaires (AttrakDiff, SUS, SUMI and SASSI) on their ability to evaluate multi-modal systems, specifically a personal digital assistant (PDA) and a tablet computer, with a uni-modal computer employed as a control system. They found that there was no agreement between the four questionnaires regarding which was the most preferable of the three systems. Their results indicated that the results of SUMI were negatively correlated with the results of the other three questionnaires, that the SASSI, AttrakDiff and SUS results were correlated on their analysis of the computer, and that SASSI and AttrakDiff were correlated for the PDA results. They concluded that the results of the SUMI analysis were the most inconsistent with the other questionnaires.

Tools Needed

SUMI is typically applied using pen and paper. The software system under analysis is also required for the task performance component of the SUMI analysis. The SUMI technique is available as an MS windows compliant software application.

The System Usability Scale (SUS)

Background and Applications

SUS offers a very quick and simple to use questionnaire designed to assess the usability of a particular device or product. It consists of 10 usability statements that are rated on a Likert scale of 1 (strongly agree with statement) to 5 (strongly disagree with statement). Answers are coded and a total usability score is derived for the product or device under analysis.

Domain of Application

Generic.

Procedure and Advice

Step 1: Create Exhaustive Task List for the Device under Analysis
Initially, the analyst should develop an exhaustive task list for the product or device under analysis. This should include every possible action associated with the operation of the device. If this is not possible due to time constraints, then the task list should be as representative of the full functionality of the device as possible. An HTA is normally used for this purpose.

Step 2: User Trial
Next, the participant should complete a thorough user trial for the device or product under analysis. He or she should be instructed to perform every task on the task list given to them.

Step 3: Complete SUS Questionnaire
Once the participant has completed the appropriate task list, he or she should be given the SUS questionnaire and instructed to complete it, based on his or her opinions of the device under analysis.

Step 4: Calculate SUS Score for the Device under Analysis
Once completed, the SUS questionnaire score is calculated in order to derive a usability score for the device under analysis. Scoring an SUS questionnaire is a very simple process. Each item in the SUS scale is given a score between 0 and 4. The items are scored as follows (Stanton and Young, 1999a):

- The score for odd numbered items is the scale position, e.g. 1, 2, 3, 4 or 5 minus 1.
- The score for even numbered items 5 minus the associated scale position.
- The sum of the scores is then multiplied by 2.5.
- The final figure derived represents a usability score for the device under analysis and should range between 0 and 100.

Advantages

- SUS is very easy to use, requiring only minimal training.
- It offers an immediately useful output in the form of a usability 'rating' for the device under analysis.
- It is very useful for canvassing user opinions of devices or products.
- The scale is generic and so can be applied in any domain.
- It is very useful when comparing two or more devices in terms of usability.
- Its simplicity and speed of use mean that it is a very useful technique to use in conjunction with other usability assessment techniques.
- It is very quick in its application.
- It can be adapted to make it more suitable for other domains.
- It is freely available (Finstad, 2010).

Disadvantages

- The output of the SUS is very limited.
- It requires an operational version of the device or system under analysis.
- It is unsophisticated.
- Non-English speakers have experienced problems understanding the terminology (Finstad, 2010).
- Finstad (2010) questions the adequacy of using a five-point Likert scale, suggesting that a seven-point scale may be more appropriate.
- According to Finstad (2010), the method does not adequately align with the ISO usability standards.

Example

Stanton and Young (1999a) conducted a study comparing 12 ergonomics methods, one of which was the SUS technique. The SUS scale was used to rate the usability of two in-car radio cassette players: the Ford 7000 RDS-EON and the Sharp RG-F832E. SUS results for both devices are presented below.

Ford radio SUS scoring

Odd-numbered items score = Scale position – 1	Even-numbered items score = 5 – scale position
Item 1. 5 – 1 = 4	Item 2. 5 – 2 = 3
Item 3. 5 – 1 = 4	Item 4. 5 – 1 = 4
Item 5. 4 – 1 = 3	Item 6. 4 – 1 = 3
Item 7. 4 – 1 = 3	Item 8. 4 – 2 = 3
Item 9. 5 – 1 = 4	Item 10. 5 – 3 = 2
Total for odd-numbered items = 18	Total for even-numbered items = 15
Grand total = 33	
SUS usability score = grand total X 2.5	

= 34 X 2.5
= 85

Sharp radio SUS scoring

Odd numbered items score = Scale position – 1

Item 1. 4 – 1 = 3
Item 3. 5 – 1 = 4
Item 5. 3 – 1 = 2
Item 7. 5 – 1 = 4
Item 9. 4 – 1 = 3
Total for odd-numbered items = 16
Grand total = 34
SUS usability score = grand total X 2.5
  = 35 X 2.5
  = 87.5

Even numbered items score = 5 – scale position

Item 2. 5 – 1 = 3
Item 4. 5 – 1 = 4
Item 6. 5 – 2 = 3
Item 8. 5 – 1 = 4
Item 10. 5 – 1 = 4
Total for even-numbered items = 18

Flowchart

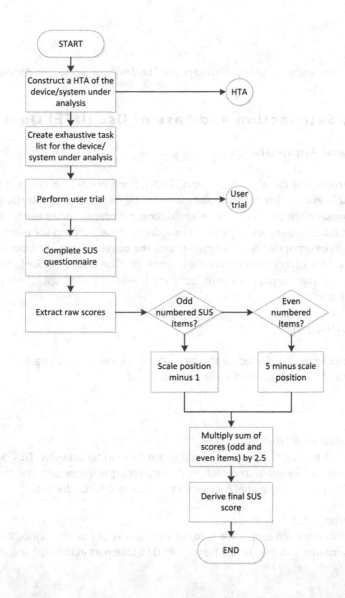

Related Methods

There are a number of other usability questionnaires available, such as SUMI and QUIS.

Approximate Training and Application Times

Both the training and application times for the SUS technique are very low. Since the SUS scale uses only 10 questions, it is very quick to use and apply. Stanton and Young (1999a) reported that questionnaire techniques such as SUS are the quickest to use and apply.

Reliability and Validity

In a comparison of 12 ergonomics techniques, SUS was tested in an analysis of two in-car radio cassettes (Stanton and Young, 1999a). It failed to achieve a statistically significant level for intra-rater reliability and predictive validity. Inter-rater reliability was also rated as moderate (on the technique's second application trial).

Tools Needed

The SUS technique can be applied using pen and paper. The device or product under analysis is also required.

Usefulness, Satisfaction and Ease of Use (USE) Questionnaire

Background and Applications

The USE questionnaire was developed by Lund (2001). Lund wanted to create a method that would enable a transferable and comparable 'usability score' to be derived for all interfaces. The method aims to incorporate user satisfaction into usability evaluation in order to utilise users' ability to subjectively evaluate what is usable. Lund argues that current usability methods neglect the user's subjective opinions, whereas USE tries to incorporate these judgments into the development of transferable usability scores.

The method involves asking users to rate their agreement, on a seven-point Likert scale, with a set of 30 statements. The 30 statements map onto four high level categories: usefulness, ease of use, ease of learning and satisfaction.

Domain of Application

The method was originally developed for the evaluation of software, hardware and services with a core emphasis on its transferability across domains (Lund, 2001).

Procedure and Advice

Step 1: Define the Task under Analysis
The initial stage of any analysis involves defining the task under analysis. This definition guides the analysis and ensures that relevant information is captured. It is important that the analyst presents a clear definition of the goals of the analysis before the next step is conducted.

Step 2: Identify Participants
The next stage of this interface evaluation is to define the user population. Once it is clear who the end-users are, a representative sample should be recruited to take part in the USE analysis.

Step 3: Design Task Scenario
At this stage, the analyst should decide and formulate participant instructions regarding which aspects of the interface should be explored and how this exploration will proceed. It is important that participants experience the system in the same manner in order for their results to be comparable. The set of tasks should be representative of the system as a whole and must be large enough to represent all aspects of the system, but not so large as to be overwhelming for the participant. It is recommended that an HTA for the software system under analysis should be used to develop the task list.

Step 4: Briefing
Before the analysis is undertaken, the analyst should ensure that all participants understand what is required of them. It may be useful for the analyst to provide an example of how to complete the questionnaire. At this point, the participants should be encouraged to question anything they are confused or unsure about to ensure that they fully understand the task, the system and the questionnaire.

Step 5: Task Performance
At this stage in the analysis, the participants are asked to undertake the task list in the order that it is listed. They should not be able to help one another or receive assistance from the analyst during this stage. This stage will continue until all participants have completed all tasks on the task list.

Step 6: Complete USE Questionnaire
The next stage of the analysis involves the participants completing the USE questionnaire. Once the task list is completed, participants should be given the questionnaire and told to complete all items based on the task list experience with the system under analysis. Again, participants should not be conferring with one another in this stage, although the analyst can aid them in completing the questionnaire.

Step 7: Analyse Data
On completion of the questionnaires, the analyst can begin to score the responses and analyse the data. Calculating average scores may be used for summary analysis.

Advantages

- The USE questionnaire provides metrics to reliably measure user reactions to a product.
- It is a very quick method to apply.
- It enables the definition of an acceptable level of user satisfaction (Lund, 1998).
- It provides a baseline evaluation that can measure progress through design iterations (Lund, 2001).
- It allows the identification of priority areas for usability improvement (Lund, 1998).
- It allows insight into design aspects that can increase usability (Lund, 1998).

Disadvantages

- Every question in the questionnaire is positively phrased, which means that the results it gives are biased towards positive responding.

Approximate Training and Application Times

Both training and application times for the method are low.

Validity and Reliability

There seems to be a degree of validity to the method (Lund, 2001), although no data is currently available.

Related Methods

There are many similar usability questionnaires, such as QUIS, repertory grid analysis, SUMI and SUS.

Tools Needed

A prototypical version of the questionnaire is available online at: http://usesurvey.com.
    The method can be conducted using pen and paper.

Flowchart

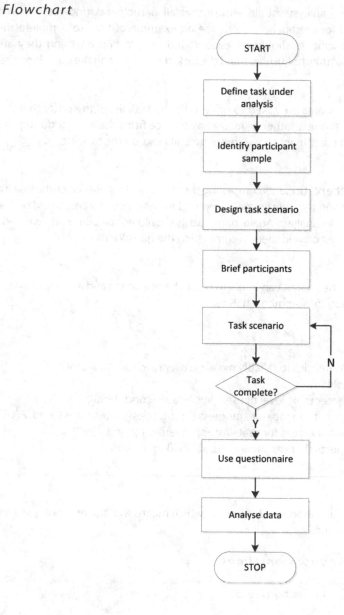

The Purdue Usability Testing Questionnaire (PUTQ)

Background and Applications

The PUTQ (Lin, Choong and Salvendy, 1997) is a usability measurement method specifically designed to enable end-users to evaluate and compare software. First proposed within a special issue of the *Behaviour and Information Technology* journal on usability evaluation, the aim of the method is to evaluate the usability of a number of products in order to assist the end-user in choosing between alternative products. Although the main focus of the method was initially proposed as a guide for software purchases, Lin, Choong and Salvendy (1997) suggest that the method could be used for a variety of roles, such as in-house evaluations by software companies.

The method was developed out of Proctor and Van Zandt's (1994) model of information processing. This model divides information processing into the perceptual stage, the action stage and the cognition stage. Figure 10.24 represents the factors of the PUTQ and the way in which they map onto the three dimensions of perception, action and cognition. Lin, Choong and Salvendy used this model as a basis to explain the way in which users interact with computer interfaces and employed it to identify the most important usability criteria.

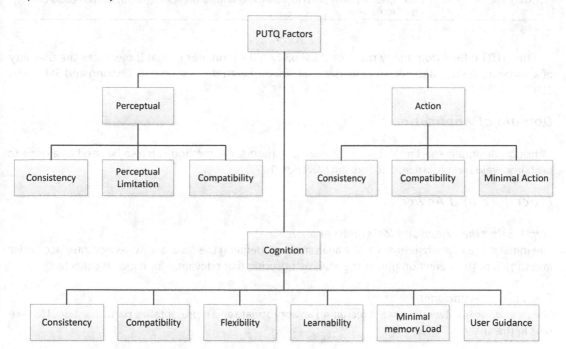

Figure 10.24   PUTQ usability factors and levels of information processing (derived from Lin, Choong and Salvendy, 1997)

The PUTQ is based around eight usability factors and three stages of information processing, as presented in Table 10.14 on the next page. The method contains a set of 100 items which are rated on a seven-point Likert scale ranging from 1 (poor) to 7 (excellent).

Table 10.14   PUTQ items (derived from Lin, Choong and Salvendy, 1997)

1. Compatibility	When stimuli naturally match appropriate responses, users' responses are more accurate and occur at a faster rate (Fitts and Seeger, 1953). It is important that the designer ensures that displayed information and required responses are compatible
2. Consistency	Consistency refers to both internal consistency within the system under analysis and external consistency with other systems in the world. A system that is both internally and externally consistent improves a user's performance and satisfaction
3. Flexibility	Flexibility refers to the system's ability to adapt to users' requirements. In order to account for the range of user requirements, user ability and user experience systems should be flexible (Gong and Salvendy, 1994)
4. Learnability	Learnability refers to the ease at which a system can be learnt. Good design and organisation improve the user's ability to learn
5. Minimal action	Minimal action refers to the positivity of a low level of actions required by the user to conduct tasks improving users' performance and satisfaction
6. Minimal memory load	Minimal memory load is a similar concept referring to the positivity of low memory requirements to reduce the mental workload of the interacting with the system
7. Perceptual limitation	Perceptual limitation refers to the user's perceptual limitations in terms of colour resolution, spatial resolution and so on
8. User guidance	User guidance refers to the improvements on performance afforded by an appropriate user guidance schema

The PUTQ differs from many traditional usability questionnaires in that it measures the usability of a software system and not just the user's satisfaction with the system (Lin, Choong and Salvendy, 1997).

Domain of Application

Although designed specifically to aid software purchasers, the method can also be used as a generic usability evaluation tool (Lin, Choong and Salvendy, 1997).

Procedure and Advice

Step 1: Select the System and Tasks under Analysis
The initial stage of analysis involves the analyst clearly defining the tasks and system or interface under investigation. This definition guides the analysis, ensuring that relevant information is collected.

Step 2: Select Participants
The next stage involves the analyst recruiting an appropriate and representative participant pool to take part in the analysis.

Step 3: Design Users' Task Scenario
At this point, the analyst must refer back to the definition of the analysis and draw out the key tasks that the participants need to undertake. The analyst should create a task scenario which covers all of these tasks and presents an accurate image of the interface to the participant. The analyst must ensure that the task scenario is not overly time-consuming for the participants.

Step 4: Users Conduct Task
The analyst then must instruct the participants to undertake the task scenario. The participants must complete each stage of the task scenario with no assistance from the analyst and without conferring with each other.

Step 5: Participants Complete the Questionnaire
This stage of the analysis involves the participants rating each statement in accordance with their experience with the system under analysis.

Step 6: Results are Analysed
The final stage of analysis involves the analysis of the participants' responses in order to identify patterns and themes in the data. Additionally, statistics can be drawn from the data to illustrate average responses.

Advantages

- The method has both construct and criterion validity and reliability (Lin, Choong and Salvendy, 1997).
- It has a strong theoretical grounding (Lin, Choong and Salvendy, 1997).

Disadvantages

- Lin, Choong and Salvendy (1997) and Kim, Proctor and Salvendy (2011) highlight the method's inability to explore the level of enjoyment the user felt whilst interacting with the system.
- The questionnaire is only applicable to traditional interfaces which utilise a mouse, keyboard and visual display (Lin, Choong and Salvendy, 1997; Kim, Proctor and Salvendy, 2011).

Reliability and Validity

The method's basis in experiments and theory ensures high levels of content and construct validity. The reliability of the method was found to be high in a study comparing the PUTQ to QUIS (Lin, Choong and Salvendy, 1997).

Tools Needed

The questionnaire items are available online from: http://oldwww.acm.org/perlman/question.cgi?form=PUTQ.
Microsoft Excel and statistical packages such as SPSS are useful for analysing the data.

Related Methods

There are many similar usability questionnaires such as QUIS, repertory grid analysis, SUMI and SUS. Lin, Choong and Salvendy (1997) argue that although the method is similar to QUIS, it represents a broader scope as it is not focused solely on user satisfaction.

Flowchart

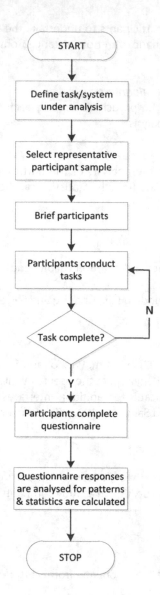

The Post-Study System Usability Questionnaire (PSSUQ)

Background and Applications

The PSSUQ (Lewis, 1991, 1992) is a method to assess user satisfaction with a computer system. The questionnaire is made up of 19 items adapted from the SUMS questionnaire which was developed by IBM in the 1980s. The questionnaire's items measure three core factors: system usefulness, information quality and interface quality (Lewis, 1993). These three factors are measured through the items, using a seven-point Likert scale with ratings ranging from 'Strongly Disagree' to 'Strongly Agree'.

The method is designed for application after a usability study has taken place to provide usability measurements. Table 10.15 on the following page presents the items included in the PSSUQ.

Domain of Application

The method is generic and can be applied to any interface or system. The variations of the method mean that it is suitable for laboratory-based, or postal, application.

Procedure and Advice (Lewis, 1993)

Step 1: Define the Task and System under Analysis

The first stage of analysis is to clearly define the task and system under analysis. This is an important stage of the analysis as it ensures that the analysis is focused on collecting appropriate and relevant information.

Step 2: Design Test Scenario

Once the task is clearly defined, the analyst should develop a test scenario which incorporates all of the tasks under analysis. The test scenario should ensure sufficient and representative interaction with the system whilst ensuring that it does not become overly time-consuming.

Table 10.15   PSSUQ items

System usability	1.	Overall, I am satisfied with how easy it is to use this system
	2.	It was simple to use this system
	3.	I could effectively complete the tasks and scenarios using this system
	4.	I was able to complete the tasks and scenarios quickly using this system
	5.	I was able to efficiently complete the tasks and scenarios using this system
	6.	I felt comfortable using this system
	7.	It was easy to learn to use this system
	8.	I believe I could become productive quickly using this system
Information quality	9.	The system gave error messages that clearly told me how to fix problems
	10.	Whenever I made a mistake using the system, I could recover easily and quickly
	11.	The information (such as online help, on-screen messages and other documentation) provided with this system was clear
	12.	It was easy to find the information I needed
	13.	The information provided was effective in helping me complete the tasks and scenarios
	14.	The organisation was effective in helping me complete the tasks and scenarios
	15.	The organisation of information on the system screens was clear
Interface quality	16.	The interface of this system was pleasant
	17.	I liked using the interface of this system
	18.	This system has all the functions and capabilities I expect it to have
	19.	Overall, I am satisfied with this system

Source: Lewis, 1993.

Step 3: Recruit Participants
An appropriate sample of end-users should be recruited to take part in the analysis. This sample should be representative of the end-user population under analysis.

Step 4: Brief Participants
The analyst should now brief the participants on the test scenario and how to complete the PSSUQ.

Step 5: Participants Undertake the Test Scenario
Participants should now undertake the test scenario. The participants should not receive any help from the analysts or other participants whilst they undertake the task scenario. This step is completed once all participants have completed all phases of the test scenario.

Step 5: Participants Complete the PSSUQ
The analyst should ask participants to complete the questionnaire whilst considering the usability test scenario they undertook.

The participants are asked to read each statement and respond by stating how strongly they agree or disagree to each statement, using the Likert scale, or state 'non-applicable' if the statement does not apply to their experience.

Step 6: Elaboration
Participants are asked to elaborate upon their ratings with written comments; spaces are provided underneath each rating to enable them to do so.

Step 7: Review
The analyst should now review the participants' responses with the participants to ensure they are correctly interpreting the responses and elaboration.

Step 8: Analysis
The compiled results of the PSSUQ can be used to calculate a number of metrics. Overall satisfaction, system usefulness, information quality and interface quality can all be derived from the results. Overall satisfaction is calculated from a sum of the ratings – as the statements are all positive and the rating scale begins with 1 (strongly agree) – a low score equates to a high level of usability. Items 1–8 represent system usability, items 9–15 represent information quality and items 16 onwards represent interface quality.

Reliability and Validity

The overall scale and the individual sub-scales all have high reliability. The validity of the scale has also been proven through correlation analyses (Lewis, 1993: 2002).

Approximate Training and Application Times

The method is simple to train and apply and, according to Lewis (1993), usually takes around 10 minutes to complete.

Advantages

- PSSUQ is a valid and reliable measure (Lewis, 1993, 2002).
- It has been shown to have high levels of sensitivity (Lewis, 1993, 2002).
- It enables the analyst to identify any usability changes attached to new iterations of the system or product (Lewis, 2002).

Disadvantages

- All of the statements are positively phrased, which presents an issue of response bias, in that people are more likely to respond positively to positive statements.

Related Methods

The computer system usability questionnaire (CSUQ) is a related method developed out of the PSSUQ for use in large-scale survey applications (Lewis, 1993).

Tools Needed

The PSSUQ can be completed using pen and paper. Access to the interface or system under analysis is required.
   The questionnaire is available online at http://drjim.0catch.com/usabqtr.pdf.

Flowchart

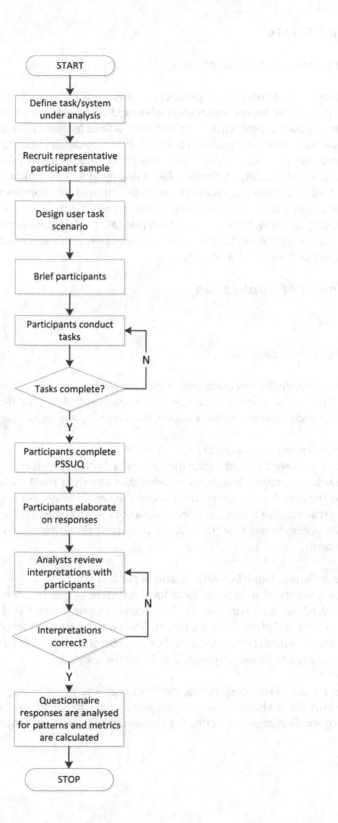

User Trials

Background and Applications

Employing user trials to test products or devices offers a simplistic and flexible means of evaluating a new product or design. User trials involve product or system end-users performing a series of tasks with a new product or device in order to evaluate various features associated with the usability of the product in question. They are appealing because they provide an indication of how the end-users will use the operational product or device. Salvendy (1997) suggests that user testing with real users is the most fundamental usability method available, as it provides direct information about how the potential end-users will utilise the interface under analysis and what problems they may encounter. The flexible nature of user trials allows them to be used to assess a wide range of features associated with a particular device, including usability, MWL, SA, error potential, task performance times and user reactions. The output of a user trial is typically utilised to generate a set of design recommendations or remedial measures for the product or device under analysis.

Domain of Application

Generic.

Procedure and Advice

Step 1: Specify Desired Outcomes of the User Trial
The first step in conducting a user trial involves specifying the desired outcomes of the analysis. The analyst should clearly define what it is that he or she wishes to assess through the user trial.

Step 2: Define the Task under Analysis
Next, the analyst should define the task (or tasks) that the user will conduct with the system or device under investigation. It is recommended that an exhaustive task list is generated, including all of the tasks that can be performed when using the device or system under analysis. If the task list becomes too great and the analysis cannot cover all of the tasks specified due to time and financial constraints, it is recommended that the task list used is as representative as possible of the device or system functions.

Step 3: Conduct an HTA for the Task under Analysis
Once a representative set of tasks for the system or device under analysis is defined, this should be described using an HTA. This involves breaking down the task under analysis into a hierarchy of goals, operations and plans. Tasks are broken down into hierarchical set of tasks, sub-tasks and plans. The HTA is useful as it gives the analyst a clear description of how the task (or tasks) should be carried out and can also be used to develop a procedural list for the user trial.

Step 4: Create Procedural List for the Task under Analysis
The HTA should be used in order to create a procedural list for the task (or tasks) under analysis. The procedural list should describe the required task steps, their sequence and the interface components used.

Step 5: Select Appropriate Participants
Once the task (or tasks) under analysis is clearly defined and described, the appropriate participants who are to take part in the user trials should be selected. The participants used should represent the potential end-users for the system or product under analysis.

Step 6: Brief Participants
The selected participants should then be briefed regarding the purpose of the analysis and also the system or product under analysis. The participants should fully understand the purpose of the user trial and the functions of the system or product under analysis before the user trial can proceed. It is useful at this stage for participants to familiarise themselves with the system or product under analysis. This might involve allowing them to consult any documentation (e.g. a user manual) associated with the system or product.

Step 7: Demonstration of Task under Analysis
Next, the participants should be given a demonstration or walkthrough of the task under analysis. It is normally useful for the analyst to walk the participants through a procedural list of the task under analysis. The analyst should verbally describe each action and physically perform any interactions with interface components. The participants should be encouraged to ask questions regarding the task during this step.

Step 8: Run User Trial
Once the participants fully understand the task under analysis and what is required of them as participants, they should be instructed to begin the first task. It is important that they are given no assistance or feedback during task performance. It is also recommended that the user trials are recorded using video and audio recording equipment. This allows the analyst to consult the recordings of the user trials during the data analysis stage in order to ensure comprehensiveness.

Step 9: Administer Appropriate Usability, Workload and SA Questionnaires
Once task performance is complete, participants should be instructed to complete appropriate MWL (e.g. the NASA-TLX), SA (e.g. SARS) and usability questionnaires (e.g. SUMI). The questionnaires used are dependent upon the nature of the analysis.

Step 10: Interview Participants
Upon completion of the trial, participant interviews should be conducted. Depending upon the nature of the analysis, the interviews can be used to assess a number of factors, such as user opinions of the system or device under analysis and errors made during the trial.

Step 11: Debrief Participants
Next, the participants should be given a debriefing interview in order to provide feedback regarding their performance during the trial and to gather user feedback regarding the system under analysis.

Step 12: Analyse Data
Once the user trial and interviews are complete, the analyst should examine the data accordingly, in line with the outcomes specified prior to the analysis. Typically, measures of usability, MWL, SA and errors are analysed statistically in order to assess their significance.

Step 13: Determine Design Recommendations
Once the data is analysed, the analyst should develop a set of design recommendations based on the findings of the user trials. These recommendations should then be used to redevelop the system or device in question.

Advantages

- User trials offer a simplistic and flexible approach to usability evaluation.
- Potentially they can be used to assess multiple features associated with a system or product's usability, including error, MWL, situation awareness and performance time.
- When employing user trials, the system is evaluated based on the potential end-user's performance. End-user opinions and advice are elicited during the user trial.
- Design recommendations are based on interviews with the system end-users.
- They give the designers a powerful insight into how the system under analysis will be used.
- If used throughout the design process, they ensure that the end-users of the system under analysis are considered.
- Once the appropriate personnel are gathered, they are simple to conduct.

Disadvantages

- User trials are time-consuming to conduct.
- Large amounts of data are collected, ensuring a lengthy data analysis process.
- It may be difficult to gain access to the required personnel or end-users. For example, when conducting a user trial for military applications, it may prove difficult to find the appropriate military personnel for the required duration.
- The end-users may often be biased towards the old system or procedure.

Related Methods

User trials are similar to heuristics evaluation. Depending upon the nature of the analysis, a user trial may utilise a number of other HF techniques, such as MWL assessment techniques (primary and secondary task performance measures, subjective rating techniques), usability metrics (SUMI, SUS, QUIS), checklists (Ravden and Johnson, 1989) and SA measurement techniques (SAGAT, SART, SARS). Interviews, questionnaires and observations are also typically used during a user trial analysis.

Flowchart

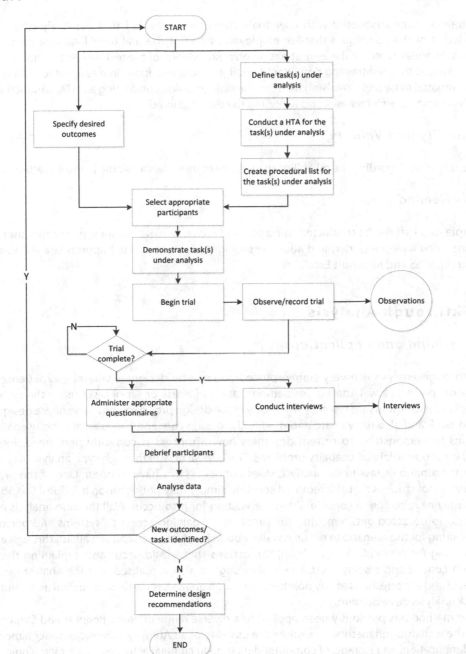

Approximate Training and Application Times

The training time associated with user trials is minimal, provided that the analyst has a working knowledge of the techniques that are employed as part of the trial (workload assessment, usability metrics, interviews, etc.). If the analyst has no prior knowledge of the techniques that are to be applied, it is estimated that the training time would be high. The application time associated with user trials is also estimated to be high, involving defining the task (or tasks), conducting an HTA, conducting the user trials and associated interviews, and analysing the data gathered.

Reliability and Validity

There is no data regarding the reliability and validity of user trials available in the literature.

Tools Needed

A simple user trial can be conducted using pen and paper. However, more sophisticated user trials may require video (video recorder) and audio recording equipment and the appropriate analysis software (Observer, SPSS and Microsoft Excel).

Walkthrough Analysis

Background and Applications

Walkthrough analysis is a very simple procedure used by designers whereby experienced system operators perform a walkthrough or demonstration of a task or set of tasks using the system under analysis. Walkthroughs are typically used early in the design process to envisage how a design concept would work and also to evaluate and modify the design concept. They can also be used on existing systems to demonstrate to system designers how a process is currently performed, highlighting flaws, error potential and usability problems. The appeal of walkthrough-type analysis lies in the fact that the scenario or task under analysis does not necessarily have to occur. One of the problems of observational study is that the required scenario simply may not occur or, if it does, the observation team may have to spend considerable time waiting for it to occur. Walkthrough analysis allows the scenario to be 'acted out', removing the problems of gaining access to systems and personnel and also waiting for the scenario to occur. A walkthrough involves an operator walking through a scenario, performing (or pretending to perform) the actions that would occur and explaining the function of each control and display used. The walkthrough is also verbalised and the analyst can stop the scenario and ask questions at any point. Walkthrough analysis is particularly useful in the initial stages of task analysis development.

The method has previously been applied to a diverse range of areas. Bennett and Stephens (2009) used the walkthrough method to explore the usability of an Autopsy Forensic Browser aimed to aid in the identification, and storage, of computer data used in criminal activities. Shah, Marchionini and Kelly (2009) used the walkthrough method to evaluate a Collaborative Information Seeking System to aid in the completion of collaborative tasks. Sheehan et al. (2009) used cognitive walkthroughs as part of a methodology to provide insights into increasing the usability of medicine prescription computerised support systems. They found that it enabled them to decipher task performance sequences and identify problems relating to the decision support system. Cognitive walkthrough was used in the evaluation of the benefits attached to implementing a computerised system to order laboratory tests in an outpatient laboratory and neurology clinic in the Netherlands (Peute and Jaspers, 2007).

Mahatody, Sagar and Kolski (2010) present a review of the walkthrough method and list 11 variations of the method, describing each in turn and comparing them:

- Heuristic Walkthrough.
- Norman Cognitive Walkthrough.
- Streamlined Cognitive Walkthrough.
- Cognitive Walkthrough for the web.
- Groupware Walkthrough.
- Activity Walkthrough.
- Interaction Walkthrough.
- Cognitive Walkthrough with users.
- Extended Cognitive Walkthrough.
- Distributed Cognitive Walkthrough.
- Enhanced Cognitive Walkthrough.

In addition to the multiple variations of the technique, recent research has developed a walkthrough environment, a computer system, to aid the analyst in the development of scenarios, general preparation of walkthroughs, in the choice of the most appropriate version to use as well as guidance for using the results of walkthrough analysis (Mahatody, Kolski and Sagar, 2009).

Domain of Application

Generic.

Procedure and Advice

There are no set rules for a walkthrough analysis. The following procedure is intended to act as a set of guidelines for conducting a such an analysis of a proposed system design concept.

Step 1: Define the Set of Representative Scenarios
First, a representative set of tasks or scenarios for the system under analysis should be defined. As a general rule, the set of scenarios used should cover every aspect of the system and its interface at least once. The personnel involved in each scenario should also be defined. If the required personnel cannot be gathered for the walkthrough, then members of the design team can be used.

Step 2: Conduct an HTA for Scenario under Analysis
Once a representative set of tasks for the system or device under analysis are defined, they should be described using an HTA. This involves breaking down the task under analysis into a hierarchy of goals, operations and plans. Tasks are broken down into hierarchical set of tasks, sub-tasks and plans. The HTA is useful as it gives the analyst a clear description of how the task (or tasks) should be carried out and also defines the component task steps involved in the scenario (or scenarios) under analysis.

Step 3: Perform Walkthrough
Next, the analyst simply takes each scenario and perform a verbalised walkthrough using the system design under analysis. It is recommended that the analyst uses the HTA to determine the component task steps involved. The scenario can be frozen at any point and questions can be asked regarding controls, displays, decisions made, situation awareness, error occurrence, etc. The walkthrough should be recorded using video recording equipment. Any problems with the design concept encountered during the walkthrough should be recorded and design remedies offered and tested.

Step 4: Analyse Data
Once the walkthrough has been performed, the data should be analysed accordingly and used with respect to the goals of the analysis. Walkthrough data is very flexible and can be used for a number of purposes, such as task analysis, constructing timelines and evaluating error potential.

Step 5: Modify Design
Once the walkthrough is complete and the data is analysed, the design can be modified based on the remedial measures proposed as a result of the walkthrough. If a new design is proposed, a further walkthrough should be conducted in order to analyse the new design.

Advantages

- When used correctly, a walkthrough can provide a very accurate description of the task under analysis and also how a proposed system design would be used.
- Walkthrough analysis allows the analyst to stop or interrupt the scenario in order to query certain points. This is a provision that is not available when using other techniques, such as observational analysis.
- It does not necessarily require the system under analysis and so can be performed in early phases of the design process (Mahatody, Kolski and Sagar, 2010).
- It is a simple, quick and low-cost technique.
- It would appear to be a very useful tool in the analysis of distributed (team-based) tasks.
- It can provide a very powerful assessment of a design concept.
- It requires little training (Mahatody, Kolski and Sagar, 2010).
- It offers a simple, guided process (Bennett and Stephens, 2009).
- It can identify errors or error potential and redundant features of a system (Liu, Tamai and Nakajima, 2009).

Disadvantages

- For the analysis to be fruitful, experienced operators for the system under analysis are required.
- The reliability of the technique is questionable.
- Peute and Jaspers (2007) argue that further guidance is needed in terms of how to develop appropriate and realistic scenarios to 'walk through'.
- Further guidance on preparation and analysis is required (Mahatody, Kolski and Sagar, 2010).

Related Methods

The walkthrough technique is very similar to VPA and observational analysis. Cognitive Walkthrough is commonly used with think-aloud methods (Peute and Jaspers, 2007; Sheehan et al., 2009).

Approximate Training and Application Times

There is no training as such for walkthrough analysis, and the associated application time is dependent upon the size and complexity of the task or scenario under analysis. The application time for walkthrough analysis is typically very low.

Reliability and Validity

Peute and Jaspers (2007) found that all of the problems that the walkthrough method identified were also present in the think-aloud procedure, providing initial levels of validation support for the method.

Flowchart

Tools Needed

A walkthrough analysis can be conducted using pen and paper. Some form of the device or system under analysis is also required (e.g. mock-up, prototype, operational device). It is also recommended that video and audio recording equipment be used to record the walkthrough.

Chapter 11

Design Methods

Introduction

The term 'design technique' is used to describe a technique that is used by designers during the early design life-cycle of a particular system, device or product. According to Wilson and Corlett (1995) ergonomics methods are used to assist the development stages of the design or re-design of equipment, workplaces, software, jobs and buildings. Wilson and Corlett (1995) suggest that the ergonomics methods should be used to develop ergonomically sound concepts, prototypes and final designs. When designing products or systems, designers may utilise a number of design techniques in order to inform and evaluate the design process. There are a number of different types of design methods available, such as interface design techniques and group design techniques. These techniques are often used to provide structure to the design process and also to ensure that the end-user of the product or system in question is considered throughout the design process. Typically, HF intervention is requested once a design is complete and problems have been unearthed by the users of the new system. Usability, error, workload and SA analyses are then conducted, and design recommendations are offered. These recommendations are often ignored due to the high costs associated with re-design. The design techniques reviewed in this document represent those techniques that are used during the actual design process of a system or product, and not those that may be used to highlight design flaws 'after the fact'. A brief description of the design techniques considered in this review is given below.

Interest in the utility and cost benefits attached to early design phase evaluation has increased in recent years. Numerous web resources are available to guide the analyst in the tools available, their strengths and weaknesses, and their appropriate application stage. One very useful website is the usability body of knowledge (see www.usabilitybok.org), which provides reviews of a number of useful design and interface analysis as well as links to numerous relevant texts on the methods. This database was an invaluable resource in updating the design and interface analysis sections of this text.

Design with Intent (DwI) is a method that can provide initial design inspiration from a varied range of domains to provide a usable interface. The method represents a toolkit to identify appropriate design templates and previous design examples to aid in the first step of system design. Contextual Inquiry is a technique to gain insights into the way in which people work in the domain under analysis. The method enables an accurate perception of the way people work before beginning the design or re-design of systems to support them. Allocation of function analysis is used by system designers to determine whether jobs, tasks, system functions, etc. are allocated to human or technological agents within a particular system. Focus group approaches use group interviews to discuss and assess user opinions and perceptions of a particular design concept. In the design process, design concepts are evaluated by the focus group and new design solutions are offered. The CARD method is a variation on the focus group method which uses cards to illustrate aspects of the system and prompt discussions, as well as dividing system aspects into three conceptual layers (observable and formal, skill and craft, description). Scenario-based design involves the use of imaginary scenarios or storyboard presentations to communicate or evaluate design concepts. A set of scenarios depicting the future use of the design concept are proposed and performed, and the design concept is evaluated. Scenarios typically use how, why and what if questions to evaluate and modify a design concept. Storyboards are a simplistic type of scenario-based design using storyboards to walk participants through the prototypical system and gain user feedback. The Wizard of Oz method represents a technologically advanced version of scenario-based design where a simulation of the system under analysis is used to explore usability issues. Mission

Table 11.1   Summary of system design techniques

Method	Type of method	Domain	Training time	App time	Related methods	Tools needed	Validation studies	Advantages	Disadvantages
Allocation of functions analysis	Systems design	Generic	High	High	Pen and paper	Pen and paper	Yes	1) Ensures that tasks are conducted by the most efficient system component. 2) Provides a structure to the automation decision-making process. 3) Ensures that automation selection is audited.	1) Time-consuming in its application. 2) A multi-disciplinary team of various experts is required.
Focus groups	Group design	Product design	Low	High	Group design methods, observation, interview	Pen and paper, video and audio recording equipment	No	1) Flexible design technique, able to consider any aspect of system design. 2) Powerful design technique	1) Assembling the focus group is difficult. 2) If the chemistry is wrong, then the data may suffer. 3) Reliability and validity is questionable.
Missions analysis	Systems design	Aviation	High	High	HTA	Pen and paper	No	1) Specifies design and information requirements. 2) Potentially exhaustive.	1) Time-consuming procedure. 2) Labour-intensive. 3) Used for cockpit design.
Scenario-based design	Design	Generic	Low	Med	Group design methods, observation, focus groups	Pen and paper, video and audio recording equipment	No	1) Flexible design technique, able to consider any aspect of system design. 2) Quick and easy, offering useful outputs. 3) Designers can see how the end product may be used.	1) Unsophisticated. 2) A multi-disciplinary team of various experts is required.
TCSD	Systems design	Generic	Low	High	Design methods, observation	Pen and paper	No	1) Simple to conduct, offering an immediately useful output. 2) Facilitates design modifications.	1) Unsophisticated. 2) Time-consuming
DwI	Design	Generic	Low	Low	N/A	Freely available toolkit	No	1) Provides cross-discipline inspiration. 2) Adaptable for different timescales.	1) Only provides design inspiration rather than a prescriptive method of development.

Table 11.1   Continued

Method	Type of method	Domain	Training time	App time	Related methods	Tools needed	Validation studies	Advantages	Disadvantages
Wizard of Oz	System design	Generic	High	Low	Storyboards	Mock-up system	No	1) Enables early design phase evaluation. 2) Provides insights into usability problems and understanding of user interaction	1) Requires a high degree of training and a detailed script. 2) Reliability and validity are questionable.
Storyboards	Design	Generic	Low	Low	Wizard of Oz	Pen and paper	No	1) Allow early design phase evaluation. 2) Provide detailed data	1) Reliability and validity are questionable. 2) Poor exploration of context.
Rich pictures	System design	Generic	Low	Low	Storyboards, walkthroughs	Pen and paper	No	1) Provide a simple and quick summary. 2) Enable insights into users' concept models	1) The method is not fully defined. 2) Requires subjective interpretation.
Contextual inquiry	Design	Generic	Medium	High	HTA, data collection	Pen and paper, video and audio recording equipment	No	1) Observational evaluation in a naturalistic environment. 2) Enables identification of non-conscious features.	1) Time-consuming. 2) Requires access to multiple participants in their work context.
CARD	Group design	Generic	Low	Low	Focus groups	Pen and paper	No	1) Enables early design phase evaluation. 2) Allows insights into users' goals and motivations.	1) The focus on the group consensus means that individual differences may be overlooked.

analysis is a technique that is employed during the design of military cockpit environments. End-user tasks and requirements are evaluated and translated into a set of design requirements for the cockpit in question. Task-centred system design (TCSD) is a quick and easy approach to evaluating system design involving the identification of the potential users and tasks associated with the design concept and evaluating the design using design scenarios and a walkthrough-type analysis. The technique offers a re-design of the interface or system design under analysis as its output. Rich Pictures enables users to draw out and illustrate their interpretation of the system under analysis, providing a common language of design. A summary of the system design techniques reviewed is presented in Table 11.1 on the previous pages.

Allocation of Function Analysis

Background and Applications

The emergence of system automation and an increase in technological capability has resulted in agent and artefact roles within complex, dynamic systems becoming ill-defined and somewhat opaque. It is now entirely feasible that human operators and technological artefacts can perform a variety of tasks within complex, dynamic systems equally as well as each other. Allocation of function analysis is used during the design process in order to allocate jobs, tasks, function and responsibility to the human or machine for the system in question (Marsden and Kirby, 2005). Allocation of function involves the design team considering each task and the relative advantages and disadvantages associated with that task being performed by the human or by the machine. Allocation of function analysis is particularly important when considering system automation.

Domain of Application

Generic.

Procedure and Advice

Step 1: Define the Task under Analysis
The first step in an allocation of function analysis is to define the task (or tasks) that is to be considered during the analysis. It is recommended that an exhaustive set of tasks for the system under analysis are considered. However, it may be that a number of the tasks are already allocated to either the human or the machine and so only those tasks that require functional allocation should be considered.

Step 2: Conduct an HTA for the Task under Analysis
Once the tasks under analysis are defined, an HTA should be conducted for each task or scenario. This involves breaking down into the task under analysis into a hierarchical set of tasks, sub-tasks and plans. It is recommended that each bottom-level task step in the HTA is considered during the allocation of function analysis.

Step 3: Conduct Stakeholder Analysis for Allocation of Functions
According to Marsden and Kirby (2005), a stakeholder analysis is conducted in order to identify stakeholder satisfaction and dissatisfaction caused by changes in the computer systems in the system or type of system under analysis. Observational study is required in order to conduct the stakeholder analysis. This involves determining the current knowledge and skills of the existing stakeholders and the potential of stakeholders to develop new knowledge and skills (Marsden and Kirby, 2005). Marsden and

Kirby also suggest that the analyst should consider a number of aspects of work that are important to the stakeholders involved, such as the development of new skills, enjoying interaction with other people and having a variety of work to do.

Step 4: Consider Human and Computer Capabilities

Next, the analyst should consider each bottom-level task step in the HTA and the associated advantages and disadvantages of allocating that task to the human operator or to the machine or system. The capability of the personnel and the technological artefacts involved should be considered with respect to the each task step in the HTA. Marsden and Kirby (2005) recommend that each task step should be allocated to the human only (H), the human and computer with the human in control (H-C), the human and computer with the computer in control (C-H) or the computer only (C).

Step 5: Assess Impact of Allocation of Function on Task Performance and Job Satisfaction

Once the tasks have been allocated, the analyst should review each allocation and determine the effects upon task performance and job satisfaction (Marsden and Kirby, 2005). The analyst should consider error potential, performance time gains/losses, impact upon cost, MWL and the job satisfaction criteria highlighted earlier in the analysis. For any allocations that have a significant negative effect upon task performance and job satisfaction, the analyst should determine an alternative allocation of function. The alternative allocation of functions for the task step in question should then be compared and the most suitable allocation selected.

Example

The following example for a decision support system in a brewery context is taken from Marsden and Kirby (2005):

1. check the desirability of trying to meet a potential increase in demand
    1.1 forecast demand
        1.1.1 review regular sales
        1.1.2 review demand from pub chains
        1.1.3 review potential demand from one-off events
    1.2 produce provisional resource plan
        1.2.1 calculate expected demand for each type of beer
        1.2.2 make adjustment for production minima and maxima
    1.3 check feasibility of plan
        1.3.1 do materials explosion of ingredients
        1.3.2 do materials explosion of casks and other packaging
        1.3.3 check material stocks
        1.3.4 calculate materials required
        1.3.5 negotiate with suppliers
        1.3.6 check staff availability
        1.3.7 check ability to deliver beer to customers
    1.4 review potential impact
        1.4.1 review impact of plan on cash flow
        1.4.2 review impact of plan on staff
        1.4.3 review impact on customer relations
        1.4.4 review impact on supplier relations
2. check the desirability of trying to meet a potential increase in demand
    2.1 forecast demand (H)

 2.1.1   *review regular sales (H)*
 2.1.2   *review demand from pub chains (H)*
 2.1.3   *review potential demand from one-off events (H)*
 2.2   *produce provisional resource plan (H-C)*
 2.2.1   *calculate expected demand for each type of beer (H-C)*
 2.2.2   *make adjustment for production minima and maxima (C)*
 2.3   *check feasibility of plan (H-C)*
 2.3.1   *do materials explosion of ingredients (H-C)*
 2.3.2   *do materials explosion of casks and other packaging (C)*
 2.3.3   *check material stocks (H-C)*
 2.3.4   *calculate materials required (C)*
 2.3.5   *negotiate with suppliers (H)*
 2.3.6   *check staff availability (H)*
 2.3.7   *check ability to deliver beer to customers (H)*
 2.4   *review potential impact (H)*
 2.4.1   *review impact of plan on cash flow (H)*
 2.4.2   *review impact of plan on staff (H)*
 2.4.3   *review impact on customer relations (H)*
 2.4.4   *review impact on supplier relations (H)*

Advantages

- Allocation of function analysis is a simplistic procedure that allows tasks to be allocated appropriately within the system or device under analysis.
- It allows the designers to ensure that the tasks are carried out by the most efficient system component.
- According to Marsden and Kirby (2005), it provides a structure to the automation decision-making process and also ensures that automation decisions are traceable.
- Provided that the appropriate personnel are used, the procedure is a simple and straightforward one.

Disadvantages

- The procedure can be laborious and time-consuming, particularly for complex systems or devices.
- A multi-disciplinary team of HF specialists, potential end-users and designers are required in order to conduct the analysis properly. It may be difficult to assemble such a team.

Related Methods

An allocation of function analysis uses HTA as its primary input. A stakeholder analysis is also conducted during the allocation of function analysis.

Approximate Training and Application Times

According to Marsden and Kirby (2005), an allocation of function analysis requires several skills on behalf of the analyst. The analyst should be proficient in task analysis techniques and stakeholder analysis techniques. It is therefore estimated that the training time for the technique is considerable in cases where the analyst has no prior experience of the techniques used. The application time for an allocation of function analysis is also estimated to be high.

Flowchart

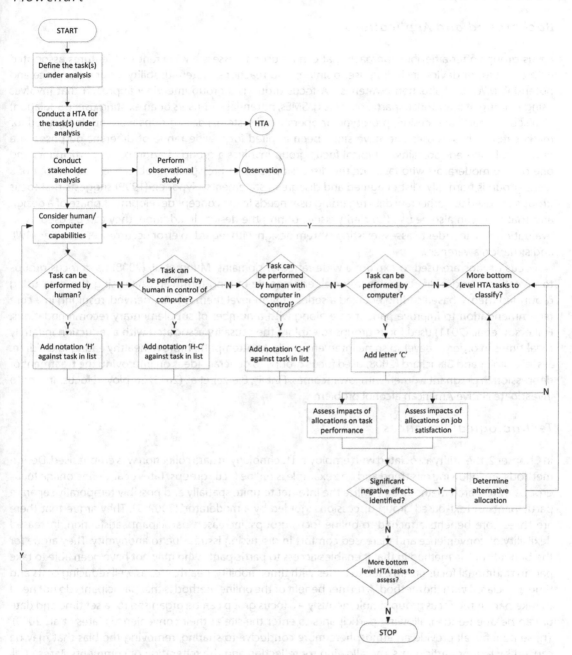

Reliability and Validity

There is no data regarding the reliability and validity of the allocation of function analysis.

Tools Needed

Allocation of function analysis can be conducted using pen and paper. It is also useful to have some form of the system under analysis (e.g. mock-up, functional diagrams, prototype, operational system).

Focus Groups

Background and Applications

Focus groups offer a flexible approach that can be used to assess a wide range of features associated with a system or device, including user opinions and reactions, system usability, error occurrence and potential, MWL and situation awareness. A focus group is a group interview approach that involves using a group of appropriate participants (e.g. SMEs, potential end-users or an existing user population) to discuss a particular design, prototype or operational system. Focus groups were originally used for market research purposes and have since been applied for a wide range of different purposes in a number of different domains. A typical focus group involves a group of appropriate participants and one or two moderators who facilitate the discussion to meet pre-specified objectives. The output of a focus group is normally a list of agreed and disagreed statements. Hyponen (1999) suggests that focus groups are used to gather raw data regarding user needs in the concept development phase of a design and that they can also be used to clarify issues during the design. In addition, they can be used as an evaluation tool in order to assess existing system design with regard to error occurrence, usability, MWL and situation awareness.

Focus groups are used to explore a wide range of domains. Munn et al. (2008) used focus groups to investigate end-of-life experience for elderly people in long-term care. Analysing the data using grounded theory-based coding enabled a series of high-level themes to be derived, resulting in a core recommendation to improve patient care along with a number of supplementary recommendations. Fulkerson et al. (2011) used focus groups to explore the causality associated with a reduction in family meal times in order to develop some interventions in an attempt to improve healthy eating in children. Jesse, Dolbier and Blanchard (2008) used focus groups to generate ideas on improving the treatment of depression in pregnant women with low incomes. Finally, Bletzer et al. (2011) employed focus groups to investigate Native American alcohol problems.

Technological Advances

In Chapter 2, the utility associated with employing technology in data collection was emphasised. Design methods are following this trend and one example is online focus groups. Tates et al. define online focus groups as 'research methods that utilize the Internet to unite spatially and possibly temporally separate participants in text-based group discussions, guided by a moderator' (2009: 2). They argue that there are three core benefits attached to online focus groups: increased participant satisfaction, increased flexibility of convenience and increased comfort in discussing issues due to anonymity. They argue for the benefits of the method in that it enables access to participants who may not have been able to take part in traditional focus groups due to issues with time, mobility, health, etc., as well reducing costs and time associated with the method. A further benefit of the online method is that participants do not need to take part in the focus group simultaneously – a focus group can be organised for a set time and date or can be open-ended, allowing participants to enter the site at their convenience (Tates et al., 2009). These benefits all provide an atmosphere more conducive to sharing, removing the bias that may be caused by louder participants and allowing for reflection and the reiteration of comments. Tates et al. cite a number of studies in which traditional and online focus groups have been compared, resulting in broad comparability between the two. In the Tates et al. (2009) study, 22 out of 31 participants said they preferred the online version and 11 out of 31 said they would not have attended a traditional focus group, with a further 9 out of 31 stating that they were unsure as to whether they would have done so or not. Arguments against the use of online focus groups include problems associated with misinterpretation due to a lack of personal and non-verbal cues, as well as the bias caused by only recruiting a computer-literate population.

Domain of Application

Generic.

Procedure and Advice

There are no set rules for conducting a focus group-type analysis. The following procedure is merely intended to act as a set of guidelines to consider when conducting such an analysis.

Step 1: Define Aims and Objectives
The first step in conducting a focus group is to clearly define its overall aims and objectives. This involves stating explicitly the purpose of the focus group, such as to discuss the C4i Gold command interface design concept.

Step 2: Determine Key Discussion Topics
Once the overall aim of the focus group has been defined, it should be divided into specific areas that are to be the topic of discussion during the focus group. Using the 'C4i gold command interface design concept' example cited above, this could be spilt into the following key discussion areas: interface layout, probability of error, task times, usability, design flaws and design remedies. The key discussion points should be placed in a logical order and this order should be adhered to during the focus group.

Step 3: Assemble Focus Group
Assembling the correct personnel for a focus group is crucial. For the example outlined above, the focus group would require a number of different personnel, including the following:

- HF experts;
- military personnel;
- experienced command and control system operators;
- a project manager;
- an HRA/HEI specialist;
- a usability specialist;
- designers;
- a data recorder;
- controllers from different domains (such as ATC or the emergency services).

Focus group participants are normally made up of end-users of the device or system under analysis. It is often useful to recruit participants via advertising or a group email.

Step 4: Administer Demographic Questionnaire
A simple demographic questionnaire is normally administered at the beginning of a focus group in order to gather information regarding participant age, gender, occupation, experience, etc.

Step 5: Introduce Design Concept
Once the demographic questionnaires have all been completed, the starting point of the focus group session is to introduce to the group the design concept that is to be the topic of discussion. This would normally take the form of a presentation. Once the presentation is finished, the focus group leader should introduce the first topic of discussion. It is recommended that the focus group is recorded using either audio or video recording equipment.

Step 6: Introduce First/Next Topic
The first topic of discussion should be introduced clearly to the group, including what the topic is, why it is important and what is hoped to be achieved by discussing that particular topic. The actual topic should be discussed thoroughly until it is exhausted and a number of points are agreed upon. Step 6 should be repeated until all of the chosen discussion points have been discussed fully.

Step 7: Transcribe Data
Once the focus group session has been completed, the data needs to be transcribed. In order to do this, the analyst should use an audio or video recording of the focus group session.

Step 8: Analyse Data
Once transcribed, the data should then be analysed accordingly. Focus group data can be analysed in a number of ways and is dependent upon the focus of the analysis. Typically, the data output from a focus group session is a set of agreed-upon statements regarding the design concept.

Advantages

- Focus groups offer a flexible approach that can be used for a wide range of purposes, ranging from user reactions and opinions to the error potential of a particular system or device.
- The make-up of the focus group is entirely up to the analyst involved. A correctly assembled focus group can provide a very powerful input into the design process.
- The analyst has complete control of the focus and direction of the analysis, and can change this at any time.
- Very powerful data can be elicited from a focus group-type analysis.
- Focus group-type interviews allow the analyst to quickly survey a large number of opinions.
- Participants discuss issues more freely in a group context.
- Focus groups can be used when the level of current knowledge on a topic is too low to develop questionnaires/interview probes (Lovejoy and Handy, 2008).
- The method reduces the amount of time required per participant compared to interviews (Bletzer et al., 2011).

Disadvantages

- Assembling the desired focus group can be a very difficult thing to do; getting such a diverse group of experts together at the same location and at the same time is not always easy. Similarly, recruiting participants is also difficult.
- Focus group data is difficult to treat statistically.
- The chemistry within the focus group has a huge effect upon the data collected.
- The reliability and validity of focus groups are questionable.
- Large amounts of data is gathered. This is time-consuming to transcribe and analyse.
- Focus group data can be subject to bias.
- The method requires a skilled moderator (Bletzer et al., 2011).
- Outspoken individuals could bias the results (Munn et al., 2008).

Related Methods

Focus groups use a semi-structured group interview approach and also typically employ questionnaires or surveys as part of the data collection procedure.

Approximate Training and Application Times

There are no training times associated with a focus group-type analysis. A typical focus group session duration is between 90 minutes and two hours. However, this is dependent upon the requirements of the focus groups and it is not unheard of for focus group sessions to last for days at a time.

Reliability and Validity

Whilst no data regarding the reliability and validity of focus groups is available in the literature, it is apparent that it could be questionable.

Tools Needed

The tools required conduct a focus group analysis include pen and paper, a video recording device, such as a video recorder and/or an audio recording device, such as a cassette recorder. A computer with a word processing package such as Microsoft Word is required to transcribe the data collected.

Flowchart

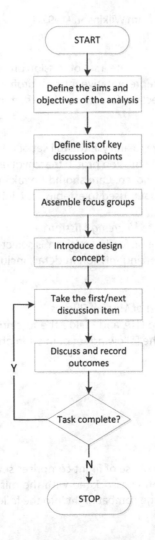

Missions Analysis

Background and Application

The mission analysis technique (Wilkinson, 1992) is a cockpit design methodology that is used to generate cockpit design requirements based on an analysis of operational procedures and requirements. The technique was developed by BAE Systems and has been used on the European Fighter Aircraft (EFA) project. It involves the breakdown of representative flight missions into flight segments and operational modes and the specification of function, information and control requirements within the cockpit. Whilst it was developed for use in the aviation domain, the actual procedure used is generic, allowing it to be applied to the interface design process in other domains.

Domain of Application

Military aviation.

Procedure and Advice

The following procedure is adapted from Wilkinson (1992).

Step 1: Compile Mission Profiles List
The first step in a mission analysis is to create a set of mission profiles for the system under investigation. The analyst should identify a set of representative mission profiles for the system. These profiles should be comprehensive, covering all aspects of future use of the system.

Step 2: Select Forcing Mission
As it would be too resource-intensive to analyse the full set of missions outlined in the mission profiles, a 'forcing mission' is selected. A single mission profile that involves the use of all of the potential design elements of the cockpit should be chosen. Care should be taken in selecting the appropriate mission profile, as it is this 'forcing mission' that is used to establish the initial cockpit design.

Step 3: Conduct an HTA for the Selected 'Forcing Mission'
Once the appropriate mission profile or 'forcing mission' is selected, an HTA should be conducted. This allows the analyst to describe the forcing mission in detail, including each of the tasks and task steps involved.

Step 4: Break Down Mission into a Set of Mission Phases
Next, the analyst should consult the HTA and divide the forcing mission profile into a set of mission phases. Wilkinson (1992) proposes the following set of mission phases for a mission profile:

- ground procedures;
- take-off;
- navigation;
- combat.

Step 5: Identify Operation Modes
According to Wilkinson (1992), each phase of flight comprises several modes of operation. The analyst should identify the modes of operation associated with the mission phases identified during step 4. Wilkinson divided beyond visual range combat flight into the following modes:

- target detection and identification;
- evaluation, prioritisation and decision;
- pre-launch manoeuvre;
- launch weapons;
- post-launch manoeuvre.

Step 6: Divide Each Flight Mode into a List of Task Steps

Next, the analyst should take each flight segment or mode and, using the HTA, describe the tasks that are involved in each flight segment. These tasks should then be divided into the following categories:

- Primary tasks: those that characterise each segment and are performed sequentially, requiring the pilot's foreground attention.
- Intermittent tasks: those that are performed by the pilot as and when required or at the request of the system.
- Continuous tasks: those that are performed continuously and concurrently (mainly monitoring) and are preferably carried out by the system, alerting the pilot only when necessary.

Step 7: Determine Task Function Requirements

For each of the tasks identified in steps 5 and 6, the analyst is then required to determine a set of task function requirements. In order to do this, the functions required to perform the task should be specified. The function categories used are presented below (Wilkinson, 1992):

- manual – purely visual, verbal or mental;
- manual augmented – e.g. Fly-by-Wire;
- manual augmented – automatically limited, e.g. anti-skid braking;
- automatic – manually limited, e.g. autopilot attitude hold mode;
- automatic – manual sanction, e.g. target nomination;
- automatic autonomous – e.g. system status monitoring.

Step 8: Determine Task/Control Requirements

Finally, the analyst should specify the information presentation and control function requirements for each task. The requirements depict how the pilot would perform the task or monitor the automated performance of the task. Therefore, the controls and displays required should be specified. It is these requirements that act as the primary output of the mission analysis and that the design aims to cater for. It is recommended that the information presentation requirements include a specification of content (what information is required), format (in what format would the information best be presented) and type of display used. The control function requirements should at least specify the function of control required, the location of the control and the type of control required.

Advantages

- The output of a mission analysis clearly specifies the system requirements to the designers.
- An exhaustive analysis of the potential user requirements of the system is conducted, including an analysis of user requirements and an analysis of the system's future use.
- The design team can use the mission analysis output to guide the design, ensuring that all requirements are catered for.

Disadvantages

- The procedure involved in a mission analysis appears to be very time-consuming and laborious.

- The selection of a representative mission profile is crucial. It may be that elements of system usage are not catered for by the analysis due to the selection of an inappropriate mission profile.
- No data regarding the reliability and validity of the technique is available in the literature.

Flowchart

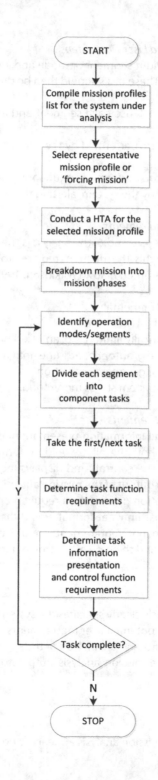

Related Methods

The mission analysis technique uses task analysis techniques such as HTA in its application.

Approximate Training and Application Times

It is estimated that the training time for the mission analysis technique would be low, provided that the analyst in question possesses sufficient domain expertise. It is apparent that a considerable amount of knowledge regarding the system under analysis is required. For example, when applying the mission's analysis in an aviation context, knowledge regarding the types of mission, the tasks involved and the level of automation available in the cockpit is required. The application time for the technique is estimated to be high.

Reliability and Validity

No data regarding the reliability and validity of the technique is available in the literature.

Tools Needed

The technique can be applied using pen and paper.

Scenario-Based Design

Background and Applications

Design scenarios offer a flexible approach to system or device design by adopting a storybook-style approach to help designers and design teams propose, evaluate and modify design concepts. According to Go and Carroll (2003), a scenario is a description that contains actors, assumptions about the environment, goals and objectives, sequences of actions and events. Scenario analyses are used throughout the design cycle to develop and present new system designs in future contexts. They typically involve the use of sketch storyboards depicting a proposed future operation of the device/system in question. At its most basic level, a scenario-type analysis involves proposing a design concept and querying the design using who, what, when, why and how-type questions (Go and Carroll, 2003). Once a scenario is created, design ideas and changes can be added to the storyboard and the design is modified as a result. Scenarios are also used to communicate design concepts to other organisations or design teams. One of the main reasons for using scenario analysis is that it is much cheaper to sketch and act out a future scenario than it is to develop a simulation, mock-up or prototype version of one. It is a powerful design tool that has been applied to the design process in a number of different domains, such as HCI, requirements engineering, object-oriented design, systems design and strategic planning (Go and Carroll, 2003). The appeal of the technique lies in its flexibility, whereby the focus and nature of the analysis is based entirely upon the analyst's requirements, and the direction of the analysis is entirely up to the analysis team.

Domain of Application

Generic.

Procedure and Advice

There are no set rules for scenario-type analysis. A rough guide proposed by Go and Carroll (2003) is presented below.

Step 1: Determine Representative Set of Scenarios

The first step in a scenario analysis is to develop and describe a representative set of scenarios for the system under investigation. Each scenario should be described fully, including its aims, objectives and activities as well as any input devices, displays or interfaces used in it. The personnel involved, the context within which the scenario may take, and individual goals, actions and possible outcomes should also be stipulated. A scenario description table should be constructed at this point, containing all of the relevant information regarding the scenario, such as goals, objectives, task steps, input devices, output devices, etc.

Step 2: Scenario Observation

Scenarios are normally based on an observation of similar scenarios to that under analysis. The analyst should record and observe the scenario under investigation. If the system or design concept does not yet exist, the scenario should be 'made up' using techniques such as group brainstorming. Any novel scenarios observed or elicited that were not expressed in step 1 should also be added to the scenario description table. Interviews and questionnaires may also be used to elicit information regarding potential scenarios.

Step 3: Act Out the Scenario

The analyst or team of analysts should then create the scenario in the form of a storyboard. The scenario should be based on the system being designed, with future contexts and situations being added to the scenario as the analysis progresses. Team members should offer intervention, proposing different contexts and events, such as 'what would happen if…?' and 'how would the operator cope if…?'. This allows the team to evaluate every possibility that might occur with the design concept. Problem scenarios are particularly useful for evaluating a design concept. This part of the scenario analysis is the most crucial and should involve maximum experimentation with the proposed design concept. All assumptions and resultant design modifications should be recorded. The process should continue until the design team is satisfied that all possible scenarios have been exhausted and the end design is complete.

Advantages

- Scenario analyses offer a quick and easy approach of evaluating a particular design concept in future contexts. This can help highlight any design flaws and future problems associated with the initial design.
- It is a very flexible technique.
- It can provide a format for communicating design concepts and issues between designers and design teams.
- It is a quick, low-cost technique that is easy to apply.
- Its output is immediately useful, giving an impression of the design in action and also highlighting any problems that may be encountered.
- Any number of scenarios can be evaluated, ranging from 'normal' to 'worst case' scenarios.

Disadvantages

- Scenarios are not very precise and many potential scenarios may be missed or left out by the analysis team.
- Scenario analysis can be time-consuming for large scenarios.

- To reap the full benefit of a scenario analysis, a multi-discipline team needs to be put together. This is often difficult to achieve.

Flowchart

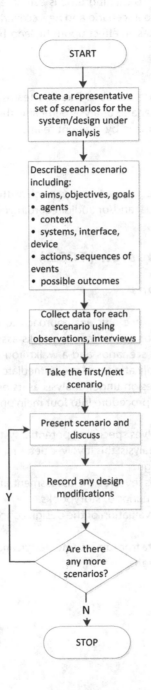

Related Methods

Scenario analysis involves the collection of data using traditional HF data collection procedures such as observational study, interviews and questionnaires. Scenario techniques are also similar to role-play techniques, which are also used by designers to visualise potential product use.

Approximate Training and Application Times

The technique is simple to use and so training time is estimated to be very low. Application time can vary, as there are no set end points to a scenario and new scenarios can be added to existing ones at any point. The size of the scenario also has an effect upon the length of the analysis.

Reliability and Validity

The reliability of the technique is questionable. Scenario teams may fail to capture all of the potential future scenarios of a design in a scenario analysis. Similarly, the technique may produce inconsistent results for the same design when applied by different teams.

Tools Needed

Scenarios are typically conducted using pen and paper. For the data collection element of scenario analysis, it is recommended that visual and/or audio recording equipment is used.

Task-Centred System Design (TCSD)

Background and Applications

TCSD is a simple, low-cost and resource-efficient approach to evaluating system design concepts. It involves the identification of the potential users, the tasks associated with the design concept, and evaluating the design using design scenarios and a walkthrough-type analysis. The technique's main appeal lies in its quick and easy application and the immediate usefulness of its output. It offers a re-design of the interface or system design under analysis as its output and is both easy to learn and to apply. Greenberg (2003) divides the procedure into four main phases:

- Identification phase: this involves specifying potential system users and example tasks.
- User-centred requirements analysis: this involves determining which user groups and which tasks will be catered for by the design.
- Design through scenarios: this involves the assessment and modification of the design concept through the use of design scenarios or storybooks.
- Evaluation: this involves the evaluation of the design concept via walkthrough-type analysis.

A typical TCSD involves gathering data from an existing design and re-designing the system using design scenarios and system task walkthroughs.

Domain of Application

Generic.

Procedure and Advice

Step 1: Identification of Potential Users
The first step in a TCSD analysis is to identify the potential end-users of the design under analysis. Specific user groups should be described. Observation and interviews are normally used to gather this data. The analyst should produce a representative list of user groups.

Step 2: Specification of Example Tasks
Once the specific user groups have been defined, a representative set of tasks for the system under analysis should be defined. This data is also collected through observation and interviews. The data for steps 1 and 2 are normally collected at the same time, i.e. observing different users performing different tasks. Once the set of representative tasks is defined fully, each individual task should be given a task description. Greenberg (2003) suggests that each task description should adhere to the following five rules:

- the description should describe what the user wants to do, but not how he or she will do it;
- the description should be very specific;
- the description should describe a complete job;
- the description should identify the users and reflect their interests; and
- when put together as a set of task descriptions, a wide range of users and task types should be described.

Once the list of tasks is complete, this should be checked and verified by the system end-users. Task descriptions that are incomplete should be rewritten.

Step 3: Determine System Users
The next step forms the first part of phase 2, the user-centred requirements analysis. Typically, system design cannot cater for all possible users. Step 3 involves determining which users or user groups the proposed design will cater for. Greenberg (2003) suggests that users should be put into typical user types or groups and also proposes that the different user types or groups should be categorised as 'absolutely must include', 'should include if possible' and 'exclude'. For example, for a military command and control system design concept, the user groups falling into the absolutely must include group would be gold command users, silver command users and bronze command personnel (foot soldiers, infantrymen).

Step 4: Determine System Tasks
The next task in the process involves clearly specifying which tasks the system design will cater for. Similar criteria to those used in step 3 (absolutely must include, should include if possible and exclude) are employed (with the addition of a 'could include' category) to categorise each task described in step 2.

Step 5: Generate Design Scenarios
Once steps 1–4 are complete, the analyst should have a set of clearly defined end-users and a set of tasks that the design will cater for. The actual design of the system can now begin. To do this, the TCSD informs the design process via the use of design scenarios or storybooks. A number of different design scenarios should be created, each exploring how the design could cope with the scenario under analysis. Whilst no guidelines are offered regarding which scenarios and how many, it is recommended that a scenario involving each of the absolutely must include, should include if possible and could include tasks identified in step 4 should be created.

Step 6: Evaluate and Modify Design Concept using Scenario
Once a set of design scenarios has been specified, this should be used to continually evaluate and modify the design concept. Each scenario should be taken individually and applied to the system design, with team members questioning the efficiency of the design with respect to the events that unfold during each scenario. This is a continuous process, with each design scenario effectively testing the design concept. This process should continue until the team is happy with the system design.

Step 7: Perform Task Walkthrough

Once all of the scenarios have been applied to the design and the design team are happy with the end design concept, the design is tested further and more thoroughly using a walkthrough analysis. Depending upon the resources available (time, money), SMEs or members of the design team can be used. However, walkthroughs using SMEs or system operators would produce more valid results. Essentially, the walkthrough involves role-playing, putting oneself in the mind and context of the user (Greenberg, 2003). Lewis and Reiman (1993) propose the following procedure for performing task-centred walkthroughs:

- Select one of the task scenarios.
- For each of the users/actions in the task:
  - Can you build a believable story that motivates the user's actions?
  - Can you rely on the user's expected knowledge and training about the system?
  - If you cannot, you have located a problem in the interface.
  - Note the problem and any comments or solutions that come to mind.
  - Once a problem is identified, assume it has been repaired.
  - Go to the next step in the task.
  - Once all of the scenarios have been subjected to a walkthrough, the end design should be complete.

Advantages

- TCSD is a simple technique to use that immediately informs system design.
- Design modifications occur naturally throughout the analysis.
- It considers the end-users and the set of tasks that the design is required to support.
- The use of design scenarios allows the design to be evaluated as it would be used.
- Correctly assembled TCSD teams can be very powerful.
- The design concept is evaluated and modified as a result of a TCSD analysis.
- It is not as resource-intensive as other techniques.

Disadvantages

- The validity and reliability of the technique are questionable.
- The use of such a simplistic technique in the design of a miltary command and control system may be questioned.
- Whilst the technique's simplicity is the main advantage associated with its use, this leads to criticisms regarding the depth of the analysis.
- Although it is not as resource-intensive as other techniques, it is still a time-consuming technique to apply.
- Assembling a TCSD team may prove difficult. For example, a TCSD analysis for the design of a military command and control system would require numerous specialists (HF, military, design, system operators, etc.). Getting such a team together in one place at one time could prove very difficult.
- It generates huge amounts of data.

Example

The following example (Tables 11.2–11.4) is adapted from a TCSD analysis of a catalogue-based department store (Greenberg, 2003). As the end output of TCSD is typically very large, only extracts of the analysis are shown below. The example is based on the evaluation and re-design of an in-store computer ordering system. For a more detailed example, see Greenberg (2003).

Table 11.2   User types

Customers	Sales clerks
First-time vs repeat customers	Experienced and trained
Computer knowledgeable vs computer naive	New staff member; has passed introductory training session
Typists vs non-typists	
Willing to use the computer vs unwilling	
People with disabilities who may have trouble with fine motor control	

Table 11.3   Tasks to be catered for by the end design

Choosing merchandise	Pay by	Reviewing cost	Merchandise pickup
One item	Cash	Individual item cost	Immediate
Multiple items	Credit or debit card	Total costs	Delivery
Modifying the selected list of items	Invoice	Comparison shopping	

Table 11.4   Example TCSD walkthrough

Task step	Knowledge? Believable? Motivated?	Comments/solutions?
a. Enters store.	OK	Finding paper catalogues is not a problem in the current store.
b. Looks for catalogue.	OK if paper catalogue is used, but what if the catalogue is online?	However, we were not told if the paper catalogue would still be used or if the catalogue would be made available online. Note – ask cheap shop about this. If they are developing an electronic catalogue, we will have to consider how our interface will work with it. For now, we assume that only a paper catalogue is used.
c. Finds red JPG stroller in catalogue.	OK	The current paper catalogue has proven itself repeatedly as an effective way for customers to brows cheap shop merchandise and to locate products.
d. Looks for computer.	Modest problem.	As a first-time customer, Fred does not know that he needs to order through the computer. Unfortunately, we do not know how the store plans to tell customers that they should use the computer. Is there is a computer next to every catalogue or are there a limited number of computers on separate counters? Are there signs telling Fred what to do? Note – ask cheap shop about the store layout and possible signage. Possible solution: instead of screen 1, a start-up screen can clearly indicate what the computer is for (e.g. 'Order your items here' in large letters).

Flowchart

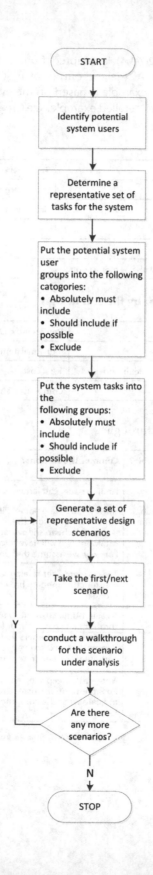

Related Methods

In conducting a TCSD analysis, a number of different HF techniques can be utilised. Observational techniques and interviews are typically used to collect data regarding the system users and the type of tasks that the system caters for. Design scenarios and walkthrough analysis are also used to evaluate the design concept. Greenberg (2003) suggests that to make a TCSD analysis more comprehensive, heuristic-type analysis is often used.

Approximate Training and Application Times

The training time for TCSD would be minimal. However, the application time, including observations, interviews, the generation of scenarios and the application of walkthrough-type analysis, would be high.

Reliability and Validity

The reliability of the TCSD technique is questionable. Greenberg (2003) suggests that it is not a precise technique and that task or user groups are likely to be overlooked. Indeed, it is apparent that when used by different analysts, the technique may offer strikingly different results. As such, the validity of such a technique is hard to define.

Tools Needed

TCSD can be conducted using pen and paper. However, for the observational analysis, it is recommended that visual and/or audio recording devices are used.

The Wizard of Oz Technique

Background and Applications

The Wizard of Oz (Kelley, 1985) is a method developed to involve the end-user population in the design process with the aim of increasing the usability of systems. The method posits that appropriate design is dependent upon accurate knowledge of the user. In order to develop a system that is simple and straightforward to navigate through, an analysis of the user is necessary. The Wizard of Oz technique is a design method that utilises experimental simulation to enable the testing of systems at the initial stages of the design process. It was originally named by Kelly in 1985, although Bevan et al. (2010) highlight issues relating to discrepancies over the origins of the technique, suggesting that many argue that it had been used informally for years before Kelly gave it this label. Once designers have an idea of the functionality required for a prototypical system, they can utilise this technique to explore any associated usability issues before spending resources creating the system (Maulsby, Greenberg and Mander, 1993; Bevan et al., 2010). The technique involves a user interacting with what he or she believes to be a real system; however, the inputs he or she makes into the system are actually intercepted by an experimenter 'behind the curtain' who responds in line with the functionality of the proposed system. It can be used continuously throughout the design process in order to test various iterations of the system with the end-user population (Kelley, 1985).

Bevan et al. (2010) emphasise the deception involved in this method, in the sense that users are unaware of the interception by the experimenter and that the analyst should ensure that he or she gains approval from an ethical committee before implementing the technique.

Domain of Application

The Wizard of Oz is a generic method that can be used to design and evaluate systems in any domain. An example of the varied range of possible applications is the current use of the technique in the development of a military cultural role-play trainer (Barber, Schatz and Nicholson, 2010).

Procedure and Advice

Step 1: Define the Task under Analysis
The initial stage of analysis involves clearly defining the task under investigation. It is important to ensure that this definition is clear, as it will guide the remaining analysis.

Step 2: Data Collection
The next stage of the method involves developing an indepth understanding of the domain under analysis. Research techniques such as interviews and observations should be utilised.

Step 3: Conduct Task Analysis
In order to identify the appropriate functionality, the analyst should conduct a task analysis, such as an HTA, on the system under investigation.

Step 4: Identify Initial Functionality
Based on the data collection and task analysis steps, the initial functionality of the system should be identified. This functionality should be defined in an unambiguous and comprehensible manner, as any variation from this will affect the validity of the results. In line with this functionality, an interface needs to be developed to simulate the proposed system.

Step 5: Develop Appropriate Task Scenarios
In order to test the applicability of the proposed design, a set of tasks needs to be created. These tasks need to be representative of those the users would undertake. Enough tasks should be included to ensure that a range of task types are explored, yet the analyst must also ensure that the task scenario is not overly taxing for the participants.

Step 6: Seek Ethical Approval for the Study
According to Bevan et al. (2010), if the participants are to be deceived into believing that they are interacting with a real system, ethical approval for the study should be sought.

Step 7: Develop a Response Script for the Wizard
In order to ensure the consistency of the Wizard's responses to participant inputs, a comprehensive response script should be developed. This script should contain a response to any possible user input in order to ensure that the Wizard's responses match the proposed system functionality and do not vary between participants.

Step 8: Recruit Appropriate Participants
The analyst must now recruit a representative sample of appropriate end-users to take part in the analysis.

Step 9: Conduct Pilot Tests and Make Any Amendments
A small sample should be taken from the recruited participants in order to pilot the study. These participants should undertake the task scenario in order to identify any issues or problems with the procedure, task

design or response script. Any issues identified at this stage should be resolved before the trial sessions begin; if a large number of changes are required, it may be appropriate to conduct a second pilot study.

Step 10: Conduct Trial Sessions
The participants can now begin the trial session. Each participant should work through a set of tasks and the trial will continue until each task has been conducted.

Step 11: Debrief
Once the trials are complete, the analyst must ensure that participants are adequately debriefed, particularly on any aspect of deception. The analyst should also question the participants on their experience with the system.

Step 12: Synthesis of Responses and Design Recommendation Development
After the trial sessions are complete, the analyst should collate the responses and identify design recommendations based on these.

Step 13: Refine Prototype
At this stage, the analyst should attempt to rectify any usability problems identified in the process.

Step 14: Repeat Test Sessions
In order to ensure that the refinements are appropriate, the analyst should conduct further usability trials. Each trial should involve an increased level of functionality so that the Wizard is gradually 'phased out'.

Step 15: Cross-validation
The fully functional system including all implemented technology is tested on participants with no help from the Wizard, followed by interviews to evaluate the system.

Advantages

- Maulsby, Greenberg and Mander (1993) posit that valuable information can be gained through the method, especially for those in the role of the Wizard.
- It is a method that enables 'rapid prototyping' of future systems (Maulsby, Greenberg and Mander, 1993; Bevan et al., 2010).
- It enables evaluation before resources are spent on, and commitments made to, developing systems (Maulsby, Greenberg and Mander, 1993; Bevan et al., 2010).
- Maulsby, Greenberg and Mander (1993) posit that in addition to evaluating future systems, the method can be used to understand the way in which users interact with an existing system.
- Maulsby, Greenberg and Mander (1993) also claim that the method enables the identification of any usability problems.
- The method allows user feedback to be incorporated into system design (Maulsby, Greenberg and Mander, 1993).
- It can provide an initial evaluation of usability and can map the improvements associated with design iterations.

Disadvantages

- According to Maulsby, Greenberg and Mander (1993) and Bevan et al. (2010), the main criticism of the method is that it allows the Wizard to produce functionality that is unrealistic or not in line with the system under development.

- The Wizard's behaviour should be heavily scripted in order to ensure consistent responses (Maulsby, Greenberg and Mander, 1993; Bevan et al., 2010).
- The Wizard requires a high level of training and expertise (Bevan et al., 2010).
- Bevan et al. (2010) argue that the variations in human and computer behaviour mean that validity and reliability of the method is questionable.
- Bevan et al. (2010) also emphasise the participant deception involved in the method.
- A believable interface must be designed and produced in order to convince users that they are interacting with a real system.

Approximate Training and Application Times

Training times for the Wizard will be high, as will the development of a system interface capable of deceiving participants. Application times for the method will be relatively low as, from the perspective of the participant, the method is simple and straightforward to undertake.

Reliability and Validity

Both the validity and reliability of the method are dependent upon the skill of the Wizard and the accuracy of the response script. Bevan et al. (2010) argue that where there are differences in human and computer actions, any such discrepancies between the proposed system functionality and the response from the experimenter will negatively impact the validity of the results.

Tools Needed

In order to conduct the method, a realistic system interface is required along with a Wizard who is competent enough to match the responses of the proposed system without the knowledge of the user.

Example

Revell, Stanton and Bessell (2010) applied the Wizard of Oz technique to the design of a command and control planning tool. Here we will discuss one function of this interface – predictive line drawing. Figure 11.1 presents a screen shot of one of the tool's interfaces: an interactive bird table.

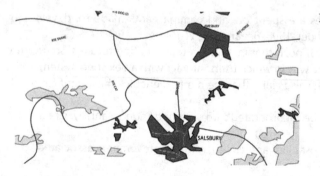

Figure 11.1   Interactive bird table

Figure 11.2 presents a participant attempting to draw a line along a boundary using an interactive pen on the bird table.

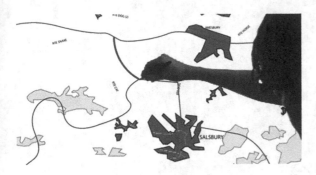

Figure 11.2   Participant drawing a line

The interface, or in this case the Wizard simulating possible functionality, predicts where the participant might want the line to go, based on the notion that the participant is following the boundary line. Figure 11.3 presents the predicted line drawn by the Wizard, which is shown as a dashed red line.

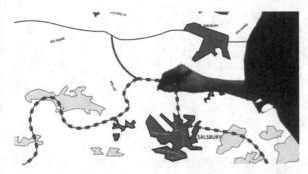

Figure 11.3   Predictive line developed by the Wizard

Of course, this may not be what the participant intended, so he or she can then use the interactive pen to gesture up to which point he or she would like to keep this predicted line. Figures 11.4 and 11.5 present screenshots of the participant gesturing to where he or she wants the line to stop.

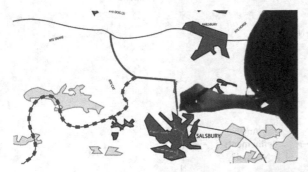

Figure 11.4   Participant gesturing where the line should finish

Figure 11.5   Wizard removing excess predicted line based on participant's gestures

Flowchart

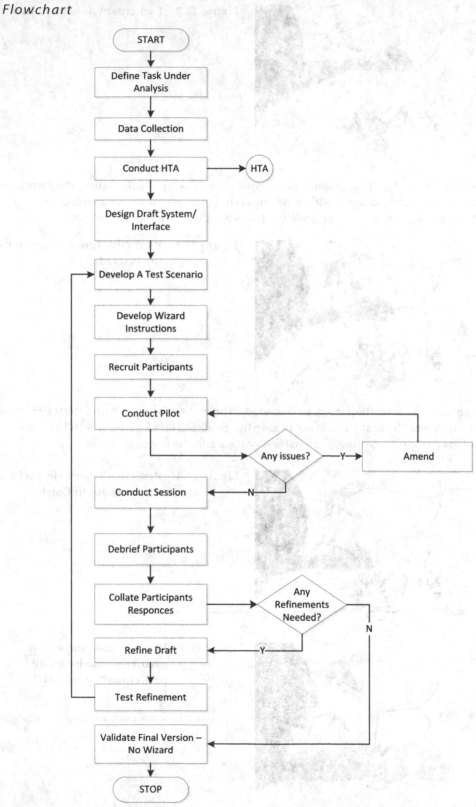

Design with Intent (DwI)

Background and Applications

The DwI method (Lockton, Harrison and Stanton, 2010) was developed to provide guidance for designers on influencing user behaviour. Lockton, Harrison and Stanton argue that all products influence behaviour to some extent and that this influence should be recognised and utilised in order to reduce errors and improve performance. The method aims to illustrate a range of design techniques to successfully influence user behaviour and provide examples of best practice for these techniques, resulting in a cross-domain toolkit of influence techniques.

DwI provides suggestions on design solutions to influence behaviour in specific situations; it is not meant to provide a comprehensive list of solutions, but rather to inspire the analyst to possible design solutions (Lockton, Harrison and Stanton, 2010). It is not a static toolkit, but is rather a constantly evolving and growing collection. Its creators encourage adaptation and feedback through the provision of a blog website (see www.designwithintent.co.uk), conducting surveys, gaining direct feedback and adapting the toolkit based on these.

The method is organised into two modes of function: inspiration (a set of headline design patterns are presented to provide raid inspiration) and prescription (a range of target behaviours are identified by the designer and their associated design patterns are prescribed by the method).

The design patterns are organised into six core groups called lenses in order to categorise similar considerations and assumptions (Lockton, Harrison and Stanton, 2010). The six lenses are presented below in Figure 11.6.

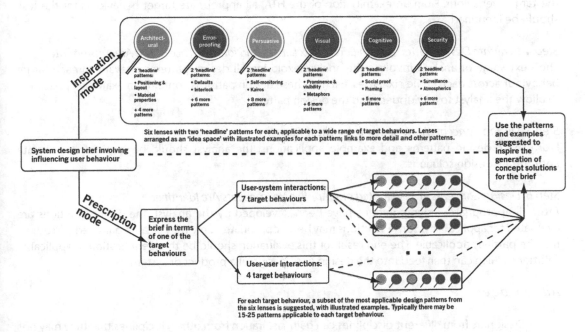

Figure 11.6   Structure of the DwI method

Each design pattern is provided with a description including examples, advantages, disadvantages, user reactions, effectiveness and procedural notes (Lockton, Harrison and Stanton, 2010).

The target behaviours utilised in the prescriptive mode of DwI are defined as 'intended outcomes, particular user behaviours, which we want to achieve through design' (Lockton, Harrison and Stanton,

2010: 385). Examples of target behaviours include 'users only get functionality when environmental criteria are satisfied' and 'multiple users are kept separate so they don't affect each other while using a system'.

Domain of Application

The initial conception behind the DwI method was to create a technique to guide the design of products to positively influence users' environmental behaviour. The design techniques incorporated with the toolkit arise from a diverse range of domains, providing the method with generic applicability.

Procedure and Advice

Step 1: Define the System or Task under Analysis
The first stage of the analysis involves clearly defining the system or system task under analysis.

Step 2: Conduct an HTA of the Goals of the Product
Once the objective of the analysis is defined, the analyst should then construct an HTA of the system.

Step 3: Define Target Behaviour
Once the HTA has been completed, appropriate target behaviours can be identified for the task or system under analysis. A pre-defined set of target behaviours is contained within the method's toolkit. Each target behaviour is listed with a code and an example in order to ensure that the analyst understands the target behaviour. From an examination of the HTA, all appropriate target behaviours for the task should be identified.

Step 4: Evaluate Design Patterns from Each of Six Lenses and Identify Those That Are Appropriate
The next stage of analysis involves the analyst exploring all design patterns relevant to each target behaviour across each of the six lenses. Each of these design patterns contains an example application to allow the analyst to fully understand the design pattern.

Step 5: Design Concepts Are Inspired from These Patterns
From these design patterns and example applications, the analyst should be inspired to generate appropriate design solutions.

Step 6: Design Concepts Are Evaluated and Applicable Concepts Are Identified
After all appropriate design solutions have been developed by the analyst, the design solutions are evaluated for applicability. Some solutions may be inappropriate, some may be amalgamated and others may be partially applicable. The end result of this evaluation should be the identification of applicable solutions which can then feed into the design or re-design of a product or system.

Advantages

- Designers from different disciplines can gain inspiration from other disciplines that they may not otherwise have been exposed to (Lockton, Harrison and Stanton, 2010).
- The method is adaptable, offering both rapid inspiration and more prescriptive design guidance.

Disadvantages

- The generic nature of the method means that many design patterns identified may be inappropriate or difficult to apply (Lockton, Harrison and Stanton, 2010).
- It relies on the individual designer's ability to develop solutions from the guidance provided.
- It only provides inspiration for design solutions, not a methodology to develop design solutions.

Approximate Application and Training Times

The method is quick to apply and requires minimal training; moreover, it is applicable to those with little design experience as well as to novices.

Reliability and Validity

As of yet, there is no reliability or validity data published.

Tools Needed

The method requires access to the Dwl workbook (which is freely available online at: http://www.danlockton.com/dwi/Main_Page [or] www.designwithintent.co.uk). These websites contain an updated and expanded version of the toolkit, which is constantly evolving based on feedback from people applying the tool.

Example

In order to explore the application of the Dwl method, an example application on reducing ATM card error is discussed which is taken from Lockton, Harrison and Stanton (2010).

In this example, the prescriptive mode of Dwl is utilised. The first stage of the analysis was to conduct an HTA of the overall goal – get cash from ATM. A previous error with the system was the user leaving his or her card in the ATM card slot after the machine had dispensed the cash. The ATM was re-designed in order to prevent this from happening, meaning that the card now has to be picked up before the cash will be dispensed to ensure users do not forget to take their card. This example explores whether Dwl would present this design solution and if it would present any other useful solutions.

The brief was 'We don't want users to leave their cards in ATMs after use'.

Using the HTA, the appropriate target behaviours were selected – in this case, target behaviour S1, "The user follows a process or path, doing things in a sequence chosen by the designer," was chosen. Additional target behaviours were also identified that were appropriate for sections of the ATM, but S1 was chosen as the main target behaviour. The design patterns for S1 across the six lenses were then consulted and 22 possible design patterns were identified. From these design patterns, applicable design concepts were inspired. Within this application, applicable design patterns included solutions from all six lenses. Examples from the architectural lens include changing the spacing between interface elements so that the card slot is adjacent to the cash dispensing slot, and moving the card visually to draw users' attention.

Flowchart

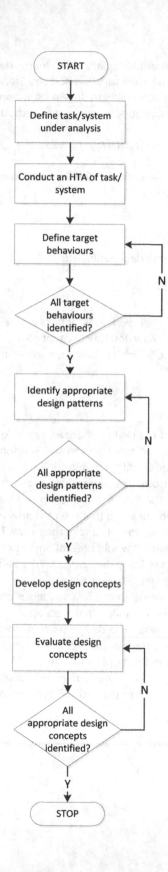

Rich Pictures

Background and Application

Rich Pictures (Checkland, 1981) is a design technique utilising graphical representations in order to illustrate and understand users' perception of and interaction with a system, task or concept. The method is part of a wider methodology called Soft Systems Methodology (Checkland, 1981), which applies the principles of complexity and evolution from systems theory thinking to the design of systems. The methodology posits that systems are comprised of multiple interacting parts which evolve in unanticipated and complex ways. Checkland (2000) asserts that a graphical image is able to represent the system as a whole rather than deconstructing it into separate components.

The Rich Pictures method involves asking users to sketch out the system under analysis in terms of their understanding of it. In this way, users are able to depict complex systems and interactions in a simple and easily comprehensible manner. The resulting illustration is then utilised to provide an understanding of users' perceptions, but can also be used to stimulate conversations around the system under analysis (Checkland, 2000), to ensure analysts' interpretations of interaction are correct (Checkland, 2000) and to provide a common language between users and those who are unfamiliar with the system (Stanton and McIlroy, 2012).

Domain of Application

The method is generic and can be applied to any domain and numerous stages of the design cycle from the development of a concept design to the evaluation of a working system.

Procedure and Advice

Step 1: Define the System or Task under Analysis
The first stage of analysis involves clearly defining the task or system under analysis. This is dependent upon the focus of analysis and should be comprehensively defined in order to ensure that appropriate information is gathered.

Step 2: Recruit Participants
The next stage of the analysis involves recruiting appropriate participants. A representative sample of the population under analysis should be selected.

Step 3: Brief Participants on the Rich Pictures Methodology
The analyst must brief the participants with regard to the Rich Pictures methodology, ensuring that they understand exactly what is being asked of them.

Step 4: Ask Participants to Complete Rich Pictures
The next stage of analysis involves the participants producing the Rich Pictures. As there are no specific guidelines relating to what a Rich Picture should contain, the participants can utilise any amount of text or images to convey their perception of the system under analysis.

Step 5: Validation of Rich Pictures
Once the Rich Pictures are created, the analyst should discuss the content with each participant. This stage allows the analyst to ensure that their interpretation of the picture matches that of the participants. If any issues arise during this discussion, the Rich Picture should be amended appropriately.

Advantages

- Rich Pictures provides a way to summarise a system in a more simplistic manner than linear text (Checkland, 2000).
- It is simple and quick to apply.
- It provides insight into how users perceive a system (Stanton and McIlroy, in press).
- It can be utilised to develop recommendations for system re-design (Stanton and McIlroy, in press).
- It enables system design to align with users' conceptual models, increasing the system's usability (Stanton and McIlroy, in press).

Disadvantages

- According to Lewis (1992), there is little agreement as to what is meant by the term 'Rich Picture'.
- The analysis of Rich Pictures involves a high level of interpretation; in order to ensure that this interpretation is valid, they must be discussed with participants.

Related Methods

Rich Pictures are similar to storyboards in their use of graphical illustrations and are similar to walkthrough analysis in the manner in which participants step through the task under analysis.

Approximate Training and Application Times

The training and application times for the method are very low.

Reliability and Validity

There is no data on the reliability or validity of the method. The criticism put forward by Lewis (1992) that there is little agreement on what Rich Pictures actually are suggests that the method has low levels of reliability. In addition, Checkland (2000) highlights the differing abilities of people to construct Rich Pictures, which could cause further reliability problems.

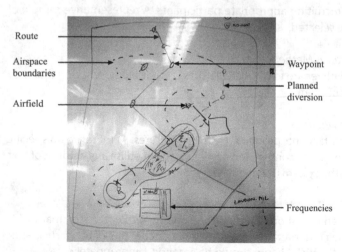

Figure 11.7   Rich Picture of air-to-ground communications

Example

Stanton and McIlroy (2012) applied the Rich Pictures method to the concept of air-to-ground communications – an example of this application is presented in Figure 11.7 on the left. The route is displayed by the solid line with waypoints and circles on the line. The dashed circles indicate airspace boundaries associated with an airfield (the cross on a circle in the centre of the airspaces), the dashed line indicates the planned diversion and the boxes represent collections of frequencies associated with each airfield.

Flowchart

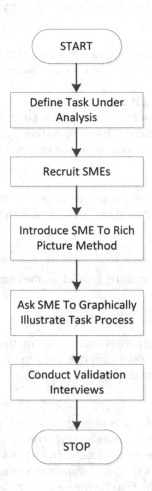

Storyboards

Background and Applications

Storyboards are a simplistic technique designed to extract and represent information in an easy-to-understand format. The method was originally used in the entertainment industry to illustrate the plot of a film or animation, but is now commonly utilised within the exploration of system interfaces (Troung, Hayes and Abowd, 2006). It involves displaying the core features of a process in a simple pictorial format. According to Troung, Hayes and Abowd (2006), storyboards can be utilised in numerous phases of the design process: to illustrate the functionality of a proposed system or the in-service evaluation and re-design of the system. Landay and Meyers (1996) posit that the method can also be utilised to represent the way in which a system will react to users' inputs before the implementation of functionality. In addition to the evaluation of concepts and existing systems, storyboards provide a stimulus for discussions about the system and enable a 'common language' that people from diverse backgrounds can understand (Van der Lelie, 2006).

Domain of Application

The method is generic and can be used to represent numerous processes in a variety of domains.

Procedure and Advice (Adapted from Troung, Hayes and Abowd, 2006)

Step 1: Define the Task under Analysis
In the first stage of analysis, the analyst must clearly define the tasks or specific system aspects that are under investigation. This definition guides the analysis and ensures that appropriate information is gathered.

Step 2: Understand the Users and the System
The next stage of analysis must ensure that the analyst is familiar with the system or interface under analysis and the end-users of that system. The analyst should conduct interviews, observations and document reviews in order to develop a comprehensive knowledge of the system and its users.

Step 3: Recruit Participants
Next, the analyst must select and recruit a representative sample of the end-user population to take part in the research. The participants should meet the requirements for the study in terms of role, age, gender, experience, etc.

Step 4: Brief Participants
Once participants are recruited, they must be briefed on the purpose of the analysis and also on the storyboard technique.

Step 5: Brainstorm
The next stage of analysis involves the participants discussing their interaction with the system under analysis. This stage of the analysis should continue until the participants have identified common perceptions and experiences relating to the system.

Step 6: Build the Storyboard
At this stage of the analysis, the participants should focus on the development of the storyboard. A storyboard is made up of a series of smaller panels; there is no limit to the number of panels, but Troung, Hayes and Abowd (2006) found that normally around three to six frames are utilised. The authors also state that each panel should focus on a specific feature of the system under analysis and should consist of four key features: a short sentence to describe the panel, text, people, time and detail.

Step 7: Feedback
In order to ensure that the storyboards developed are representative of the participants' perceptions, the analyst should hold discussions with the participants. During these discussions, the analyst should voice his or her own interpretation of the storyboard to ensure that his or her understanding matches those of the participants.

Step 8: Amendments
Based on the feedback from the participants, any necessary amendments should be made.

Step 9: Design Implications
The completed storyboard will present an accurate depiction of the way in which users interact with the system under analysis. Utilising this knowledge, the analyst should develop a series of design or re-design recommendations.

Advantages

- The method can be used to represent the functionality of a system, allowing user evaluation before it has been developed (Troung, Hayes and Abowd, 2006: Landay and Myers, 1996).

- It provides insight into users' responses and how these are affected by system functionality (Troung, Hayes and Abowd, 2006).
- Storyboards provide a greater level of detail than flowcharts or technical graphics (Sharon et al., 2010).
- It does not require any technical skills (Sharon et al., 2010).

Disadvantages

- There is little data on the effectiveness of storyboards for the design of interfaces (Sharon et al., 2010).
- The method is not able to represent the context of a situation (Troung, Hayes and Abowd, 2006).
- Knowing when to start and end the storyboard procedure can be hard to ascertain (Sharon et al., 2010: Troung, Hayes and Abowd, 2006).

Tools Required

In order to create storyboards, a pen and white board are required. Troung, Hayes and Abowd (2006) identify a number of generic software packages such as Microsoft PowerPoint or Adobe illustrator which can be utilised in the storyboard process. In addition to this, Sharon et al. (2010) posit that numerous specific storyboard software applications can be used, such as Comic Life, StoryBoard Artist and StoryBoard Quick.

Approximate Application and Training Times

The method is a simple process that is quick to apply.

Reliability and Validity

There is no data regarding the reliability or validity of storyboards.

Related Methods

Storyboards can be related to the Wizard of Oz method in that both provide a means for system evaluation before the implementation of functionality.

Example

Storyboards were applied to a design problem relating to the re-design of a helicopter pilot's communication planning system (Stanton and McIlroy, 2012). A number of helicopter pilots familiar with the system worked with the analysts and designers to create storyboards illustrating idealised processes and interfaces which would enable them to plan their communications more effectively. Figures 11.8 and 11.9, on the following pages, represent detailed storyboards of the process undertaken to identify and insert required frequencies into a communication planning system. In this way, the analysts and designers were able to gain an accurate understanding of the processes, tasks and sequences that must be conducted within this sequence, as well as the reference material required. These storyboards were then updated by the analysts and designers to include updated functionality and tools to aid the pilots, enabling them to provide feedback on these amendments.

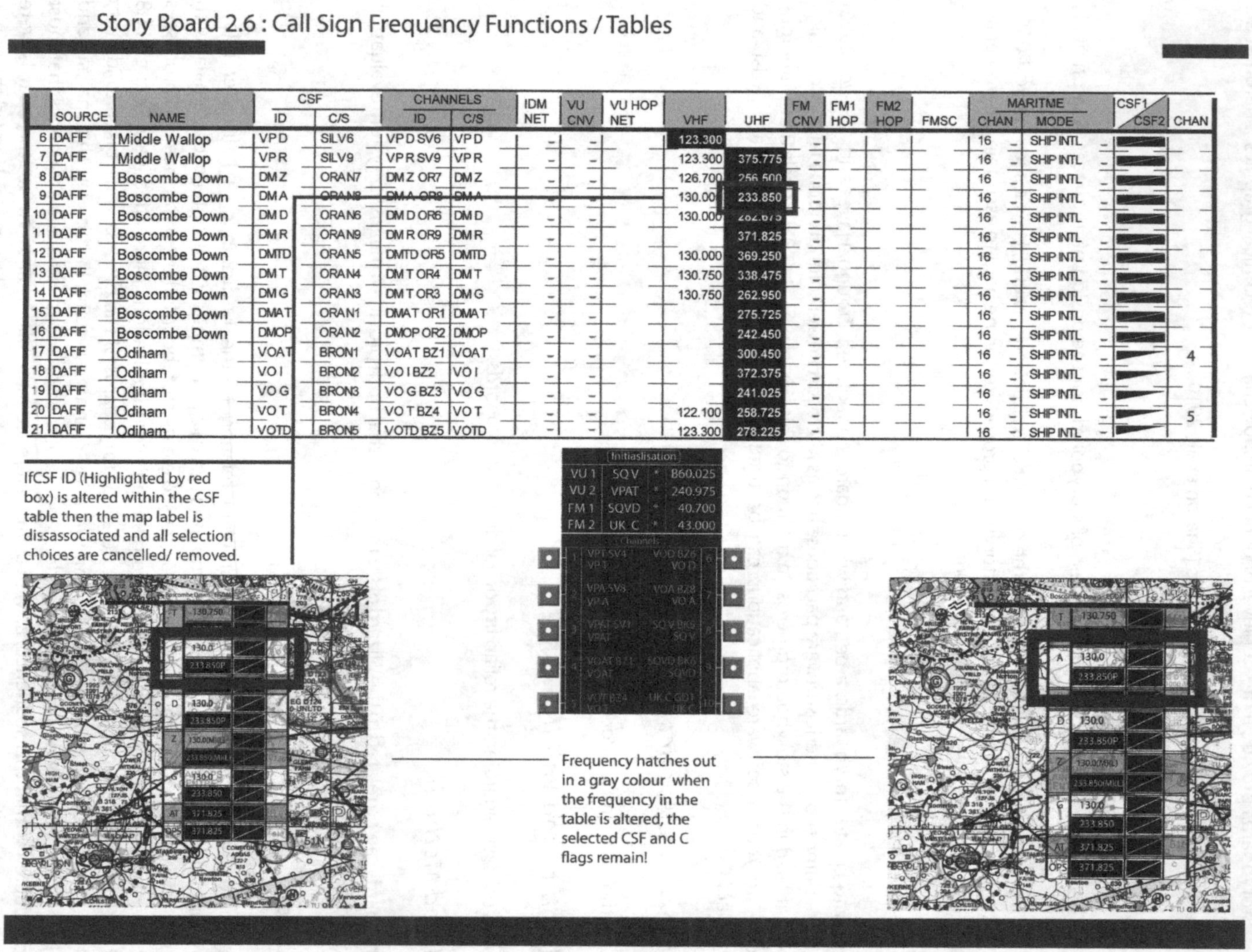

Story Board 2.6 : Call Sign Frequency Functions / Tables

		CSF		CHANNELS		IDM NET	VU CNV	VU HOP NET	VHF	UHF	FM CNV	FM1 HOP	FM2 HOP	FMSC	MARITIME		CSF2 / CSF1	CHAN
SOURCE	NAME	ID	C/S	ID	C/S										CHAN	MODE		
6 DAFF	Middle Wallop	VPD	SILV6	VPD SV6	VPR	-	-	-	123.300	375.775	-	-	-	-	16	SHP NTL		
7 DAFF	Middle Wallop	VPR	SILV9	VPR SV9	VPR	-	-	-		256.600	-	-	-	-	16	SHP NTL		
8 DAFF	Boscombe Down	DMZ	ORAN7	DM Z OR7	DM Z	-	-	-	126.700	256.600	-	-	-	-	16	SHP NTL		
9 DAFF	Boscombe Down	DMA	ORAN8	DMA OR8	DMA	-	-	-	130.00	233.850	-	-	-	-	16	SHP NTL		
10 DAFF	Boscombe Down	DM D	ORAN6	DM D OR6	DM D	-	-	-	130.000	232.675	-	-	-	-	16	SHP NTL		
11 DAFF	Boscombe Down	DM R	ORAN9	DM R OR9	DM R	-	-	-		371.825	-	-	-	-	16	SHP NTL		
12 DAFF	Boscombe Down	DMTD	ORAN5	DMTD OR5	DMTD	-	-	-		369.250	-	-	-	-	16	SHP NTL		
13 DAFF	Boscombe Down	DM T	ORAN4	DM T OR4	DM T	-	-	-	130.000	338.475	-	-	-	-	16	SHP NTL		
14 DAFF	Boscombe Down	DM G	ORAN3	DM T OR3	DM G	-	-	-	130.750	262.950	-	-	-	-	16	SHP NTL		
15 DAFF	Boscombe Down	DMAT	ORAN1	DMAT OR1	DMAT	-	-	-	130.750	275.725	-	-	-	-	16	SHP NTL		4
16 DAFF	Boscombe Down	DMOP	ORAN2	DMOP OR2	DMOP	-	-	-		242.450	-	-	-	-	16	SHP NTL		
17 DAFF	Odiham	VOAT	BRON1	VOAT BZ1	VOAT	-	-	-		300.450	-	-	-	-	16	SHP NTL		
18 DAFF	Odiham	VO I	BRON2	VO I BZ2	VO I	-	-	-		372.375	-	-	-	-	16	SHP NTL		
19 DAFF	Odiham	VO G	BRON3	VO G BZ3	VO G	-	-	-		241.025	-	-	-	-	16	SHP NTL		
20 DAFF	Odiham	VO T	BRON4	VO T BZ4	VO T	-	-	-	122.100	258.725	-	-	-	-	16	SHP NTL		
21 DAFF	Odiham	VOTD	BRON5	VOTD BZ5	VOTD	-	-	-	123.300	278.225	-	-	-	-	16	SHP NTL		5

If CSF ID (Highlighted by red box) is altered within the CSF table then the map label is disassociated and all selection choices are cancelled/ removed.

Initialisation

VU 1	SQ V	860.025
VU 2	VPAT	240.975
FM 1	SQVD	40.700
FM 2	UK C	43.000

Frequency hatches out in a gray colour when the frequency in the table is altered, the selected CSF and C flags remain!

Figure 11.8   Example storyboard

Story Board 2.8 : Call Sign Frequency Functions / Tables

SOURCE	NAME	CSF ID	CSF C/S	CHANNELS ID	CHANNELS C/S	IDM NET	VU CNV	VU HOP NET	VHF	UHF	FM CNV	FM1 HOP	FM2 HOP	FMSC	MARITIME CHAN	MARITIME MODE	CSF1 CSF2 CHAN
6 DAFF	Middle Wallop	VPD	SILV6	VPD SV6	VPD				123.300						16	-	SHP INTL
7 DAFF	Middle Wallop	VPR	SILV9	VPR SV9	VPR				123.300	375.775					16	-	SHP INTL
8 DAFF	Boscombe Down	DM Z	ORANZ	DMZ OR7	DM Z				128.700	365.500					16	-	SHP INTL
9 DAFF	Boscombe Down	DM A	ORAN8	DMA OR8	DM A				130.000	233.850					16	-	SHP INTL
10 DAFF	Boscombe Down	DM R	ORAN9	DMR ORG	DM R					371.825					16	-	SHP INTL
11 DAFF	Boscombe Down	DM R	ORAN9	DMR ORG	DMR				130.000	369.250					16	-	SHP INTL
12 DAFF	Boscombe Down	DMTD	ORAN5	DMTD OR5	DMTD				130.750	338.475					16	-	SHP INTL
13 DAFF	Boscombe Down	DM T	ORAN4	DMT OR4	DM T				130.750	262.950					16	-	SHP INTL
14 DAFF	Boscombe Down	DM G	ORAN3	DMT OR3	DM G				130.750	275.725					16	-	SHP INTL
15 DAFF	Boscombe Down	DMAT	ORAN1	DMAT OR1	DMAT					242.450					16	-	SHP INTL
16 DAFF	Boscombe Down	DMOP	ORAN2	DMOP OR2	DMOP					300.450					16	-	SHP INTL
17 DAFF	Odiham	VOAT	BRON1	VOAT BZ1	VOAT					372.375					16	-	SHP INTL
18 DAFF	Odiham	VO I	BRON2	VO I BZ2	VO I					241.025					16	-	SHP INTL
19 DAFF	Odiham	VO G	BRON3						122.100	258.725					16	-	SHP INTL
20 DAFF	Odiham	VOTD	BRON5	VOTD BZ5	VOTD				122.100	258.725					16	-	SHP INTL
21 DAFF	Odiham	VOTD	BRON5	VOTD BZ5	VOTD				123.300	278.225					16	-	SHP INTL

Initialisation (floating table)

VU 1	SQV	*	B60.025
VU 2	VPAT		240.975
FM 1	SQVD		40.700
FM 2	UK C		43.000

Initialisation (floating Radio bar)

VU 1	SQV	*	B60.025
VU 2	DM A		233.850
FM 1	SQVD		40.700
FM 2	UK C		43.000

On 'Click and Drag' mouse command, within a row, a small box representing that row, with Name of place and ID channel will display grayed out and slightly transparent as shown above.

Boscombe Down — DMZ OR7

If it is dragged and dropped onto the floating Initialisation Radio bar then the Channel C/S will be displayed in that particular slot, showing if squelch (*) is on and the frequency.

Note: it is also possible to click / select any C/S entry and then click on the radio Initialisation button to do the same as above. This is also applicable to the ID channels entrys (And Channels floating table).

Figure 11.9   Second example storyboard

Flowchart

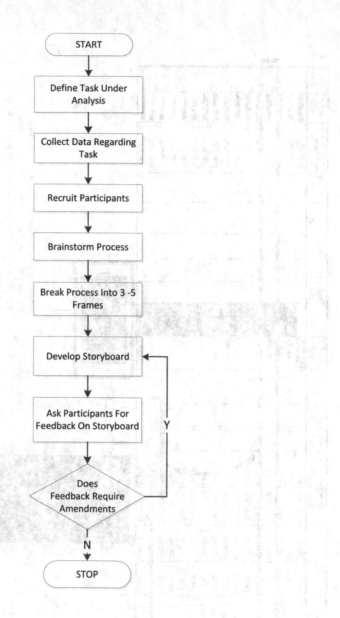

Contextual Inquiry

Background and Applications

Contextual inquiry (Whiteside, Bennett and Holtzblatt, 1988) is a method based on the principles of participatory design whereby developing an accurate perception of the way in which users work is vital (Holtzblatt and Jones, 1993). The technique utilises a hybrid of observation and interview practices in order to develop an accurate understanding of participants' experiences, tasks and processes within the system under analysis. Holtzblatt and Jones (1993) posit that understanding the nature of a participant's interaction with the system or task aids in the design, or re-design, of systems or processes to support them.

The method highlights the importance of context, positing that task performance is affected by the environment in which that task is performed and the artefacts utilised within task performance

(Holtzblatt and Jones, 1993). In light of this, interviews and observations are undertaken within the environment in which the participants normally conduct the task under analysis (Herzon et al., 2010). In utilising observation and interview techniques, contextual inquiry provides an accurate insight into task performance and the constraints associated with this, whereas interviews alone could only extract an idealised description (Holtzblatt and Jones, 1993).

Domain of Application

The method is generic and can be applied in any setting.

Procedure and Advice (Holtzblatt and Jones, 1993)

Step 1: Define the Task under Analysis
The first stage of analysis involves the analyst clearly defining the task or system under investigation. This definition provides a facous for the analysis and ensures that the data gathered is relevant and appropriate.

Step 2: Recruit Appropriate Participants
Once there is a clear definition of the focus of the study, the analyst must identify and recruit a representative sample of participants It is important that these participants have recent experience performing the task or using the system under analysis.

Step 3: Brief Participants
At this stage, the analyst must brief the participants on the objectives of the research and exactly what they will be participating in. The analyst must ensure that all participants are clear on the purpose of the study and their role in it.

Step 4: Conduct Initial Interview
The analyst must now conduct the initial interview with the participants. The participants should be asked to provide an overview of the task or system under analysis and their interaction with the task or system. Herzon et al. (2010) posit that this interview should be recorded or, if this is not viable, that two analysts should be present – one to conduct the interview and the second to transcribe the participants' responses.

Step 5: Observation
The next stage of contextual inquiry involves the analyst entering the participants' workplace and observing their normal work routine and task performance.

Step 6: Post-observation Interview
Once the observation phase of the analysis is complete, the analyst should conduct a second interview with the participants. This interview should be utilised to clarify the analyst's interpretations of task performance. Herzon et al. (2010) point out that it is important to ensure that this interview is recorded or that there are two analysts present.

Step 7: Summation
At this stage, the analyst must collate and summarise all of the information gained throughout the analysis.

Step 8: Review
In order to ensure that the analyst's interpretation of the participants' work tasks and task performance is accurate, a further interview should be conducted with the participants in which the research summary is discussed.

Step 9: Refine Interpretation Based on Participants' Feedback
Based on step 8, any amendments that are necessary should be made to the results summary.

Step 10: Development of an Affinity Diagram
The information gained from this analysis is compiled by the analyst into an affinity diagram, which provides a summary of the task under analysis organised into appropriate categories.

According to Holtzblatt and Jones (1993), in addition to the affinity diagram, depending upon the depth of analysis, a number of models can be developed from the data collected:

- Physical model: this represents a 'caricature' of the work environment illustrating the manner in which components (people and equipment) move within the work environment, and any constraints placed upon this movement.
- Flow model: this represents the roles and responsibilities within the work environment and how these are supported, for example, the communication which flows through the work environment.
- Sequence model: this represents the tasks involved in the work environment and the manner in which these should be undertaken, as well as the relationships between tasks.
- Cultural model: this represents the prevailing mindset of the work environment and the values, policies and influences which affect people within this environment.
- Object/artefact model: this represents the artefacts utilised within task performance in the work domain, as well as exploring the way in which these artefacts are created and modified through task performance.

These models then represent the key aspects and issues associated with the different components of the work environment.

Further analysis could also be conducted including HTA and timeline analysis.

Advantages

- The method enables the identification of aspects of work that the participant may not be consciously aware of.
- Participants are interviewed and observed within their work environment (Herzon et al., 2010).
- It enables the development of accurate work profiles for participants.

Disadvantages

- The method requires access to multiple participants and their work environment.
- Contextual inquiry can be time-consuming.
- It is quite intrusive for participants.
- It is subjective.

Tools Needed

The method can be conducted using pen and paper, but video recording equipment is recommended for the interview and observation stages (Herzon et al., 2010).

Approximate Application and Training Times

Both the application of the method and the analysis of the resulting data are time-consuming. Holtzblatt and Jones (1993) suggest that a quick analysis would equate to spending a day developing each of the models and an additional day developing the affinity diagram.

Reliability and Validity

There is no data in the literature regarding the reliability and validity of this method.

Related Methods

The method utilises the data collection techniques of interviews and observations. HTA can be used alongside contextual inquiry.

Flowchart

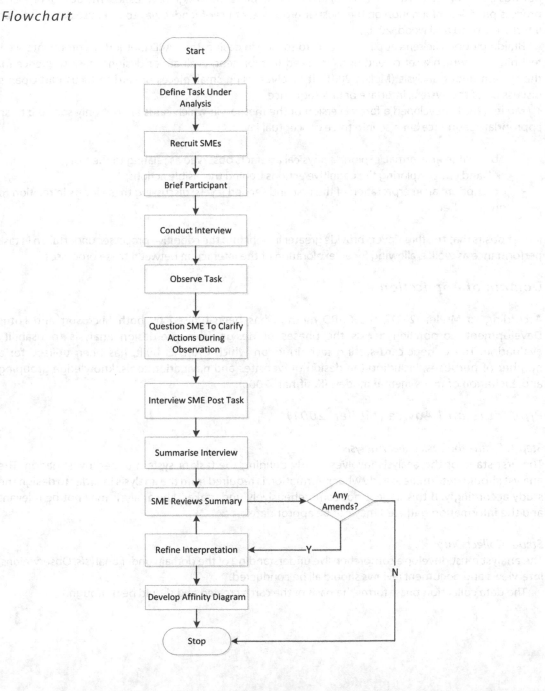

Collaborative Analysis of Requirements and Design (CARD)

Background and Applications

CARD (Tudor et al., 1993) is a method aimed at eliciting information on the design of systems based on the principles of participatory design. It is built upon the generic card-sorting technique in which participants are asked to sort a series of cards, labelled with concepts, into a sequence or pattern. According to Killam et al. (2010), card sorting was originally developed to explore abstract reasoning, but was later utilised by Nielsen (1994b) in order to explore the design of websites. The output of such a process provides information on the most appropriate, or user-friendly, way to organise data within an interface (Wood and Wood, 2008).

Building upon Nielsen's application of card sorting in design, the CARD method is a group interview technique in which a set of end-users is asked to represent, evaluate or design a task sequence for the system under analysis (Muller, 2001). It involves using small pieces of card to focus participants' discussion of the system, interface or task sequence.

Muller (2001) developed a formal version of the method in which cards are not only sorted into an appropriate sequence but also into three conceptual layers:

- observable and formal: exploring physical artefacts observed as relating to the task;
- skill and craft: exploring the cognitive actions behind the visible actions;
- description: an interpretation of the first and second layers, discussing these layers in relation to one another.

Muller posits that the three layers provide greater insight into the cognitive processes undertaken in task performance as well as allowing for an exploration of the interaction between these processes.

Domain of Application

According to Muller (2001), the CARD method has been utilised by both Microsoft and Lotus Development Corporation across the phases of design from pre-design analysis to usability evaluation. The generic card-sorting technique, on which CARD is built, has been utilised for a number of purposes, including the design of websites and navigation tools, knowledge grouping and elicitation of users' mental models (Kaufman, 2006).

Procedure and Advice (Muller, 2001)

Step 1: Define the Task under Analysis
The first stage of the analysis involves clearly defining the task or system under investigation. The analyst should determine exactly what information is required from the analysis in order to design the study accordingly. If this stage is not comprehensively undertaken, the analysis may not be relevant and the information gathered may not be appropriate.

Step 2: Collect Data
The analyst must develop a comprehensive understanding of the domain under analysis. Observations, interviews and document reviews should all be conducted.

The data collection phase forms the basis of the cards created and should be thorough.

Step 3: Create Cards
The analyst must then utilise the domain knowledge gained in order to create appropriate cards. A card should be created to represent each task step in the system under analysis. Muller (2001) suggests that additional blank cards should be produced in order to allow participants to add or modify task steps.

Step 4: Recruit Appropriate Participants
Assembling the correct participants for the CARD procedure is crucial. Participants should provide a representative sample of the end-users of the system under analysis.

At this stage, the analyst must also decide on sample size; this is dependent upon the scope of the study being conducted and the resources available.

Step 5: Brief Participants and Introduce Materials
Before the CARD sorting process begins, the analyst should brief the participants on what they will be asked to do, explaining the process to them and presenting them with the cards.

Step 6: Current Procedure
At this stage, the participants are asked to sequentially place the cards in an order which represents the task steps currently involved in the system under analysis. This is carried out as a group activity in which participants are encouraged to discuss their interaction with the system.

Once the cards have been placed in a relevant sequence, the participants are asked to separate the cards into the three conceptual layers: observable and formal; skill and craft; and description.

Step 7: Critique Sequence
The next stage of analysis involves a group discussion on the positive and negative aspects associated with the current procedure in the system under analysis. In light of this discussion, the participants are asked to consider different ways in which they could re-structure the cards in a more appropriate, or user-friendly, sequence.

Step 8: Determine a New Sequence
The participants should agree on a new card sequence (which may include new or modified cards) in which the tasks are more appropriately sequenced and the cards should be laid out accordingly. As in step 6, the participants are then asked to separate the cards into the three conceptual layers.

Step 9: Analysis
The output of the CARD analysis is a visual representation of the current, and ideal, sequence of task steps for the system under analysis.

Wood and Wood (2008) posit that traditional card-sorting data can be analysed visually, searching for themes, or through the use of cluster analysis to statistically analyse similarities and differences between sorts. We posit that these techniques could also be utilised within the specific CARD method.

Advantages

- The method allows for early and pre-design evaluation.
- It provides insights into end-users' goals, strategies and motivations (Muller, 2001).
- It allows interpretation of end-users' work processes across three levels: observed actions, cognitive actions and the analyst's interpretation (Muller, 2001).
- As a variation of card sorting, it is a simple method for participants and analysts (Kaufman, 2006; Killam et al., 2010).

Disadvantages

- Users may have very different perceptions of task sequences and therefore the results may vary between participants (Killam et al., 2010).

Tools Needed

The method can be conducted using pen, paper and card. The analysis also requires access to the population under analysis.

There are a number of tools to aid in the generic card-sorting method which could be utilised for CARD. Chaparro, Hinkle and Riley (2008) highlight three of the most prominent tools – CardZort, WebSort and OpenSort – each of which is commercially available.

Approximate Training and Application Times

CARD is a simple method with low training and application times, although the application time is dependent upon the complexity of the task sequence under analysis.

Reliability and Validity

There is no data available in the literature regarding the reliability and validity of the method.

Related Methods

The method is similar to focus groups in that a group of appropriate participants is utilised in order to evaluate the design of a system.

Example

The addition of the three conceptual levels enables a deeper insight into the evaluation of the users' perception of a task sequence. An example of this deeper insight was presented by Muller (2001) with respect to the process of creating a file on a computer and then turning the computer off. At the level of observable and formal behaviour, a child creates a file, turns off the computer, turns the computer on, searches through the files and a parent opens the file the child had created earlier. This tells us little about the exact nature or reason for the task. At the level of skill and craft, insights into the motivations for tasks are highlighted. In this example, at this level, it becomes clear that the child could not understand the computer file system and had lost the file created earlier, as the parent found the file for the child in an unintuitive location. The additional level of analysis not only provides greater insight into the motivations behind actions but also as to which aspects of the interface were difficult for the user – in this case, the computer filing system. Insights such as these enable the analyst to identify and prioritise improvements to the interface.

Flowchart

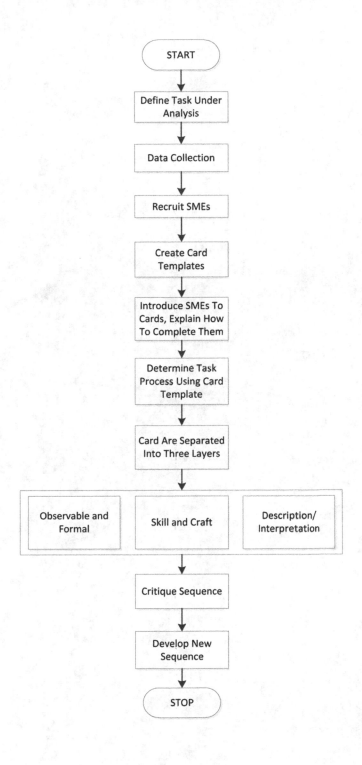

Chapter 12

Performance Time Prediction Methods

The temporal nature of task performance is an important feature of activity in complex systems. Data regarding the duration of the component task steps involved in activity is used for a number of reasons, including the design and development of processes and procedures, performance evaluation and performance prediction. Task performance time prediction is used in the design of systems and processes in order to determine whether proposed design concepts offer performance time reductions, and also to offer performance times associated with a particular task or set of tasks. Predicted task performance times are compared to existing performance times in order to evaluate the impact of proposed design concepts. They are also evaluated in order to ensure that task performance with the proposed design meets the associated performance time constraints or requirements. According to Card, Moran and Newell (1983), it is useful for system designers to possess a model enabling the prediction of how much time it takes to accomplish a given task. The prediction of performance times associated with operator tasks was first attempted in the HCI domain. The Goals, Operators, Methods and Selection rules (GOMS) family of techniques included the Keystroke Level Model (KLM), which offered a set of standard times for operator actions, such as button press, mental operation and homing. Operator tasks are broken down into unit tasks and standard times are assigned to each unit task. These unit-task times are then summed to calculate the total performance time. Although initially developed for the HCI domain, the technique has been used elsewhere; For example, Stanton and Young (1998) used the KLM to predict the performance time for the operation of two in-car stereo/radio devices. Baber (2004) describes the potential of CPA for predicting task performance times. Timeline analysis techniques have also been used to predict performance time. According to Kirwan and Ainsworth (1992), the American National Standards Institute defines timeline analysis as:

> *An analytical technique for the derivation of human performance requirements which attends to both the functional and temporal loading for any given combination of tasks.*

Typically, observational data is used to construct graphically the performance times associated with operator tasks. Timeline-type analysis seems to be potentially suited to analysing team performance times. Kirwan and Ainsworth (1992) also suggest that timelines are useful in assessing task allocation and identifying communications requirements. A summary of the performance time assessment techniques reviewed is presented on the following page in Table 12.1.

CPA

Background and Applications

CPA is a popular technique in project management (Lockyer and Gordon, 1991) and is used to estimate the duration of a project in which some activities can be performed in parallel. The assumption is that a given task cannot start until all preceding tasks that contribute to it are complete. This means that some tasks might be completed and the process is waiting for other tasks before it is possible to proceed. The tasks which are completed but are waiting for others are said to be 'floating', i.e. they can shift their start times with little impact on the overall process. On the other hand, tasks that the others wait for are said

Table 12.1   Summary of performance time assessment techniques

Method	Type of method	Domain	Training time	App time	Related methods	Tools needed	Validation studies	Advantages	Disadvantages
CPA	Task analysis	HCI	Med	Med	KLM	Pen and paper	Yes	1) Considers parallel task activity. 2) Can be used to assess or predict task performance times. 3) More efficient than KLM	1) Can be tedious and time-consuming for large, complex tasks. 2) Only models error-free performance. 3) Times are not available for all actions
KLM	Performance time assessment + prediction	HCI	Low	Low	NGOMSL, CMN-GOMS, CRM-GOMS	Pen and paper	Yes	1) Quick and easy to use, requiring very little training. 2) Can be used to compare task times for two or more devices. 3) Output is immediately useful	1) Designed specifically for use in HCI. 2) Only caters for expert, error-free performance. 3) Does not take context into account
Timeline analysis	Performance time assessment + prediction	Generic	Low	Low	KLM, CPA	Pen and paper	No	1) Quick and easy to use, requiring very little training. 2) Could be used to represent team-based activity. 3) Workload can be mapped onto the timeline graph	1) Predictive use is questionable. 2) Reliability and validity questionable. 3) Of limited use

to lie on the critical path, and any change to these tasks will have an impact on the overall process time. It is possible to apply these ideas to any time-based activity, including human performance.

In order to calculate CPA, it is necessary to know the order in which tasks are performed, their duration and their dependency. The notion of dependency is, for traditional CPA, based on the question of what tasks need to be completed before another task is allowed to commence. However, when applied to human performance models, dependency offers a richer conceptual framework in that it allows consideration of parallel activity. Traditional methods for modelling human response time are constrained because they do not represent parallelism. For example, the KLM method offers a simple additive method for calculating response times in computing tasks (Card, Moran and Newell, 1983). This assumes that all tasks are performed in series and that the total process time is simply the sum of all the task times. However, it is apparent that people are able to perform some tasks in parallel. Models based on CPA can be constructed to represent some aspects of parallel activity, which can provide more accurate estimates of performance time (Schweickert, 1978; Gray, John and Atwood, 1993; Baber and Mellor, 2001).

Describing Dependency

In order to introduce the concept of dependency, it is necessary to make assumptions about the order in which tasks are performed and the nature of the tasks themselves. Clearly, some tasks need to be completed before others can start (which is central to traditional CPA modelling). This means that we can consider temporal dependency as the first stage in constructing a CPA model. However, temporal dependency tells us nothing about *why* some tasks can be performed in parallel. In order to consider this issue, we turn to notions of multiple resources.

Multiple Resources

For the HF community, it is convenient to assume that tasks involving different modalities, such as speaking and looking, can be performed with little interference. This assumption is not without criticism and there are several experiments that we will not consider here which suggest that interference can occur at the stage of central processing of information. This means that, like many assumptions within HF, what serves as a useful aid in engineering applications is not necessarily supported as a generalisable component of human cognition (although my feeling is that for many contexts, the assumption is sufficiently well supported to be treated as robust). Wickens (1992) amalgamated a considerable amount of research on multiple task performance to propose a theory of multiple attentional resources. The theory proposes a general pool (or reservoir) of attentional resources which is shared across stages of human information processing: as the demands of one stage increase, so the resources available to other stages diminish. In order to manage this distribution of resource, the theory assumes that there are two sub-pools: one for visio-spatial resources and one for verbal-acoustic resources. Such a model would help to determine the possibility of tasks being performed in series or parallel, i.e. two 'visual' tasks would need to be performed in series (for the simple reason that one cannot look in two places at the same time), but an 'auditory' and 'visual' task could possibly be performed in parallel, e.g. the (visual) monitoring of displays could be performed in parallel with the (auditory) hearing of an alarm. The suggestion is that, as tasks draw from the same sub-pool, their interference requires serial processing, but if they use different sub-pools, they can be performed in parallel. A complication with this assumption is that the various stages of processing might draw on different versions of the sub-pool, e.g. at the input stage, the 'sub-pool' could be constrained by sensory limitations (for example, you cannot look at two places at once, but need to move your eyes between the places), and at the output stage, the 'sub-pool' would be constrained by response mechanisms, e.g., speaking or pressing buttons. Thus, at the observable stages of human response, it is possible to make certain assumptions relating to the manner in which information is presented to the person or responses are made. However, the central processing stage is not so amenable to reductionism and it is not entirely clear what 'codes' are used to represent information. While this could be a problem for experimental psychology, HF tends to stick with the observable aspects of input/output and uses these notions for characterizing tasks. So, we would consider 'input' in terms of vision or hearing and 'output' in terms of speech or manual response (left or right hand). For the purposes of this approach, we also include a generalised 'cognition' component (it would be possible to assume that cognition is performed using different codes and to include some additional components, but this is neither substantiated by research nor particularly necessary). The list of codes used in this analysis is as follows:

- visual;
- auditory;
- spoken response;
- manual response (left);
- manual response (right);
- cognition.

Domain of Application

Primarily HCI, but also generic.

Example

In order to illustrate the procedure, the following example will be used: a security guard is watching a bank of closed-circuit television (CCTV) displays that receive images from cameras around a building. If anything suspicious occurs, the guard uses a joystick to manipulate the camera and issues a spoken notification that an intruder has been seen.

Procedure and Advice

In this chapter, construction of a CPA model is based upon a method initially developed by Gray, John and Atwood (1993) and further refined by Baber and Mellor (2001). The method may be proceduralised as follows.

Step 1: Analyse the Tasks to be Modelled
The tasks need to be analysed in fine detail if they are to be modelled by multimodal CPA. HTA can be used (Figure 12.1), but it needs to be conducted down to the level of individual task units. This fine-grain level of analysis is essential if reasonable predictions of response times are to be made.

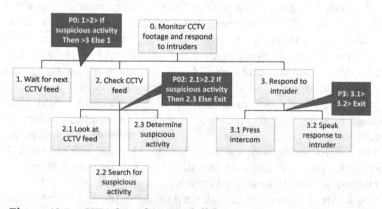

Figure 12.1   HTA based on modalities

Step 2: Order the Tasks
This requires an initial sketch (drawn as a flowchart) of the task sequences in terms of temporal dependency (Figure 12.2). At this stage, the analyst is considering whether more than one task might feed into subsequent tasks.

Step 3: Allocate Sub-tasks to Modality

Each unit task then needs to be assigned to a modality (Table 12.2 below). For the purposes of control room tasks, these modalities are as follows:

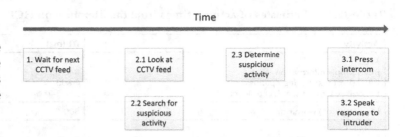

Time

| 1. Wait for next CCTV feed | 2.1 Look at CCTV feed | 2.3 Determine suspicious activity | 3.1 Press intercom |
| 2.2 Search for suspicious activity | | | 3.2 Speak response to intruder |

Figure 12.2   Representation based on temporal dependency

Table 12.2   Defining modalities

Visual	Auditory	Manual -L	Manual-R	Spoken	Cognition
Look at CCTV feed			Press intercom	Speak response	Determine suspicious activity
Search for suspicious activity					

- Visual tasks: for example, looking at a displays, or written notes and procedures.
- Auditory tasks: for example, listening for an auditory warning or to a verbal request.
- Cognition: for example, making decisions about whether or not to intervene and selecting intervention strategies.
- Manual tasks: for example, typing codes on the keyboard, pressing buttons and moving a cursor with a mouse or a tracker ball. Typically, a distinction is made between tasks performed using the left and right hand because this can be used to define the opportunity for serial or parallel performance.
- Speech tasks: for example, talking to colleagues or using a speech recognition system.

Step 4: Sequence the Sub-tasks in a Multimodal CPA Diagram

The tasks are put into the order of occurrence, checking the logic for parallel and serial tasks. For serial tasks, the logical sequence is determined by the task analysis. For parallel tasks, the modality determines their placement in the representation (Figure 12.3 below).

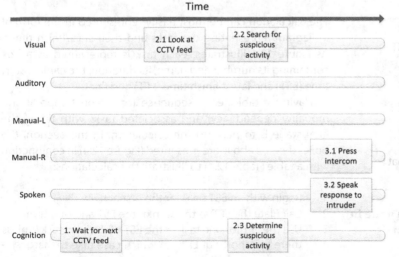

Figure 12.3   Representation based on modalities

Step 5: Allocate Timings to the Sub-tasks

Timings for the tasks are derived from a number of sources. For the purposes of this exercise, the timings used are based on the HCI literature and are presented in Table 12.3.

Step 6: Determine the Time to Perform the Whole Task

The time in which the task may be performed can be found by tracing through

Table 12.3   Estimates of activity times from the literature on HCI

Activity	RT (ms)	Source
Read Read simple information Read short textual descriptions Recognise familiar words or objects	340 1,800 314-340	Baber and Mellor (2001) John and Newell (1990) Olson and Olson (1990)
Hear (auditory warning)	300	Graham (1999)
Search Checking or monitoring or searching Scanning, storing and retrieving	2,700 2,300–4,600	Baber and Mellor (2001) Olson and Olson (1990)
Diagnosis or decision Mental preparation for response Choosing between alternative responses Simple problem solving	1,350 1,760 990	Card et al. (1983) John and Newell (1990) Olson and Nielson (1988)
Speak	100 per phoneme or space	Hone and Baber (2001)
Move hand to tracker ball or keyboard	214-400 320	Card et al. (1983) Baber and Mellor (2001)
Move tracker ball to target item Move cursor via tracker ball 100 mm	1,500 1,245	Olson and Olson (1990) Baber and Mellor (2001)
Press key (e.g. ACK or CANCEL key)	200 80–750 230	Baber and Mellor (2001) Card et al. (1983) Olson and Olson (1990)
Type Average typist (40 wpm) Typing random letters Typing complex codes	280 500 750	Card et al. (1983) Card et al. (1983) Card et al. (1983)
Auditory processing (e.g. speech)	2,300	Olson and Olson (1990)
Switch attention from one part of a visual display to another	320	Olson and Olson (1990)

the CPA using the longest node-to-node values. The calculations in CPA are fairly simple, providing two basic rules are followed:

1.   on the 'Forward-pass', take the longest time;
2.   on the 'Backward-pass', take the shortest time.

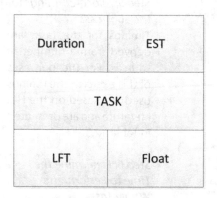

Figure 12.4   Key for each node in CPA diagram

The calculation can be most easily represented in the form of a diagram representing the tasks and their start/finish times. As Figure 12.4 illustrates, each task is represented as a box containing its number and name, its duration, the earliest start time (EST) and latest finish time (LFT) and float.

Having established a sequence (based on temporal and modality dependency) and associated tasks with times, the final stage is to perform the calculation. In this section, the boxes defined above are presented (Figure 12.4) in conjunction with a table (Table 12.4) to illustrate the calculations:

1.   Begin with an EST on 0 for the first activity.
2.   Calculate the EFT as the sum of the EST and duration.
3.   Use the EFT for one task as the EST of the next task (unless there is a choice of EFTs, in which case take the largest – see value marked [a]).

4. Continue calculating the EFT until the end.
5. Set the LFT to equal the EFT of the final task.
6. Subtract duration from LFT to get LST.
7. Insert LST as LFT on previous task (unless there is a choice, in which case take the smallest: see value marked [b]).
8. Continue until first task reached.

Table 12.4   Summary analysis

Task	Duration	EST	EFT	LST	LFT	Float
Wait for CCTV feed	500	0	500	0	500	0
Look at CCTV	280	500	780	500	780	0
Search for suspicious activity	2,700	780	3,480	780	3,480	0
Determine suspicious activity	1,350	780	2,130	2,130	3,480	1,350
Press intercom	200	3,480	3,680	3,480	3,680	0
Speak response	2,000**	3,680	5,680	3,680	5,680	0

*Assume feed cycles every half-second or so.
**Assume the phrase 'This is Security. You are under surveillance' is spoken.

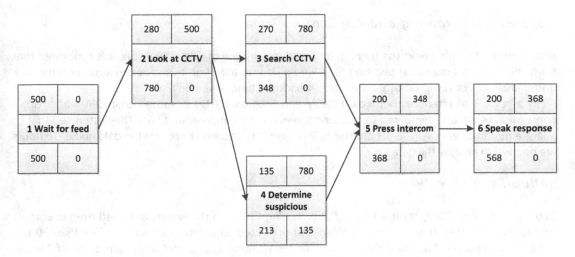

Figure 12.5   Summary analysis

Advantages

- CPA allows the analyst to gain a better understanding of the task via splitting the task into the activities that need to be carried out in order to ensure successful task completion.
- It allows the consideration of parallel unit task activity (Baber and Mellor, 2001), which the KLM does not.
- It gives predicted performance task times for the full task and also for each task step.
- It determines a logical, temporal description of the task in question.
- It does not require a great deal of training.

- It is a structured and comprehensive procedure.
- It can accommodate parallelism in user performance.
- It provides reasonable fit with observed data.
- Olson and Olson (1990) suggest that CPA can be used to address the shortcomings of the KLM.

Disadvantages

- CPA can be tedious and time-consuming for complex tasks.
- It only models error-free performance and cannot deal with unpredictable events such as those seen in man–machine interactions.
- Its modality can be difficult to define.
- It can only be used for activities that can be described in terms of performance times.
- Times are not available for all actions.
- It can be overly reductionistic, particularly for tasks that are mainly cognitive in nature.

Related Methods

CPA is one of a number of performance time prediction methods which also include the KLM (Card, Moran and Newell, 1983) and timeline analysis.

Approximate Training and Application Times

Although no data regarding the training and application time of CPA is available, it is suggested that both the training time the application time would be low, although this is dependent upon the task under analysis. For complex, larger tasks, the application time would be high.

In a review of interface methods, Harvey and Stanton (2012) proposed that CPA was a time-consuming technique compared to heuristic analysis, layout analysis and HTA. They estimated that to analyse multimodel in-vehicle car interfaces, 2–4 hours are required to collect the data and 8–10 hours are needed to analyse the data.

Reliability and Validity

Baber and Mellor (2001) compared predictions using CPA with the results obtained from user trials and found that the 'fit' between observed and predicted values had an error of less than 20 per cent. This suggests that the approach can provide robust and useful approximations of human performance.

Tools Needed

CPA can be conducted using pen and paper.

Flowchart

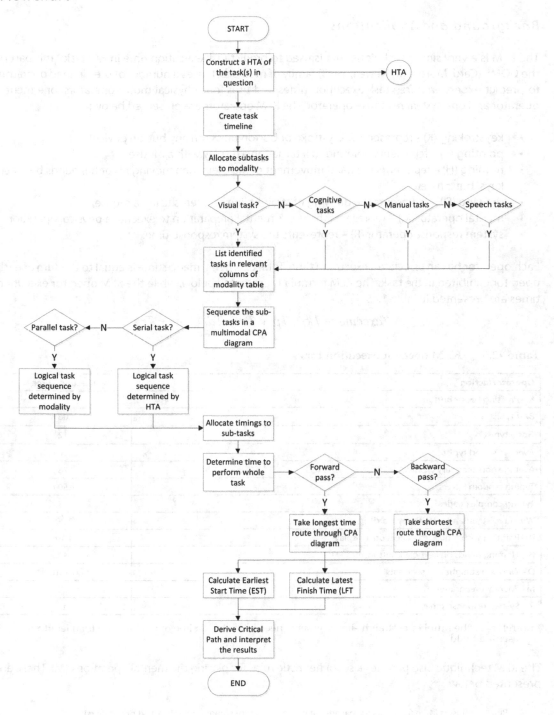

The Keystroke Level Model (KLM)

Background and Applications

The KLM is a very simple technique that is used to predict task execution time in HCI tasks. It is part of the GOMS (Card, Moran and Newell, 1983) family of methods. It uses a number of pre-defined operators to predict expert error-free task execution times and uses four physical motor operators, one mental operator and one system response operator. The KLM operators are presented below:

- keystroking (K) – represents a keystroke or button press (on any button device);
- pointing (P) – represents pointing to a target on a display with a mouse;
- homing (H) – represents the hand movement of the user when moving his or her hands between keys, buttons, etc.;
- drawing (D) – represents the drawing of straight line segments using a mouse;
- mental operator (M) – represents the user's mental preparation to execute a physical operation;
- system response operator (R) – represents the system response time.

Each operator has an associated execution time. Total task performance time is equal to the sum of each operator exhibited in the task. The KLM formula is presented below, while the KLM operator execution times are presented in Table 12.5.

$$Texecute = Tk + Tp + Th + Td + Tm + Tr$$

Table 12.5   KLM operator execution times

Operator/action	Execution time
K – Pressing key or button	
Best typist	.08
Good typist	.12
Average skilled typist	.20
Average non-secretary typist	.28
Typing random letters	.50
Typing complex codes	.75
Worst typist (unfamiliar with keyboard)	1.20
P – Pointing with mouse to a target on a display	1.10
H – Homing hands on keyboard, button, etc.	.40
D – Drawing straight-line segments	.9nd + .16ld*
M – Mental preparation	1.35
R – System response time	t

* nd refers to the number of straight-line segments needed to draw a line of total length; total length is represented by ld

The KLM technique also provides a set of heuristic rules for placing the mental operations (M). These are presented below.

Rule 0:   Insert Ms in front of all Ks that are not part of argument strings proper (e.g. text or numbers)

Rule 1:   If an operator following an M is fully anticipated in an operator previous to M, then delete the M

Rule 2:   If a string of MKs belongs to a cognitive unit (e.g. the name of a command), then delete all Ms but the first

Rule 3:   If a K is a redundant terminator (e.g. the terminator of a command immediately following the terminator of its argument), then delete the M in front of it

Rule 4:   If a K terminates a constant string (e.g. a command name), then delete the M in front of it, but if the K terminates a variable string (e.g. an argument string), then keep the M in front of it

The KLM technique has been used in collaboration with a number of other HF techniques. Pettitt, Burnett and Stevens (2007) adapted the KLM method to enable the modelling of task performance whilst employing the occlusion technique, while Saitwal et al (2010) used a combination of GOMS and the KLM to assess the usability of the Armed Forces Health Longitudinal Technology Application interface.

Recent research has focused on evaluating the applicability of the KLM for use on mobile devices due to their surge in popularity. Holleis et al. (2007) explored extensions to the technique for mobile devices, while Luo and John (2005) also looked at extending the KLM for use with mobile devices, discussing the use of CogTool, a tool to aid in the development of KLM. Luo and John described CogTool as a software application which enables interface mock-ups to be created in an interactive storyboard format. A KLM is automatically populated from this storyboard. They concluded that the results of their study provided evidence for the need to update the KLM for stylus-based activities and for finger swipes. Further research into the KLM and mobile devices was conducted by Li et al. (2010), who also argue that the KLM is inappropriate for a range of tasks focused on interacting with mobile phones. They claim that the device-dependent method requires an extension in order to consider the entire range of mobile phone interactions now available. They also propose an extended version of the KLM in which an additional 14 operators, including tapping with stylus (T), homing finger to somewhere (HF) and homing stylus to somewhere (HS), are included for use with GPS applications on mobile phones. GPS applications were chosen as Li et al. argue that they represent the majority of technical restrictions in mobile devices. Li et al. (2010) argue that such blocks will aid the analyst in building a KLM, will allow for the development of simplified models and will define user norms.

Domain of Application

HCI.

Procedure and Advice

Step 1: Compile Task List and Determine the Scenario to be Analysed
First, the analyst should compile an exhaustive task list for the device or system under investigation. Once the task list is complete, the analyst should select the particular task (or set of tasks) that is to be analysed.

Step 2: Determine the Component Operations Involved in the Task
Once the task under analysis has been defined, the analyst should determine the component operations involved in the task. The KLM calculates task performance time by summing the component operations involved in the task.

Step 3: Insert Physical Operations
Any homing or button presses involved in the task should be recorded. The time for each component should be recorded.

Step 4: Insert System Response Time
Next, the analyst should insert the appropriate system response time. This is normally determined from manufacturer specifications (Stanton and Young, 1999a). If these are not readily available, a domain expert estimate is sufficient.

Step 5: Insert Mental Operations
Finally, the mental operation times should be inserted. The analyst should use the KLM heuristic rules to place the mental operations.

Step 6: Calculate the Total Task Time

To calculate the total task time, the analyst should add each associated component operation time. The sum of the operation times equals the total task performance time (error-free performance). For maximum accuracy, the final sum should be multiplied by 1.755.

Advantages

- KLM is very easy and quick to use.
- It requires very little training (Stanton and Young, 1999a).
- Although it was developed specifically for HCI, it has been applied successfully in alternative domains, such as driving (Stanton and Young, 1999a) and also 'bank deposit reconciliation systems' (John and Kieras, 1994).
- It can be used to quickly compare the task times for two different devices or systems.
- It has proved to be effective at predicting transaction time, within acceptable limits of tolerance, e.g. usually within 20 per cent of the mean time observed from human performance (Card, Moran and Newell, 1983; Olson and Olson, 1990).
- It provides an immediately useful output of estimated task performance time.
- It offers encouraging reliability and validity data (Stanton and Young, 1999a).

Disadvantages

- KLM was designed specifically for computer-based tasks (HCI). New operators may have to be developed for the technique to be used in other domains.
- It only models error-free expert performance.
- It does not take context into account.
- There is limited validation evidence associated with its use outside of HCI.
- It assumes that all performance is serial and cannot deal with parallel activity.
- It ignores other unit task activity and also variation in performance.
- It ignores flexible human activity (Baber and Mellor, 2001).

Related Methods

The GOMS family of methods include NGOMSL (Natural Goals, Operators, Methods and Selection rules Language), the KLM, CMN-GOMS (Card, Moran and Newell – Goals, Operators, Methods and Selection rules) and CPM-GOMS (Cognitive Perceptual Model – Goals, Operators, Methods and Selection rules). An HTA for the system or device under analysis is also very useful when conducting a KLM analysis.

Approximate Training and Application Times

Stanton and Young (1999a) suggested that KLM is requires a moderately long training time. Execution time is dependent upon the size of the task under analysis, but is generally low. Stanton and Young also reported that KLM execution times improve considerably on the second application.

Reliability and Validity

Stanton and Young (1999a) reported outstanding reliability and validity measures for KLM. Out of the 12 HF techniques tested, the KLM was the only technique to achieve acceptable levels across the three ratings of inter-rater reliability, intra-rater reliability and validity.

Luo and John (2005) also found positive validation results in their analysis of KLM. Their study compared a KLM developed around a PDA with actual participant times, illustrating a high degree of correlation between the two (average prediction error of 3.7 per cent).

Tools Needed

KLM is pen-and-paper-based method. The analyst should also have access to the device or system under analysis and the KLM operator times.

Example

The following example (Stanton and Young, 1999a) is taken from a KLM analysis of a Ford in-car stereo system.

When using the Ford 7000 RDS EON in-car stereo, in order to switch the device on, the user has to push the on/off button. For the KLM analysis, this would be presented as:

Task | Execution time(s)
Switch on | MHKR = 2.65 + 1 = 3.65

where M = the driver thinking about pressing the on/off button, H = the driver positioning his or her finger over the button, K = the driver actually pressing the button and R = the time it takes for the radio to turn on (system response time).

The above example is a very simple one. A more complicated one, again for the Ford 7000 RDS EON, would be to adjust the treble on the system. In order to do this, the driver would have to push the bass button twice and then use the volume knob. Using a KLM analysis, this would be presented as:

Task | Execution time(s)
Adjust treble | MHKKHKR = 4.15+0.3 = 4.45

where M = the driver thinking about the following actions, H = the driver positioning his or her finger over the BASS button, KK = the driver pressing the BASS button twice, H = the driver positioning his or her finger over the volume button, K = the driver turning the volume button and R = the system response time.

The full KLM analysis of the Ford and Sharp in-car radios performed by Stanton and Young (1999a) is presented in Table 12.6.

As a result of the KLM analysis, it can be concluded that when performing the set of tasks outlined above, it takes around five seconds longer to complete using the Ford design.

Table 12.6   KLM output

Task	Time – Ford	Time - Sharp	Difference +/-
Switch unit on	MHKR = 2.65+1 = 3.65	MHKR = 2.65+1 = 3.65	0
Adjust volume	MHKR = 2.65+0.1 = 2.75	MHKR = 2.65+0 = 2.65	+0.1
Adjust bass	MHKHKR = 3.95+0.2 = 4.15	MHKR = 2.65+0 = 2.65	+1.5
Adjust treble	MHKKHKR = 4.15+0.3 = 4.45	MHKR = 2.65+0 = 2.65	+1.8
Adjust balance	MHKKHKR = 4.15+0.3 = 4.45	MHKKR = 2.85+0.1 = 2.95	+1.5
Choose new preset	MHKR = 2.65+0.2 = 2.85	MHKR = 2.65+0.2 = 2.85	0
Use seek	MHKR = 2.65+1 = 3.65	MHKR = 2.65+1 = 3.65	0
Use manual search	MHKHKR = 3.95+1 = 4.95	MHKR = 2.65+1 = 3.65	1.3

Table 12.6   Continued

Task	Time – Ford	Time - Sharp	Difference +/-
Store station	MHKR = 2.65+1 = 3.65	MHKR = 2.65+3 = 5.65	-2
Insert cassette	MHKR = 2.65+1 = 3.65	MHKR = 2.65+1 = 3.65	0
Autoreverse and FF	MHKRHKRKR = 4.15+5 = 9.15	MHKRKRK = 3.05+5 = 8.05	1.1
Eject cassette	MHKR = 2.65+0.5 = 3.15	MHKR = 2.65+0.3 = 2.95	0.2
Switch off	MHKR = 2.65+0.5 = 3.15	MHKR = 2.65+0.7 = 3.35	-0.2
Total time	53.65	48.35	5.3

Flowchart

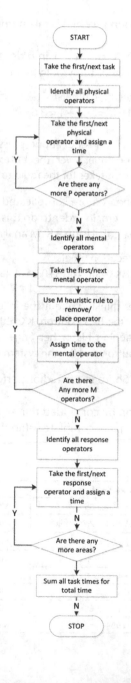

Timeline Analysis

Background and Applications

Although not a set methodology, timeline analysis is an approach that can be used in order to depict scenarios in terms of tasks and their associated task performance times. It can be used to display the functional and temporal requirements of a task and can be employed both predictively and retrospectively. The output is typically a graph. It can also be combined with workload analysis to represent the workload associated with each task step (Kirwan and Ainsworth, 1992). In terms of analysing command and control and team-based tasks, the appeal of timeline analysis lies in the fact that it can potentially depict individual and team task steps over time.

Domain of Application

Generic.

Procedure and Advice

Step 1: Data Collection
The first step in any timeline analysis is to collect specific data from the system under investigation. Task performance times should be recorded for all of the behaviours exhibited in the system. Typically, observational analysis is used during the data collection phase. If the technique is being applied retrospectively, then the analyst should observe the scenario under analysis. If a predictive timeline is required, similar scenarios in similar systems should be observed.

Step 2: Conduct an HTA
Once sufficient data regarding the task under analysis is collected, an HTA should be conducted. HTA (Annett et al., 1971; Shepherd, 1989; Kirwan and Ainsworth, 1992) is based upon the notion that task performance can be expressed in terms of a hierarchy of goals (what the person is seeking to achieve), operations (the activities executed to achieve the goals) and plans (the sequence in which the operations are executed). The hierarchical structure of the analysis enables the analyst to progressively re-describe the activity in greater levels of detail. The analysis begins with an overall goal of the task, which is then broken down into subordinate goals. At this point, plans are introduced to indicate in which sequence the sub-activities are performed. When the analyst is satisfied that this level of analysis is sufficiently comprehensive, the next level may be scrutinised. The analysis proceeds downwards until an appropriate stopping point is reached (see Annett et al., 1971 and Shepherd, 1989 for a discussion of the stopping rule).

Step 3: Determine Performance Times
Step 3 allows the analyst to create a performance time database for the analysis. Each task step in the HTA should be assigned a performance time. If the analysis is retrospective, this involves sifting through the data gathered during observations and recording the task performance times for each task. If a predictive timeline is required, the analyst should record the performance times for similar tasks to that involved in the predicted scenario.

Step 4: Construct the Timeline Graph
The timeline graph normally flows from left to right, with time running along the Y-axis and the tasks running along the X-axis.

Advantages

- Timeline analysis can be used to compare the performance times associated with two different systems or designs.
- It can be used to represent team-based tasks and parallel activity.
- It can be used to highlight problematic tasks or task sequences in the design of systems and processes.
- Workload analysis can be mapped directly onto a timeline graph. This makes for a very powerful analysis.
- It is a simple technique and requires little training.
- It requires very few resources once the data collection phase is complete.

Disadvantages

- The reliability and validity of the technique is questionable.
- Observation data is often flawed by a number of biases.
- When used predictively, it can only model error-free performance.
- The initial data collection phase is time-consuming and resource-intensive.

Approximate Training and Application Times

The amount of training time required for timeline analysis is very low. The application time is minimal once the initial data collection is complete. The data collection time involved is dependent upon the scenario under analysis. For large, complex scenarios, the data collection time associated with timeline analysis is very high.

Reliability and Validity

Kirwan and Ainsworth (1992) report that the technique possesses a high face validity. No data regarding the reliability and validity of the technique is available in the literature.

Tools Needed

Once the data collection phase is complete, timeline analysis can be conducted using pen and paper. The data collection phase (observation) typically requires using video and audio recording devices.

Flowchart

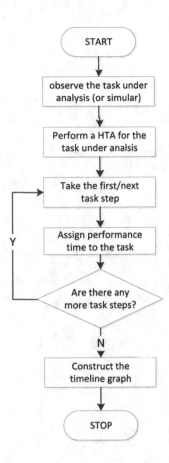

Human Factors Methods Integration: Applications of the Event Analysis of the Systemic Teamwork Framework

Introduction

The aim of this chapter is to demonstrate how different HF methods, originally developed for the study of distinct concepts, can be integrated to provide exhaustive analyses of performance in complex socio-technical systems. Although applying the methods described in this book in isolation is perfectly acceptable, the complexity of modern-day systems is such that HF methods are increasingly being applied together as part of frameworks or integrated suites of methods (e.g. Stanton, Baber and Harris, 2008; Walker et al., 2006). Scenarios are often so complex and multi-faceted, and analysis requirements so diverse, that various methods need to be applied as one method in isolation cannot cater for the scenario and analysis requirements. The EAST framework (Stanton, Baber and Harris, 2008) described in this chapter is one such example of how a combination of methods can be useful for examining performance in complex systems.

The Event Analysis of Systemic Teamwork

Two case studies are presented which focus on command and control activities in the rail and energy distribution domains. Command and control scenarios are characterised by multiple individuals and teams working together in pursuit of a common goal (comprising multiple interacting sub-goals). High levels of communication and coordination are required, and there is often considerable onus placed on technologies to facilitate this. Various sub-constructs are also evident within command and control, including planning, directing, coordinating and control of resources, SA, etc. When examining command and control, the descriptive constructs of interest can be distilled down to simply:

- *why* (the goals of the system, sub-system(s) and actor(s));
- *who* (the actors performing the activity are, including humans and technologies);
- *when* (activities take place and which actors are associated with them);
- *where* (activities and actors are physically located);
- *what* (activities are undertaken, what knowledge/decisions/processes/devices are used and what levels of workload are imposed); and
- *how* (activities are performed and how actors communicate and collaborate to achieve goals).

More than likely, none of the HF methods described in this book can independently cover all of these constructs. Using an integrated suite of methods, however, allows scenarios to be analysed exhaustively from the perspectives described, but more importantly enables the effects of constructs on other constructs to be considered; for example, how communications influence the way in which tasks are performed and in turn how the way in which the tasks are being performed influences the knowledge/decisions/processes being used. There are also further advantages associated with the integration of existing methods, because not only does this bring reassurance in terms of a validation history, but it also enables the same data to be analysed from multiple perspectives. These multiple perspectives, as well as being inherent in the scenario being described and measured, also provide a form of internal validity. Assuming that the separate methods integrate on a theoretical level, their application to the same data set offers a form of 'analysis triangulation'.

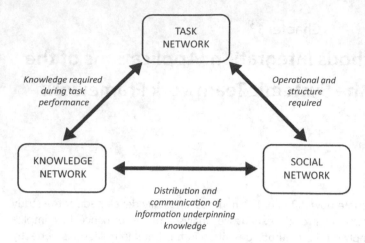

Figure 13.1   Network of networks approach

The Event Analysis of Systemic Teamwork (EAST; Stanton et al., 2008) provides one such framework of methods that allows collaborative performance to be comprehensively described and evaluated. Since its conception, the framework has been applied in many domains, including land and naval warfare (Stanton et al., 2006), aviation (Stewart et al., 2008), air traffic control (Walker et al., 2010), railway maintenance (Walker et al., 2006) and the emergency services (Houghton et al., 2008). It is underpinned by the notion that complex collaborative systems can be meaningfully understood through a network of networks approach (see Figure 13.1). Specifically, three networks are considered: task, social and knowledge networks. Task networks describe the goals and subsequent tasks being performed within the system. Social networks analyse the organisation of the system (i.e. communications structure) and the communications taking place between the actors working in the team. Finally, knowledge networks describe the information and knowledge (distributed SA) that the different actors use and share during task performance.

The process begins with the conduct of an observational study (see Chapter 2) of the scenario under analysis. HTA (see Chapter 3) is then used to describe the goals, sub-goals and operations involved during the scenario. The resultant task network identifies the actors involved, what tasks

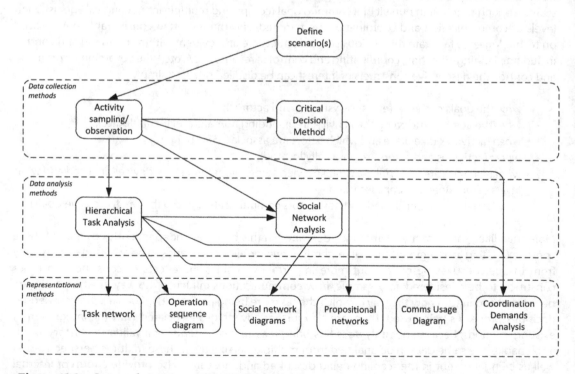

Figure 13.2   Internal structure of EAST framework

are being performed, the temporal structure of tasks and the interrelations between tasks. The HTA output is then used to construct an OSD (see Chapter 5) of the task, detailing all activities and interactions between the actors involved. The social network is embodied by SNA (see Chapter 9), which considers the associations between agents during the scenario. It is important to note here that agents may be human or technological, and so SNA caters too for human-machine interactions. The CDM (see Chapter 4) focuses on the decision-making processes used during task performance and the information and knowledge underpinning decision-making. Based on this data, the propositional network approach (see Chapter 7) represents SA during the scenario, from both the point of view of the overall system and the individual agents performing activities within the system. Thus, the type, structure and distribution of knowledge throughout the scenario are represented. The overlap between methods and the constructs they access is explained by the multiple perspectives provided on issues such as the 'Who' and the 'What'. For example, the HTA deals with 'what' tasks and goals, the CDM deals with 'what' decisions are required to achieve goals, and the propositional networks deal with 'what' knowledge or SA is underpinning the tasks being performed and the decisions being made. Each of these is a different but complementary perspective on the same descriptive construct and a different but complementary perspective on the same data derived from observation and interview, which is an example of analysis triangulation. The structure of the EAST framework is represented in Figure 13.2 on the left.

Procedure and Advice

Step 1: Define Analysis Aims
First, the aims of the analysis should be clearly defined so that appropriate scenarios are used and relevant data is collected. In addition, not all components of the EAST framework may be required, so it is important to clearly define the aims at this point to ensure that the appropriate EAST methods are applied.

Step 2: Define the Task under Analysis
Next, the task (or tasks) or scenario (or scenarios) under analysis should be clearly defined. This is dependent upon the aims of the analysis and may include a range of tasks or one task in particular. It is normally standard practice to develop an HTA for the task under analysis if sufficient data and SME access are available. This is useful later on in the analysis and is also enlightening, allowing the analyst to gain an understanding of the task before the observation and analysis begins.

Step 3: Conduct Observational Study of the Task or Scenario under Analysis
The observation step is the most important part of the EAST procedure. Typically, a number of analysts are used in scenario observation. All activities involved in the scenario under analysis should be recorded along an incident timeline, including a description of the activity undertaken, the agents involved, any communications made between agents and the technology involved. Additional notes should be made where required, including the purpose of the activity observed, any tools, documents or instructions used to support activity, the outcomes of activities, any errors made and also any information that the agent involved feels is relevant. In addition, it is useful to video record the task and record verbal transcripts of all communications if possible.

Step 4: Conduct CDM Interviews
Once the task under analysis is complete, each 'key' agent (e.g. scenario commander, agents performing critical tasks) involved should be subjected to a CDM interview. This involves dividing the scenario into key incident phases and then interviewing the actor involved in each phase using a set of pre-defined CDM probes (e.g. O'Hare et al., 2000; see also Chapter 4 for more information on the CDM).

Step 5: Transcribe Data

Once all of the data is collected, it should be transcribed in order to make it compatible with the EAST analysis phase. An event transcript should then be constructed. This should describe the scenario over a timeline, including descriptions of activity, the actors involved, any communications made and the technology used. In order to ensure the validity of the data, the scenario transcript should be reviewed by one of the SMEs involved.

Step 6: Reiterate HTA

The data transcription process allows the analyst to gain a deeper and more accurate understanding of the scenario under investigation. It also allows any discrepancies between the initial HTA scenario description and the actual activity observed to be resolved. Typically, activities in complex socio-technical systems do not run entirely according to protocol, and certain tasks may have been performed during the scenario that were not described in the initial HTA description. The analyst should compare the scenario transcript to the initial HTA and add any changes as required.

Step 7: Conduct CDA

The CDA method (see Chapter 9 for a full description) involves extracting teamwork tasks from the HTA and rating them against the associated CDA taxonomy. Each teamwork task is rated against each CDA behaviour on a scale of 1 (low) to 3 (high). The total coordination for each teamwork step can be derived by calculating the mean across the CDA behaviours. The mean total coordination figure for the scenario under analysis should also be calculated.

Step 8: Construct CUD

The CUD (see Chapter 9 for a full description) is used to describe the communications between teams of actors spread across different geographical locations. A CUD output describes how and why communications between actors occur, which technology is involved in the communication, and the advantages and disadvantages associated with the technology used. A CUD analysis is typically based upon observational data of the task or scenario under analysis, although walk/talkthrough analysis and interview data can also be used (Watts and Monk 2000).

Step 9: Conduct SNA

SNA (See Chapter 9 for a full description) is used to analyse the relationships between the actors involved in the scenario under investigation. It is normally useful to conduct a series of SNAs representing different phases of the task under analysis (using the task phases defined during the CDM part of the analysis). It is recommended that the Applied Graph and Network Analysis (AGNA) SNA software package is used for the SNA phase of the EAST methodology.

Step 10: Construct OSD

The OSD (see Chapter 5 for a full description) represents the activity observed during the scenario under analysis. The analyst should construct the OSD using the scenario transcript and the associated HTA as inputs. Once the initial OSD is completed, the analyst should then add the results of the CDA to each teamwork task step.

Step 11: Construct Propositional Networks

The final step of the EAST analysis involves constructing propositional networks (see Chapter 7 for a full description) for each scenario phase identified during the CDM interviews. The WESTT software package should be used to construct the propositional networks. Following construction, information usage should be defined for each actor involved via shading of the information elements within the propositional networks.

Step 12: Validate Analysis Outputs
Once the EAST analysis is complete, it is pertinent to validate the outputs using appropriate SMEs and recordings of the scenario under analysis. Any problems identified should be corrected at this point.

Advantages

- The analysis produced is extremely comprehensive and activities are analysed from various perspectives.
- The framework approach allows methods to be chosen based on analysis requirements.
- EAST has been applied in a wide range of different domains.
- The approach is generic and can be used to evaluate activities in any domain.
- A number of HF concepts are evaluated, including distributed SA, cognition, decision-making, teamwork and communications.
- It uses structured and valid HF methods and has a sound theoretical underpinning.
- The outputs can be used to examine and inform system, technology, procedure and training design.
- It avoids bias by focusing on objective and manifest phenomena.
- Its results are comparable across domains.
- Its results are graphical and easily interpreted, yet are amenable to further summarisation using tables and numerical indices.
- The summary level is underpinned by considerable detail that can be explored further, for example, in the context of system design.

Disadvantages

- When undertaken in full, the EAST framework is a very time-consuming approach.
- The use of various methods ensures that the framework incurs a high training time.
- In order to conduct an EAST analysis properly, a high level of access to the domain, task and SMEs is required.
- Some parts of the analysis can become overly time-consuming and laborious to complete.
- Some of the outputs can be large, unwieldy and difficult to present in reports, papers and presentations.

Application 1: Railway Maintenance Example

Three examples of complex and dynamic resource systems are taken from the UK rail industry. The scenarios serve as a source of live data to demonstrate the capability of the method. The data is sourced, in this case, from written transcripts based on communications between parties in the scenarios and interviews with SMEs.

Broadly speaking, the C4i activities under consideration are those involved in the setting up of safety systems required when carrying out railway track maintenance (for example, see Figure 13.3 on the next page). Safety systems are required so that workers on the track do not come into conflict with moving trains and so that trains do not travel over railway infrastructure that is rendered unsafe by the maintenance work or the requirement for it. The strict procedures underpinning these systems are specified nationally in the UK railway industry *Master Rule Book* (Rail Safety and Standards Board, 2003).

Railway operations are an example of civilian C4i, where a 'management infrastructure' is required and in place. Maintenance activities on the railways possess all the essential ingredients of C4i, including

Figure 13.3   Example of track maintenance activities (Rail Safety and Standards Board, 2003)

a common goal, individuals and teams coordinating to reach it but who are dispersed geographically, and numerous systems, procedures and technology to support their endeavours.

Background

Under normal conditions, a signaller has the key responsibility for controlling train movements and maintaining safety for an area of railway line. This control occurs remotely from the line at a control centre (a signalbox or signalling centre). These can be located many miles from where the activity could be taking place.

During maintenance, another person takes responsibility for an area of the line (sometimes referred to as a 'possession') and/or for preventing trains passing over the possession (referred to as 'protection'). These individuals are normally termed the person in charge of possession (PICOP) or controller of site safety (COSS). Communication and coordination are required to transfer responsibility between the signaller and the PICOP/COSS. The PICOP/COSS also has to communicate and coordinate with various other personnel, who include personnel carrying out maintenance within the ICOP/COSS's areas of control, drivers of trains and on the track-plant, and personnel implementing aspects of the possession. All of these people (and equipment) may be spread over a sizeable geographical area. Three specific maintenance scenarios are briefly described below, with Figure 13.4 providing additional clarity on the general layout and relative geographical positions of personnel.

Scenario 1: Planned Maintenance Activities

This scenario describes the processes and activities for setting up a possession for a stretch of track so that planned maintenance can take place. This requires coordination between multiple parties, including communication between the signaller and the PICOP (so that appropriate 'protecting' signals are set to danger) and the provision of instructions to a 'competent person' to place a form of protection against oncoming trains at the limits of the possession (these take the form of explosive charges called detonators that emit a loud noise to alert drivers who may have just run over them). Additional complexity comes in the form of a number of engineering work sites within the possession, each of which has an engineering supervisor (ES) and a COSS responsible for setting up and managing it. The ES

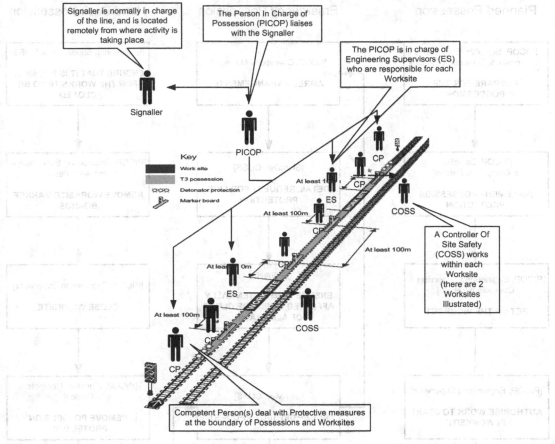

Figure 13.4 Overall diagram of the various track possession scenarios (adapted from Rail Safety and Standards Board, 2003)

will also use 'competent personnel' to place marker boards as a form of additional protection at the ends of the individual worksites.

Scenario 2: Emergency Engineering Work

When railway personnel are required to carry out unplanned emergency engineering work on the line, such as when track or infrastructure has been damaged or has suddenly degraded, the passage of trains must be stopped and an emergency protection procedure called a T2(X) must be applied. For this procedure, a portion of the railway which is normally under the control of a signaller working remotely from a signalbox becomes the responsibility of a COSS, who will work on the line. The workers will be protected from train movements by the signaller placing signals at the limits of the work zone to danger. Emergency protection can be arranged between the COSS and the signaller following discussions with the Network Rail Area Operations Manager. It should be noted in the scenario above that *non*-emergency engineering work involves greater advanced planning and protection, whereas in emergency scenarios, organisation tends to occur 'on the day'.

Scenario 3: Ending a Track Possession

When the possession is ended, the 'set-up' procedure outlined in scenario 1 is largely reversed. First, the ES of a worksite has to check that the worksite can be closed. This requires agreement between the

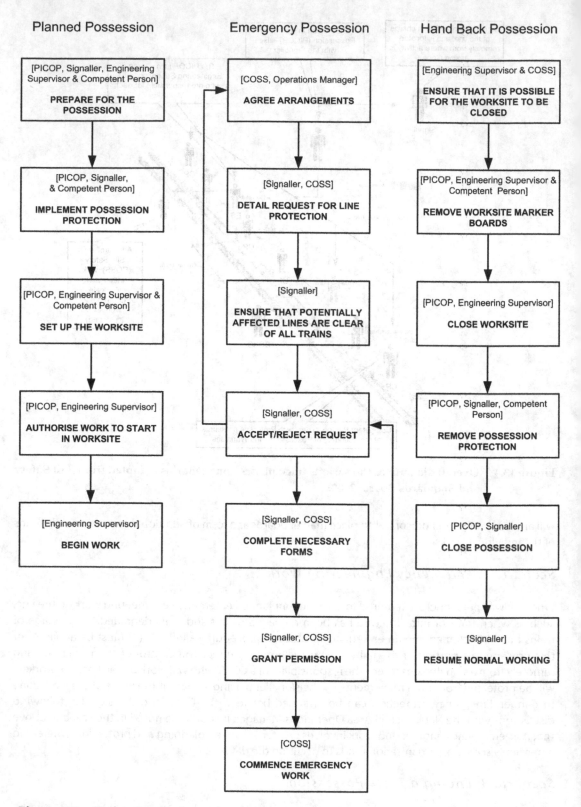

Planned Possession

[PICOP, Signaller, Engineering Supervisor & Competent Person]

PREPARE FOR THE POSSESSION

↓

[PICOP, Signaller, & Competent Person]

IMPLEMENT POSSESSION PROTECTION

↓

[PICOP, Engineering Supervisor & Competent Person]

SET UP THE WORKSITE

↓

[PICOP, Engineering Supervisor]

AUTHORISE WORK TO START IN WORKSITE

↓

[Engineering Supervisor]

BEGIN WORK

Emergency Possession

[COSS, Operations Manager]

AGREE ARRANGEMENTS

↓

[Signaller, COSS]

DETAIL REQUEST FOR LINE PROTECTION

↓

[Signaller]

ENSURE THAT POTENTIALLY AFFECTED LINES ARE CLEAR OF ALL TRAINS

↓

[Signaller, COSS]

ACCEPT/REJECT REQUEST

↓

[Signaller, COSS]

COMPLETE NECESSARY FORMS

↓

[Signaller, COSS]

GRANT PERMISSION

↓

[COSS]

COMMENCE EMERGENCY WORK

Hand Back Possession

[Engineering Supervisor & COSS]

ENSURE THAT IT IS POSSIBLE FOR THE WORKSITE TO BE CLOSED

↓

[PICOP, Engineering Supervisor & Competent Person]

REMOVE WORKSITE MARKER BOARDS

↓

[PICOP, Engineering Supervisor]

CLOSE WORKSITE

↓

[PICOP, Signaller, Competent Person]

REMOVE POSSESSION PROTECTION

↓

[PICOP, Signaller]

CLOSE POSSESSION

↓

[Signaller]

RESUME NORMAL WORKING

Figure 13.5   Task networks for each scenario

ES and each controller of site safety (COSS) within the worksite (there is a COSS responsible for each piece of work being undertaken within the worksite). Once this has been checked and the PICOP has been informed, the ES can instruct a competent person (CP) to remove the worksite marker boards. The PICOP is informed when this is completed and then, when all the worksites within a possession are closed, the possession itself can be closed. The PICOP can then instruct a CP to remove the possession protection and will inform the signaller that the lines are now safe and clear for trains to run on. Control of the line is then passed from the PICOP to the signaller and the normal running of trains over the lines can resume.

Outputs from Component Methods

Task Networks

The first step in the EAST methodology, subsequent to collecting the data, is to model the goal structure of the scenario using HTA. The output of this step is represented as task networks. These are graphical representations of the 'plan 0' within the HTA and depict the task structure in terms of how tasks relate to each other functionally and temporally. In the present case, the highly proceduralised and rigid nature of the activity is seen a more or less linear task flow (see Figure 13.5 on the opposite page).

The following is an extract from the verbal transcripts and shows a typical communication occurring between the Signaller (S) and the PICOP:

S:	Wimbledon… [answers phone with name of signalbox].
S:	…panel 1 [signalling panel, and associated geographical area being worked by the signaller].
PICOP:	Hello Wimbledon, it's [name of PICOP] at Waterloo [station].
S:	…Yes…
PICOP:	The blocks [protective measures] have now been put out mate on the down main slow and the up main slow… [referring to different lines].
S:	…right…
PICOP:	…and it's clear of 15 64 b points and 15 12 b points.
S:	It's all yours, at, er, what we on, 9:53 then.
PICO:	Oh, 9:53, cheers mate.
S:	…ok…
PICOP:	Can I take your name please, I forgot to write it down earlier?
S:	[provides name].
PICOP:	[name] thanks a lot mate…
S:	Ok…
PICOP:	Bye.

It is interesting to note the data collection transcripts which highlight that some of the required and critical steps may be implied (e.g. the PICOP's closure of communication is by saying 'bye', which is implied by the signaller to mean that there are no remaining issues or ambiguity, and that points 15 64b and 15 12b are indeed clear) and that the sequencing of communications is more flexible than the procedures may suggest (for example, the PICOP may only inquire in more detail as to the name of the signaller at the stage when it is needed for the completion of documentation, rather than at the beginning of a call). However, at the level illustrated above, any informality or flexibility occurs within the confines of a well-defined procedure.

CDA

Based on Figure 13.6, the supposition that C4 is dominated by coordination tasks appears to be justified. In the three scenarios under analysis, the tasks that fall into the 'teamwork' track and that require coordination form between 66 and 72 per cent of the total tasks undertaken. Figure 13.7 below extracts

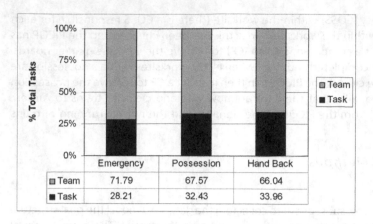

**Figure 13.6   Results of CDA analysis showing percentage
of task/teamwork activities undertaken within
each scenario**

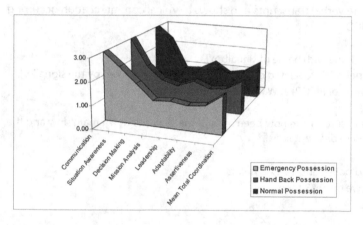

**Figure 13.7   Results of CDA analysis showing profile of
results on each of the coordination dimensions**

the teamwork tasks for further analysis. The analysis proceeds according to seven coordination dimensions and one summary total coordination score (based on the mean of the individual scores). This analysis reveals a broadly similar pattern of coordination activity within the scenarios; certainly, the total coordination figures are comparable, falling within the mid-point of the rating scale. Of more interest is the pattern of results across the seven individual dimensions where a distinctive footprint emerges. Communication, SA and decision-making are prominent dimensions, and there is also a smaller 'blip' for the leadership dimension. Leadership can be taken as perhaps a further indication of some decision-making activity. It can further be noted that the larger (and more complex) the scenario, the larger the leadership 'blip' is. Therefore, in summary, not only are the majority of total tasks dominated by coordination activities, but those activities are dominated by communications and the creation and maintenance of SA.

CUD

General observations from the CUD analysis are that the communications are entirely verbal. Given the nature of the scenario, verbal and telephone communications appear to be appropriate in most stages of the interaction. However, possible technology options could be helpful in three respects:

1. first, removing possible sources of error inherent in verbal communications;
2. second, removing the cumbersome nature of read-back procedures; and
3. third, alleviating the physical disturbance to other tasks caused by the unscheduled and ad-hoc presentation of verbal communications.

Of course, any new approaches would require fuller risk justification and assessment within the wider task context before application. The key point is that the CUD method provides a systematic way of presenting the existing situation and considering alternatives to it based on data. Figure 13.8 summarises the communications technology in use within the scenario's.

Social networks

Social Network Analysis is used in the EAST method to represent and summarise the communication/ information links between agents in the scenario. Figure 13.8 presents a graphical representation of the networks derived from each of the scenarios, and is also annotated with comms information drawn from the CUD analysis.

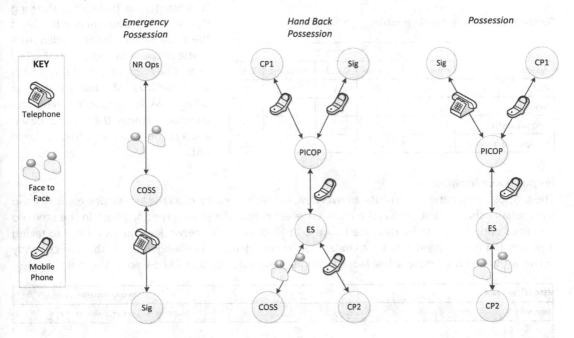

Figure 13.8   Graphical representation of social networks overlain with communications media drawn from CUD analysis

A range of mathematical metrics (derived from graph theory) can be applied. The results show that the PICOP, signaller and ES have the highest levels of sociometric status and centrality. These metrics indicate that they are key agents in the scenarios. The notion of centrality is also borne out when considering 'betweenness', that is, the PICOP and ES most frequently fall between pairs of other positions in the network.

Having identified the key agents, it is also possible to view the network as a whole under the concept of network density. Density is the degree of interconnectivity between agents, or the number of network links used compared to those that are theoretically available (the maximum being a case where all agents are linked to each other). In Table 13.1 below, it can be seen that the emergency possession scenario has the densest pattern of connectivity, with the remaining two scenarios being broadly comparable.

Although the metrics allow comparison between networks, intelligent interpretation is required. For example, a network with every agent connected to each other would permit the easy dissemination of information, but might also be inefficient. Similarly, having one central node may have advantages for coordination, but offers the potential for an information bottleneck. The main point is that the interpretation and subsequent comparison of networks has to take into account a range of contextual factors. For the time being, the networks derived from the scenarios above appear to be relatively well matched to the procedures being undertaken, with a mix of central agents and interconnectivity.

Table 13.1   Comparison of network density between scenarios

	Possession	Emergency possession	Hand back possession
Density	0.4	0.67	0.33

Representational Methods

Scenario Process Charts (OSD)

Figure 13.9 presents a sample of an enhanced OSD from scenario 1 (the planned maintenance activities) and highlights how the preceding methods are integrated with it. The operations loading is presented in Table 13.2, showing the PICOP, the signaller and the CP as the most heavily loaded individuals in the network in terms of tasks. The operations loading table provides a further level of summarisation in being able to capture, in a relatively compact manner, the process-based aspects of what is often a large OSD.

Table 13.2   Task loading table

| Agent | OPERATIONS | | | | |
	Operation	Receive	Decide	Transport	Total
PICOP	54	30	1		85
Signaller	17	16	1		34
Eng supervisor	13	17		1	31
Competent person 1	17	20		1	38

Propositional Networks

The propositional network provides an overview, for each scenario, of all the knowledge elements and their relationships. It also allows the knowledge elements related to a specific phase in the scenario (previous and current) to be described. Figure 13.10 displays the network elements related to taking a possession for emergency maintenance. Shaded cells denote knowledge objects that are currently active within the task phase, while faded cells indicate previously active knowledge. The main image in

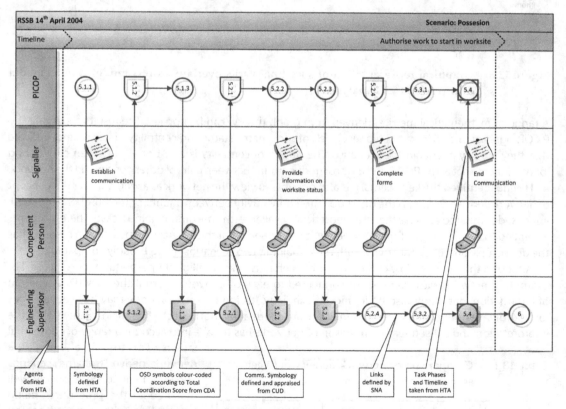

Figure 13.9   Enhanced OSD summary representation

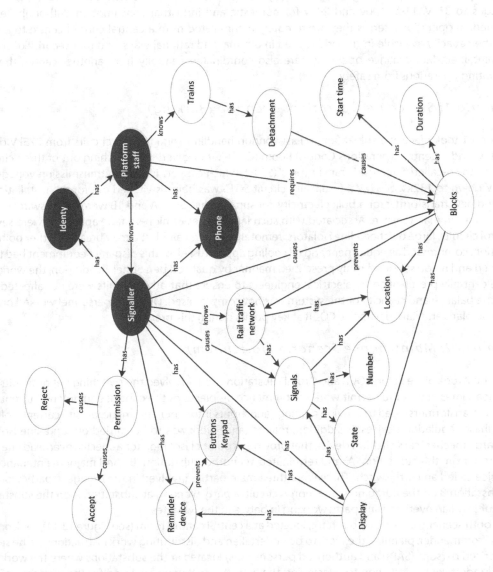

Phase 1 – Establish Communication

Phase 2 – Establish Line to be Blocked

Phase 3 – Confirm Line is Blocked

Phase 4 – Agree Detachment Complete

Figure 13.10 Illustration of propositional networks for phases within Scenario 3

Figure 13.10 is intended to provide information on knowledge objects, whereas the smaller networks alongside are merely illustrative of changing activation.

Application 2: Civilian Command and Control Example

The analysis presented in this section concerns two scenarios undertaken on a major UK electrical distribution network. The distribution grid in question consists of 341 geographically dispersed substations in England and Wales, which are used to distribute electricity to consumers. Power stations (and feeds from continental Europe) energise the grid, which uses an interconnected network of 400,000 volt (400kV), 275kV (the super grid network) and 132kV overhead lines and towers, or cables running in tunnels to carry electricity from source to substations. The substations are the national distribution company's interface with regional electricity companies that step down the grid's transmission voltages to 33kV, 11kV, 400v and 240v for domestic and industrial consumption. Although flexibly manned, in operational terms they are remotely manipulated from a central control centre to ensure that the capacity available in the grid is used in optimal and rational ways, and that security of supply is maintained. Maintenance operations are also coordinated centrally from another centre, thereby separating operations from safety.

Scenario 1: Switching Operations Scenario (Barking)

Scenario 1 took place at a substation in East London handling voltages and circuits from 275kV down to 33kV, and a Central Operations Control Room (COCR). It involved the switching out of three circuits relating to so-called 'Supergrid Transformers' (SGTs), which convert incoming transmission voltages of 275kV down to 132kV or 33kV. Specifically, circuit SGT5 was being switched out for the installation of a brand new transformer for a bulk electricity consumer, while SGT1A and 1B were being switched out for substation maintenance. Associated with such large pieces of high-voltage apparatus were several control circuits, large overhead line isolators, remotely operated air blast circuit breakers, other points of isolation, compressed air equipment and oil cooling apparatus. All of this disparate equipment had to be handled and made safe in a highly prescribed manner by qualified personnel. In addition, the work had to be centrally pre-planned by electrical engineers to ensure that other circuits were not affected and that the balance and capacity of the system was not compromised. Qualified personnel worked on-site to these plans and liaised with the COCR at key points during this process.

Scenario 2: Maintenance Scenario (Tottenham)

Scenario 2 took place at the COCR and a rural substation site. It involved the switching out of circuits and overhead lines in order to permit work to commence on pieces of the control equipment (current and voltage transformers) used to provide readings and inputs into other automatic, on-site current, voltage and phase regulation devices. In addition, maintenance work was to be carried out on a line isolator (the large mechanical switching device that provides a point of isolation for a specific overhead line that departed from this substation (A) and terminated at another substation (B)) and major maintenance on a device called an earth switch. There were three main parties involved in the outage: a party working at substation B on the outgoing substation A circuit, a party working at substation A on the substation B circuit, and an overhead line party working in between the two sites.

In both scenarios, the COCR operator, located at a central control room (see Figure 13.11), took on the role of commander, planning the work to be undertaken and distributing work instructions to the senior authorised persons (SAPs) and authorised persons (APs) located at the substations where the work was to be undertaken. In addition to overseeing the activities that were analysed for the purposes of this research, the COCR operator was also involved in other activities being undertaken elsewhere on the grid and so had other responsibilities and tasks to attend to during the study. The other agents involved in the scenarios included the Central Command (CC) Operator and Overhead Line Party (OLP) personnel working on the overhead lines. The COCR operator communicated with the other agents via landline telephone and mobile phone, and also had access to substation diagrams, work logs and databases and the Internet. The structure of personnel for Scenario 1 is presented in Figure 13.12. The structure of personnel for Scenario 2 is presented in Figure 13.13.

Figure 13.11 Central control room

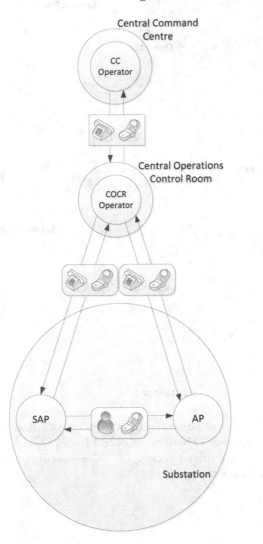

Figure 13.12 Scenario 1 network structure: CC Operator = Central Command Operator; COCR Operator = Central Operations Control Room Operator; SAP = Senior Authorised Person; AP = Authorised Person

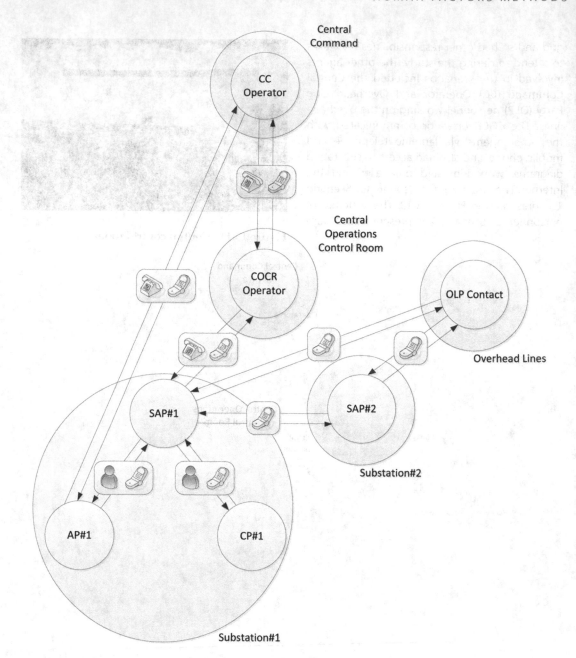

Figure 13.13 Scenario 2 network structure: CC Operator = Central Command Operator; COCR Operator = Central Operations Control Room Operator; SAP = Senior Authorised Person; AP = Authorised Person; CP = Competent Person; OLP Contact = Overhead Line Party Contact

Methodology

Design

The study was an observational study that involved direct observation of the activities undertaken during the scenarios analysed. Three researchers acted as observers, with one located in the field at the substation involved and two located at the central control room.

Participants

This study involved 11 participants who worked for the energy distribution organisation in question. Scenario 1 involved the following four participants: a CC operator, a COCR operator, a SAP and an AP. Scenario 2 involved the following seven participants: a CC operator, a COCR operator, a SAP, an AP and a CP at one substation, a SAP at another substation and an overhead line party contact. Due to access restrictions and the nature of the study (observation during real work activities), it was not possible to collect demographic data for the participants involved.

Materials

The observers used pen and paper as well as video and audio recording equipment to collect data during the observations. All observational and verbal transcripts were transcribed using Microsoft Word. A set of pre-defined interview probes were used for the CDM interviews, along with a CDM pro-forma to record participants' responses. A Dictaphone was also used to record the CDM interviews. Various other software programs were utilised to support the analysis presented, including the AGNA network analysis and WESTT software packages and Microsoft Visio.

Procedure

The analyses were based on data collected during live observational study of the two scenarios. In each scenario, two observers were located at the COCR observing the COCR operator, and one observer was located in the field with the SAPs/APs at the substation involved. The analysts located at the COCR observed all of the COCR operator's activities and were able to discuss the activities being undertaken and query different aspects of the scenarios as they unfolded. The analyst located at the substations observed the SAP/APs undertaking the work required and was also able to discuss aspects of the scenario with them. Observational transcripts were constructed and audio recordings were used to record the communications between those involved. The data recorded via observational transcripts included a description of the activity (specific task steps, e.g. issue instructions to SAP at substation) performed by each of the agents involved, transcripts of the communications that occurred between agents during the scenarios, the technology used to mediate these communications, the artefacts used to aid task performance (e.g. tools, computers, instructions, substation diagrams, etc.), time and additional notes relating to the tasks being performed (e.g. why the task was being performed, what the outcomes were, etc.). CDM interviews were conducted with the key agents involved (the COCR operator and the SAPs) upon the completion of the scenario. This involved breaking down the scenario into a series of key decision points and administering CDM probes in order to examine the decision-making processes used at each point. For validation purposes, an SME from the energy distribution company reviewed the data collected and the subsequent analysis outputs.

Results

Task Networks

HTAs were constructed for each scenario based on standard operating procedures and direct observation of the scenarios. The HTAs described each scenario in terms of a structured hierarchy of goals and sub-goals, along with feedback loops. One useful way of summarising large and complex HTA outputs is through the construction of a task network, which provides a summary of the main higher-level goals and tasks involved, and the interaction between them. Task networks for each scenario are presented in Figure 13.14.

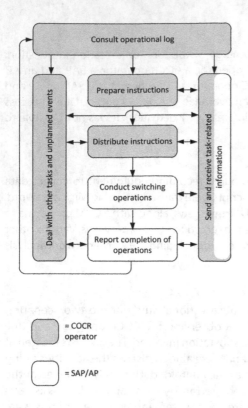
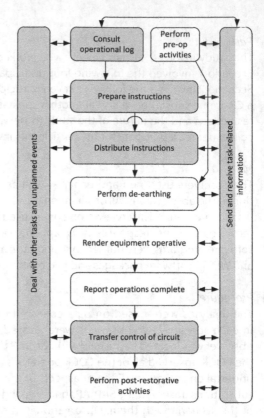

Figure 13.14 Task models for Scenarios 1 and 2. Shading denotes which agents undertook activities relating to each goal

Social Networks

Social network analysis is used to examine individuals or teams that are linked to each other by communications and to subject that network to mathematical analysis using Graph Theory (Driskell and Mullen, 2004). Social networks were constructed for each scenario based on the observed communications between agents during task performance. The structure of each network was then analysed using three network analysis metrics: density, sociometric status and centrality. Network *density* represents the level of interconnectivity of the network in terms of communications links between agents. Density is expressed as a value between 0 and 1, with 0 representing a network with no connections between agents and 1 representing a network in which every agent is connected to every other agent (Kakimoto et al., 2006, cited in Walker et al., 2012). *Sociometric status* provides a measure of how 'busy' each agent is relative to the total number of agents within the network under analysis (Houghton et al., 2008). *Centrality* is also a metric of the standing of each agent within a network (Houghton et al., 2008), but here this standing is in terms of its 'distance' from all other agents in the network. A central agent is one that is close to all other agents in the network, and a message conveyed from that agent to an arbitrarily selected other agent in the network would, on average, arrive via the least number of relaying hops (Houghton et al., 2008). The social network analysis for scenarios 1 and 2 is presented in Figure 13.15.

The social network analysis outputs enable the identification of the key communications 'hubs' within each network, as well as a comparison of the communications network apparent in each scenario. Key agents, specifically those that act as communications hubs within each network, are identified through examination of the centrality and sociometric status values. Agents with values above the centrality and sociometric status mean values for the overall network are defined as key agents. For Scenario 1, the key agents are the COCR operator and the SAP/AP at the substation, while for Scenario 2, the COCR operator

Scenario 1

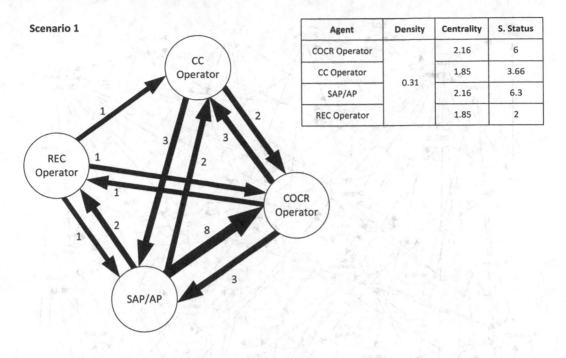

Agent	Density	Centrality	S. Status
COCR Operator		2.16	6
CC Operator	0.31	1.85	3.66
SAP/AP		2.16	6.3
REC Operator		1.85	2

Scenario 2

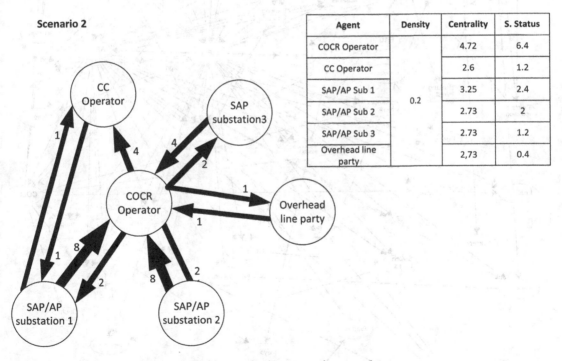

Agent	Density	Centrality	S. Status
COCR Operator		4.72	6.4
CC Operator		2.6	1.2
SAP/AP Sub 1	0.2	3.25	2.4
SAP/AP Sub 2		2.73	2
SAP/AP Sub 3		2.73	1.2
Overhead line party		2,73	0.4

Figure 13.15   Social network analysis outputs for Scenarios 1 and 2

is the key agent. The network density values show that the communications network in Scenario 1 was denser than that of Scenario 2. In other words, there are fewer agents involved in the scenario but a similar number of communications between them.

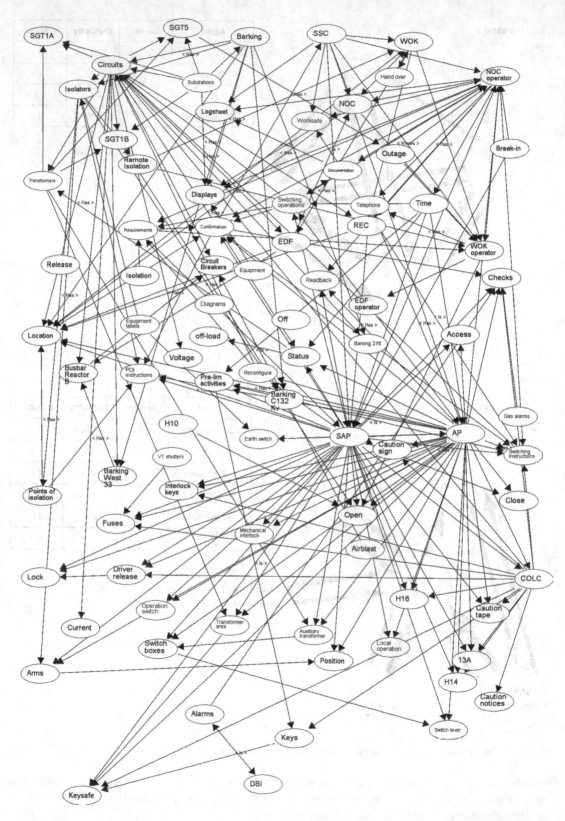

Figure 13.16 Propositional network for Scenario 1

Propositional Networks

Propositional networks were constructed for each scenario phase (as defined by the task networks) using the observational transcripts, CDM interview responses and the HTA of the tasks performed. The propositional network for Scenario 1 is presented in Figure 13.16. Propositional networks depict distributed SA based on the notion that SA comprises concepts and the relationships between them. Within Figure 13.16, each node represents a concept, and the directional arrows show the relationships between the concepts (e.g. substation 'has' location, display 'shows' circuit). The propositional networks are then examined in order to identify the usage and ownership of concepts by each agent involved and also the key concepts underpinning SA. The usage of concepts by each agent is presented in Figure 13.17 (Scenario 1) and Figure 13.18 (Scenario 2). Further, the usage of concepts by each agent per scenario phase is presented in Table 13.3.

The key concepts under-pinning SA were identified through the use of the centrality and sociometric status (see Chapter 9). Key concepts are defined as those that have salience for each scenario phase, salience being defined as those concepts that act as hubs to other concepts. Those concepts with a sociometric status and centrality value above the mean values for the overall network are taken to be key concepts.

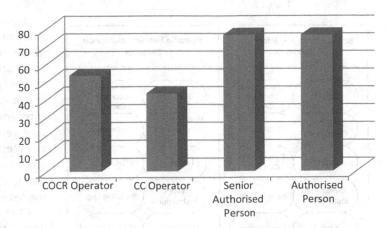

Figure 13.17 Information element usage for Scenario 1 (overall)

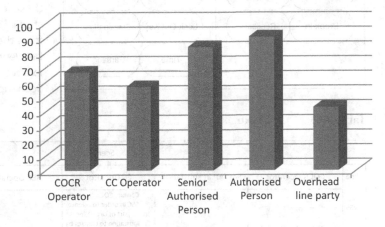

Figure 13.18 Information element usage for Scenario 2 (overall)

Table 13.3   Information element usage by agents throughout each phase of Scenario 2

Agent	Information element usage							
	Scenario phase							
	1	2	3	4	5	6	7	Total
COCR operator	22	41	24	0	20	30	0	**67**
SAP	34	36	0	29	18	37	0	**84**
AP	33	29	0	24	18	3	14	**91**
CC operator	26	0	0	0	0	0	0	**57**
OLP contact	15	0	0	9	13	0	0	43

Table 13.4   Key information elements for Scenarios 1 and 2

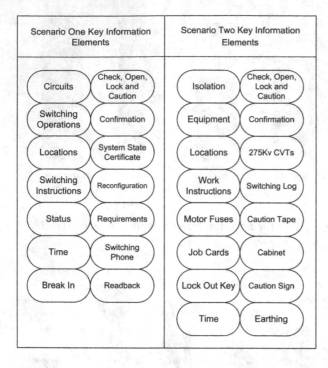

Scenario One Key Information Elements		Scenario Two Key Information Elements	
Circuits	Check, Open, Lock and Caution	Isolation	Check, Open, Lock and Caution
Switching Operations	Confirmation	Equipment	Confirmation
Locations	System State Certificate	Locations	275Kv CVTs
Switching Instructions	Reconfiguration	Work Instructions	Switching Log
Status	Requirements	Motor Fuses	Caution Tape
Time	Switching Phone	Job Cards	Cabinet
Break In	Readback	Lock Out Key	Caution Sign
		Time	Earthing

Additional EAST Analyses

A deeper level of analysis is provided through additional methods from the framework, including OSDs, CDA (Burke, 2004) and CUDs (Watts and Monk, 2000). In the present study, the task, social and knowledge network analyses were supplemented by OSD, CDA and CUD analyses. The purpose of these additional analyses was to examine levels of workload and teamwork during the scenarios and also to evaluate the technologies used to mediate communications.

CUD (Watts and Monk, 2000) is used within the EAST framework to describe the communications between agents and evaluate the technologies used to mediate these communications. For the present study, those bottom-level task steps from the HTAs involving communications between agents were examined. An extract from the CUD analysis from Scenario 1 is presented in Table 13.5.

Table 13.5   CUD extract

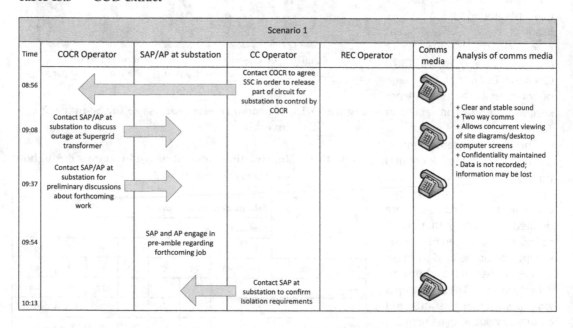

Scenario 1						
Time	COCR Operator	SAP/AP at substation	CC Operator	REC Operator	Comms media	Analysis of comms media
08:56			Contact COCR to agree SSC in order to release part of circuit for substation to control by COCR			+ Clear and stable sound + Two way comms + Allows concurrent viewing of site diagrams/desktop computer screens + Confidentiality maintained - Data is not recorded; information may be lost
09:08	Contact SAP/AP at substation to discuss outage at Supergrid transformer					
09:37	Contact SAP/AP at substation for preliminary discussions about forthcoming work					
09:54		SAP and AP engage in pre-amble regarding forthcoming job				
10:13			Contact SAP at substation to confirm isolation requirements			

CDA (Burke, 2004) is used to derive a rating of the level of coordination exhibited between team members during collaborative scenarios. Again using the HTA as its primary input, teamwork tasks were first identified, following which coordination between team members on each teamwork task step was rated using a taxonomy of key teamwork behaviours. An extract of the CDA for Scenario 1 is presented in Table 13.6.

Table 13.6   CDA extract

Teamwork = 64% of HTA tasks							Task-work = 36% of HTA tasks			
Mean co-ordination across all teamwork tasks = 1.57										
Task	Agent	Task type	Comms	SA	DM	MA	Lead	Adapt	Assert	Mean Co-ord
1.1.1. Use phone to contact COCR	CC Operator	Task work	-	-	-	-	-	-	-	-
1.1.2. Exchange IDs	CC Operator & COCR Operator	Team work	3	3	1	1	1	1	1	1.57
1.1.3. Agree SSC documentation	CC Operator & COCR Operator	Team work	3	3	3	1	1	1	1	1.86
1.1.4.1. Agree SSC with CC Operator	COCR Operator	Team work	3	3	3	1	1	1	1	1.86
1.1.4.1. Agree time CC Operator	COCR Operator	Team work	3	3	3	1	1	1	1	1.86
1.1.5.1. Record details onto log sheet	COCR Operator	Task work	-	-	-	-	-	-	-	-
1.1.5.2. Enter details into worksafe system	COCR Operator	Task work	-	-	-	-	-	-	-	-
1.2.1. Ask for isolators to be opened remotely	COCR Operator	Team work	3	3	1	2	2	1	1	1.86
1.2.2. Perform remote isolation	CC Operator	Task work	-	-	-	-	-	-	-	-
1.2.3. Check substation diagram on screen	COCR Operator	Task work	-	-	-	-	-	-	-	-
1.2.4. Terminate comms	CC Operator & COCR Operator	Team work	3	1	1	1	1	1	1	1.29

For Scenario 1, the CDA analysis revealed that 64 per cent of the bottom-level HTA tasks were teamwork tasks. The mean level of coordination across all teamwork tasks was 1.57, which represents a medium level of coordination. For Scenario 2, 67.7 per cent of the tasks represented teamwork tasks and the mean level of coordination achieved was 1.55.

Summary

The aim of this chapter has been to demonstrate an integrated framework of HF methods for analysing performance in complex socio-technical systems. Although only a summary of the outputs derived was presented, the utility of taking a framework approach to the analysis of performance in complex socio-technical systems has been demonstrated. In this case, EAST enables the description and examination of the task and social and knowledge networks underpinning complex socio-technical system performance, along with a deeper analysis of various specific constructs. For example, in both scenarios, the methods within the EAST framework supported examination of the following elements of command and control.

Table 13.7   Elements of command and control examined through EAST

	HTA	Social network analysis	Propositional networks	CDM	CUD	CDA	OSD
Tasks	■			■		■	■
Communications		■					■
Teamwork		■				■	
Situation awareness			■				
Decision making			■	■			
Technologies	■		■			■	■
Workload	■			■			

Table 13.7 demonstrates how, when taken individually, the outputs offer insight into various components of command and control (e.g. SA, teamwork, communications); for example, the propositional networks alone can be used to make judgments on the levels of SA attained throughout the scenarios and also to determine what knowledge underpins SA during energy distributed maintenance scenarios. However, integration of the outputs provides a much deeper insight. For example, viewing the task, social and knowledge networks together enables judgments to be made on how the procedures used and how the organisation of the social network supported SA during operations. In the case of the task networks, the standard operating procedures engendered high levels of closed-loop communication between the agents involved throughout the scenario, including issuing of work instructions, regular updates on work progress and reporting of completed operations. In the case of the social networks, in both scenarios, each of the agents involved had one or more ways of communicating with all other agents, (e.g. landline telephone, mobile phone and emails). Further, the hierarchical structure of the network meant that the COCR operator, effectively as network commander, would regularly contact (or be contacted by) agents in the field in order to gather or provide work progress updates, a process which served to update SA throughout the activities. Combining the analysis outputs thus allows the examination of not only *how* SA was developed and maintained and *what* it comprised, but also *why* efficient levels of SA were achieved.

The EAST framework lends itself to indepth evaluations of complex socio-technical system performance, examination of specific constructs within complex socio-technical systems (e.g. SA, decision-making, teamwork) and also system, training, procedure and technology design. Whilst not providing direct recommendations, the analyses produced are often very useful in pinpointing specific issues limiting performance or generating system re-design recommendations.

Conclusion

This study presents an example of the integration of HF methods. The analysis outputs provide a number of distinct but overlapping perspectives on the two scenarios analysed. The case study applications demonstrate the utility of applying integrated suites of HF methods to tackle complex socio-technical system performance. The approach presented summarises the task, social and knowledge structures involved into a form that enables comparisons to be made on key metrics across actors and scenarios. Beyond this, the framework approach presents demonstrates how the individual HF methods described in this book can be combined to fully encompass the boundaries of complex socio-technical systems.

Bibliography and References

Adams, M. J., Tenney, Y.J. and Pew, R. W. (1995) Situation awareness and the cognitive management of complex systems. *Human Factors* 37(1), 85–104.

Ainsworth, W. (1988) Optimization of string length for spoken digit input with error correction. *International Journal of Man Machine Studies* 28, 573–81.

Ainsworth, L. and Marshall, E. (1998) Issues of quality and practicality in task analysis: preliminary results from two surveys. *Ergonomics* 41(11), 1604–17.

Almeida, I. M. and Johnson, C. W. (2005). Extending the borders of accident investigation: applying novel analysis techniques to the loss of the Brazilian space programme's launch vehicle VLS-1 V03. Available at: http://www.dcs.gla.ac.uk/~johnson/papers/Ildeberto_and_Chris.PDF.

Anderson, J. (1980) *Cognitive Psychology and its Implications*. San Francisco, CA: Freeman.

Anderson, J. (1983) *The Architecture of Cognition*. Cambridge, MA: Harvard University Press.

Anderson, L. and Wilson, S. (1997). Critical incident technique. In D. L. Whetzel and G. R. Wheaton (eds), *Applied Measurement Methods in Industrial Psychology*. Palo Alto, CA: Davis-Black, pp. 89–112.

Andow, P. (1990) *Guidance on HAZOP Procedures for Computer-Controlled Plants*. UK Health and Safety Executive Contract Research Report No. 26/1991.

Andrews, J. D. and Moss, T. R. (1993) *Reliability and Risk Assessment*. Harlow: Longman Scientific and Technical.

Annett, J. (1997) Analysing team skills. In R. Flin, E. Salas, M. Strub and L. Martin (eds), *Decision Making under Stress*. Aldershot: Ashgate.

Annett, J. (2000) Theoretical and pragmatic influences on Task Analysis methods. In J. M. Schraagen, S. F. Chipman and V. L. Shalin (eds), *Cognitive Task Analysis*. Mahwah, NJ: Lawrence Erlbaum.

Annett, J. (2002) A note on the validity and reliability of ergonomics methods. *Theoretical Issues in Ergonomics Science* 3(2), 229–32.

Annett, J. (2003) Hierarchical Task Analysis. In D. Diaper and N. A. Stanton (eds), *Handbook of Task Analysis in Human-Computer Interaction*. Mahwah, NJ: Lawrence Erlbaum.

Annett, J. (2004) Hierarchical Task Analysis. In N. A. Stanton, A. Hedge, K. Brookhuis, E. Salas and H. Hendrick (eds), *Handbook of Human Factors and Ergonomics Methods*. Boca Raton, FL: CRC Press, pp. 329–37.

Annett, J. (2005) Conclusions. In J. R. Wilson and E. N. Corlett (eds), *Evaluation of Human Work*, 3rd edn. Boca Raton, FL: CRC Press.

Annett, J., Cunningham, D. J. and Mathias-Jones, P. (2000) A method for measuring team skills. *Ergonomics* 43(8), 1076–94.

Annett, J., Duncan, K. D., Stammers, R. B. and Gray, M. J. (1971) Task analysis. *Training Information No. 6*. London: HMSO.

Annett, J. and Stanton, N. A. (2000) *Task Analysis*. London: Taylor & Francis.

Arnold, R. (2009) A qualitative comparative analysis of SOAM and STAMP in ATM occurrence investigation. Unpublished MSc thesis. Available at: http://sunnyday.mit.edu/Arnold-Thesis.pdf.

Arroyo, E., Selker, T. and Wei, W. (2006) Usability tool for analysis of web designs using mouse tracks. *Extended Abstracts of the ACM Conference on Human Factors in Computing Systems*, 484–9.

Artman, H. and Garbis, C. (1998) Situation awareness as distributed cognition. In T. Green, L. Bannon, C. Warren and J. Buckley (eds), *Cognition and Cooperation. Proceedings of Ninth Conference of Cognitive Ergonomics*. Limerick: Ireland.

Asimakopoulos, S., Dix, A. and Fildes, R. (2011). Using hierarchical task decomposition as a grammar to map actions in context: application to forecasting systems in supply chain planning. *International Journal of Human-Computer Studies* 69(4), 234–50.

Astley, J. A. and Stammers, R. B. (1987) Adapting Hierarchical Task Analysis for user-system interface design. In J. R. Wilson, E. N. Corlett and I. Manenica (eds), *New Methods in Applied Ergonomics*. London: Taylor & Francis.

Atterer, R., Wnuk, M. and Schmidt, A. (2006) Knowing the user's every move: user activity tracking for website usability evaluation and implicit interaction. In *Proceedings of the 15th International Conference on the World Wide Web*, 2006, pp. 203–12.

Baber, C. (1991) *Speech Technology in the Control Room Systems: A Human Factors Perspective*. Chichester: Ellis Horwood.

Baber, C. (1996) Repertory Grid theory and its application to product evaluation. In P. Jordan et al. (eds), *Usability in Industry*. London: Taylor & Francis.

Baber, C. (2004) Critical Path Analysis. In N. A. Stanton, A. Hedge, K. Brookhuis, E. Salas and H. Hendrick (eds), *Handbook of Human Factors and Ergonomics Methods*. Boca Raton, FL: CRC Press, pp. 41.1–41.8.

Baber, C. (2005a) Evaluating Human-Computer Interaction. In J. R. Wilson and E. N. Corlett (eds), *Evaluation of Human Work*, 3rd edn. Boca Raton, FL: CRC Press.

Baber, C. (2005b) Repertory Grid for product evaluation. In N. A. Stanton, A. Hedge, K. Brookhuis, E. Salas and H. Hendrick (eds), *Handbook of Human Factors and Ergonomics Methods*. Boca Raton, FL: CRC Press, p. 31-7.

Baber, C., Houghton, R. J., McMaster, R., Salmon, P., Stanton, N. A., Stewart, R. J. and Walker, G. (2004) *Field Studies in Emergency Services*. HFI DTC Technical Report/WP1.1.1/01.

Baber, C. and Mellor, B. A. (2001) Modelling multimodal Human-Computer Interaction using Critical Path Analysis. *International Journal of Human Computer Studies* 54, 613–36.

Baber, C. and Stanton, N. A. (1994) Task analysis for error identification: a methodology for designing error-tolerant consumer products. *Ergonomics* 37, 1923–41.

Baber, C. and Stanton, N. A. (1996a) Human error identification techniques applied to public technology: predictions compared with observed use. *Applied Ergonomics* 27(2), 119–31.

Baber, C. and Stanton, N. A. (1996b) Observation as a technique for usability evaluations. In P. W. Jordan, B. Thomas, B. A. Weerdmeester and I. McClelland (eds), *Usability Evaluation in Industry*. London: Taylor & Francis.

Baber, C. and Stanton, N. A. (1999) Analytical prototyping. In J. M. Noyes and M. Cook (eds), *Interface Technology: The Leading Edge*. Baldock: Research Studies Press.

Baber, C. and Stanton, N. A. (2002) Task analysis for error identification: theory, method and validation. *Theoretical Issues in Ergonomics Science* 3(2), 212–27.

Baber, C. and Stanton, N. A. (2004) *Methodology for DTC-HFI WP1 Field Trials*. HFI DTC Technical Report/WP.2.1.

Baber, C., Walker, G., Stanton, N. A. and Salmon, P. (2004) *Report on Initial Trials of WP 1.1 Methodology Conducted at Fire Service Training College*. HFI DTC Technical Report/WP 1.1.1/1.1.

Bainbridge, L. (1982) Ironies of automation. In J. Rasmussen, K. Duncan and J. Neplat (eds), *New Technology and Human Error*. New York: Wiley.

Baker, D. (2004) Behavioural Observation Scales (BOS). In N. A. Stanton, A. Hedge, K, Brookhuis, E. Salas and H. Hendrick. (eds), *Handbook of Human Factors and Ergonomic Methods*. London: Taylor & Francis.

Baker, D., Salas, E. and Cannon-Bowers, J. A. (1998) Team Task Analysis: lost but hopefully not forgotten. *Industrial and Organizational Psychologist* 35(3), 79–83.

Baker, D. and Krokos, K. (2007) Development and validation of aviation causal contributors for error reporting systems (ACCERS). *Human Factors* **49**(2), 185–99.

Baldauf, D., Burgard, E. and Wittman, M. (2009) Time perception as a workload measure in simulated car driving. *Applied Ergonomics* 40, 929–35.

Barber, D., Schatz, S. and Nicholson, D. (2010) AVATAR: developing a military cultural role play trainer. *Proceedings of the 3rd International Conference on Applied Human Factors and Ergonomics (AHFE)*, 17–20 July, Miami, Florida.

Bartlett, F.C. (1932) *Remembering: A Study in Experimental and Social Psychology*. Cambridge: Cambridge University Press.

Baysari, M. T., Cappnecchia, C. and McIntosh, A. S. (2011) A reliability and usability study of TRACEr-RAV: the technique for the retrospective analysis of cognitive errors – for rail, Australian version. *Applied Ergonomics* 42, 852–9.

Baysari, M. T., Caponecchia, C., McIntosh, A. S. and Wilson, J. R. (2009) Classification of errors contributing to rail incidents and accidents: a comparison of two human error identification techniques. *Safety Science* 47, 948–57.

Baysari, M. T., McIntosh, A. S. and Wilson, J. R. (2008) Understanding the Human Factors contribution to railway accidents and incidents in Australia. *Accident Analysis and Prevention* 40(5), 1750–1757.

Bedny, G. and Meister, D. (1999) Theory of activity and situation awareness. *International Journal of Cognitive Ergonomics* 3(1), 63–72.

Bell, H. H. and Lyon, D. R. (2000) Using observer ratings to assess situation awareness. In M. R. Endsley (ed.), *Situation Awareness Analysis and Measurement*. Hillsdale, NJ: Lawrence Erlbaum.

Bennett, D.J. and Stephens, P. (2009) A cognitive walkthrough of autopsy forensic browser. *Information Management and Computer Security* 17(1), 20–29.

Bernardin, H. J. and Buckley, M. R. (1981). Strategies in rater training. *Academy of Management Review* 6, 205–12.

Bergqvist, P., Gustafsson, F., Haraldsson, J. and Laden, G. (2007) 'FRAM visualiser'. Available at: http://code. google.com/p/framvisualizer.

Bevan, N. (2006) 'Contextual inquiry. UsabilityNet'. Available at: http://www.usabilitynet.org/tools/ contextualinquiry.htm.

Bevan, N., Chauncey, W., Chung, S., DeBoard, D. and Herzon, C. (2010) 'Wizard of Oz. Usability body of knowledge'. Available at: http://www.usabilitybok.org/wizard-of-oz.

Beyer, H. and Holtzblatt, K. (1998) *Contextual Design: Defining Customer-Centered Systems*. San Francisco, CA: Morgan Kaufmann Publishers.

Billings, C. E. (1995) Situation awareness measurement and analysis: a commentary. *Proceedings of the International Conference on Experimental Analysis and Measurement of Situation Awareness*. Daytona Beach, FL: Embry-Riddle Aeronautical University Press.

Bisantz, A. M., Roth, E. M., Brickman, B., Gosbee, L. L., Hettinger, L. and McKinney, J. (2003) Integrating cognitive analyses in a large-scale system design process. *International Journal of Human Computer Studies* 58, 177–206.

Blandford, A. and Wong, B. L. W. (2004) Situation awareness in emergency medical dispatch. *International Journal of Human-Computer Studies* 61(4), 421–52.

Bletzer, K. V., Yuan, N. P., Koss, M. P., Polacca, M., Eaves, E. R. and Goldman, D. (2011) Taking humor seriously: talking about drinking in Native American focus groups. *Medical Anthropology* 30(3), 295–318.

Bodker, S. (1988). Understanding representation in design. *Human Computer Interaction* 13(2), 107–25.

Boehm, B. (2006) Some future trends and implications for systems and software engineering processes. *Systems Engineering* 9(1), 1–19.

Bolstad, C. A., Riley, J. M., Jones, D. G. and Endsley, M. R. (2002) Using goal directed task analysis with Army brigade officer teams. In *Proceedings of the 46th Annual Meeting of the Human Factors and Ergonomics Society, Baltimore, MD*. Santa Monica, CA: Human Factors and Ergonomics Society, 472–6.

Bowers, C. A. and Jentsch, F. (2005) Team Workload. In N. A. Stanton., A. Hedge., K. Brookhuis., E. Salas and H. Hendrick (eds), *Handbook of Human Factors Methods*. London: Taylor & Francis.

Braband, J. and Brehmke, B. (2002) Application of why-because graphs to railway near-misses. In C. W. Johnson (ed.), *Investigation and Reporting of Incidents and Accidents*. GIST Technical Report G2002-2, Department of Computing Science, University of Glasgow, http://www.dcs.gla.ac.uk/~johnson/iria2002/IRIA_2002.pdf.

Braband, J., Evers, B. and Stefano, E. (2003) Towards a hybrid approach for incident root cause analysis. Siemens transportation systems – rail automation system development integrity. In *Proceedings of the 21st International System Safety Conference*, Ottawa, Canada, 4–8 August.

Bradbury-Jones, C. and Tranter, S. (2008) Inconsistent use of the critical incident technique in nursing research. *Journal of Advanced Nursing* 64(4), 399–407.

Brewer, W.F. (2000) Bartlett's concept of the schema and its impact on theories of knowledge representation in contemporary cognitive psychology. In A. Saito (ed.), *Bartlett, Culture and Cognition*. London: Psychology Press.

Brookhuis, K.A. and de Waard, D. (2010) Monitoring drivers' mental workload in driving simulators using physiological measures. *Accident Analysis and Prevention* 42, 898–903.

Brostrom, R., Bengtsson, P., Axelsson, J. (2011) Correlation between safety assessments in the driver-car interaction design process. *Applied Ergonomics* 42, 575–82.

Brown, J. L., Vanable, P. A. and Eriksen, M. D. (2008) Computer-assisted self-interviews: a cost effectiveness analysis. *Behavior Research Methods* 40, 1–7.

Burford, B. (1993) Designing adaptive ATMs. Unpublished MSc thesis, University of Birmingham.

Burke, S. C. (2004) Team Task Analysis. In N. A. Stanton et al. (eds), *Handbook of Human Factors and Ergonomics Methods*. Boca Raton, FL: CRC Press.

Butterfield, L. D., Borgen, W. A., Amundson, N. E. and Maglio, A. S. T. (2005) Fifty years of the critical incident technique: 1954–2004 and beyond. *Qualitative Research* 5, 475–97.

Byers, J. C., Bittner, A. C. and Hill, S. G. (1989) Traditional and Raw Task Load Index (TLX) correlations: are paired comparisons necessary? In A. Mital and B. Das (eds), *Advances in Industrial Ergonomics and Safety*, vol. 1. London: Taylor & Francis, pp. 481–8.

Caffiau, S., Scapin, D., Girard, P., Baron, M. and Jambon, F. (2010) Increasing the expressive power of task analysis: systematic comparison and empirical assessment of tool-supported task models. *Interacting with Computers* 22, 569–93.

Card, S. K., Moran, T. P. and Newell, A. (1983) *The Psychology of Human-Computer Interaction*. Hillsdale, NJ: Lawrence Erlbaum.

Carswell, C. M., Lio, C. H., Grant, R., Klein, M. I., Clarke, D., Seales, W. B. and Strup, S. (2010) Hands-free administration of subjective workload scales: acceptability in a surgical training environment. *Applied Ergonomics* 42, 138–45.

Casali, J. G. and Wierwille, W. W. (1983) A comparison of rating scale, secondary task, physiological and primary task workload estimation techniques in a simulated flight task emphasising communications load. *Human Factors* 25, 623–41.

Celik, M. and Cebi, S. (2009) Analytical HFACS for investigating human errors in shipping accidents. *Accident Analysis and Prevention* 41(1), 66–75.

CCPS (Centre for Chemical Process Safety) (1994) *Guidelines for Preventing Human Error in Process Safety*. New York: American Institute of Chemical Engineers.

Cha, D. W. (2001) Comparative study of subjective workload assessment techniques for the evaluation of ITS-orientated human-machine interface systems. *Journal of Korean Society of Transportation* 19(3), 45–58.

Cha, D. and Park, P. (1997) User required information modality and structure of in-vehicle navigation system focused on the urban commuter. *Computers and Industrial Engineering* 33, 517–20.

Chamarro, A. and Fernandez- Castro, J. (2009) The perception of causes of accidents in mountain sports: a study based on the experience of victims. *Accident Analysis and Prevention* 41, 197–201.

Chaparro, B. S., Hinkle, V. D. and Riley, S, K. (2008) The usability of computerized card sorting: a comparison of three applications by researchers and end users. *Journal of Usability Studies* 4(1), 31–48.

Chau, M. and Wong, C.H. (2010) Designing the user interface and functions of a search engine development tool. *Decision Support Systems* 48, 369–82.

Checkland, P. (1981) *Systems Thinking, Systems Practice*. London: John Wiley.

Checkland, P. (2000) Soft systems methodology: a thirty year perspective. *Systems Research and Behavioural Science* 17, 11–58.

Chen, M.-C., Anderson, J. R. and Sohn, M.-H. (2001) Can a mouse cursor tell us more?: correlation of eye/mouse movements on web browsing. In *CHI '01 Extended Abstracts*, pp. 281–2.

Chin Li, W., Harris, D. and San Yu, C. (2008) Routes to failure: analysis of 41 aviation accidents from the Republic of China using the Human Factors analysis and classification system. *Accident Analysis and Prevention* 40, 426–34.

Chin, J. P., Diehl, V. A. and Norman, K. L. (1988) Development of an instrument measuring user satisfaction of the Human-Computer Interface. In *CHI'88*.

Chin, M., Sanderson, P. and Watson, M. (1999) Cognitive work analysis of the command and control work domain. In *Proceedings of the 1999 Command and Control Research and Technology Symposium (CCRTS), June 29 - July 1, Newport, RI, USA*, vol. 1. Newport, RI: U.S. Naval War College, pp. 233–48.

Ciavarelli, A. and Sather, T. (2002) Human Factors checklist: an aircraft accident investigation tool, School of Aviation Safety, California. Available at: *https://www.netc.navy.mil/nascweb/sas/files/hfchklst.pdf*.

Collier, S. G. and Folleso, K. (1995) SACRI: a measure of situation awareness for nuclear power control rooms. In D. J. Garland and M. R. Endsley (eds), *Experimental Analysis and Measurement of Situation Awareness*. Daytona Beach, FL: Embry-Riddle University Press.

Collins, A. M. and Loftus, E. F. (1975) A spreading-activation theory of semantic processing. *Psychological Review* 82, 407–28.

Cooke, N. J. (2004) Measuring team knowledge. In N. Stanton, A. Hedge, K. Brookhuis, E. Salas and H. Hendrick (eds), *Handbook of Human Factors and Ergonomics Methods*. Boca Raton, FL: CRC Press.

Cooper, G. E. and Harper, R. P. (1969) *The Use of Pilot Rating in the Evaluation of Aircraft Handling Qualities*. Report No. ASD-TR-76-19.

Couper, M. P. and Rowe, B. (2010) Evaluation of a Computer-Assisted Self-Interview (CASI) component in a CAPI Survey. Available at: http://www.amstat.org/sections/srms/Proceedings/papers/1995_177.pdf.

Crandall, B., Klein, G. and Hoffman, R. (2006) *Working Minds: A Practitioner's Guide to Cognitive Task Analysis*. Cambridge, MA: MIT Press.

Crawford, J. O., Taylor, C. and Po, N. L. W. (2001) A case study of on-screen prototypes and usability evaluation of electronic timers and food menu systems. *International Journal of Human Computer Interaction* 13(2), 187–201.

Crichton, M. (2005) Attitudes to teamwork, leadership and stress in oil industry drilling teams. *Safety Science* 43, 679–96.

Cronholm, S. (2009) The usability of usability guidelines – a proposal for meta-guidelines. In *Proceedings of the 21st Australasian Computer-Human Interaction Conference (OZCHI)*, pp. 233–40.

Crutchfield, J. P. and Shalizi, C. R. (1999) Thermodynamic depth of causal states: objective complexity via minimal representations. *Physical Review E* 59, 275–83.

Dean, T. F. (1997) *Directory of Design Support Methods, Defence Technical Information Centre, DTIC-AM*. MATRIS Office, ADA 328 375, September.

Dekker, S. W. A. (2002a) Reconstructing human contributions to accidents: the new view on human error and performance. *Journal of Safety Research* 33, 371–85.

Dekker, S. W. A. (2002b) The re-invention of human error. Technical Report 2002-01. Available at: http://www.lu.se/upload/Trafikflyghogskolan/TR2002-01_ReInventionofHumanError.pdf.

Dennehy, K. (1997) *Cranfield – Situation Awareness Scale, User Manual*. Applied Psychology Unit, College of Aeronautics, Cranfield University, COA Report No. 9702, Bedford, January.

Diaper, D. (1989) *Task Analysis in Human Computer Interaction*. Chichester: Ellis Horwood.

Diaper, D. (2004) Understanding task analysis for human-computer interaction. In D. Diaper and N. A. Stanton (eds), *The Handbook of Task Analysis for Human-Computer Interaction*. Mahwah, NJ: Lawrence Erlbaum.

Diaper, D. and Stanton, N. A. (2004) *Handbook of Task Analysis in Human-Computer Interaction*. Mahwah, NJ: Lawrence Erlbaum.

Dijkstra, A. (2011) KLM Royal Dutch Airlines/TU Delft. Resilience engineering and safety management systems in aviation. Available at: http://systemssafety.net/Dijkstra%20Resilience%20and%20SMS.pdf.

Dillon, A. and McKnight, C. (1990) Towards a classification of text types: a Repertory Grid approach. *International Journal of Man-Machine Studies* 33, 623–36.

Dinadis, N. and Vicente, K. J. (1999) Designing functional visualisations for aircraft systems status displays. *International Journal of Aviation Psychology* 9, 241–69.

Djokic, J., Lorenz, B. and Fricke, H. (2010) Air traffic control complexity as workload driver. *Transportation Research Part C* 18, 930–936.

Dominguez, C. (1994) Can SA be defined? In M. Vidulich, C. Dominguez, E. Vogel and G. McMillan (eds), *Situation Awareness: Papers and Annotated Bibliography*. Report AL/CF-TR-1994-0085) Wright-Patterson Airforce Base, OH: Air Force Systems Command.

Doytchev, D. E. and Szwillus, G. (2009) Combining task analysis and fault tree analysis for accident and incident analysis: a case study from Bulgaria. *Accident Analysis and Prevention* 41, 1172–9.

Driskell, J. E. and Mullen, B. (2005) Social Network Analysis. In N. A. Stanton et al. (eds), *Handbook of Human Factors and Ergonomics Methods*. Boca Raton, FL: CRC Press.

Drury, C. G. (1990) Methods for direct observation of performance, In J. Wilson and E. N. Corlett (eds), *Evaluation of Human Work: A Practical Ergonomics Methodology*, 2nd edn. London: Taylor & Francis.

Dunjo, J., Fthenakis, V., Vilchez, J. A. and Arnaldos, J. (2010) Hazard and operability (HAZOP) analysis. A literature review. *Journal of Hazardous Materials* 173, 19–32.

Durga Rao, K., Gopika, V., Sanyasi Rao, V. V. S., Kushwaha, H. S., Verma, A. K. and Srividya, A. (2009) Dynamic fault tree analysis using Monte Carlo simulation in probabilistic safety assessment. *Reliability Engineering and System Safety* 94, 872–83.

Durso, F. T., Hackworth, C. A., Truitt, T., Crutchfield, J. and Manning, C. A. (1998) Situation awareness as a predictor of performance in en route air traffic controllers. *Air Traffic Quarterly* 6, 1–20.

Durso, F. T., Truitt, T. R., Hackworth, C. A., Crutchfield, J. M., Nikolic, D., Moertl, P. M., Ohrt, D. and Manning, C. A. (1995) Expertise and chess: a pilot study comparing situation awareness methodologies. In D. J. Garland and M. Endsley (eds), *Experimental Analysis and Measurement of Situation Awareness*. Daytona Beach, FL: Embry-Riddle Aeronautical University Press.

Easterby, R. (1984) Tasks, processes and display design. In R. Easterby and H. Zwaga (eds), *Information Design*. Chichester: Wiley.

El Bardissi, A. W., Wiegmann, D. A., Dearani, J. A., Daly, R. C. and Sundt, T. M. (2007) Application of the Human Factors analysis and classification system methodology to the cardiovascular surgery operating room, *Annals of Thoracic Surgery* **83**(4), 1412–19.

Embrey, D. E. (1986) SHERPA: a systematic human error reduction and prediction approach. Paper presented at the *International Meeting on Advances in Nuclear Power Systems*, Knoxville, Tennessee.

Embrey, D. E. (1993) Quantitative and qualitative prediction of human error in safety assessments. *Institute of Chemical Engineers Symposium Series* 130, 329–50.

Endsley, M. R. (1988) Situation Awareness Global Assessment Technique (SAGAT). In *Proceedings of the National Aerospace and Electronics Conference*. New York: IEEE.

Endsley, M. R. (1989) *Final Report: Situation Awareness in an Advanced Strategic Mission*. Northrop Document 89-32. Northrop Corporation.

Endsley, M. R. (1990) Predictive utility of an objective measure of situation awareness. In *Proceedings of the Human Factors Society 34th Annual Meeting*. Santa Monica, CA: Human Factors Society.

Endsley, M. R. (1993) A survey of situation awareness requirements in air-to-air combat fighters. *International Journal of Aviation Psychology* 3, 157–68.

Endsley, M. R. (1995a) Toward a theory of situation awareness in dynamic systems. *Human Factors* 37(1), 32–64.

Endsley, M. R. (1995b) Measurement of situation awareness in dynamic systems. *Human Factors* 37(1), 65–84.

Endsley, M.R. (2000) Theoretical underpinnings of situation awareness: a critical review. In M. R. Endsley and D. J. Garland (eds), *Situation Awareness Analysis and Measurement*. Mahwah, NJ: Lawrence Erlbaum.

Endsley, M. R. (2001) Designing for situation awareness in complex systems. In *Proceedings of the Second International Workshop on Symbiosis of Humans, Artifacts and Environment, Kyoto, Japan*.

Endsley, M. R., Bolte, B. and Jones, D. G. (2003) *Designing for Situation Awareness: An Approach to User-Centred Design*. London: Taylor & Francis.

Endsley, M. R. and Garland, D. G. (2000) *Situation Awareness Analysis and Measurement*. Mahwah, NJ: Lawrence Erlbaum.

Endsley, M. R., Holder, C. D., Leibricht, B. C., Garland, D. C., Wampler, R. L. and Matthews, M. D. (2000) *Modelling and Measuring Situation Awareness in the Infantry Operational Environment*. Alexandria, VA: Army Research Institute.

Endsley, M. R. and Jones, W. M. (1997) *Situation Awareness, Information Dominance and Information Warfare*. Technical Report 97-01. Belmont, MA: Endsley Consulting.

Endsley, M.R. and Kiris, E.O. (1995) *Situation Awareness Global Assessment Technique (SAGAT) TRACON Air Traffic Control Version User Guide*. Lubbock, TX: Texas Tech University.

Endsley, M. R. and Robertson, M. M. (2000) Situation awareness in aircraft maintenance teams. *International Journal of Industrial Ergonomics* 26(2), 301–25.

Endsley, M. R. and Rogers, M. D. (1994) Situation awareness information requirements for en route Air Traffic Control. DOT/FAA/AM-94/27. Washington, DC, Federal Aviation Administration Office of Aviation Medicine.

Endsley, M. R., Selcon, S. J., Hardiman, T. D. and Croft, D. G. (1998) A comparative evaluation of SAGAT and SART for evaluations of situation awareness. In *Proceedings of the Human Factors and Ergonomics Society Annual Meeting*. Santa Monica, CA: Human Factors and Ergonomics Society.

Endsley, M.R. and Smolensky, M. (1998) Situation awareness in air traffic control: the picture. In M. Smolensky and E. Stein (eds), *Human Factors in Air Traffic Control*. New York: Academic Press.

Endsley, M. R., Sollenberger, R. and Stein, E. (2000) Situation awareness: a comparison of measures. In *Proceedings of the Human Performance, Situation Awareness and Automation: User-Centered Design for the New Millennium*. Savannah, GA: SA Technologies, Inc.

Engelmann, C., Schneider, M., Kirschbaum, C., Grote, G., Dingemann, J., Schoof, S. and Ure, B. M. (2011) Effects of intraoperative breaks on mental and somatic operator fatigue: a randomised clinical trial. *Surgical Endoscopy* 25, 1245–50.

Ericsson, K. A. and Simon, H. A. (1980) Verbal reports as data. *Psychological Review* 87, 215–51.

Ericsson, K. A. and Simon, H. A. (1993) *Protocol Analysis: Verbal Reports as Data*, revised edn. Cambridge, MA: MIT Press.

Erlandsson, M. and Jansson, A. (2007) Collegial verbalisation – a case study on a new method of information acquisition. *Behaviour and Information Technology* 26, 535–43.

EUROCONTROL (European Organisation for the safety of air navigation) (2005) ESARR Advisory material/ guidance document (EAM/ GUI) EAM2/ GUI8. Guidelines on the systemic occurrence analysis methodology (SOAM). Available at: http://www.eurocontrol.int/src/gallery/content/public/documents/deliverables/esarr2_awareness_package/eam2_gui8_e10_ri_web.pdf.

Evan, W. M. and Miller, J. R. (1969) Differential effect of response bias of computer vs. traditional administration of a social science questionnaire: an exploratory methodological experiment. *Behavioral Science* 14, 216–27.

Evans, L. (2002) Transportation safety. In R. W. Hall (ed.), *Handbook of Transportation Science*, 2nd edn. Norwell, MA: Kluwer Academic Publishers.

Eyferth, K., Niessen, C. and Spath, O. (2003) A model of air traffic controllers conflict detection and conflict resolution. *Aerospace Science and Technology* 7(6), 409–16.

Eysenck, M. W. and Keane, M. T. (1990) *Cognitive Psychology: A Student's Handbook*. Hove: Lawrence Erlbaum.

Fackler, J. C., Watts, C., Grome, A., Miller, T., Crandall, B. and Pronovost, P. (2009) Critical care physician cognitive task analysis: an exploratory study. *Critical Care* 13 (R33). Available at: http://ccforum.com/content/13/2/R33.

Farmer, E. W., Jordan, C. S., Belyavin, A. J., Bunting, A. J., Tattersall, A. J. and Jones D. M. (1995) *Dimensions of Operator Workload*. Defence Evaluation and Research Agency, Report DRA/AS/MMI/CR95098/1.

Federal Aviation Administration, (1996) *Report on the Interfaces between Flightcrews and Modern Flight Deck Systems*, Federal Aviation Administration, Washington DC, 1996.

Ferdous, R., Khan, F., Sadiq, R., Amyotte, P. and Veitch, B. (2009) Handling uncertainties in event tree analysis. *Process Safety and Environmental Protection* 87, 283–92.

Finstad, K. (2010) The usability metric for user experience. *Interacting with Computers* 22, 323–7.

Fitts, P. M. and Seeger, C. M. (1953) S - R compatibility: spatial characteristics of stimulus and response codes. *Journal of Experimental Psychology* 46, 199–210.

Flanagan, J. C. (1954) The Critical Incident Technique. *Psychological Bulletin* 51, 327–58.

Flin, R., Goeters, K. M., Hormann, H. J. and Martin, L. (1998) A generic structure of non-technical skills for training and assessment. *Proceedings of the 23rd Conference of the European Association for Aviation Psychology, Vienna*, 14–18 September.

Fowlkes, J. E., Lane, N. E., Salas, E., Franz, T. and Oser, R. (1994) Improving the measurement of team performance: the TARGETs methodology. *Military Psychology* 6(1), 47–61.

Fracker, M. (1991) *Measures of Situation Awareness: Review and Future Directions*. Report No. AL-TR-1991-0128. Wright Patterson Air Force Base, Ohio: Armstrong Laboratories.

Fulkerson, J. A., Kubik, M. Y., Rydell, S., Boutelle, K. N., Garwick, A., Story, M., Neumark-Sztainer, D. and Dudovitz, B. (2011) Focus groups with working parents of school aged children: what's needed to improve family meals. *Journal of Nutrition Education and Behaviour* 43(3), 189–94.

Gertmann, D. I. and Blackman, H. S. (1994) *Human Reliability and Safety Analysis Data Handbook*. New York: John Wiley & Sons.

Gertmann, D. I., Blackman, H. S., Haney, L. N., Seidler, K. S. and Hahn, H. A. (1992) INTENT: a method for estimating human error probabilities for decision based errors. *Reliability Engineering and System Safety* 35, 127–37.

Gilbreth, F. B. (1911) *Motion Study*. Princeton, NJ: Van Nostrand.

Gillan, D. J. and Cooke, N. J. (2001) Using Pathfinder Networks to analyze procedural knowledge in interactions with advanced technology. In E. Salas (ed.), *Advances in Human Performance and Cognitive Engineering Research*. Amsterdam: JAI Press.

Glaser, B. G. and Strauss, A. L. (1967) *The Discovery of Grounded Theory*. Piscataway, NJ: Aldine Publishing Co.

Glendon, A. I. and McKenna, E. F. (1995) *Human Safety and Risk Management*. London: Chapman & Hall.

Go, K. and Carroll, J., M. (2003) Scenario-based task analysis. In D. Diaper and N. Stanton (eds), *The Handbook of Task Analysis for Human-Computer Interaction*. Mahwah, NJ: Lawrence Erlbaum.

Goillau, P. J. and Kelly, C. (1996) Malvern Capacity Estimate (MACE) – a proposed cognitive measure for complex systems. *First International Conference on Engineering Psychology and Cognitive Ergonomics*.

Golightly, D., et al. (2010) The role of situation awareness for understanding signalling and control in rail operations. *Theoretical Issues in Ergonomics Science* 11(1), 84–98.

Gong, Q. and Salvendy, G. (1994) Design of skill-based adaptive interface: the effect of a gentle push. *Proceedings of the Human Factors and Ergonomics Society, 38th Annual Meeting* (HFES, Santa Monica, CA).

Goodyear, R., Reynolds, P. and Both Gragg, J. (2010) University faculty experiences of classroom incivilities: a critical incident study. Paper presented at the AERA annual meeting, Denver.

Gorman, J. C., Cooke, N. and Winner, J. L. (2006) Measuring team situation awareness in decentralised command and control environments. *Ergonomics* 49(12–13), 1312–26.

Goschnick, B. and Sonenberg, L. (2008) From task to agent oriented meta-models and back again. In P. Forbrig and F.Paterno (eds), *HCSE/TAMODIA*, LNCS 5247.

Graham, R. (1999) Use of auditory icons as emergency warnings: evaluation within a vehicle collision avoidance application. *Ergonomics* 42(9), 1233–48.

Grandjean, E. (1988) *Fitting the Task to the Man*. London: Taylor & Francis.

Gray, W. D., John, B. E. and Atwood, M. E. (1993) Project Ernestine: validating a GOMS analysis for predicting and explaining real-world performance. *Human Computer Interaction* 8(3), 237–309.

Green, D., Stanton, N., Walker, G. and Salmon, P. M. (2005) Using wireless technology to develop a virtual reality command and control centre. *Virtual Reality* 8, 147–55.

Green, W. S. and Jordan, P. W. (1999) *Human Factors in Product Design: Current Practice and Future Trends*. London: Taylor & Francis.

Greenberg, S. (2003) Working through task-centred system design. In D. Diaper and N. Stanton (eds), *The Handbook of Task Analysis for Human Computer Interaction*. Mahwah, NJ: Lawrence Erlbaum.

Gregorich, S. E., Helmreich, R. L. and Wilhelm, J. A. (1990) The structure of cockpit management attitudes. *Journal of Applied Psychology* 75(6), 682–90.

Griffin, T. G. C., Young, M. S. and Stanton, N. A. (2010) Investigating accident causation through information network modelling. *Ergonomics*, 53(2), 198–210.

Gugerty, L.J. (1997) Situation awareness during driving: explicit and implicit knowledge in dynamic spatial memory. *Journal of Experimental Psychology: Applied* 3, 42–66.

Hahn, A. H. and DeVries, J. A. (1991) Identification of human errors of commission using SNEAK analysis. In W. Karwowski and W. S Marras (eds), *The Occupational Ergonomics Handbook*. Boca Raton, FL: CRC Press.

Halley, J. D. and Winkler, D. A. (2008) Classification of emergence and its relation to self-organization. *Complexity* 13(5), 10–15.

Harris, D. and Li, W-C. (2011) An extension of the Human Factors analysis and classification system for use in open systems. *Theoretical Issues in Ergonomics Science* 12(2), 108–28.

Harris, C. J. and White, I. (1987) *Advances in Command, Control and Communication Systems*. London: Peregrinus.

Harris, D., Stanton, N. A., Marshall, A. Young, M. S., Demagalski, J. and Salmon, P. (2005) Using SHERPA to predict design-induced error on the flight deck. *Aerospace Science and Technology Journal* 9(6), 525–32.

Harrison, A. (1997) *A Survival Guide to Critical Path Analysis*. London: Butterworth-Heinemann.

Hart, S. G. (2006) NASA-Task Load Index (NASA-TLX); 20 years later. In *Proceedings of the Human Factors and Ergonomics Society 50th Annual Meeting*. Santa Monica, CA: Human Factors and Ergonomic Society.

Hart, S. G. and Staveland, L. E. (1988) Development of a multi-dimensional workload rating scale: results of empirical and theoretical research. In P. A. Hancock and N. Meshkati (eds), *Human Mental Workload*. Amsterdam: Elsevier.

Harvey, C. and Stanton, N.A. (2012) Trade-off between context and objectivity in an analytic approach to the evaluation of in-vehicle interfaces. *IET Intelligent Transport Systems* 6(3), 243–58 .

Hauss, Y. and Eyferth, K. (2003) Securing future ATM-concepts' safety by measuring situation awareness in ATC. *Aerospace Science and Technology* 7(6), 417–27.

Hazlehurst, B., McMullen, C. K. and Gorman, P. N. (2007) Distributed cognition in the heart room: how situation awareness arises from coordinated communications during cardiac surgery. *Journal of Biomedical Informatics* 40(5), 539–51.

Helander, M. (1978) Applicability of drivers' electrodermal responses to the design of the traffic environment. *Journal of Applied Psychology* 63, 481–8.

Helmreich, R. L. (1984) Cockpit management attitudes. *Human Factors* 26, 63–72.

Helmreich, R. L. (2000) On error management: lessons from aviation. *British Medical Journal* 320, 781–5.

Helmreich, R. L., Foushee, H., Benson, R. and Russini, W. (1986) Cockpit resource management: exploring the attitude–performance linkage. *Aviation, Space and Environmental Medicine* 57, 1198–200.

Helmreich, R. L., Wilhelm, J. and Gregorich, S. (1988) Revised versions of the Cockpit Management Attitudes Questionnaire (CMAQ) and CRM Seminar evaluation form. NASA/University of Texas Technical Report 88-3.

Herzon, C., DeBoard, D., Wilson, C. and Bevan, N. (2010) Contextual inquiry. Usability body of knowledge. Available online at http://www.usabilitybook.org/contextual-inquiry.

Hess, B. (1999) Graduate student cognition during information retrieval using the world wide web: a pilot study. *Computers and Education* 33(1), 1–13.

Hilburn, B. G. (1997) Free flight and air traffic controller mental workload. Presented at the *9th Symposium on Aviation Psychology*, Columbus, Ohio, USA.

Hoffman, K. A., Aitken, L. M. and Duffield, C. (2009). A comparison of novice and expert nurses' cue collection during clinical decision-making: verbal protocol analysis. *International Journal of Nursing Studies* 46(10), 1335–44.

Hogg, D. N., Folleso, K., Strand-Volden, F. and Torralba, B. (1995) Development of a situation awareness measure to evaluate advanced alarm systems in nuclear power plant control rooms. *Ergonomics* 38(11), 2394–413.

Höhl, M. and Ladkin, P. (1997) Analysing the 1993 Warsaw accident with a WB-Graph. Report RVS-Occ- 97-09, 8 September, Faculty of Technology, Bielefeld University. Available at: http://www.rvs.uni-bielefeld.de.

Holleis, P., Otto, F., Hufsmann, H. and Schmidt, A. (2007) Keystroke Level Model for advanced mobile phone interaction. In *CHI 2007 Proceedings: Models of Mobile Interaction*, 28 April–3 May, San Jose, California, USA.

Hollnagel, E. (1993) *Human Reliability Analysis: Context and Control*. London: Academic Press.

Hollnagel, E. (1998) *Cognitive Reliability and Error Analysis Method – CREAM*. Oxford: Elsevier Science.

Hollnagel, E. (2003) *Handbook of Cognitive Task Design*. Mahwah, NJ: Lawrence Erlbaum.

Hollnagel, E. (2004) *Barriers and Accident Prevention*. Aldershot: Ashgate.

Hollnagel, E., Kaarstad, M. and Lee, H-C. (1999) Error mode prediction. *Ergonomics* 42, 1457–71.

Hollnagel, E., Pruchnicki, S., Woltjer, R. and Etcher, S. (2008) A functional resonance accident analysis of Comair flight 5191. Paper presented at the *8th International Symposium of the Australian Aviation Psychology Association*, Sydney, Australia.

Holtzblatt, K. and Jones, S. (1993) Contextual Inquiry: a participatory technique for system design. In D. Schuler and A. Namioka (eds), *Participatory Design: Principles and Practices*. Mahwah, NJ: Lawrence Erlbaum.

Holtzblatt, K., Wendell, J. B. and Wood, S. (2005) Rapid contextual design: a how-to guide to key techniques for user-centered design. San Francisco, CA: Morgan Kaufmann.

Hone, K. S. and Baber, C. (1999) Modelling the effect of constraint on speech-based human computer interaction. *International Journal of Human Computer Studies* 50(1), 85–107.

Hone, K. S. and Baber, C. (2001) Designing habitable dialogues for speech-based interaction. *International Journal of Human Computer Studies* 54, 637–62.

Hong, E. S., Lee, I. M., Shin, H. S., Nam, A. W., Kong, J. S. (2009) Qualitative risk evaluation based on event tree analysis technique: application to the design of shield TBM. *Tunnelling and Underground Space Technology* 24, 269–77.

Hopkins, A. (2005) *Safety, Culture and Risk: The Organisational Causes of Disasters*. Sydney, CCH.

Horan, P. (2002) A new and flexible graphic organiser for IS learning: the rich picture. In E. and A. Zaliwska (eds), *IS'2002 Proceedings of the Informing Science and IT Education Conference*, Cork, Ireland, Informing Science Institute, pp. 133–8.

Hornby, G. S. (2007) Modularity, reuse and hierarchy: measuring complexity by measuring structure and organization. *Complexity* 13(2), 50–61.

Horrey, W. J., Lesch, M. F. and Garabet, A. (2008) Assessing the awareness of performance decrements in distracted drivers. *Accident Analysis and Prevention* 40, 675–82.

Houghton, R. J., Baber, C., Cowton, M., Stanton, N. A. and Walker, G. H. (2008) WESTT (Workload, Error, Situational Awareness, Time and Teamwork): an analytical prototyping system for command and control. *Cognition Technology and Work* 10(3), 199–207.

Houghton, R. J., Baber, C., McMaster, R., Stanton, Salmon, P., Stewart, R. and Walker, G. (2006) Command and control in emergency services operations: a social network analysis. *Ergonomics* 49(12–13), 1204–25.

Houghton, R. J., Baber, C., Cowton, M., Walker, G. H and Stanton, N. (2008). WESTT (Workload, Error, Situational Awareness, Time and Teamwork): an analytical prototyping system for command and control. *Cognition Technology and Work* 10, 199–207.

Hutchins, E. (1995) *Cognition in the Wild*. Cambridge, MA: MIT Press.

Hyponen, H. (1999) Focus groups. In H. A. Williams, J. Bound and R. Coleman (eds), *The Methods Lab: User Research for Design. Design for Ageing Network*. Geneva: International Standards Organisation.

Isaac, A., Shorrock, S.T. and Kirwan, B., (2002) Human error in European air traffic management: the HERA project. *Reliability Engineering and System Safety* 75, 257–72.

ISO 13407 (1996) *Human-Centred Design Processes for Interactive Systems*. Geneva: International Standards Organization.

ISO 9126 (2000) *Software Engineering – Product Quality*. Geneva: International Standards Office.

ISO 9241 (1998) *Ergonomics of Office Work with VDTs – Guidance on Usability*. Geneva: International Standards Office.

ISO 9241:11 (1998) *Ergonomics Requirements for Office Work with Visual Display Terminals – Part 11: Usability*. Geneva: International Standards Organization.

James, N. and Patrick, J. (2004) The role of situation awareness in sport. In S. Banbury and S. Tremblay (eds), *A Cognitive Approach to Situation Awareness: Theory and Application*. Aldershot: Ashgate.

Jansson, A., Olsson, E. and Erlandsson, M. (2006) Bridging the gap between analysis and design: improving existing driver interfaces with tools from the framework of cognitive work analysis. *Cognition, Technology and Work*, 8, 41–9.

Jarmasz, J., Zobarich, R., Bruyn-Martin, L. and Lamourex, T. (2009) Team cognition during a simulated close air support exercise: results from a new behavioural rating instrument. In D. Andrews, R. P. Hertz and M. B. Wolf (eds), *Human Factors in Combat Identification*. Aldershot: Ashgate.

Jeannott, E., Kelly, C. and Thompson, D. (2003) *The Development of Situation Awareness Measures in ATM Systems*. EATMP report. HRS/HSP-005-REP-01.

Jenkins, D. P. (2012) Using cognitive work analysis to describe the role of UAVs in military operations. *Theoretical Issues in Ergonomics Science* 13(3), 335–57.

Jenkins, D. P., Salmon, P. M., Stanton, N. A. and Walker, G. H. (2010a) A systemic approach to accident analysis: a case study of the Stockwell shooting. *Ergonomics* 53(1), 1–17.

Jenkins, D. P., Stanton, N. A., Salmon, P. M. Walker, G. H. (2009a) *Cognitive Work Analysis: Coping with Complexity*. Aldershot: Ashgate.

Jenkins, D. P., Stanton, N. A., Salmon, P. M., Walker, G. H. (2010b) Using the decision-ladder to add a formative element to naturalistic decision-making research. *International Journal of Human Computer Interaction* 26(2–3), 132–46

Jenkins, D. P., Stanton, N. A., Walker, G. H., Salmon, P. M. and Young, M. S. (2009b) Using cognitive work analysis to explore system flexibility, *Theoretical Issues in Ergonomics Science* 9(4), 273–95.

Jensen, R. S. (1997) The boundaries of aviation psychology, Human Factors, aeronautical decision making, situation awareness and crew resource management. *International Journal of Aviation Psychology* 7(4), 259–67.

Jentsch, F. and Bowers, C. A. (2004) Team communications analysis. In N. A. Stanton, A. Hedge, K. Brookhuis, E. Salas and H. Hendrick (eds), *Handbook of Human Factors Methods*. London: Taylor & Francis.

Jesse, D. F., Dolbier, C.L. and Blanchard, A. (2008) Barriers to seeking help and treatment suggestions from prenatal depressive symptoms: focus groups with rural low income women. *Issues in Mental Health Nursing* 29, 3–19.

John, B. A and Kieras, D. E. (1994) The GOMS family of analysis techniques. Report CMU-CS-94-181. Pittsburgh, PA: Carnegie-Mellon University.

John, B. A. and Newell, A. (1990) Toward an engineering model of stimulus-response compatibility. In R. W. Proctor and T. G. Reeve (eds), *Stimulus-Response Compatibility*. Amsterdam: North-Holland.

Johnson, A. M., Copas, A. J., Erens, B., Mandalia, S., Fenton, K., Korovessis, C., Wellings, K. and Field, J. (2001) Effect of computer-assisted self-interviews on reporting of sexual HIV risk behaviours in a general population sample: a methodological experiment. *Epidemiology and Social: Concise Communication* 15(1), 111–15.

Johnson, C. W. and de Almeida, I. M. (2008) Extending the borders of accident investigation: applying novel analysis techniques to the loss of the Brazilian Space Programme's Launch Vehicle VLS-1 V03. *Safety Science* 46(1), 38–53.

Johnson, P., Diaper, D. and Long, J. (1984) Tasks, skills and knowledge: task analysis for knowledge-based descriptions. In B. Shackel (ed.), *Interact '84 – First IFIP Conference on Human-Computer Interaction*. Amsterdam: Elsevier.

Jones, D. G. (2000) Subjective measures of situation awareness. In M. R. Endsley and D. J. Garland (eds), *Situation Awareness Analysis and Measurement*. Mahwah, NJ: Lawrence Erlbaum.

Jones, D. G. and Endsley, M. R. (2000) Can real-time probes provide a valid measure of situation awareness? *Proceedings of the Human Performance, Situation Awareness and Automation: User Centred Design for the New Millennium Conference*, October.

Jones, D. G. and Kaber, D. B. (2005) Situation awareness measurement and the Situation Awareness Global Assessment Technique. In N. A. Stanton, A. Hedge, K. Brookhuis, E. Salas and H. Hendrick (eds), *Handbook of Human Factors and Ergonomics Methods*. Boca Raton, FL: CRC Press.

Jordan, C. S., Farmer, E. W. and Belyavin, A. J. (1995) The DRA Workload Scales (DRAWS): a validated workload assessment technique. *Proceedings of the 8th International Symposium on Aviation Psychology* 2, 1013–18.

Jorna, P.G.A. (1992) Spectral analysis of heart rate and psychological state: a review of its validity as a workload index. *Biological Psychology* 34, 1043–54.

Kakimoto, T., Kamei, Y., Ohira, M. and Matsumoto, K. 2006. Social network analysis on communications for knowledge collaboration in OSS communities. In *Proceedings of the 2nd International Workshop on Supporting Knowledge Collaboration in Software Development (KCSD'06)*, Tokyo, Japan, 35–41.

Karwowski, W. (1998) *The Occupational Ergonomics Handbook*. Boca Raton, FL: CRC Press.

Karwowski, W. (2001) *International Encyclopedia of Ergonomics and Human Factors Vols I–III*. London: Taylor & Francis.

Kath, L.M., Magley, V.J. and Marmet, M. (2010) The role of organisational trust in safety climates influence on organisational outcomes. *Accident Analysis and Prevention* 42, 1488–97.

Kaufman, J. (2006) Card sorting: an inexpensive and practical usability technique. *Intercom*, 17–19.

Kelley, J.F. (1985) CAL – a natural language program developed with the OZ paradigm: implications for supercomputing systems. *First International Conference on Supercomputing Systems (St. Petersburg, Florida, 16–20 December 1985)*. New York: ACM.

Kelly, G.A. (1955) *The Psychology of Personal Constructs*. New York: Norton.

Kennedy, G. A. L., Siemieniuch, C. E., Sinclair, M. A., Kirwan, B. A. and Gibson, W. H. (2007) Proposal for a sustainable framework process for the generation, validation and application of human reliability assessment within the engineering design lifecycle. *Reliability Engineering and System Safety* 92, 755–70.

Kennedy, R. (1995) Can Human Reliability Assessment (HRA) predict real accidents? A case study analysis of HRA. In A. I. Glendon and N. A. Stanton (eds), *Proceedings of the Risk Assessment and Risk Reduction Conference, 22nd March 1994, Aston University, Birmingham*.

Kennedy, R. and Kirwan, B. (1998) Development of a hazard and operability-based method for identifying safety management vulnerabilities in high risk systems. *Safety Science* 30, 249–74.

Kieras, D. (2003) GOMS models for task analysis. In D. Diaper and N. Stanton (eds), *The Handbook of Task Analysis for Human-Computer Interaction*. Mahwah, NJ: Lawrence Erlbaum.

Killam, B., Preston, A., McHarg, S. and Wilson, C. (2010) Card sorting. Usability body of knowledge. Available at: http://www.usabilitybok.org/card-sorting.

Kim, I. S. (2001) Human reliability analysis in the man-machine interface design review. *Annals of Nuclear Energy* 28, 1069–81.

Kim, K., Proctor, R. W. and Salvendy, G. (2012) The relation between usability and product success in cell phones. *Behaviour and Information Technology* 31(10), 969–82.

Kirakowski, J. (1996) The software usability measurement inventory: background and usage. In P. Jordan, B. Thomas and B. Weerdmeester (eds), *Usability Evaluation in Industry*. London: Taylor & Francis.

Kirwan, B. (1990) Human reliability assessment. In J. R. Wilson and E. N. Corlett (eds), *Evaluation of Human Work: A Practical Ergonomics Methodology*, 2nd edn. London: Taylor & Francis.

Kirwan, B. (1992a) Human error identification in human reliability assessment. Part 1: overview of approaches. *Applied Ergonomics* 23, 299–318.

Kirwan, B. (1992b) Human error identification in human reliability assessment. Part 2: detailed comparison of techniques. *Applied Ergonomics* 23, 371–81.

Kirwan, B. (1994) *A Guide to Practical Human Reliability Assessment*. London: Taylor & Francis.

Kirwan, B. (1996a) Human Error Recovery and Assessment (HERA) guide. Project IMC/GNSR/HF/5011, Industrial Ergonomics Group, School of Manufacturing Engineering, University of Birmingham, March.

Kirwan, B. (1996b) The validation of three human reliability quantification techniques – THERP, HEART and JHEDI: Part 1 – technique descriptions and validation issues. *Applied Ergonomics* 27(6), 359–73.

Kirwan, B. (1997a) The validation of three human reliability quantification techniques – THERP, HEART and JHEDI: Part 2 – Results of validation exercise. *Applied Ergonomics* 28(1), 17–25.

Kirwan, B. (1997b) The validation of three human reliability quantification techniques – THERP, HEART and JHEDI: Part 3 – Practical aspects of the usage of the techniques. *Applied Ergonomics* 28(1), 27–39.

Kirwan, B. (1998a) Human error identification techniques for risk assessment of high-risk systems – Part 1: review and evaluation of techniques. *Applied Ergonomics* 29, 157–77.

Kirwan, B. (1998b) Human error identification techniques for risk assessment of high-risk systems – Part 2: towards a framework approach. *Applied Ergonomics* 5, 299–319.

Kirwan, B. and Ainsworth, L. K. (1992) *A Guide to Task Analysis*. London: Taylor & Francis.

Kirwan, B., Evans, A., Donohoe, L., Kilner, A., Lamoureux., Atkinson, T. and MacKendrick, H. (1997) 'Human Factors in the ATM system design life cycle'. *FAA/Eurocontrol ATM R&D Seminar*, Paris, France. Available at: http://atm-seminar-97.eurocontrol.fr/kirwan.htm.

Klein, G. (2000) Cognitive task analysis of teams. In J. M. Schraagen, S. F. Chipman, V. L. Shalin (eds), *Cognitive Task Analysis*. Mahwah, NJ: Lawrence Erlbaum.

Klein, G. and Armstrong, A. A. (2004) Critical Decision Method. In N. A. Stanton, A. Hedge, K. Brookhuis, E. Salas and H. Hendrick (eds), *Handbook of Human Factors and Ergonomics Methods*. Boca Raton, FL: CRC Press.

Klein, G., Calderwood, R. and MacGregor, D. (1989) The Critical Decision Method for eliciting knowledge. *IEEE Transactions on Systems, Man and Cybernetics* 19(3), 462–72.

Klein, G., Schmitt, J., McCloskey, M., Heaton, J., Klinger, D. and Wolf, S. (1996) *A Decision-Centred Study of the Regimental Command Post*. Fairborn, OH: Klein Associates.

Kletz, T. (1974) HAZOP and HAZAN: notes on the identification and assessment of hazards. *Journal of Loss Prevention in the Process Industries* 8(6), 349–53.

Kletz, T. (1991) *An Engineer's View of Human Error*, 2nd edn. Rugby: Institution of Chemical Engineers.

Klinger, D. W. and Hahn, B. B. (2004) Team decision requirement exercise: making team decision requirements explicit. In N. A. Stanton, A. Hedge, K, Brookhuis, E. Salas and H. Hendrick (eds), *Handbook of Human Factors Methods*. London: Taylor & Francis.

Kvarnstrom, S. (2008) Difficulties in collaboration: a critical incident study of interprofessional healthcare teamwork. *Journal of Interprofessional Care* 22(2), 191–203.

Ladkin, P. B. and Stuphorn, J. (2003) Two causal analyses of the Black Hawk shootdown during Operation Provide Comfort. In P. Lindsay and T. Cant (eds), *Proceedings of the Eighth Australian Workshop on Safety Critical Systems and Software (SCS 2003), Canberra, Australia*, pp. 3–23 .

Ladkin, P. B. (2005b) Networks and distributed systems. Available at: http://www.rvs.uni-bielefeld.de.

Ladkin, P. B. (2005b) *Why-Because Analysis of the Glenbrook, NSW Rail Accident and Comparison with Hopkins's Accimap*. Report RVS-RR-05-05, 19 December, Faculty of Technology, Bielefeld University. Available at: http://www.rvs.uni-bielefeld.de.

Ladkin, P. B. and Loer, K. (1998) *Why-Because Analysis: Formal Reasoning about Incidents*. Technical Report RVS-BK-98-01, Faculty of Technology, Bielefeld University. Available at: http://www.rvs.uni-bielefeld.de.

Landay, J. A. and Myers, B. A. (1995) *Just Draw It! Programming by Sketching Storyboards*. Carnegie Mellon University, Human-Computer Interaction Institute Technical Report CMU-HCII-95-106 and School of Computer Science Technical Report CMU-CS-95-199. Available at: http://www.cs.cmu.edu/~landay/research/publications/storyboard-tr/storyboard.pdf.

Landay, J. A. and Myers, B. A. (1996) Sketching storyboards to illustrate interface behaviors. *Conference Companion of ACM Conference on Human Factors in Computing Systems: CHI '96*, 193–4. Available at: http://www.cs.cmu.edu/~landay/research/publications/CHI96/short_storyboard.pdf.

Lane, R., Stanton, N. A. and Harrison, D. (2006) Hierarchical Task Analysis to medication administration errors. *Applied Ergonomics* 37(5), 669-679.

Langford, J. and McDonagh, D. (2002) *Focus Groups: Supporting Effective Product Development*. London: Taylor & Francis.

Lawrence, D., Atwood, M.E., Dews, S. and Turner, T. (1995) Social interaction in the use and design of a workstation: two contexts of interaction. In P.J. Thomas (ed.), *The Social and Interactional Dimensions of Human Computer Interfaces*. Cambridge: Cambridge University Press.

Lawton, R. and Ward, N. J. (2005) A systems analysis of the Ladbroke Grove rail crash. *Accident Analysis and Prevention* 37, 235–44.

Lee, S. M., Ha, J. S. and Seong, P. H. (2011) CREAM based communication error analysis method (CEAM) for nuclear power plant operators' communication. *Journal of Loss Prevention in the Process Industries* 24, 90–97.

Leveson, N. (2001) *Evaluating Accident Models Using Recent Aerospace Accidents*. NASA Report. Available at: http://sunnyday.mit.edu/accidents/nasa-report.pdf.

Leveson, N. (2002) *A New Approach to System Safety Engineering*. Cambridge, MA: Aeronautics and Astronautics, Massachusetts Institute of Technology.

Leveson, N. (2004) A new accident model for engineering safer systems. *Safety Science* 42, 237–70.

Leveson, N., Allen, P. and Storey, M.-A. (2002) The analysis of a friendly fire accident using a systems model of accidents. In *Proceedings of the 20th International System Safety Society Conference (ISSC 2003)*. Unionville, VA: System Safety Society.

Lewis, C., Polson, P., Wharton, C. and Rieman, J. (1990) Testing a walkthrough methodology for theory-based design of walk-up-and-use interfaces. In *Proceedings of CHI'90 Conference on Human Factors in Computer Systems*. New York: Association for Computer Machinery.

Lewis, C. and Reiman, J. (1993) *Task Centred User Interface Design: A Practical Introduction*. Boulder, CO: University of Colorado. Shareware book available at: http://hcibib.org/tcuid/tcuid.pdf.

Lewis, D. (1973) Causation. *Journal of Philosophy* 70, 556–67.

Lewis, J. R. (1991) *User Satisfaction Questionnaires for Usability Studies: 1991 Manual of Directions for the ASQ and PSSUQ*. Technical Report 54.609. Boca Raton, FL: International Business Machines Corporation.

Lewis, J. R. (1992) Psychometric evaluation of the post-study system usability questionnaire: the PSSUQ. In *Proceedings of the Human Factors Society 36th Annual Meeting*. Santa Monica, CA: Human Factors Society.

Lewis, J. R. (1993) *IBM Computer Usability Satisfaction Questionnaires: Psychometric Evaluation and Instructions for Use*. Technical Report 54.786 Available at: http://drjim.0catch.com/usabqtr.pdf.

Lewis, J. R. (2002) Psychometric evaluation of the PSSUQ using data from five years of usability studies. *International Journal of Human Computer Interaction* 14(3–4), 463–88.

Lewis, P. J. (1992) Rich picture building in the soft systems methodology. *European Journal of Information Systems* 1, 351–60.

Li, H., Liu, Y., Liu, J., Wang, X., Li, Y. and Rau, P. L. P. (2010) Extended KLM for mobile phone interaction: a user study result. *CHI 2010: Work in Progress*, 12–13 April, Atlanta, Georgia.

Li, W.C., Harris, D., Hsu, Y.L. and Li, L.W. (2009) The application of the Human Error Template (HET) for redesigning standard operating procedures in aviation operations. In D. Harris (ed), *Engineering Psychology and Cognitive Ergonomics*. Oklahoma: Springer Verlag, pp. 600–605.

Li, W.C., Harris, D. and Yu, C.S. (2008) Routes to failure: analysis of 41 civil aviation accidents from the Republic of China using the Human Factors analysis and classification system. *Accident Analysis and Prevention* 40(2), 426–34.

Liang, Y. and Lee, J.D. (2010) Combining cognitive and visual distraction: less than the sum of its parts. *Accident Analysis and Prevention* 42, 881–90.

Licu, T., Cioran, F., Hayward, B. and Lowe, A. (2007) EUROCONTROL—Systemic Occurrence Analysis Methodology (SOAM)—a 'reason'-based organisational methodology for analysing incidents and accidents. *Reliability Engineering and System Safety* 92, 1162–9.

Lim, K. Y. and Long, J. (1994) *The MUSE Method for Usability Engineering*. Cambridge: Cambridge University Press.

Lin, H. X., Choong, Y. Y. and Salvendy, G. (1997) A proposed index of usability: a method for comparing the relative usability of different software systems. *Behaviour and Information Technology* 16(4–5), 267–78.

Lindhe, A., Rosen, L., Norberg, T. and Bergstedt, O. (2009) Fault tree analysis for integrated and probabilistic risk analysis of drinking water systems. *Water Research* 43, 1641–53.

Liu, S., Tamai, T. and Nakajima, S. (2009) Integration of formal specification, review and testing for software component quality assurance. *SAC'09*, 8–12 March, Honolulu, Hawaii.

Ljung, M. (2010) Generalisation of case studies in road traffic when defining pre-crash scenarios for active safety function evaluation. *Accident Analysis and Prevention* 42, 1172–83.

Lockton, D., Harrison, D. and Stanton, N. A. (2010) The Design with Intent method: a design tool for influencing user behaviour. *Applied Ergonomics* 41, 382–92.

Lockyer, K. and Gordon, J. (1991) *Critical Path Analysis and Other Project Network Techniques*. London: Pitman.

Lovejoy, K. and Handy, S. (2008) A case for measuring individuals' access to private vehicle travel as a matter of degrees: lessons from focus groups with Mexican immigrants in California. *Transportation* 35, 601–12.

Lund, A. M. (1998) USE questionnaire resource page. Available at:http://www.stcsig.org/usability/ newsletter/0110_measuring_with_use.html.

Lund, A. M. (2001) Measuring usability with the USE questionnaire. *Usability Interface Usability SIG Newsletter*.

Luo, L. and John, B.E. (2005) Predicting task execution time on handheld devices using the Keystroke Level Model. *CHI 2005*, 2–7 April, Portland, Oregon.

Luximon, A. and Goonetilleke, R. S. (2001) A simplified subjective workload assessment technique. *Ergonomics* 44(3), 229–43.

Ma, R. and Kaber, D. B. (2005) Situation awareness and workload in driving while using adaptive cruise control and a cell phone. *International Journal of Industrial Ergonomics* 35, 939–53.

Ma, R. and Kaber, D. B. (2007). Situation awareness and driving performance in a simulated navigation task. *Ergonomics* 50, 1351–64.

MacMillan, J., Paley, M. J., Entin, E. B. and Entin, E. E. (2004) Questionnaires for distributed assessment of team mutual awareness. In N. A. Stanton., A. Hedge., K. Brookhuis., E. Salas and H. Hendrick (eds), *Handbook of Human Factors Methods*. London: Taylor & Francis.

Mahatody, T., Kolski, C. and Sagar, M. (2010) CWE: assistance environment for the evaluation operating a set of variations of the cognitive walkthrough ergonomic inspection method. In D. Harris (ed.), *Engineering Psychology and Cognitive Ergonomics*. Oklahoma: Springer Verlag.

Mahatody, T., Sagar, M. and Kolski, C. (2010) State of the art on the cognitive walkthrough method, its variants and evolutions. *International Journal of Human Computer Interaction* 26(8), 741–85.

Marsden, P. and Kirby, M. (2005) Allocation of functions. In N. A. Stanton, A. Hedge, K, Brookhuis, E. Salas and H. Hendrick (eds), *Handbook of Human Factors Methods*. London: Taylor & Francis.

Marshall, A., Stanton, N., Young, M., Salmon, P., Harris, D., Demagalski, J., Waldmann, T. and Dekker, S. (2003) Development of the human error template – a new methodology for assessing design induced errors on aircraft flight decks. ERRORPRED Final Report E!1970, August.

Matheus, C. J., Kokar, M. M. and Baclawski, K. (2003) A core onotology for situation awareness. In *Proceedings of the 6th International Conference on Information Fusion*, Cairns, Australia.

Matthews, M. D. and Beal, S. A. (2002) *Assessing Situation Awareness in Field Training Exercises*. U.S. Army Research Institute for the Behavioural and Social Sciences. Research Report 1795.

Matthews, M. D., Pleban, R. J., Endsley, M. R. and Strater, L. D. (2000) Measures of infantry situation awareness for a virtual MOUT environment. In *Proceedings of the Human Performance, Situation Awareness and Automation: User Centred Design for the New Millennium Conference*.

Matthews, M.D., Strater, L.D. and Endsley, M. R. (2004) Situation awareness requirements for infantry platoon leaders. *Military Psychology* 16, 149–61.

Maulsby, D., Greenberg, S. and Mander, R. (1993) Prototyping an intelligent agent through Wizard of Oz. *InterChi 1993*.

Maxiom, R.A. and Reeder, R.W. (2005) Improving user-interface dependability through mitigation of human error. *International Journal of Human-Computer Studies* 63, 25–50.

McDonnell, P., Smallenburg, J., Spiewela, J., Wilson, B. and Newman, M. (2009) Heuristic evaluation of the interdepartmental program in classical art and archaeology website. Available at: http://www.jacekspiewla.com/projects/ipcaa/ipcaa_heuristic.pdf.

McFadden, K. L. and Towell, E. R. (1999) Aviation Human Factors: a framework for the new millennium. *Journal of Air Transport Management* 5, 177–84.

McGuinness, B. (1999) Situational Awareness and the Crew Awareness Rating Scale (CARS). *Proceedings of the 1999 Avionics Conference*. Heathrow, London, 17–18 November. ERA Technology report 99-0815 (paper 4.3).

McGuinness, B. and Ebbage, L. (2000) Assessing Human Factors in command and control: workload and situational awareness metrics. In *Proceedings of 2002 Command and Control Research and Technology Symposium, Monterey, California*.

McGuinness, B. and Foy, L. (2000) A subjective measure of SA: the Crew Awareness Rating Scale (CARS). Presented at the *Human Performance, Situational Awareness and Automation Conference*, Savannah, Georgia, 16–19 October.

McIlroy, R.C. and Stanton, N.A. (2011) Observing the observer: non-intrusive verbalisations using the Concurrent Observer Narrative Technique. *Cognition, Technology and Work* 13, 135–49.

Medina, A. L., Lee, S. E., Wierwille, W. W. and Hanowski, R. J. (2004) Relationship between infrastructure, driver error and critical incidents. *Proceedings of the Human Factors and Ergonomics Society 48th Annual Meeting.*

Megaw, E. D. (2005) The definition and measurement of mental workload. In J. R. Wilson and E. N. Corlett (eds), *Evaluation of Human Work*, 3rd edn. Boca Raton, FL: CRC Press.

Mercurio, D., Podofillini, L., Zio, E. and Dang, V.N. (2009) Identification and classification of dynamic event tree scenarios via possibilistic clustering: application to a steam generator. *Accident Analysis and Prevention* 41, 1180–1191.

Militello, L. G. and Hutton, J. B. (2000) Applied Cognitive Task Analysis (ACTA): a practitioner's toolkit for understanding cognitive task demands. In J. Annett and N. S Stanton (eds), *Task Analysis*. London: Taylor & Francis.

Miller, G. A., Galanter, E. and Pribram, K. H. (1960) *Plans and the Structure of Behaviour*. New York: Holt.

Miller, J. E., Patterson, E. S. and Woods, D. D. (2006) Elicitation by critiquing as a cognitive task analysis methodology. *Cognition, Technology and Work* 8(2), 90–102.

Miller, R. B. (1953) *A Method for Man-Machine Task Analysis*. Report WADC-TR-53-137, Wright Air Development Center, Wright-Patterson AFB, Ohio.

Milligan, F. J. (2007) Establishing a culture for patient safety—the role of education, *Nurse Education Today* **27**(2), 95–102.

Mills, S. (2007) Contextualising design: aspects of using usability context analysis and Hierarchical Task Analysis for software design. *Behaviour & Information Technology* 26(6), 499–506.

Ministry of Defence (2000) *Human Factors Integration: An Introductory Guide*. London: HMSO.

Morgan, J. F. and Hancock, P. A. (2011) The effect of prior task loading on mental workload: an example of hysteresis in driving. *Human Factors* 53(1), 75–86.

Muller, M. J. (2001) Layered participatory analysis: new developments in the CARD Technique. *Proceedings of ACM Conference on Human Factors in Computing Systems*, Seattle, Washington.

Munn, J. C., Dobbs, D., Meier, A., Williams, C. S., Biola, H. and Zimmerman, S. (2008) The end of life experience in long term care: five themes identified from focus groups with residents, family members and staff. *The Gerontologist* 48(4), 485–94.

Naikar, N. (2006) A comparison of the decision ladder template and the recognition primed decision model. Submitted for proposed publication of papers resulting from the *International Workshop on Intelligent Decision Support Systems: Retrospect and Prospects*, 29 August–2 September 2005, Siena, Italy; subsequently unpublished.

Naikar, N., Lintern, G. and Sanderson, P. M. (2002) Cognitive work analysis for air defense applications in Australia. In M. D. McNeese and M. A. Vidulich (eds), *Cognitive Systems Engineering in Military Aviation Environments: Avoiding Cogminutia Fragmentosa!* Human Systems Information Analysis Center: Wright Patterson Air Force Base, Dayton, Ohio.

Naikar, N., Moylan, A. and Pearce, B. (2006) Analysing activity in complex systems with cognitive work analysis: concepts, guidelines and case study for control task analysis. *Theoretical Issues in Ergonomics Science* 7(4), 371–94.

Naikar, N., Pearce, B., Drumm, D. and Sanderson, P. (2003) Designing teams for first-of-a-kind, complex systems using the initial phases of cognitive work analysis: case study. *Human Factors* 45(2), 202–17.

Naikar, N. and Sanderson, P. M. (1999) Work domain analysis for training-system definition and acquisition. *International Journal of Aviation Psychology* 9, 271–90.

Naikar, N. and Sanderson, P. M. (2001) Evaluating design proposals for complex systems with work domain analysis. *Human Factors* 43(4), 529–42.

Naikar, N. and Saunders, A. (2003) Crossing the boundaries of safe operation: an approach for training technical skills on error management. *Cognition Technology and Work* 5, 171–80.

Neerincx, M. A. (2003) Cognitive Task Load Analysis: allocating tasks and designing support. In E. Hollnagel (ed.), *Handbook of Cognitive Task Design*. Mahwah, NJ: Lawrence Erlbaum.

Neisser, U. (1976) *Cognition and Reality: Principles and Implications of Cognitive Psychology*. San Francisco, CA: Freeman.

Nelson, W. R, Haney, L. N, Ostrom, L. T. and Richards, R. E. (1998) Structured methods for identifying and correcting potential human errors in space operations, *Acta Astronautica* 43, 211–22.

Newman, J. F., LaDue, D. S. and Heinselman, P. L. (2009) Identifying critical strengths and limitations of current radar systems. *25th Conference on International Interactive Information and Processing Systems (IIPS) for Meteorology, Oceanography and Hydrology*.

Nielsen, J. (1994a) Heuristic evaluation. In J. Nielsen and R. L. Mack (eds), *Usability Inspection Methods*. New York: John Wiley & Sons.

Nielsen, J. (1994b) *Usability Engineering*. Boston: Academic Press.

Nielsen, J. and Molich, R. (1990) Heuristic evaluation of user interfaces. In J. C Chew and J. Whiteside (eds), *Empowering People: CHI 90 Conference Proceedings*. Monterey, CA: ACM Press.

Nofi, A. (2000) Defining and Measuring Shared Situational Awareness, DARPA. Available at: http://www.thoughtlink.com/publications/DefiningSSA00Abstract.htm.

Norman, D. A. (1988) *The Design of Everyday Things*. Cambridge, MA: MIT Press.

Nouvel, D., Travadel, S. and Hollnagel, E. (2007) Introduction of the concept of functional resonance in the analysis of a near-accident in aviation. *33rd ESReDA Seminar: Future Challenges of Accident Investigation*.

Noyes, J. and Baber, C. *Applied Computing. User Centred Design of Systems*. London: Springer Verlag.

Nygren, T. E. (1991) Psychometric properties of subjective workload measurement techniques: implications for their use in the assessment of perceived mental workload. *Human Factors* 33(1), 17–33.

O'Brian, K. S. and O'Hare, D. (2007) Situational awareness ability and cognitive skills training in a complex real world task. *Ergonomics* 50(7), 1064–91.

O'Hare, D., Wiggins, M., Williams, A. and Wong, W. (2000) Cognitive task analysis for decision centred design and training. In J. Annett and N.A. Stanton (eds), *Task Analysis*. London: Taylor & Francis.

O'Reilly, J. M, Hubbard, M., Lessler, J., Biemer, P. P. and Turner, C. F. (1994) Audio and video computer assisted self-interviewing: preliminary tests of new technologies for data collection. *Journal of Official Statistics* 10(2), 197–214.

Ockerman, J. (2010) Contextual design for mobile technology. Available at: http://www.iswc.ethz.ch/events/tutorials/slides_ockerman.ppt#425,2,Contextual Design Timeline.

Ogden, G. C. (1987) Concept, knowledge and thought. *Annual Review of Psychology* 38, 203–27.

Olsen, N. S. and Shorrock, S. T. (2010) Evaluation of the HFACS ADF safety classification system: inter-coder consensus and inter-order consistency. *Accident Analysis and Prevention* 42, 437–44.

Olson, J. R. and Nilsen, E. (1988) Analysis of the cognition involved in spreadsheet software interaction. *Human-Computer Interaction*, 3, 309–50.

Olson, J. R. and Olson, G. M. (1990) The growth of cognitive modelling in human-computer interaction since GOMS. *Human-Computer Interaction* 5(2), 221–65.

Olsson, G. and Lee, P. L. (1994) Effective interfaces for process operators. *Journal of Process Control* 4, 99–107.

Omodei, M. M. and McLennan, J. (1994) Studying complex decision making in natural settings: using a head-mounted video camera to study competitive orienteering. *Perceptual and Motor Skills* 79(3), 1411–25.

Oppenheim, A. N. (2000) *Questionnaire Design, Interviewing and Attitude Measurement*. London: Continuum.

Ormerod, T. C. (2000) Using task analysis as a primary design method: the SGT approach. In J. M. Schraagen, S. F. Chipman and V. L. Shalin (eds), *Cognitive Task Analysis*. Mahwah, NJ: Lawrence Erlbaum.

Ormerod, T. C., Richardson, J. and Shepherd, A. (1998) Enhancing the usability of a task analysis method: a notation and environment for requirements. *Ergonomics* 41(11), 1642–63.

Ormerod, T. C. and Shepherd, A. (2003) Using task analysis for information requirements specification: the Sub-Goal Template (SGT) method. In D. Diaper and N. A. Stanton (eds), *The Handbook of Task Analysis for Human-Computer Interaction*. Mahwah, NJ: Lawrence Erlbaum.

Paletz, S. B. F., Bearman, C., Orasanu, J. and Holbrook, J. (2009) Socializing the Human Factors analysis and classification system: incorporating social psychological phenomena into a Human Factors error classification system. *Human Factors* 51(4), 435–45.

Patrick, J., Gregov, A. and Halliday, P. (2000) Analysing and training task analysis. *Instructional Science* 28(4), 51–79.

Patrick, J., James, N., Ahmed, A. and Halliday, P. (2006) Observational assessment of situation awareness, team differences and training implications. *Ergonomics* 49(12–13), 393–417.

Patrick, J. and Morgan, P. L. (2010) Approaches to understanding, analysing and developing situation awareness. *Theoretical Issues in Ergonomics Science* 11(1), 41–57.

Patterson, J. M. and Shappell, S. A. (2010) Operator error and system deficiences: analysis of 508 mining incidents and accidents from Queensland, Australia using HFACS. *Accident Analysis and Prevention* 42(4), 1379–85.

Pennycook, W. A. and Embrey, D. E. (1993) An operating approach to error analysis. In W. Karwowski and W. S Marras (eds), *The Occupational Ergonomics Handbook*. Boca Raton, FL: CRC Press.

Pettitt, M., Burnett, G. and Stevens, A. (2007) An extended Keystroke Level Model (KLM) for predicting visual demand of in-vehicle information systems. *CHI 2007 Proceedings Models of Mobile Interaction*, 28 April–3 May, San Jose, CA.

Peute, L. W. P. and Jaspers, M. W. M. (2007) The significance of a usability evaluation of an emerging laboratory order entry system. *International Journal of Medical Informatics* 76, 157–68.

Pew, R. W., Miller, D. C. and Feehrer, C. S. (1981) *Evaluation of Proposed Control Room Improvements through Analysis of Critical Operator Decisions, NP 1982*. Palo Alto, CA: Electric Power Research Institute.

Phipps, D., Meakin, G. H., Beatyy, P. C. W., Nsoedo, C. and Parker, D. (2008) Human Factors in anaesthesia practice: insights from a task analysis. *British Journal of Anaesthesia* 100(3), 333–43.

Pocock, S., Fields, B. Harrison, M. and Wright, P. (2001a) THEA – a reference guide. Unpublished.

Pocock, S., Harrison, M.D., Wright, P. C. and Johnson, P. (2001b) THEA: a technique for Human Error Assessment early in design. In M. Hirose (ed.), *Interact01*. Amsterdam: IOS Press.

Polson, P. G., Lewis, C., Rieman, J. and Wharton, C. (1992) Cognitive walkthroughs: a method for theory-based evaluation of user interfaces. *International Journal of Man-Machine Studies* 36, 741–73.

Prasanna, R., Yang, L. and King, M. (2009) GDIA: a cognitive task analysis protocol to capture the information requirements of emergency first responders. In *Proceedings of the 6th International ISCRAM Conference*, Gothenberg, Sweden.

Pretorius, A. and Cilliers, P. J. (2007) Technical Note. Development of a Mental Workload Index: a systems approach. *Ergonomics* 50(9), 1503–15.

Proctor, R. and Van Zandt, T. (1994) *Human Factors in Simple and Complex Systems*. Needham Heights, MA: Allyn and Bacon.

Quillian, R. (1969) The teachable language comprehender: a simulation program and theory of language. *Communications of the ACM*, 12459–76.

Qureshi, Z. H. (2007) A review of accident modelling approaches for complex socio-technical systems. In *Proceedings of the Defence and Systems Institute*, University of South Australia. Available at: http://crpit.com/confpapers/CRPITV86Qureshi.pdf.

Rafferty, L. A., Stanton, N. A. and Walker, G. H. (2012) *The Human Factors of Fratricide*. Aldershot: Ashgate.

Rail Safety and Standards Board. (2003) *Master Rule Book*. Newark, NJ: Willsons..

Ramos, M., Sedivi, B. M. and Sweet, E. M. (1998) Computerised self-administered questionnaires. In M. P. Couper, R. P. Baker, J. Bethlehem, C. Z. Clark, J. Martin, W. L. Nicholls and M. O. O'Reilly (eds), *Computer Assisted Survey Information Collection*. New York: John Wiley.

Rasmussen, J. (1974) *The Human Data Processor as a System Component: Bits and Pieces of a Model* (Report No. Risø-M-1722). Roskilde, Denmark.

Rasmussen, J. (1985) The role of hierarchical knowledge representation in decision making and system management. *IEEE Transactions on Systems, Man and Cybernetics* 15, 234–43.

Rasmussen, J. (1986) *Information Processing and Human-Machine Interaction*. Amsterdam: North-Holland.

Rasmussen, J. (1997) Risk management in a dynamic society: a modelling problem. *Safety Science* 27(2–3), 183–213.

Rasmussen, J., Pedersen, O. M., Carnino, A., Griffon, M., Mancini, C. and Gagnolet, P. (1981) *Classification Scheme for Reporting Events Involving Human Malfunctions*. Ris_-M-2240, SINDOC(81)14. Riso National Laboratories, Roskilde, Denmark.

Rasmussen, J., Pejtersen, A. and Goodstein, L. P. (1994) *Cognitive Systems Engineering*. New York: Wiley.

Ravden, S. J. and Johnson, G. I. (1989) *Evaluating Usability of Human-Computer Interfaces: A Practical Method*. Chichester: Ellis Horwood.

Rea, L. M. and Parker, R. A. (2005) *Designing and Conducting Survey Research: A Comprehensive Guide*, 3rd edn. San Francisco, CA: Jossey-Bass.

Reason, J. (1990) *Human Error*. Cambridge: Cambridge University Press.

Reason, J., Manstead, A., Stradling, S., Baxter, J. and Campbell, K. (1990) Errors and violations on the roads: a real distinction? *Ergonomics* 33, 1315–332.

Redding, R. E. (1989) Perspectives on cognitive task-analysis: the state of the state of the art. In *Proceedings of the Human Factors Society*. Santa Monica, CA: Human Factors Society.

Reid, G. B. and Nygren, T. E. (1988) The subjective workload assessment technique: a scaling procedure for measuring mental workload. In P. S. Hancock and N. Meshkati (eds), *Human Mental Workload*. Amsterdam: Elsevier.

Reinach, S. and Viale, A. (2006) Application of a human error framework to conduct train accident/incident investigations. *Accident Analysis and Prevention* 38(2), 396–406.

Revell, K. M. A., Stanton, N. A. and Bessell, K. (2010) Digitising command and control. HFI-DTC Report. BAE Systems, Yeovil, Somerset.

Robie, C., Brown, D. J. and Beaty, J. C. (2007). Do people fake on personality inventories? A verbal protocol analysis. *Journal of Business and Psychology* 21(4), 489–509.

Rodden, K. and Fu, X. (2007) Exploring how mouse movements relate to eye movements on web search results pages. SIGIR 2007 Workshop Web Information Seeking and Interaction held in conjunction with the *30th Annual International ACM SIGIR Conference*, 27 July, Amsterdam.

Roscoe, A. H. (1992) Assessing pilot workload: why measure heart rate, HRV and respiration? *Biological Psychology* 34, 259–88.

Roscoe, A. and Ellis, G. (1990) *A Subjective Rating Scale for Assessing Pilot Workload in Flight*. Farnborough: RAE.

Roth, E. M., Patterson, E. S. and Mumaw, R. J. (2002) Cognitive engineering: issues in user-centred system design. In J. J Marciniak (ed.), *Encyclopedia of Software Engineering*, 2nd edn. New York: John Wiley & Sons.

Royal Australian Aviation Force. (2001) *F111 Deseal/ Reseal Board of Inquiry Report*. Available at: http://www.airforce.gov.au/docs/Volume1.htm.

Rubio, S., Diaz, E., Martin, J. and Puente, J. M. (2004) Evaluation of subjective mental workload: a comparison of SWAT, NASA-TLX and workload profile methods. *Applied Psychology: An International Review* 53(1), 61–86.

Rudnicky, A.I. and Hauptmann, A.G. (1991) Models for evaluating interaction protocols in speech recognition. *Proceedings of the CHI Conference*, April.

Saitwal, H., Feng, X., Walji, M., Patel, V. and Zhang, J. (2010) Assessing performance of an electronic health record (HER) using cognitive task analysis. *International Journal of Medical Informatics* 79, 501–6.

Salas, E., Prince, C., Baker, P. D. and Shrestha, L. (1995) Situation awareness in team performance. *Human Factors* 37, 123–36.

Salas, E. (2004) Team methods. In N. A. Stanton, A. Hedge, E. Salas, H. Hendrick and K. Brookhaus (eds), *Handbook of Human Factors and Ergonomics Methods*. London: Taylor & Francis.

Salmon, P. M., Jenkins, D. P., Stanton, N. A. and Walker, G. H. (2010a) Hierarchical Task Analysis versus cognitive work analysis: comparison of theory, methodology and contribution to system design. *Theoretical Issues in Ergonomics Science* 11(6), 504–31.

Salmon, P. M., Regan, M. A. and Johnston, I. (2005) *Human Error and Road Transport. Phase One Literature Review*. Monash University Accident Research Centre Report no. 256. Available at: http://www.monash.edu.au/muarc/reports/muarc256.pdf.

Salmon, P. M., Stanton, N. A., Gibbon, A. C., Jenkins, D. P. and Walker, G. H. (2010b) *Human Factors Methods and Sports Science: A Practical Guide*. Abingdon: Taylor & Francis.

Salmon, P. M., Stanton, N. A., Jenkins, D. P. and Walker, G.H. (2011) Coordination during multi-agency emergency response: issues and solutions. *Disaster Prevention and Management* 20(2), 140–158.

Salmon, P. M., Stanton, N. A., Lenne, M., Jenkins, D. P., Rafferty, L. A. and Walker, G. H. (2011) *Human Factors Methods and Accident Analysis: Practical Guidance and Case Study Application*. Aldershot: Ashgate

Salmon, P. M, Stanton, N. A., Walker, G. H., Baber, C., Jenkins, D. P. and McMaster, R. (2008a) What really is going on? Review of situation awareness models for individuals and teams. *Theoretical Issues in Ergonomics Science* 9(4), 297–323.

Salmon, P. M., Stanton, N. A., Walker, G. and Green, D. (2004) Future battlefield visualisation: investigating data representation in a novel C4i system. In *Proceedings of the Land Warfare 2004 Conference*, Defence Science and Technology Organisation, Melbourne, September.

Salmon, P. M., Stanton, N. A., Walker, G. and Green, D. (2006) Situation awareness measurement: a review of applicability for C4i environments. *Applied Ergonomics* 37(2), 225–38.

Salmon, P. M., Stanton, N. A., Walker, G. H. and Jenkins, D. P. (2009a) *Distributed Situation Awareness: Advances in Theory, Measurement and Application to Teamwork*. Aldershot: Ashgate.

Salmon, P. M., Stanton, N. A., Walker, G., Jenkins, D. and Baber, C. (2008) Representing situation awareness in collaborative systems: a case study in the energy distribution domain. *Ergonomics* 51(3), 367–84.

Salmon, P. M., Stanton, N. A., Walker, G. H., Jenkins, D. P., Ladva, D., Rafferty, L. and Young, M. S. (2009b) Measuring situation awareness in complex systems: comparison of measures study. *International Journal of Industrial Ergonomics* 39, 490–500.

Salmon, P. M., Stanton, N. A., Walker, G., McMaster, R. and Green, D. (2005), Command and control in the energy distribution domain. Unpublished manuscript.

Salmon, P., Stanton, N. A., Young, M. S., Harris, D., Demagalski, J., Marshall, A., Waldman, T. and Dekker, S. (2002) Using existing HEI techniques to predict pilot error: a comparison of SHERPA, HAZOP and HEIST. In *Proceedings of HCI-Aero 2002*. Cambridge, MA: MIT Press.

Salmon, P., Stanton, N. A., Young, M. S., Harris, D., Demagalski, J., Marshall, A., Waldmann, T. and Dekker, S. (2003a) Using existing HEI techniques to predict pilot error: a comparison of SHERPA, HAZOP and HEIST. In S. Chatty, J. Hansman and G. Boy (eds), *Proceedings of International Conference on Human-Computer Interaction in Aeronautics – HCI-Aero 2002*. Menlo Park, CA: AAAI Press.

Salmon, P. M., Stanton, N. A., Young, M. S., Harris, D., Demagalski, J., Marshall, A., Waldmann, T. and Dekker, S. (2003b) Predicting design induced pilot error: a comparison of SHERPA, Human Error HAZOP, HEIST and HET, a newly developed aviation specific HEI method. In D. Harris, V. Duffy, M. Smith and C. Stephanidis (eds), *Human-Centred Computing – Cognitive, Social and Ergonomic Aspects*. Mahwah, NJ: Lawrence Erlbaum Associates.

Salvendy, G. (1997) *Handbook of Human Factors and Ergonomics*, 2nd edn. New York: John Wiley & Sons.

Sanders, M. S. and McCormick, E. J. (1993) *Human Factors in Engineering and Design*. New York: McGraw-Hill.

Sandin, J. (2009) An analysis of common patterns in aggregated causation charts from intersection crashes. *Accident Analysis and Prevention* 41, 624–32.

Sarter, N. B. and Woods, D. D. (1991) Situation awareness – a critical but ill-defined phenomenon. *International Journal of Aviation Psychology* 1(1), 45–57.

Scarborough, A. and Pounds, J. (2001) Retrospective Human Factors analysis of ATC operational errors. *11th International Symposium on Aviation Psychology*, Columbus, OH, 5–8 March.

Schaafstal, A. and Schraagen, J. M. (2000) Training of troubleshooting: a structured, task analytical approach. In J. M. Schraagen, S. F. Chipman and V. L. Shalin (eds), *Cognitive Task Analysis*. Hillsdale, NJ: Erlbaum.

Schneiderman, B. (1998) *Designing the User Interface,* 3rd edn. Boston: Addison-Wesley.

Schraagen, J. M., Chipman, S. F. and Shalin, V. L. (2000) *Cognitive Task Analysis*. Mahwah, NJ: Lawrence Erlbaum.

Schuler, D., Russo, P., Boose J. and Bradshaw J. (1990) Using personal construct techniques for collaborative evaluation, *International Journal of Man-Machine Studies* **33,** 521–36.

Schweickert, R. (1978) A critical path generalization of the additive factor method. *Journal of Mathematical Psychology* 18, 105–39.

Scott, J. (1991) *Social Network Analysis: A Handbook*. London: Sage Publications.

Seagull, F. J. and Xiao, Y. (2001) Using eye-tracking video data to augment knowledge elicitation in cognitive task analysis. In *Proceedings of the Human Factors and Ergonomics Society 45th Annual Meeting*. Minneapolis, MN: HFES.

Seamster, T. L., Redding, R. E. and Kaempf, G. L. (2000) A skill-based cognitive task analysis framework. In J. M. Schraagen, S. F. Chipman and V. L. Shalin (eds), *Cognitive Task Analysis*. Hillsdale, NJ: Erlbaum.

Seigal, A. I., Bartter, W. D., Wolf, J. J., Knee, H. E. and Haas, P. M. (1984) *Maintenance Personnel Performance Simulation (MAPPS) Model: Summary Description*. NUREG/CR-3626, U.S Nuclear Regulatory Commission, Washington DC.

Selcon, S. J. and Taylor, R. M. (1989) Evaluation of the Situational Awareness Rating Technique (SART) as a tool for aircrew system design. *Proceedings of AGARD Symposium on Situational Awareness in Aerospace Operation*, Copenhagen, October.

Sexton, J. B., Helmreich, R. L., Neilands, T. B., Rowan, K., Vella, K., Boyden, J., Roberts, P. R. and Thomas, E. J. (2006) The Safety Attitudes Questionnaire: psychometric properties, benchmarking data and emerging research. *BMC Health Services Research* 6(44). Available at: http://www.biomedcentral.com/content/pdf/1472-6963-6-44.pdf.

Sexton, J. B., Thomas, E. J. and Grillo, S. P. (2003) *The Safety Attitudes Questionnaire (SAQ) Guidelines for Administration*. Technical Report 03-02. The University of Texas Center of Excellence for Patient Safety Research and Practice (AHRQ grant # 1PO1HS1154401 and U18HS1116401).

Shadbolt, N. R. and Burton, M. (1995) Knowledge elicitation: a systemic approach. In J. R. Wilson and E. N. Corlett (eds), *Evaluation of Human Work: A Practical Ergonomics Methodology*. London: Taylor & Francis.

Shah, C., Marchionini, G. and Kelly, D. (2009) Learning design principles for a collaborative information seeking system. *CHI 2009*, 4–9 April, Boston, MA.

Shappell, S. A. and Wiegmann, D. A. (2000) *The Human Factors Analysis and Classification System*. Report Number DOT/FAA/AM-00/07. Washington DC: Federal Aviation Administration.

Shappell, S., Detwiler, C. Holcomb, K., Hackworth, C., Boquet, A. and Wiegman, D. A. (2007) Human error and commercial aviation accidents: an analysis using the Human Factors analysis and classification system. *Human Factors* 49(2), 227–42.

Shappell, S. A. and Wiegmann, D. A. (2003) *A Human Error Analysis of General Aviation Controlled Flight into Terrain Accidents Occurring between 1990–1998* (Report Number DOT/FAA/AM-03/4). Washington DC: Office of Aerospace Medicine.

Shappell, S. and Wiegmann, D. (2009) A methodology for assessing safety programs targeting human error in aviation. *International Journal of Aviation Psychology* 19(3), 252–69.

Sharon, T., Ekhteraei, S., Mc Harg, S. and Chauncey, W. (2010) Storyboard. Usability body of knowledge. Available at: http://www.usabilitybok.org/methods/storyboard.

Sheehan, B., Kaufman, D., Stetson, P. and Currie, L.M. (2009) Cognitive analysis of decision support for antibiotic prescribing at the point of ordering in a neonatal intensive care unit. *AMIA Annual Symposium Proceedings*.

Shepherd, A. (1989) Analysis and training in information technology tasks. In D. Diaper (ed.), *Task Analysis for Human-Computer Interaction*. Chichester: Ellis Horwood.

Shepherd, A. (2002) *Hierarchical Task Analysis*. London: Taylor & Francis.

Shorrock, S. T. (2007) Errors of perception in air traffic control. *Safety Science* 45, 890–904.

Shorrock, S. T. and Kirwan, B. (1999) The development of TRACEr: a technique for the retrospective analysis of cognitive errors in ATC. In D. Harris (ed.), *Engineering Psychology and Cognitive Ergonomics*, vol. 3. Aldershot: Ashgate.

Shorrock, S. T. and Kirwan, B. (2002) Development and application of a human error identification tool for air traffic control. *Applied Ergonomics* 33, 319–36.

Shu, Y. and Furuta, K. (2005) An inference method of team situation awareness based on mutual awareness. *Cognition Technology & Work* 7(4), 272–87.

Siemieniuch, C. E. and Sinclair, M. A. (2006) Systems integration. *Applied Ergonomics* 37(1), 91–110.

Singh, A. and Wesson, J. (2009) Evaluation criteria for assessing the usability of ERP systems. *SAICSIT '09 Proceedings of the 2009 Annual Research Conference of the South African Institute of Computer Scientists and Information Technologists*, New York.

Skillicorn, D. B. (2004) Finding unusual correlations using matrix decompositions. *Second Symposium on Intelligence and Security Informatics, ISI 2004*, Tucson, June, available in *Springer-Verlag Lecture Notes in Computer Science* 3073, 83–99.

Sklet, S. (2002) *Methods for Accident Investigation*. Reliability, Safety and Security Studies at NTNU – Norwegian University of Science and Technology. Report number ROSS (NTNU) 200208. Available at: http://www.ntnu.no/ross/reports/accident.pdf.

Smalley, J. (2003) Cognitive factors in the analysis, design and assessment of command and control systems. In E. Hollnagel (ed.), *Handbook of Cognitive Task Design*. Mahwah, NJ: Lawrence Erlbaum.

Smith, A.F., Casey, K., Wilson, J. and Fischbacher-Smith, D. (2011) Wristbands as aids to reduce misidentification: an ethnographically guided task analysis. *International Journal for Quality in Health Care*, 23(5), 590–599.

Smith, K. and Hancock, P. A. (1995) Situation awareness is adaptive, externally directed consciousness. *Human Factors* 37(1), 137–48.

Smith, P. A., Baber, C., Hunter, J. and Butler, M. (2008) Measuring team skills in crime scene investigation: exploring ad hoc teams. *Ergonomics* 51(10), 1463–88.

Smolensky, M. W. (1993) Toward the physiological measurement of situation awareness: the case for eye movement measurements. In *Proceedings of the Human Factors and Ergonomics Society 37th Annual Meeting*, Santa Monica, Human Factors and Ergonomics Society.

Snook, S.A. (2000) *Fratricide: The Accidental Shootdown of U.S. Black Hawks over Northern Iraq*. Princeton, NJ: Princeton University Press.

Sonnenwald, D. H., Maglaughlin, K. L. and Whitton, M. C. (2004) Designing to support situation awareness across distances: an example from a scientific collaboratory. *Information Processing and Management* 40(6), 989–1011.

Stanton, N. A. (1995) Analysing worker activity: a new approach to risk assessment? *Health and Safety Bulletin* 240, 9–11.

Stanton, N. A. (2002) Human error identification in human computer interaction. In J. Jacko and A. Sears (eds), *The Human Computer Interaction Handbook*. Mahwah, NJ: Lawrence Erlbaum.

Stanton, N.A. (2003) Human error identification in human computer interaction. In J. Jacko and A. Sears (eds), *The Human Computer Interaction Handbook*. Mahwah, NJ: Lawrence Erlbaum.

Stanton, N. A. (2004) The psychology of task analysis today. In D. Diaper and N. A. Stanton (eds), *Handbook of Task Analysis in Human-Computer Interaction*. Mahwah, NJ: Lawrence Erlbaum.

Stanton, N. A. (2005a) Hierarchical Task Analysis: developments, extensions and applications. *Applied Ergonomics* 37(1), 55–79.

Stanton, N. A. (2005b) Human Factors and ergonomics methods. In N. A. Stanton, A. Hedge, K. Brookhuis, E. Salas and H. Hendrick (eds), *Handbook of Human Factors and Ergonomics Methods*. Boca Raton, FL: CRC Press.

Stanton, N. A. (2006) Hierarchical Task Analysis: developments, applications and extensions. *Applied Ergonomics* 37(1), 55–79.

Stanton, N. A. and Annett, J. (2000) Future directions for task analysis. In J. Annett and N. A. Stanton (eds), *Task Analysis*. London: Taylor & Francis.

Stanton, N. A. and Baber, C. (1996a) A systems approach to human error identification. *Safety Science* 22, 215–28.

Stanton, N. A. and Baber, C. (1996b) Task Analysis for error identification: applying HEI to product design and evaluation. In P. W. Jordan, B. Thomas, B. A. Weerdmeester and I. L. McClelland (eds), *Usability Evaluation in Industry*. London: Taylor & Francis.

Stanton, N. A. and Baber, C. (1998) A systems analysis of consumer products. In N. A. Stanton (ed.), *Human Factors in Consumer Products*. London: Taylor & Francis.

Stanton, N. A. and Baber, C. (2002) Error by design: methods to predict device usability. *Design Studies* 23(4), 363–84.

Stanton, N. A., Baber, C. and Harris, D. (2008) *Modelling Command and Control: Event Analysis of Systemic Teamwork*. Aldershot: Ashgate.

Stanton, N. A., Baber, C. and Young, M. S. (2005) Observation. In N. A. Stanton, A. Hedge, K. Brookhuis, E. Salas and H. Hendrick (eds), *Handbook of Human Factors and Ergonomics Methods*. Boca Raton, FL: CRC Press.

Stanton, N. A., Chambers, P. R. G. and Piggott, J. (2001) Situational awareness and safety. *Safety Science* 39, 189–204.

Stanton, N. A., Harris, D., Salmon, P. M., Demagalski, J. M., Marshall, A., Young, M. S. and Dekker, S. (2006a) Predicting design induced pilot error using HET (Human Error Template) – a new formal human error identification method for flight decks. *Aeronautical Journal* 110(1104), 107–15.

Stanton, N. A., Hedge, A., Brookhaus, K., Salas, E. and Hendrick, H. (2005) *Handbook of Human Factors and Ergonomics Methods*. Boca Raton, FL: CRC Press.

Stanton, N. A. and McIlroy, R. C. (2012) Designing mission communication planning: the role of Rich Pictures and cognitive work analysis. *Theoretical Issues in Ergonomics Science* 13(2), 146–68.

Stanton, N. A., Rafferty, L. A. and Forster, M. (2012) Contemporising the combat estimate: Soft effects and influence operations. *Journal of Battlefield Technology* 15(1), 35–40.

Stanton, N. A., Rafferty, L. A., Salmon, P. M., Revell, K. M. A., McMaster, R., Caird-Daley, A. and Cooper-Chapman, C. (2010a) Distributed decision making in multi-helicopter teams: a case study of mission planning and execution from a non-combatant evacuation operation training scenario. *Journal of Cognitive Engineering and Decision Making* 4(4), 328–53.

Stanton, N. A., Salmon, P. M., Walker, G. H. and Jenkins, D. P. (2009) Genotype and phenotype schemata as models of situation awareness in dynamic command and control teams. *International Journal of Industrial Ergonomics* 39(3), 480–489.

Stanton, N. A., Salmon, P. M., Walker, G. H. and Jenkins, D. P. (2010b) Is situation awareness all in the mind? *Theoretical Issues in Ergonomics Science* 11(1–2), 29–40.

Stanton, N. A. and Stevenage, S. V. (1998) Learning to predict human error: issues of acceptability, reliability and validity. *Ergonomics* 41(11), 1737–56.

Stanton, N. A., Stewart, R., Harris, D., Houghton, R. J., Baber, C., McMaster, R., Salmon, P., Hoyle. G., Walker, G., Young. M. S., Linsell, M., Dymott, R. and Green, D. (2006b) Distributed situation awareness in dynamic systems: theoretical development and application of an ergonomics methodology. *Ergonomics* 22(49), 1288–311.

Stanton, N. A. and Young, M. S. (1998) Is utility in the mind of the beholder? A review of ergonomics methods. *Applied Ergonomics* 29(1), 41–54.

Stanton, N. A. and Young, M. (1999a) *A Guide to Methodology in Ergonomics: Designing for Human Use*. London: Taylor & Francis.

Stanton, N. A. and Young, M. S. (1999b) What price ergonomics? *Nature* 399, 197–8.

Stanton, N. A. and Young, M. S. (2003) Giving ergonomics away? The application of ergonomics methods by novices. *Applied Ergonomics* 34(5), 479–90.

Stanton, N. A. and Wilson, J. (2000) Human Factors: step change improvements in effectiveness and safety. *Drilling Contractor* Jan/Feb, 36–41.

Staples, L. J. (1993) The task analysis process for a new reactor. *Proceedings of the Human Factors and Ergonomics Society 37th Annual Meeting – Designing for Diversity*. Seattle, Washington, 11–15 October. The Human Factors and Ergonomics Society, Santa Monica, California.

Stefanidis, D., Wang, F., Korndorffer Jr., J. K., Dunne, J. B. and Scott, D. J. (2010) Robotic assistance improves intracorporeal suturing performance and safety in the operating room while decreasing operator workload. *Surgical Endoscopy* 24, 377–82.

Stewart, R., Stanton, N. A., Harris, D., Baber, C., Salmon, P. M., Mock, M., Tatlock, K., Wells, L. and Kay, A. (2008) Distributed situation awareness in an airborne warning and control system: application of novel ergonomics methodology. *Cognition Technology and Work* 10(3), 221–9.

Stoner, H. A., Wiese, E. E. and Lee, J. D. (2003) Applying ecological interface design to the driving domain: the results of an abstraction hierarchy analysis. *Proceedings of the Human Factors and Ergonomics Society 47th Annual Meeting*, 2003.

Stuve, T. P. (2005) A practical guide to the why because analysis method: performing a why because analysis. Available at: http://www.rvs.uni-bielefeld.de/research/WBA.

Svedung, J. and Rasmussen, J. (2002) Graphic representation of accident scenarios: mapping system structure and the causation of accidents. *Safety Science* 40, 397–417.

Swain, A. D. and Guttmann, H. E. (1983) *A Handbook of Human Reliability Analysis with Emphasis on Nuclear Power Plant Applications*. NUREG/CR-1278, USNRC, Washington DC-20555.

Swann, C. D. and Preston, M. L. (1995) Twenty-five years of HAZOPs. *Journal of Loss Prevention in the Process Industries* 8(6), 349–53.

Swezey, R. W., Owens, J. M., Bergondy, M. L. and Salas, E. (2000) Task and Training Requirements Analysis Methodology (TTRAM): an analytic methodology for identifying potential training uses of simulator networks in teamwork-intensive task environments. In J. Annett and N. Stanton (eds), *Task Analysis*. London: Taylor & Francis.

Tates, K., Zwaanswijk, M., Otten, R., van Dulmen, S., Hoogerbrugge, P. M., Kamps, W. A. and Bensing, J.M. (2009) Online focus groups as a tool to collect data in hard to include populations: examples from paediatric oncology. *BMC Medical Research Methodology* 9(15), 1–8. Available at: http://www.biomedcentral.com/1471-2288/9/15.

Tay, R. (2009) Drivers' perceptions and reactions to roadside memorials. *Accident Analysis and Prevention* 41, 663–9.

Taylor, R. M. (1990) Situational Awareness Rating Technique (SART): the development of a tool for aircrew systems design. In *Situational Awareness in Aerospace Operations* (AGARD-CP-478) 3/1 –3/17, Neuilly Sur Seine: NATO-AGARD.

Thornton, G.E. and Zorich, S. (1980) Training to improve observer accuracy. *Journal of Applied Psychology* 65, 351–4.

Troung, K. N., Hayes, G. R. and Abowd, G. D. (2006) Storyboarding: an empirical determination of best practices and effective guidelines. *Designing Interactive Systems*, 12-21.

Tsang, P. S. and Velazquez, V. L. (1996) Diagnosticity and multidimensional subjective workload ratings. *Ergonomics* 39, 358–81.

Tsui, K. M., Abu-Zahra, K., Casipe, R., M'Sadoques, J. and Drury, J. L. (2009) A process for developing specialised heuristics: case study in assistive robotics. University of Massachusetts Lowell Technical Report 2009-11, Department of Computer Science.

Tudor, L. G., Muller, M. J., Dayton, T. and Root, R. W. (1993) A participatory design technique for high-level task analysis, critique and redesign: the CARD method. In *Proceedings of Human Factors and Ergonomics Society Annual Meetings*, Human Factors and Ergonomics Society.

Tullis, T. S. and Stetson, J. N. (2004) A comparison of questionnaires for assessing website usability. UPA presentation. Available at: http://home.comcast.net/~tomtullis/publications/UPA2004TullisStetson.pdf.

Uhlarik, J. and Comerford, D. A. (2002) *A Review of Situation Awareness Literature Relevant to Pilot Surveillance Functions* (DOT/FAA/AM-02/3). Washington DC: Federal Aviation Administration, U.S. Department of Transportation.

USAF Accident Investigation Board (1994) *U.S. Army Black Hawk Helicopters 87-26000 and 88-26060: Volume 1, Executive Summary: UH-60 Black Hawk Helicopter Accident*, 14 April. Available at: http://www.dod.mil/pubs/foi/operation_and_plans/PersianGulfWar/973-1.pdf.

United States General Accounting Office (1997) *Operation Provide Comfort: Review of U.S. Air Force Investigation of Black Hawk Fratricide Incident*. Report to Congressional Requesters.

Van der Lelie, C. (2006) The value of storyboards in the product design process. *Personal and Ubiquitous Computing* 10(2–3), 159–62.

Van Welie, M. and van der Veer, G. C. (2003) Groupware task analysis. In E. Hollnagel (ed.), *Handbook of Cognitive Task Design*. Mahwah, NJ: Lawrence Erlbaum.

Vatrapu, R., Suthers, D. and Medina, R. (2008) Usability, sociability and learnability: a CSCL design evaluation framework. In *Proceedings of the 16th International Conference on Computers in Education*. Available at: http://www.apsce.net/ICCE2008/papers/ICCE2008-paper242.pdf.

Verplank, B., Fulton, J., Black, A. and Moggridge, B. (1993) Observation and invention – use of scenarios in interaction design. Tutorial notes for *InterCHI 93*, Amsterdam.

Vicente, K. J. (1999) *Cognitive Work Analysis: Towards Safe, Productive and Healthy Computer-Based Work*. Mahwah, NJ: Lawrence Erlbaum.

Vicente, K.J. and Christoffersen, K. (2006) The Walkerton E.coli outbreak: a test of Rasmussen's framework for risk management in a dynamic society. *Theoretical Issues in Ergonomics Science* 7(2), 93–112.

Vidulich, M. A. (1989) The use of judgement matrices in subjective workload assessment: the Subjective Workload Dominance (SWORD) technique. In *Proceedings of the Human Factors Society 33rd Annual Meeting*. Santa Monica, CA: Human Factors Society.

Vidulich, M. A. and Hughes, E. R. (1991) Testing a subjective metric of situation awareness. In *Proceedings of the Human Factors Society 35th Annual Meeting*. Santa Monica, CA: Human Factors Society.

Vidulich, M. A. and Tsang, P. S. (1985) Assessing subjective workload assessment. A comparison of SWAT and the NASA bipolar methods. *Proceedings of the Human Factors Society 29th Annual Meeting*. Santa Monica, CA: Human Factors Society.

Vidulich, M. A. and Tsang, P. S. (1986a) *Collecting NASA Workload Ratings*. Moffett Field, CA: NASA Ames Research Center.

Vidulich, M. A. and Tsang, P. S. (1986b) Technique of subjective workload assessment: a comparison of SWAT and the NASA bipolar method. *Ergonomics* 29(11), 1385–98.

Vidulich, M. A., Ward, G. F. and Schueren, J. (1991) Using the Subjective Workload Dominance (SWORD) technique for projective workload assessment. *Human Factors* 33(6), 677–91.

Waag, W. L. and Houck, M. R. (1994) Tools for assessing situational awareness in an operational fighter environment. *Aviation, Space and Environmental Medicine* 65(5), A13–A19.

Walker, G. H. (2004) Verbal Protocol Analysis. In N. A. Stanton, A. Hedge, K. Brookhuis, E. Salas and H. Hendrick (eds), *Handbook of Human Factors and Ergonomics Methods*. Boca Raton, FL: CRC Press.

Walker, G. H., Gibson, H., Baber, C. and Stanton, N. A. (2004) *EAST Analysis of Railway Data: Possession Scenario*. HFI DTC Technical Report/ W.P. 1.1.3.

Walker, G. H., Gibson, H., Stanton, N. A., Baber, C., Salmon, P. and Green, D. (2006) Event analysis of systemic teamwork (EAST): a novel integration of ergonomics methods to analyse C4i activity. *Ergonomics* 49(12), 1345–69.

Walker, G. H., Stanton, N. A., Baber, C., Wells, L., Jenkins, D. P., Salmon, P. M. (2010) From ethnography to the EAST method: a tractable approach for representing distributed cognition in air traffic control. *Ergonomics* 53(2), 184–97.

Walker, G. H., Stanton, N. A., Kazi, T. A., Salmon, P. M. and Jenkins, D. P. (2009a) Does advanced driver training improve situation awareness? *Applied Ergonomics* 40(4), 678–87.

Walker, G. H., Stanton, N. A. and Salmon, P. M. (2011) Cognitive compatibility of motorcyclists and car drivers. *Accident Analysis and Prevention* 43(3), 878–88.

Walker, G.H., Stanton, N.A., Salmon, P.A. and Jenkins, D.J. (2009b) How can we support the commander's involvement in the planning process? An exploratory study into remote and co-located command planning. *International Journal of Industrial Ergonomics* 39, 456–64.

Walker, G. H., Stanton, N. A., Salmon, P. M., Jenkins, D. P., Monnan, S. and Handy, S. (2012) Communications and cohesion: a comparison between two command and control paradigms. *Theoretical Issues in Ergonomics Science* 13(5), 508–27.

Walker, G. H., Stanton, N. A. and Young, M. S. (2001a) Hierarchical Task Analysis of driving: a new research tool. In M. A. Hanson (ed.), *Contemporary Ergonomics*. London: Taylor & Francis.

Walker, G. H., Stanton, N. A. and Young, M. S. (2001b) An on-road investigation of vehicle feedback and its role in driver cognition: implications for cognitive ergonomics. *International Journal of Cognitive Ergonomics* 5(4), 421–44.

Walker, G. H., Stanton, N. A. and Young, M. S. (2008) Feedback and driver situation awareness (SA): a comparison of SA measures and contexts. *Transportation Research Part F: Traffic Psychology and Behaviour* 11, 282–99.

Wasserman, S. and Faust, K. (1994) *Social Network Analysis: Methods and Applications*. Cambridge: Cambridge University Press.

Watts, L. A. and Monk, A. F. (2000) Reasoning about tasks, activities and technology to support collaboration. In J. Annett and N. Stanton (eds), *Task Analysis*. London: Taylor & Francis.

Weber, R. P. (1990) *Basic Content Analysis*. London: Sage Publications.

Wechsung, I. and Naumann, A.B. (2008) Evaluation methods for multimodal systems: a comparison of standardised usability questionnaires. *PIT*, 276–84.

Wei, J. and Salvendy, G. (2006) Development of a human information processing model for cognitive task analysis and design. *Theoretical Issues in Ergonomics Science* 7(4), 345–70.

Wellens, A. R. (1993) Group situation awareness and distributed decision-making: from military to civilian applications. In N. J. Castellan (ed.), *Individual and Group Decision Making: Current Issues*. Mahwah, NJ: Lawrence Erlbaum.

Wester, A. E., Bocker, K. B. E., Volkerts, E. R., Verster, J. C. and Kenemans, J. L. (2008) Event-related potentials and secondary task performance during simulated driving. *Accident Analysis and Prevention* 40, 1–7.

Westman, A., Sjoling, M., Lindberg, A. and Bjornstig, U. (2010) The SKYNET data: demography and injury reporting in Swedish skydiving. *Accident Analysis and Prevention* 42, 778–83.

Whalley, S. J. (1988) Minimising the cause of human error. In B. Kirwan and L. K. Ainsworth (eds), *A Guide to Task Analysis*. London: Taylor & Francis.

Whalley. S. J. and Kirwan, B. (1989) An evaluation of five human error identification techniques. Paper presented at the *5th International Loss Prevention Symposium*, Oslo, June.

Whiteside, J. Bennett, J. and Holtzblatt, H. (1988) Usability engineering: our experience and evaluation. In M. Helander (ed.), *Handbook of Human-Computer Interaction*. New York: Elsevier Science.

Wickens, C. D. (1987a) Attention. In P. A. Hancock (ed.), *Human Factors Psychology*. Amsterdam: North-Holland.

Wickens, C. D. (1987b) Information processing, decision-making and cognition. In G. Salvendy (ed.), *Handbook of Human Factors*. New York: John Wiley & Sons.

Wickens, C.D. (1992) *Engineering Psychology and Human Performance*. New York: HarperCollins.

Wickens, C. D., Gordon, S. E. and Lui, Y. (1998) *An Introduction to Human Factors Engineering*. New York: Longman.

Wiegmann, A. and Shappell, S. (2001a) A human error analysis of commercial aviation accidents using the Human Factors analysis and classification system (HFACS). *Aviation, Space and Environmental Medicine* 72(11), 1006–16.

Wiegmann, D. and Shappell, S. (2001b) Human error analysis of commercial aviation accidents: application of the Human Factors Analysis and Classification System (HFACS). In *Proceedings of the Eleventh International Symposium on Aviation Psychology*, Ohio State University.

Wiegmann, D. and Shappell, S. (2003) *A Human Error Approach to Aviation Accident Analysis: The Human Factors Analysis and Classification System*. Aldershot: Ashgate.

Wierwille, W. W. and Eggemeier, F. T. (1993) Recommendations for mental workload measurement in a test and evaluation environment. *Human Factors* 35, 263–82.

Wilkinson P.R. (1992) The integration of advanced cockpit and systems design. *AGARD Avionics Panel Symposium*, May, Madrid, AGARD-CP-521 Paper 26.

Williams, J. C. (1986) HEART – a proposed method for assessing and reducing human error. In *9th Advances in Reliability Technology Symposium*, University of Bradford.

Williams, J. C. (1988) A data based method for assessing and reducing human error to improve operational performance. *IEEE Fourth Conference on Human Factors and Power Plants*, 5–9 June, Monterey, CA.

Williams, J. C. (1989) Validation of human reliability assessment techniques. *Reliability Engineering* 11, 149–62.

Wilson, B (1990) *Systems: Concepts, Methodologies and Applications*. Chichester, John Wiley.

Wilson, J. R. (1995) A framework and context for ergonomics methodology. In J. R. Wilson and E. N. Corlett (eds), *Evaluation of Human Work*, 2nd edn. London: Taylor & Francis.

Wilson, J. R. and Corlett, N. E. (1995) *Evaluation of Human Work: A Practical Ergonomics Methodology*. London: Taylor & Francis.

Woltjer, R. (2010) Resilience assessment based on models of functional resonance. Available at: http://www.rvs.uni-bielefeld.de/research/WBA.

Woo, D.M. and Vicente, K.J. (2003) Sociotechnical systems, risk management and public health: comparing the North Battleford and Walterton outbreaks. *Reliability Engineering and System Safety* 80, 253–69.

Wood, J. and Wood, L. (2008) Card sorting: current practices and beyond. *Journal of Usability Studies* 4(1), 1–6.

Woodson, W., Tillman, B. and Tillman, P. (1992) *Human Factors Design Handbook*. New York: McGraw-Hill.

Wu, R. C., Orr, M. S., Chignell, M. and Straus, S. E. (2008) Usability of a mobile electronic medical record prototype: a verbal protocol analysis. *Informatics for Health & Social Care* 33(2), 139–49.

Xiao, T., Sanderson, P.M., Mooij, M. and Fothergill, S. (2008) Work domain analysis for assessing simulated worlds for ATC studies. *Proceedings of the 52nd Human Factors and Ergonomics Society Annual Meeting*. Santa Monica, CA: Human Factors and Ergonomics Society.

Yamaoka, T. and Baber, C. (2000) Three point task analysis and human error estimation. *Proceedings of the Human Interface Symposium*, Tokyo, Japan

Young, M. and Stanton, N. A. (1997) Automobile automation. *International Journal of Cognitive Ergonomics* 1(4), 325–36.

Young, M. S. and Stanton, N. A. (2001) Mental workload: theory, measurement and application. In W. Karwowski (ed.), *International Encyclopedia of Ergonomics and Human Factors, Volume 1*. London: Taylor & Francis.

Young, M. S. and Stanton, N. A. (2004) Taking the load off: investigations of how adaptive cruise control affects mental workload. *Ergonomics* 47(8), 1014–35.

Young, M. S. and Stanton, N. (2005) Mental workload. In N. A. Stanton, A. Hedge, K, Brookhuis, E. Salas and H. Hendrick (eds), *Handbook of Human Factors Methods*. Boca Raton, FL: CRC Press.

Zachary, W. W., Ryder, J. M. and Hicinbothom, J. H. (1998) Building cognitive task analyses and modelling of decision-making in complex environments. In J. A. Cannon-Bowers and E. Salas (eds), *Making Decisions Under Stress: Implications for Individual and Team Training*. Washington DC: American Psychological Association.

Zakariasen Victoroff, K. and Hogan, S. (2005) Students' perceptions of effective learning experiences in dental school: a qualitative study using a critical incident technique. *Journal of Dental Education* 70(2), 124–32.

Subject Index

Author Index